PVP-Vol. 458

COMPUTER TECHNOLOGY AND APPLICATIONS

presented at

THE 2003 ASME PRESSURE VESSELS AND PIPING CONFERENCE
CLEVELAND, OHIO
JULY 20–24, 2003

sponsored by

THE PRESSURE VESSELS AND PIPING DIVISION, ASME

edited by

WOLF REINHARDT
BABCOCK & WILCOX CANADA

JOHN MARTIN
LOCKHEED MARTIN, INC.

THE AMERICAN SOCIETY OF MECHANICAL ENGINEERS
Three Park Avenue / New York, N.Y. 10016

ISBN No. 0-7918-1699-0

FOREWORD

The application of the computer continues to expand in the design, manufacturing and life assessment of pressure vessels. The Computer Technology Committee of the ASME Pressure Vessel and Piping Division that sponsors the present volume aims to help engineers utilize present computer capabilities and to keep them up-to-date with their expanding potential. To this end, the Committee organizes paper sessions where technical advances can be presented and discussed, panel sessions that bring together experts to elucidate current issues, and a software demonstration forum where vendors present the latest software.

This volume includes papers that were presented at the 2003 ASME Pressure Vessel and Piping Conference in Cleveland, Ohio, July 20-24. The compilation showcases novel applications of existing methods and presents new research into the improvement of computer-based analysis. The technical papers in this volume are grouped into the following areas:

- Efficient Computational Models for Elastic-Plastic and Limit Load Analysis of Pressure Vessel Components
- Finite Element Methodology for Manufacturing Simulations and Field Applications
- Non-Linear FEA and Emerging Methods
- New and Emerging Computational Methods
- Risk and Reliability

A separate volume sponsored by the Computer Technology Committee contains papers related to the design of bolted joints.

The papers in this volume have been formally reviewed, and the resulting collection of high-quality papers speaks for itself. The time and efforts that the presenters, authors and reviewers and their organizations invested into these papers are greatly appreciated. Special thanks are due to the session developers N. Badie, J. A. Farquharson, D. Jones, G. Morandin, Y. H. Park, B. N. Rao, R. Sauvé and J. Tang, who generously donated their time collecting the papers and reviews and making this volume possible. Thanks also go to the session chairs and vice chairs who ensure the paper presentations run smoothly. Finally, we would like to express our gratitude for the support from Don Metzger, Chair of the Computer Technology Committee, and from Ismael Kisisel, the Technical Program Chair of the conference.

John Martin
Lockheed Martin Inc.

Wolf Reinhardt
Babcock & Wilcox Canada

CONTENTS

EFFICIENT COMPUTATIONAL MODELS FOR ELASTIC-PLASTIC AND LIMIT LOAD ANALYSIS OF PRESSURE VESSEL COMPONENTS

Introduction

David P. Jones
Bechtel Bettis Inc.

Wolf D. Reinhardt
Babcock & Wilcox Canada

There continues to be a strong interest in the pressure vessel community to take advantage of plastic design methods for weight saving and economical reasons. The drawback of these methods lies in the higher computational cost involved in a nonlinear analysis. Finding more efficient methods to solve the underlying problem numerically can offset such drawbacks. The subsequent collection of papers addresses these problems in various ways.

Two of the subsequent papers focus on efficient methods to obtain an inelastic stress-strain field for higher temperature problems where both plasticity and creep phenomena are important. In one case, this is achieved by matching an inhomogenous linear-elastic solution to the nonlinear one. The other paper explores the potential of strain energy matching through Neuber's rule for the prediction of near-notch tip stresses and strains.

Methods to accelerate the convergence of linear matching methods are proposed by one researcher using the concept of redistribution nodes. In another paper the same objective is achieved with the help of a variational approach and the use of upper and lower bound solutions.

Two of the papers address the solution of the plastic shakedown problem. One of these gives an overview and example applications of existing methods, while the other proposes a non-cyclic method for establishing the shakedown boundary.

One paper discusses bellows design with plastic limit analysis and its simplification through the use of design equations. The equations are verified over a large parameter range.

The efficient evaluation of the constraint effect in fracture mechanics through weight functions is demonstrated in a final paper.

Both authors and reviewers devoted a substantial amount of time and effort to the papers in this session. The session developers wish to use this opportunity to express their appreciation and gratitude for these contributions to all involved.

PVP-Vol. 458, Computer Technology and Applications

PVP2003-1884

APPLICATION OF THE LINEAR MATCHING METHOD TO THE INTEGRITY ASSESSMENT FOR THE HIGH TEMPERATURE RESPONSE OF STRUCTURES

Haofeng Chen and Alan R.S. Ponter
Department of Engineering, University of Leicester, Leicester, LE1 7RH, UK
Email: hfc3@le.ac.uk, asp@le.ac.uk
Tel: 0044-1162522549 Fax: 0044-1162522525

ABSTRACT

The paper describes a first attempt to produce a complete system of calculations that cover the entire range of assessments required in the Life Assessment method R5 based on a new programming method, the Linear Matching Method, and using shakedown and related concepts. We show that two solutions types are possible, the first assuming a constant residual stress field that provides shakedown and related limits. The second method involves the evaluation of the amplitude of the changing residual stress field. This provides the first stage for the ratchet limit and the amplitude of plastic strain. By adaptation the elastic follow-up factor corresponding to creep dwell periods may also be evaluated

NOTATION

$\lambda P_i(x_j,t)$	mechanical loads
$\lambda\theta(x_j,t)$	temperature history
λ, Z	load parameter and elastic follow-up factor
V, S	volume and surface of the body
$\lambda\hat{\sigma}_{ij}$	linear elastic stress field
σ, ε	stress and strain
λ^E, λ^S, λ^P	elastic, shakedown and ratchet limit multipliers
E, S, P, R	elastic, shakedown, reverse plasticity and ratchetting region
$\bar{\rho}_{ij}$, $\rho_{ij}^{pr}(t)$	constant and changing residual stress field
$\dot{\varepsilon}_{ij}^{pr}$, $\Delta\varepsilon_{ij}^{pr}$	plastic strain rate history and accumulated strain
Δu_i^{pr}	displacement increment
t_c, Δt	material time scale and cycle time
δt	creep dwell time
$\theta, \theta_0, \Delta\theta$	temperatures
σ_P, σ_t	effective tension and thermal stress
D, L	diameter of the hole and length of the plate
σ_Y, E, ν	yield stress, elastic modulus and Poison's ratio
σ_{t0}	effective elastic thermal stress
a, r	radius of the hole and distance to the centre of hole
$\sigma_c(t_f,\theta)$, σ_y^{LT}	creep rupture stress and yield stress
t_f	the time to creep rupture
R, g	functions of creep rupture time and temperature
σ_0, $\dot{\varepsilon}_0$, n	creep materials data
$\Delta\varepsilon^p$, $\Delta\varepsilon^c$	plastic strain range and accumulated creep strain
$\Delta\rho^p$, $\Delta\rho^c$	varying residual stress associated with reverse plasticity mechanism and creep relaxation
D_c, N_0	creep endurance limit and cycles to failure
σ^c, $\dot{\varepsilon}^c$	creep stress and creep strain rate

INTRODUCTION

The integrity assessment procedure R5 [1] has been widely used for the high temperature response of the structures. In R5 procedures, for the assessment of creep and fatigue damage, the Reference Stress technique [1] is widely used on the basis of the shakedown calculations.

For perfect plasticity when fluctuating temperature fields are significant, the shakedown limit is often a reverse plasticity limit. If the fluctuation of temperature is raised above this level, the shakedown condition is exceeded but the cyclic plastic strains form a reverse plasticity mechanism in a confined region and the body does not suffer strain growth. Such a situation is often acceptable in design and the significant limit then become a ratchet limit.

When the load history is in excess of shakedown but less than a ratchet limit there are two properties required in low temperature design, i.e. design below the creep range. The amplitude of plastic strain provides information concerning fatigue crack initiation in low cycle fatigue and the capacity of the body to withstand additional constant mechanical load indicates the proximity to a ratchet limit. A varying residual stress field associated with the reverse plasticity occurs at this stage.

There exists other condition. Engineering structures exposed to high temperature environment exhibit time dependent behaviour [1]. In order to avoid full inelastic analyses, simplified inelastic analysis methods are desirable. The loadings applied to high temperature structures often consist of severe cyclic thermal stresses, possibly beyond yield, and relatively smaller, steadier mechanical loads. Under these circumstances, behaviour during periods of steady operation at high temperature results in creep, and involves the relaxation of initially high stresses as creep strain replaces elastic strain. The concept of "elastic follow-up" was introduced to allow determination of quantities serving as measure for life assessment of structures subjected to creep conditions without performing full time dependent structural analysis [2].

The Elastic Compensation Method [3], the matching of the non-linear material behaviour to a linear material, forms the basis for a powerful upper bound programming method that may be applied to a significant class of problems concerned by R5 procedures. This generalized method, now called the Linear Matching Method [4-13], has been applied with considerable rigor to cyclic loading problems where the residual stress field remains constant. This includes classical limit loads, shakedown limits, creep rupture and rapid cycle creep problems, where the cycle time is small compared with material time scales [4-8]. For the steady cyclic behaviour associated with complex histories of load and temperature where the residual stress field changes during a cyclic state, the Linear Matching Method can still be used. This includes the plastic strain amplitude and ratchet limit associated with reverse plasticity mechanism, the creep strain accumulation and elastic follow-up over creep dwell associated with the creep-reverse plasticity mechanism [9-13].

The purpose of this paper is to not only present the theoretical backgrounds of the application of Linear Matching Method on the assessment of the high temperature response of structures concerned by R5, but also demonstrate that the Linear Matching Method have both the advantages of programming methods and the capacity to be implemented easily within commercial finite element code ABAQUS [14]. The summary of solution sequence for structural integrity assessment based on the Linear Matching Method is given in Table 1, which concerns a range of cyclic problems when the residual stress field remains *constant* (shakedown, creep rupture and creep rapid cycle solutions) and also when the residual stress field *varies* during the cycle (plastic strain amplitude, ratchet limit and elastic follow-up over creep dwell).

In this paper, a 3-D holed plate subjected to cyclic thermal load and mechanical load is assessed in detail as a typical example to confirm the applicability of the Linear Matching Method on the assessment procedures concerned by R5. Numerical examples of the application of the methods to practical problem, a 3D tubeplate, are presented in the separated paper [15].

DEFINITION OF THE PROBLEM

We consider the problem of a body with volume V subjected to a history of cyclic load $\lambda P_i(x_j,t)$ on S_T and a temperature $\lambda\theta(x_j,t)$ within *V*. λ is a load parameter. Hence $\lambda\hat{\sigma}_{ij}$ is the elastic stress field corresponding to cyclic thermal and mechanical loads. In our methods we first generate the elastic solution for $\lambda = 1$.

PLASTICITY LIMITS

The core of the methodology involves iterative methods capable of obtaining the classical load limits for an elastic-perfectly plastic material. We assume the von Mises yield condition, i.e $\overline{\sigma} \le \sigma_y$ where $\overline{\sigma}$ is the von Mises effective stress and σ_y is the uniaxial yield stress.

If we define λ^E, λ^S, λ^P as the elastic limit multiplier, shakedown limit multiplier and ratchet limit multiplier respectively, the properties that defines these limits are as follows:

E - Elastic region - $0 \le \lambda \le \lambda^E$, where $\overline{\sigma}(\lambda\hat{\sigma}_{ij}) \le \sigma_y$ throughout V

S - Shakedown - $\lambda^E \le \lambda \le \lambda^S$, where $\overline{\sigma}(\lambda\hat{\sigma}_{ij} + \overline{\rho}_{ij}) \le \sigma_y$ and $\overline{\rho}_{ij}$ is a constant residual stress field.

P -Reverse Plasticity- $\lambda^S \le \lambda \le \lambda^P$, where $\overline{\sigma}(\lambda\hat{\sigma}_{ij} + \overline{\rho}_{ij} + \rho_{ij}^{pr}) \le \sigma_y$, and $\rho_{ij}^{pr}(t)$ is a changing residual stress field, derived from a plastic strain rate history $\dot{\varepsilon}_{ij}^{pr}$ that satisfies the *zero growth* condition $\int_0^{\Delta t} \dot{\varepsilon}_{ij}^{pr} dt = 0$ everywhere in V.

R - Ratchetting - $\lambda^P \le \lambda$, where $\sigma(\lambda\hat{\sigma}_{ij} + \overline{\rho}_{ij} + \rho_{ij}^{pr}) \le \sigma_y$, and $\rho_{ij}^{pr}(t)$ is a changing residual stress field, derived from a plastic strain rate history $\dot{\varepsilon}_{ij}^{pr}$ that satisfies the *growth* condition $\int_0^{\Delta t} \dot{\varepsilon}_{ij}^{pr} dt = \Delta\varepsilon_{ij}^{pr}$ where $\Delta\varepsilon_{ij}^{pr}$ is a compatible accumulated strain giving rise to non-zero displacement increment Δu_i^{pr}.

The behaviour progresses, for increasing λ, from the most benign, the **E** region, to the most serious, the **R** region. At the transition values of λ we reach the position that, for increasing λ there no longer exists a solution of the form that characterised the exiting region. Hence, when λ increases above λ^E, somewhere $\overline{\sigma}(\lambda\hat{\sigma}_{ij}) > \sigma_y$. Similarly when λ increases above λ^S, there no longer exists a constant residual stress field $\overline{\rho}_{ij}$ so that $\overline{\sigma}(\lambda\hat{\sigma}_{ij} + \overline{\rho}_{ij}) \le \sigma_y$ everywhere.

To evaluate these limits two separate methods are required; a shakedown method that evaluates a constant residual stress field $\overline{\rho}_{ij}$ and at the same time an estimate of the largest value of load parameter λ for which the yield condition is satisfied, and ; a method for evaluating a changing residual stress field $\rho_{ij}^{pr}(t)$ associated with a local reverse plasticity mechanism $\dot{\varepsilon}_{ij}^{pr}$. The Linear Matching methods have been developed as upper bound programming methods [5-12] so that, when implemented as a finite element method, it is possible to show that the solution generated is the least upper bound associated with the finite element mesh. But the essence of the method may be understood in simpler terms. The following is a brief description of the Linear Matching Method applied to shakedown problems; a full description may be found in [8].

SHAKEDOWN

For the evaluation of a shakedown limit we solve a sequence of linear problems where the inelastic strain rate is linearly related to a stress history $\lambda\hat{\sigma}_{ij} + \bar{\rho}_{ij}$;

$$\dot{\varepsilon}_{ij}^{p} = \frac{1}{\mu(t)}(\lambda\hat{\sigma}_{ij} + \bar{\rho}_{ij})' \text{ and } \dot{\varepsilon}_{kk}^{p} = 0 \qquad (1)$$

where $\mu(t)$ is a shear modulus that varies in both time and space. Here $\mu(t)$ is chosen so that the yield condition and the linear relation (1) give rise to the same effective stress for the previous solution in the series. At the same time λ is chosen as an upper bound value also from the previous solution. Each problem is solved by integrating (1) over the cycle of loading so that we obtain a linear relationship between the accumulated strain over the cycle $\Delta\varepsilon_{ij}^{p}$ and the constant residual stress field $\bar{\rho}_{ij}$ and this can be solved as a standard finite element initial strain problem. The process needs to be started with a first solution and this is done by just choosing constant values of $\mu(t)$. General theory [8] shows that the load parameter converges to the least upper bound associated with the finite element mesh. Effectively we generate the exact shakedown limit within the approximations of the finite element method, for a linear elastic solution evaluated by the same mesh. The accuracy of the solution then relies only upon the choice of mesh geometry. For convergence something of the order of ten iterations are required for a reasonable estimate (of the order of 1% above the optimal upper bound) and about 50 iterations for no change in the fifth significant figure over five iterations λ.

The method has been implemented in ABAQUS using user routines by a method described by Engelhardt[4]. Effectively ABAQUS is set up to carry out a step by step calculation but each increment is reinterpreted as an iteration of the method. The initial value linear problems are set up in the user routine UMAT and the volume integrals required in the evaluation of the upper bound load parameter is obtained by routines designed for the evaluation of the total strain energy. The examples in this paper have been generated by the 3D method described in [5]. Before discussing the ratchet limit we first discuss examples where the basic shakedown method with some adaptations is used.

NUMERICAL APPLICATION

We consider a 3-D holed plate subjected to a temperature difference $\Delta\theta$ between the edge of the hole and the edge of the plate and uniaxial tension σ_P acting along one plate edge. The geometry of the structure and its finite element mesh are shown in Fig.1. The 20-node solid isoparametric elements with reduced integration are adopted. The ratio between the diameter D of the hole and the length L of the plate is 0.2 and the ratio of the depth of the plate to the length L of the plate is 0.05. The yield stress of the material is assumed to be 360 MPa. The elastic modulus is E= 208 GPa and Poison's ratio $\nu = 0.3$.

The variation of the temperature with radius r was assumed to be;

$$\theta = \theta_0 + \Delta\theta \ln(5a/r)/\ln(5)$$

giving a simple approximation to the temperature field, corresponding to $\theta = \theta_0 + \Delta\theta$ around the edge of the hole and $\theta = \theta_0$ at the edge of the plate. The elastic stress field and the maximum effective value, σ_t, at the edge of the holed plate due to the thermal load was calculated by ABAQUS, where $\theta_0 = 200°C$, $\Delta\theta = 400°C$ and a coefficient of thermal expansion of $1.25E-5°C^{-1}$. The cycle of loading consists of constant applied stress σ_P and cyclic variation of $\Delta\theta$ between zero and $\Delta\theta = 400°C$. The method is capable of solving for any loading history, however complex.

Fig.2 shows the shakedown limits for a set of ratios of σ_t / σ_p displayed as a Bree/Miller interaction diagram. The axes have been normalised with respect to the uniaxial yield stress σ_y where $\sigma_t = \sigma_{t0} = 2\sigma_y$ corresponds to the reverse plasticity shakedown limit. Figure 2 also shows the ratchet limit which we discuss below.

The basic method may be adapted to solve more complex problems. R5 includes limits associated with creep rupture and overall creep strain accumulation both of which may be evaluated by calculations involving a constant residual stress field. In the following two sections we summarize Linear Matching Methods for these calculations as adaptations of the basic shakedown method discussed above.

AN EXTENDED SHAKEDOWN METHOD FOR CREEP RUPTURE ASSESSMENT

The R5 method for creep rupture seeks to show that the history of cyclic loading lies within shakedown where the yield stress is given by a combination of a the material yield stress σ_y^{LT} at lower temperatures and a temperature dependent creep rupture stress $\sigma_c(t_f, \theta)$ corresponding to an acceptable creep rupture time t_f. Hence the loads are regarded as acceptable provided a history of stress of the form $\hat{\sigma}_{ij} + \bar{\rho}_{ij}$ neither exceed the yield stress or a creep rupture stress and this can be recognised as a standard shakedown problem. But the way that this requirement occurs in R5 life requires a novel interpretation. The load is prescribed, i.e. the load parameter λ is known, and there is a need to find the largest value of the creep rupture time

t_f and associated creep rupture stress $\sigma_c(t_f,\theta)$ so that shakedown occurs, thereby defining conservatively the remaining creep rupture life of the structure. This requirement has always caused computational problems as, previously no direct method for this limit has been available. The Linear Matching Method for shakedown may, however, be adapted to solve this problem. Full details are given in [8,16].

Hence the yield stress at each point of the body at instant $t = t_m$ is defined by;

$$\sigma_y(x_i,t_m) = \min\left\{\begin{matrix}\sigma_y^{LT} \\ \sigma_c(t_f,\theta(x_i,t_m))\end{matrix}\right\} \tag{3}$$

We assume the following form for $\sigma_c(t_f,\theta)$;

$$\sigma_c(t_f,\theta) = \sigma_y^{LT} R\left(\frac{t_f}{t_0}\right) g\left(\frac{\theta}{\theta_0}\right) \tag{4}$$

and $\sigma_c(t_f,\theta) = \sigma_y^{LT}$ when $R\left(\frac{t_f}{t_0}\right) g\left(\frac{\theta}{\theta_0}\right) = 1$. In order to simplify the calculation, for the example below, the form of temperature dependence for σ_c has been adopted as $g\left(\frac{\theta}{\theta_0}\right) = \frac{\theta_0}{\theta - \theta_0}$, where $\theta_0 = 200°C$. This is typical of relationships derived from creep rupture data.

Two kinds of calculations can be performed. The shakedown limit for a prescribed creep rupture parameter *R* can be evaluated by a traditional shakedown analysis with the revised yield stress by equation (3). For a structure with predefined load history, an inversion of the iterative process to optimise creep rupture time has also been proposed. Hence we wish to compute the value of R for which the shakedown limit is given by $\lambda = 1$.

The method involves the derivation, for a particular strain rate history $\dot{\varepsilon}_{ij}^p$ a relationship between a change in the load parameter $\Delta\lambda$ and a change ΔR;

$$\Delta\lambda \dot{D}_p^E = \frac{\Delta R}{R} \dot{D}_p^c \tag{5}$$

where $\dot{D}_p^E$ and $\dot{D}_p^c$ are determinate volume integrals[8,16].

We begin by choosing an initial value of $R = R_0$. The value of *R* is then change according to equation (5) at each iteration so that λ remains at the preassigned value of , say, $\lambda = 1$, and the process is repeated. At each iteration the value of R increases and converges, from below, to the value for which the shakedown limit is given by $\lambda = 1$.

Using the above numerical procedures, the shakedown limits for the 3D holed plate for five different creep parameters (R=0.1, 0.5, 1.0, 1.5 and 2.0) are shown in Fig. 3. As expected the more severe the creep behaviour the more affected the revised yield condition (3) is and the shakedown limit lowers accordingly. Note that for R=2 or more, $\sigma_y = \sigma_y^{LT}$ throughout the volume and the shakedown limit including the creep is the same as the normal shakedown limit without including the creep. This is because the material does not operate in the creep range for this particular behaviour and prescribed temperature distribution.

In order to verify the applicability of the inversion of the iterative process to optimise creep rupture time, two load points A and B with R=0.5 in Fig. 3 are chosen to evaluate the converging creep parameter R. Fig. 4 shows two solutions for a creep parameter R converging to 0.5 corresponding to load point A and B respectively. The slower convergence for the solution corresponding to point B is due to the fact that a reverse plasticity mechanism operates in the converged state.

This problem demonstrates the potential flexibility of the Linear Matching Method. It is clear from this example that the shakedown problem may be posed in other ways; in this particular problem the quantity optimised concerns a material property which enters the problem in only part of the volume and only during part of the load cycle. It is clearly possible, using the type of technique discussed in this paper, to pose a variety of optimisation problems depending upon the needs of the problem.

ASSESSMENT OF OVERALL DISPLACEMENT AND STRAIN GROWTH DUE TO CREEP

In R5 the overall growth of creep strain during the lifetime of the structure is assessed using a reference stress evaluated from the limit load or shakedown limit of the structure. In most applications fatigue and creep/fatigue limits are more significant and overall creep strain growth is not a primary problem. However it is possible to obtain a conservative estimate using a further adaptation of the shakedown method described above, and this is discussed briefly here.

For creep, the general problem is governed by two time scales, a material time scale t_c which may be taken as the time for the uniaxial creep strain to accumulate to the elastic strain at some mean stress, and cycle time Δt. The general solution is strongly dependent upon the relative values of these two time scales and, at the two extremes of $\Delta t/t_c \to 0$ and $\Delta t/t_c \to \infty$, particular solutions exist, the rapid and slow cycle solutions respectively [6]. For the rapid cycle solution the accumulation of strain during the cycle is very small compared with elastic strain and hence $\dot{\rho}_{ij}^r$ is negligible and the form of the solution is identical to a shakedown solution:

$$\sigma_{ij}(x_i,t) = \lambda\hat{\sigma}_{ij}(x_i,t) + \bar{\rho}_{ij}(x_i) \tag{6}$$

The solution also depends upon the creep constitutive relationship and the simplest and most conservative of such solutions is based on the Bailey Orowan model [6]. For this model the evaluation of the average accumulation of creep strain per cycle $\Delta\varepsilon_{ij}^c / \Delta t$ is identical in form to the shakedown solution with two exceptions. The value of λ is maintained constant and the yield stress σ_y is replaced by a flow stress $\bar{\sigma}_{\max}^f$ as:

$$\bar{\sigma}^{f}_{\max} = \sigma_0 \left\{ \frac{\int_0^{\Delta t} \bar{\dot{\varepsilon}}(\dot{\varepsilon}^c_{ij})dt}{\dot{\varepsilon}_0} \right\}^{\frac{1}{n}} \tag{7}$$

where σ_0 and $\dot{\varepsilon}_0$ are material parameters with *n* being the creep index. $\bar{\dot{\varepsilon}}(\dot{\varepsilon}^c_{ij})$ denotes the von Mises effective strain rate corresponding to the history of creep strain rate $\dot{\varepsilon}^c_{ij}$.

This solution has also been used as the beginning point for one of the methods of the evaluation of an elastic follow-up factor Z. The rapid cycle solution is used to provide the stresses at the beginning of the creep dwell period and the value of Z is found by allowing creep relaxation to take place as a step by step process [13]. There are further adaptations that allow for estimates in excess of shakedown.

PLASTIC SHAKEDOWN

The second basic numerical procedure involves the evaluation of the amplitude of plastic strain in excess of shakedown. Again the method has been developed as an upper bound method [10], based on a new minimum theorem [9] applicable to the load history in excess of a elastic shakedown limit. In the limit when the loads lie on the elastic shakedown limit, the theorem reduces to the upper bound shakedown theorem. Essentially, the method computes a closed cycle of plastic strain that gives rise to a zero accumulation of strain over the loading cycle. The changing residual stress field when added to the elastic stress history satisfies the reverse plasticity condition. The problem is solved by the Linear Matching Method through a sequence of linear problems. Although the general methodology is capable of being applied to any problem, however complex, implementation as a finite element scheme has only been completed for the case where the elastic stress history varies between two extremes. The plastic strain history then consists of two increments of plastic strain $\Delta\varepsilon^p_{ij}$ and $-\Delta\varepsilon^p_{ij}$ with corresponding changes in residual stress $\Delta\rho^r_{ij}$ and $-\Delta\rho^r_{ij}$. For this case the minimum theorem provides the methodology of computing $\Delta\varepsilon^p_{ij}$ and $\Delta\rho^r_{ij}$ as a single problem. The details of the method are given in reference [10].

The evaluation of the ratchet limit then involves a further stage. The elastic stress history $\lambda\hat{\sigma}_{ij}$ is augmented by the addition of the changing residual stress field ρ^r_{ij} to form a stress history that satisfies the reverse plasticity condition. A shakedown limit is then evaluated, again using the Linear Matching Method, for the *addition* load that will take the structure to a shakedown limit, now automatically a ratchet limit as the reverse plasticity condition is always satisfied. It is possible to show that this calculation gives the ratchet limit, evaluated exactly, except for the approximations of the finite element method.

This method forms the basis of a plastic strain amplitude estimate for variable load that feeds into a low cycle fatigue limit. The method has also been adapted to allow for cyclic hardening. The amplitude of plastic strain is sensitive to cyclic hardening and values for perfect plasticity tend to be high, giving conservative fatigue limits. The extension to cyclic hardening, given in [10] allows the yield stress to increase until the relationship between the amplitude of effective stress and amplitude of effective strain correspond to prescribed data.

The elastic-linear strain-hardening model adopted is shown in Fig. 5 where the stress and strain quantities are von Mises effective values. The perfectly plastic case corresponds to $\bar{E}_1 = 0$.

Fig.6 shows the maximum of the varying plastic strain magnitudes $\Delta\bar{\varepsilon}^p_{\max}$ at the edge of the hole in the reverse plasticity region for the different material models. The varying plastic strain magnitudes are induced by the varying thermal load rather than constant mechanical load. As mentioned above maximums of the varying plastic strain magnitudes are significantly reduced by adopting the elastic-linear strain-hardening model. In Fig.6, the calculated maximums of the varying plastic strain magnitudes for the elastic-perfectly plastic model from a step by step ABAQUS analysis is shown and corresponds very closely to the Linear Matching Method solution.

In engineering practice, the Neuber approximate values [17] for the maximum of the varying plastic strain magnitude are widely adopted. A comparison with our computed values and the Neuber approximate values are presented in Fig.7. For the cyclic hardening material model, the Neuber values are in good agreement with our method. For the elastic-perfectly plastic model, the Neuber approximate values differ significantly from our results for sufficiently high thermal stresses. The reason is that the Neuber approximation is most suitable for structures subjected to the varying mechanical load rather than varying thermal load. It can also be seen that in this case the Neuber approximate values are conservative.

Using the developed shakedown and ratchet techniques, the calculated shakedown and ratchet limit, as well as elastic boundary for the holed pate subjected to mechanical and thermal loads, are shown in Figure 2. Note that $\sigma_{t0} = 2\sigma_y$ is the value of σ_t at the reverse plasticity shakedown limit. Compared with the elastic-perfectly plastic model, the inclusion of cyclic hardening in the reverse plasticity method has the effect of increasing the ratchet limit but only for extremely high thermal loads. If the magnitude of the varying thermal load is confined to, say, twice the shakedown limit ($\sigma_t / \sigma_{t0} \le 2$) the different material models have very little influence on the ratchet limit.

RATCHET BOUNDARY FOR COMPLEX LOADING HISTORIES

The method described above has only been implemented for load histories that have two extremes. However, the observation that the ratchet boundary is insensitive to cyclic hardening allows the development of a simplified method that assumes that complete cyclic hardening occurs in the reverse plasticity regime. Details of the method may be found in reference [15] where it is applied to the thermal transient loading of a tube plate.

ELASTIC FOLLOW-UP DURING A CREEP DWELL PERIOD

When the structure is subjected to high temperature, i.e. design in the creep range, a creep-reverse plasticity mechanism may appeared in the body instead of pure reverse plasticity mechanism. We assume the structure is subjected to high temperature during a dwell period δt following time t_1, and the body is subjected to low temperature at time t_2. Hence creep relaxation occurs in the body during δt and reverse plasticity appears in the body at time t_2. There are two kinds of creep-reverse plasticity mechanisms possible. For sufficient high loads in the absence of creep a varying plastic strain $\Delta\varepsilon^p$ as well as the varying residual stress $\Delta\rho^p$ associated with pure reverse plasticity mechanism need to be calculated before the evaluation of the accumulated creep strain $\Delta\varepsilon^c$ and the corresponding residual stress $\Delta\rho^c$ for creep relaxation. The total varying strain $\Delta\varepsilon^{T\!P}$ is then the summation of varying plastic strain $\Delta\varepsilon^p$ and accumulated creep strain $\Delta\varepsilon^c$. For lower loads in the absence of creep the load point lies in the shakedown region i.e. $\Delta\varepsilon^p = 0, \Delta\rho^p = 0$, and the total varying strain $\Delta\varepsilon^{T\!P}$ equals to the accumulated creep strain $\Delta\varepsilon^c$ induced by creep relaxation followed by plastic strain increment at time t_2

A method for carrying out this calculation has been presented in the [13]. The method is an adaptation of the reverse plasticity method discussed above where, at every point in the structure we assume that a constant elastic follow-up factor Z exists during δt. Once the accumulated creep strain $\Delta\varepsilon^c$ and the corresponding residual stress $\Delta\rho^c$ due to creep relaxation are calculated, the elastic follow-up factor Z, the creep endurance limit D_c, the cycles to failure N_0 induced by low cycle fatigue mechanism can be evaluated and the total damage parameter then be determined thereafter.

NUMERICAL APPLICATION

We still use the holed plate as a typical example, where the structure is subjected to varying thermal loads shown in Fig.1. The creep material data are assumed to obey a Norton type law, $\frac{\dot{\varepsilon}^c}{\dot{\varepsilon}_0} = \left(\frac{\sigma}{\sigma_0}\right)^n$ $\sigma_0 = \frac{\sigma_y}{2}$, $\dot{\varepsilon}_0 = 2\times10^{-6}$/s and n=5.

The two load cases shown in table (2) are considered. For load case 1, the applied load domain is beyond the elastic shakedown region and the reverse plasticity mechanism appears in the body (see Fig. 2). The varying plastic strain $\Delta\varepsilon^p$ associated with reverse plasticity mechanism equals to 2.869×10^{-3}, and the total varying strain $\Delta\varepsilon^{T\!P}$ is the summation of varying plastic strain $\Delta\varepsilon^p$ and accumulated creep strain $\Delta\varepsilon^c$. For load case 2, the applied load domain is in the elastic shakedown region, where $\Delta\varepsilon^p = 0, \Delta\rho^p = 0$, and the total varying strain $\Delta\varepsilon^{T\!P}$ equals to the accumulated creep strain $\Delta\varepsilon^c$ induced by creep relaxation.

Fig.8 shows the variation of the elastic follow-up factor Z with creep dwell time for the holed plate subjected to varying thermal loads from 0 to $1.5\sigma_{t0}$ for three alternative methods all of which give similar constant values. Solution 1 consists of a single solution for a prescribed dwell time. Solution 2 assumes Z remains constant during a sequence of increments so that Z can vary during the dwell period and gives a solution that is nearly identical to the more efficient Solution 1.The remaining solution consists of generating the rapid cycle solution and then allowing the stress to relax as a step by step process. Fig. 9 shows the variation of elastic follow-up factor Z with creep dwell time varying thermal loads from 0 to $0.8\sigma_{t0}$, i.e. within shakedown.

It is worth noting that values of Z is virtually the same for all three methods of calculation and that the values are relatively low compared with Z=3 frequently used in practice. From other examples we conclude that the value of Z is rarely greater than 2 where a distinct thermal stress concentration occurs. Note also that the material data used in this calculation ignores the variation of creep rate with temperature. When this is included the Z values reduce further giving values not significantly greater than unity. The life assessments of structures under creep/fatigue conditions are sensitive to the value of Z and accurate estimates of creep strain are required for accurate life assessments.

CONCLUSIONS

The paper briefly discusses the way in which recently developed programming method, the Linear Matching Method may be applied to the various stages of a life assessment method. Although the methods have been applied to R5, the British Energy high temperature life assessment method, the general methodology may be applied to either life assessment or design calculations. The flexibility of the method indicates that it is possible to develop a sequence of calculations that are essentially concerned with two types of calculations, both associated with the cyclic behaviour of a structure under thermal and mechanical loading. The first calculation assumes that the residual stress field remains constant and finds solutions for which the accumulation of strain over the cycle is compatible. Such solutions include limit and shakedown analysis, creep rapid cycle solutions and creep rupture limits. The second calculation involves the change in residual stress associated with changes in load during the cycle. Such a calculation may be used as the first stage of a ratchet calculation, and as a method for evaluating the amplitude of plastic strain. With further adaptations the creep strain during creep dwell periods may be found and cyclic strain hardening may be included.

The particular sequence of solutions described here is by no means unique and must be regarded as a first attempt at producing a design or life assessment process that still relies upon elementary materials data but produces estimates that are specific to the geometry and loading history of the structure.

These methods have been applied to a number of specific cases to assess the advantage of this approach over conventional methods. We find, in all cases, the greatest advantage comes from the ability to produce accurate estimates of plastic strain range and elastic follow-up factors. In addition there are cases that go beyond the limits of rule based methods but these methods are able to produce good results. There are therefore good reasons to believe that the adoption of these methods will produce more accurate predictions, particularly of creep/fatigue failure and allow a much wider range of load history to be assessed.

ACKNOWLEDGEMENTS

The authors gratefully acknowledge the support of the Engineering and Physical Sciences Research Council and British Energy Ltd during the course of this work. All solutions discussed in this paper have been generated using the finite element code ABAQUS.

REFERENCES

1. Ainsworth R.A. (editor), 1997, R5: Assessment Procedure for the High Temperature Response of Structures, Issue 2, Nuclear Electric.
2. Seshadri R., 1990, The effect of multiaxiality and follow-up on creep damage. J. of Pressure Vessel Tech 112, 378-385.
3. Boyle J.T., Hamilton R., Shi J., Mackenzie D., 1997, Simple method of calculating lower-bound limit loads for axisymmetric thin shells. J. of Pressure Vessel Tech 119, 236-242.
4. Engelhardt M., 1999. Computational modelling of Shakedown. PhD thesis, Univrsity of Leicester.
5. Chen H. F. & Ponter A.R.S., 2001, Shakedown and Limit Analyses for 3-D Structures Using the Linear Matching Method, *International Journal of Pressure Vessels and Piping*, Vol. 78(6), pp. 443-451
6. Ponter A.R.S., Cocks A.C.F., 1994, Computation of shakedown limits for structural components (Brussels Diagram)-Part 2-The creep range. Nuclear Science and Technology, European Commission, Report No. EUR 15682EN, Luxembourg.
7. Ponter A. R. S., Fuschi P. and Engelhardt M., 2000, Limit analysis for a general class of yield conditions. *European Journal of Mechanics, A/Solids*. 19:401-421.
8. Ponter A.R.S., Engelhardt M., 2000, Shakedown Limits for a General Yield Condition. European Journal of Mechanics, A/Solids 19, 423-445.
9. Ponter A.R.S. & Chen H. F., 2001, A minimum theorem for cyclic load in excess of shakedown, with application to the evaluation of a ratchet limit, European Journal of Mechanics, A/Solids, 20 (4), 539-553.
10. Chen H. F. & Ponter A.R.S., 2001, A Method for the Evaluation of a Ratchet Limit and the Amplitude of Plastic Strain for Bodies Subjected to Cyclic Loading, European Journal of Mechanics, A/Solids, 20 (4), 555-571.
11. Ponter A.R.S., Chen H. F., Boulbibane M. and Habibullah M., The Linear Matching Method for the evaluation of limit loads, shakedown limits and related problems, Fifth World Congress on Computational Mechanics, July 7-12, 2002, Vienna, Austria.
12. Ponter A.R.S., Chen H. F. and Habibullah M., Computaion of Ratchet Limits for Structures subjected to Cyclic Loading and Temperature, 2002 ASME Pressure Vessel and Piping Conference, Vancouver, British Columbia, Canada, August 4 - 8, 2002
13. Chen H. F. & Ponter A.R.S., 2002, Methods for the Evaluation of Creep Relaxation and the Amplitude of Reverse Plastic Strain for Bodies Subjected to Cyclic Loading, University of Leicester.
14. ABAQUS, User's Manual, Version 5.8, 1998
15. Chen H. F. & Ponter A.R.S., 2002, Integrity Assessment for 3D Tubeplate using Linear Matching Method, University of Leicester.
16. Chen H. F. & Ponter A.R.S., Linear matching method for creep rupture assessment, *International Journal of Pressure Vessels and Piping*, submitted, 2003
17. Neuber H., 1961, Trans ASME Jnl. Appl. Mech. 28, 544-550.

Table 1 Summary of Solution Sequence based on the Linear Matching Method (LMM)

Stage	Variable	Calculation Method	Subsidiary calculation/result	Comments
1	Temperature $T(x,t)$	Transient temperature history		Same as R5
2	Elastic stresses $\hat{\sigma}(x,t)$	Transient elastic stress history		Same as R5
3	Shakedown limit	Elastic shakedown		LMM
4	Creep damage – time to creep rupture	Extended shakedown solution	Evaluate t_R	LMM
5	Creep Deformation $\Delta\varepsilon^c$	Rapid cycle creep solution, BO method, constant $\overline{\rho}$	Identify reverse plasticity region for Stage 7	LMM estimate ignoring relaxation
6	Plastic strain range $\Delta\varepsilon^p$	Reverse plasticity solution	Fatigue cycles to failure N_0 from data	LMM
7	Plastic ratchet limit	Shakedown solution assuming cyclic hardening	Factor of safety on mechanical load λ	LMM with σ_y defined by stage 5.
8	Creep damage – Elastic follow-up factor	Monotonic creep computation, starting from rapid cycle solution	Creep endurance limit D_c from data, hence N_0^*	Uses standard ABAQUS routine
		Creep-reverse plasticity solution method		LMM

Table 2 The definition of load domains for the holed plate

Case	The cyclic thermal load $\Delta\theta$	$\Delta\varepsilon^p$	$\Delta\varepsilon^{rp}$
Case a	$1.5\sigma_{t0} \to 0 \to 1.5\sigma_{t0} \to 0 \to 1.5\sigma_{t0} \cdots$	2.869×10^{-3}	$\Delta\varepsilon^p + \Delta\varepsilon^c(t)$
Case b	$0.8\sigma_{t0} \to 0 \to 0.8\sigma_{t0} \to 0 \to 0.8\sigma_{t0} \cdots$	0	$\Delta\varepsilon^c(t)$

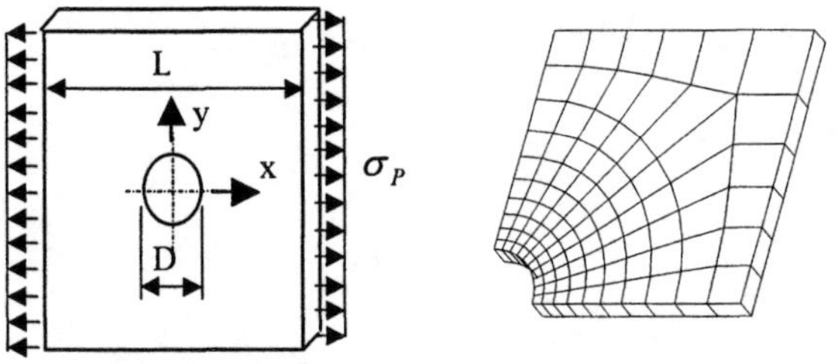

Fig 1 The geometry of the holed plate subjected to axial loading and fluctuating radial temperature distribution and its finite element mesh

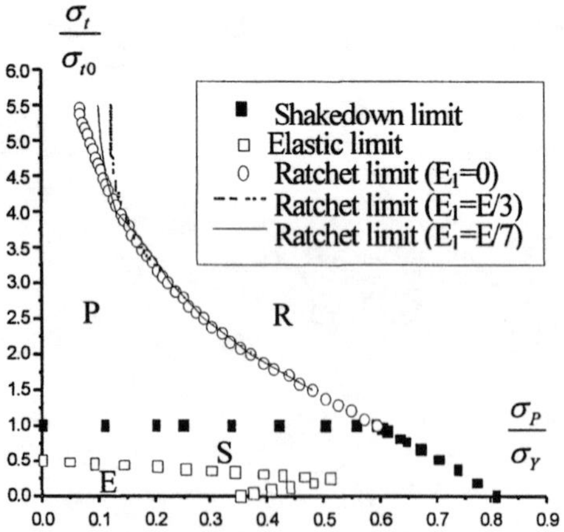

Fig. 2 The elastic, shakedown, reverse plasticity and ratchet region for the holed plate with mechanical and thermal loading

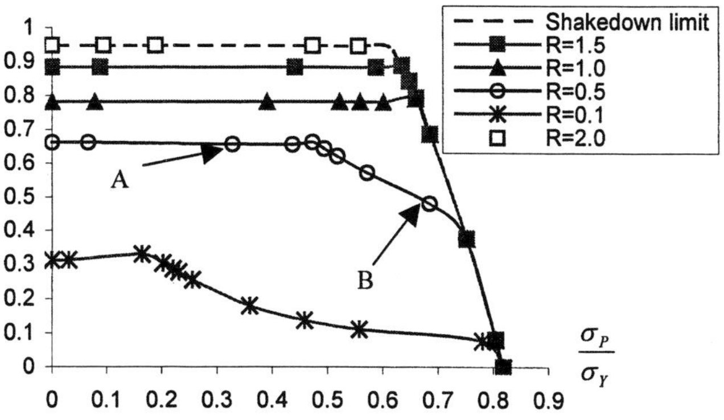

Fig.3 The shakedown limits for the 3D holed plate for five different creep parameters (R=0.1, 0.5, 1.0, 1.5 and 2.0)

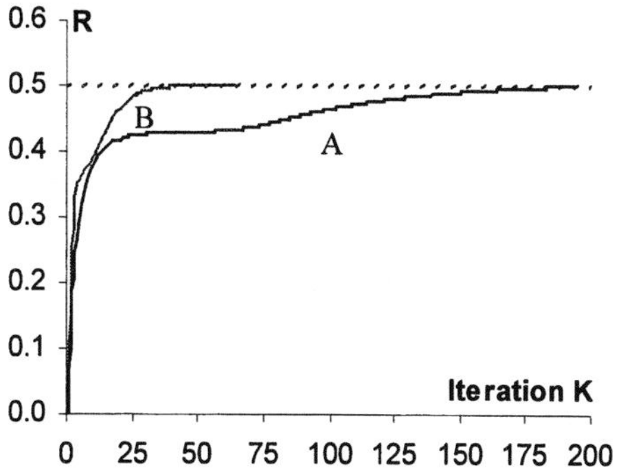

Fig.4 The convergence conditions for the solution of the optimisation of creep parameter R for shakedown to occur

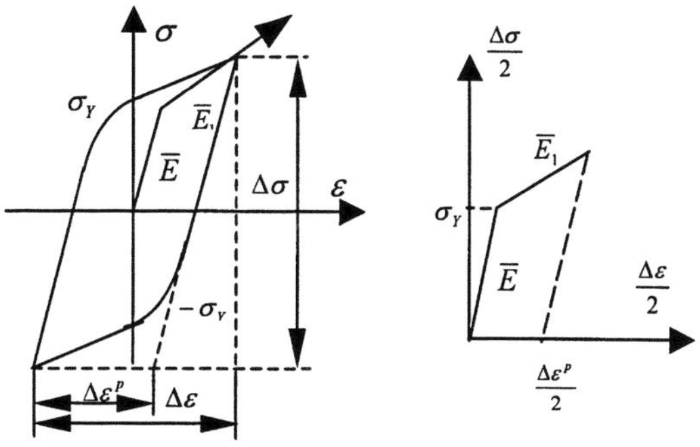

Fig. 5 The elastic-linear strain-hardening model

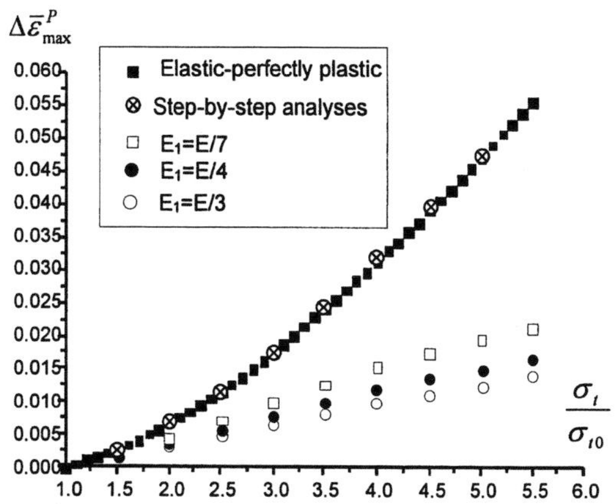

Fig.6 The maximum of the varying plastic strain magnitudes in the reverse plasticity region,

$$\Delta\bar{\varepsilon}^P_{\max} = \sqrt{\frac{2}{3}\Delta\varepsilon^P_{\max}\Delta\varepsilon^P_{\max}}$$

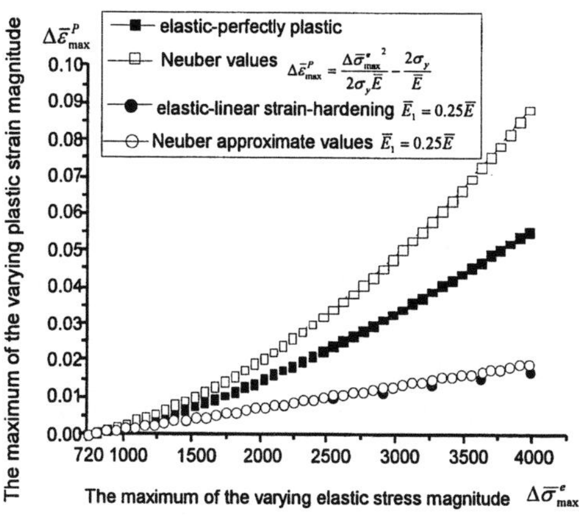

Fig.7 The comparison of the maximums of the varying plastic strain magnitudes in the reverse plasticity region by our method and Neuber approximate values

$$\left(\overline{E} = \frac{3E}{2(1+\nu)},\ E = 208GPa,\ \nu = 0.3,\ \sigma_y = 360MPa\right)$$

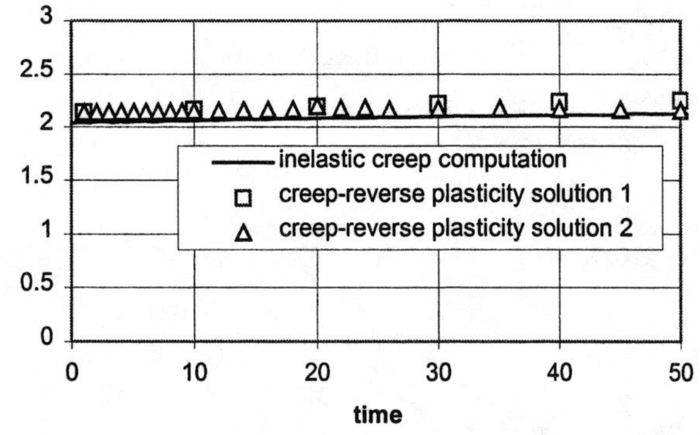

Fig .8 The elastic follow-up factor Z with creep dwell time for a holed plate subjected to varying thermal loads from 0 to 1.5 σ_{t0}

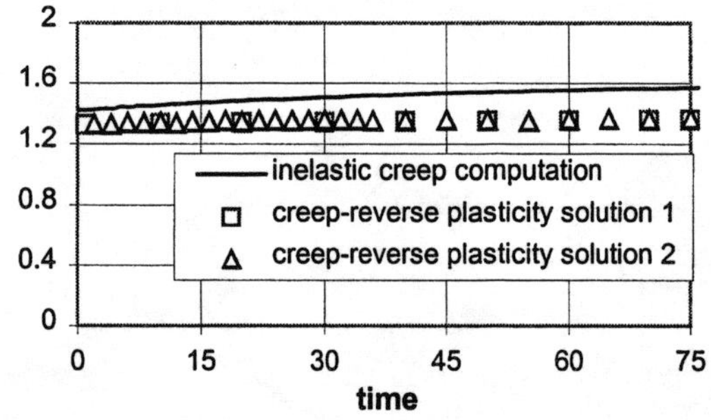

Fig .9 The elastic follow-up factor Z with creep dwell time for a holed plate subjected to varying thermal loads from 0 to 0.8 σ_{t0}

PVP-Vol. 458, Computer Technology and Applications

PVP2003-1885

DISTINGUISHING RATCHETING AND SHAKEDOWN CONDITIONS IN PRESSURE VESSELS

W. Reinhardt
Babcock & Wilcox Canada
581 Coronation Boulevard
Cambridge ON N1R 5V3
CANADA

ABSTRACT

The nature of the boundary between stable cycling and ratcheting is discussed using several illustrative example scenarios. The examples are analyzed in the context of the elastic methods currently in the ASME Code to demonstrate the conservatism of the existing approach that exists in some cases, and the unconservative estimation that exists in others. It is shown that the limit on the linearized primary plus secondary stress range can be related to conditions for elastic shakedown in certain kinematic hardening models of plasticity, while the limits on thermal stress ratchet address only scenarios similar to the Bree problem.

INTRODUCTION

Pressurized components are generally subjected to a combination of cyclic thermal and pressure loads. The resulting stress state at points inside the structure includes primary (load controlled) and secondary (deformation controlled) stresses and can, therefore, exceed the yield limit of the material without causing immediate plastic failure (collapse). However, cyclically recurring loads (transients) that exceed the yield limit of the material can cause deformations which accumulate with each cyclic load application, i.e. ratcheting. Ratcheting, which is also called incremental collapse, can render a pressure vessel unserviceable and is therefore an undesirable condition that must be taken into account in a design analysis. Design by analysis rules for pressure vessels and piping such as those contained in the ASME B&PV Code, Section III Subsection NB, Section VIII Div. 2 [1] and ASME B31.1 [2], provide stress limits for this pupose.

The classical shakedown theory consists of the theorems of Melan [3], or lower bound shakedown theorem, and of Koiter [4], which provides an upper bound for the shakedown loads. By suitable manipulations of the stress or displacement fields, the bounds converge to a unique value, at which shakedown is established for a certain load system. The classical theorems aim at a state in which the structure is eventually elastic everywhere, but this state may occur only after the cyclic load has been applied a number of times. In fact, the necessary number of load applications may be infinite as long as the plastic strains per cycle in the structure converge to zero. During this "shakedown" phase the structure may exhibit accumulating deformations, but this "transient ratcheting" is acceptable since its amount is limited.

Of the classical theorems, particularly the Melan theorem is attractive for design purposes because its application guarantees a lower bound of the shakedown loads, and there is no doubt that cumulative deformations are impossible in a structure where the stresses are elastic everywhere. In the classical context "shakedown" is always used in this sense, i.e. the structure becomes eventually completely elastic. The proponents of the classical approach have seen this as an advantage because a structure that is completely elastic is not only safe against ratcheting, but also against low cycle fatigue.

From the practical point of view, however, the classical shakedown concept is not satisfactory as a design limit because all structures contain points of stress concentration, where the yield stress is often exceeded locally. These points are justified for the intended cyclic operation with a fatigue analysis, and pose no problem otherwise. Ratcheting is not a concern if the plastic action remains confined to a local region in the structure because for the continued deformation of the structure, bulk plasticity is required. In other words, a small surface scratch in a structure may cause a local stress concentration and localized plasticity, maybe even to the initiation of a fatigue crack, but it will not lead to a failure by ratcheting. A plastic mechanism is necessary for a ratcheting failure, just like a plastic mechanism is necessary for a non-cyclic plastic collapse under primary loads. The design Codes [1], [2] recognize the irrelevance of local plasticity for shakedown by imposing limits on the linearized stress range.

The situation where cyclic plasticity, but no ratcheting, occurs in a structure will be subsequently referred to as plastic shakedown. There are unfortunately no classical plastic shakedown theorems similar to those that exist for elastic shakedown. Ponter and others [5,6] were partially successful in extending the classical theory but the full set of upper and lower bound theorems is still not known. In the absence of the complete theory, the problem of defining the load level at which ratcheting starts to occur in a structure can be solved only through an actual cyclic elastic-plastic analysis or by methods involving suitable conservative simplifications. A number of such methods have been proposed, and the present paper intends to categorize and evaluate them to aid the designer in an informed decision. All FE results presented in this paper were obtained with ANSYS [7].

DEFINITION OF SHAKEDOWN

In the following, a structure will be considered that is subject to a system of loads that is cyclic with a period T (which could include loads that are constant in time, of course). The system of loads consists of primary loads, like pressure, and secondary loads, like applied temperatures and externally applied displacements (anchor motions). The response of the structure is divided into three possible types, namely shakedown, ratcheting or plastic collapse. Of these, the safety of the structure against plastic collapse is evaluated during the (stationary) primary stress analysis and is therefore regarded as given here.

Unlike other sources that advocate the use of elastic shakedown criteria, e.g. [8], this paper will not consider plastic cycling in the structure as a mode of failure by itself. It is assumed that plastic cycling, if it occurs, is addressed in an elastic-plastic fatigue analysis and that ratcheting itself is the failure mode that needs to be evaluated. The ASME Code [1] supports this view in NB-3228.4 (b) with the statement "... and the design shall be considered to be acceptable if shakedown occurs (as opposed to continuing deformation), ...". Shakedown is therefore defined as the absence of continuing, accumulated deformation (ratcheting).

MODES OF FAILURE ASSOCIATED WITH CYCLIC LOADING

The primary stress evaluation ensures that a pressure vessel does not experience any excessive plastic deformation under a single load application. It also establishes safety against gross failure (e.g. burst), although the actual margin of safety against this failure mode is typically not known from the analysis. The fatigue analysis considers the effect of cyclic stresses that may lead to cracking, crack growth and eventually to a catastrophic failure. It is obvious that these failure modes must be prevented. The question is, however, whether the fatigue analysis by itself addresses the failure due to cyclic loads sufficiently. The ASME Codes [1,2] give a negative answer to this question.

Ratcheting is a separate failure mode since deformations accumulate as load cycles are applied. This will finally render the vessel unserviceable. The concern here is similar to that of the primary stress analysis, namely to prevent excessive permanent (plastic) deformations. Although its mechanism is of course different, ratcheting has a similar effect as creep, i.e. the failure of the component is incremental. In contrast to creep it can be avoided altogether, however.

It is known that a prolonged creep deformation of a component will eventually lead to a fracture due to the accumulation of damage within the material. The same is true for a component that ratchets. The damage caused by the ratcheting strain is not fully accounted for in the fatigue analysis [9], which deals only with the reversing part of the deformation. Only the occurrence of shakedown allows therefore the use of a fatigue analysis with one of the methods described in [1].

Demonstrating shakedown means that the cumulative deformation will end at some point, and a purely reversed strain cycling will occur from this point onward. However, some amount of permanent deformation will be incurred before the onset of stable cycling. Particularly when the loading is severe, it may be necessary to evaluate this "transient ratcheting" deformation. The accompanying plastic strains build up the residual stress field in the structure that is responsible for causing shakedown. The applied (secondary) strain ranges therefore typically limit the amount of deformation during the transient phase. Concerns are justified e.g. if the structure is subjected to very severe thermal strains or if there are very stringent deformation limits on a component. An example where the deformation of a component can be critical is at a sealing surface. In NB-3228.4, the ASME Code reminds the designer therefore explicitly of the possible existence of deformation limits: "... and the design shall be considered to be acceptable if shakedown occurs ..., and if the deformations which occur prior to shakedown do not exceed specified limits." Although this statement is only found in the section on plastic analysis, the same consideration could be necessary when elastic shakedown criteria are used.

MODELLING OF MATERIAL BEHAVIOUR

Modeling the detailed plastic response of a metallic material to general cyclic loading is very complex. The response includes directional (kinematic) hardening, cyclic (isotropic) hardening, deformations of the yield surface, and everything may depend on the non-proportionality of the loading. The development of plasticity models capable of representing the observed material shakedown and ratcheting behaviour is the topic of ongoing research. A universally applicable model does not exist currently. Hassan et al. [10] reviewed the performance of several existing plasticity models against experimental data and found that none of the models was satisfactory in its prediction of the ratcheting response. It is therefore not safe to assume that the models supplied by FEA software predict the ratcheting response of a given material with any accuracy, even if the uniaxial stress strain curve is represented well. Unsurprisingly, this applies even more to the simple linear hardening material models. These will be discussed below.

However, a full description of ratcheting is not required for a design limit. Only a conservative description of the shakedown boundary is needed, and a perfectly-plastic description is generally adequate for this purpose. In what sense such a conservatism exists opposite simple hardening models will be discussed briefly below. The advantage of the perfectly-plastic description is that it is consistent with limit analysis, and that once the plasticity model has been chosen (typically the Tresca or von Mises model), only a single material property is needed, namely the yield stress.

Which value should be used for this yield stress is important. It is not correct to use the yield that is obtained from a unidirectional tension test, except perhaps for the hydrostatic test that occurs at the beginning of the operating life and consists of only a few cycles. After being subjected to plastic cycling, most materials tend to harden or soften compared to their initial state. Very simplistically said, materials that are initially soft tend to harden while initially strong materials tend to soften. To use the initial yield stress for materials that exhibit cyclic hardening is conservative. On the other hand, experiments with softening materials have shown that these materials can shake down initially and start ratcheting after undergoing a number of stable cycles.

Therefore, if the cyclic material behaviour were known, it would be logical to use the cyclic yield stress which is also often used in a cyclic fatigue analysis. The ASME Code [1] implies that 1.5 Sm is the cyclic yield stress. This is fairly adequate, since it implies the use of the unidirectional yield stress for low strength material which are expected to harden, but a value below the unidirectional yield stress for higher strength materials where Su determines Sm, and 1.5 Sm is less than the unidirectional yield stress. For austenitic materials, 1.5 Sm could actually be above the unidirectional yield at temperature, but then these materials typically harden quite strongly, So in the absence of more detailed knowledge of the cyclic yield, the use of 1.5 Sm is conservative.

DEFINITIONS AND FUNDAMENTALS OF THE PERFECTLY PLASTIC MODEL

A small deformation formulation is adopted, and an elastic-perfectly plastic material is considered with the following properties:

1. The total strain is composed of an elastic (e) and a plastic (p) component, as well as an applied (given) thermal strain (T)

$$\varepsilon_{ij} = \varepsilon_{ij}^{e} + \varepsilon_{ij}^{p} + \varepsilon_{ij}^{T} \tag{1}$$

$$\varepsilon_{ij} = \frac{1}{2}\left(\frac{\partial u_i}{\partial x_j} + \frac{\partial u_j}{\partial x_i}\right) \tag{2}$$

2. The elastic behaviour follows Hooke's law:

$$\varepsilon_{ij}^{e} = C_{ijkl}\sigma_{kl} \tag{3}$$

where C_{ijkl} is here approximated as independent of temperature.

3. The thermal strains are given from the local temperature T by

$$\varepsilon_{ij}^{T} = \alpha\left(T - T_{ref}\right) \tag{4}$$

Again, α is taken to be independent of temperature.

4. Using the temperature independent cyclic yield stress σ_{yc}, the yield function is defined by

$$f(\sigma_{ij}) \le \sigma_{Yc} \tag{5}$$

where the yield function f has the properties of a norm, i.e. .

$$f(\sigma_{ij}) \ge 0 \text{ and } f(\sigma_{ij}) = 0 \Leftrightarrow \sigma_{ij} = 0 \quad \text{(positive definite)} \tag{6}$$

$$f(\sigma_{1\,ij} + \sigma_{2\,ij}) \le f(\sigma_{1\,ij}) + f(\sigma_{2\,ij}) \quad \text{(triangle inequality)} \tag{7}$$

$$f(\mu\,\sigma_{ij}) = |\mu|\, f(\sigma_{ij}) \quad \text{(homogenous of degree 1)} \tag{8}$$

5. The plastic flow rule is

$$\begin{aligned} \dot\varepsilon_{ij}^{p} &= g_{ij}(\sigma_{kl}) \quad \text{if} \quad f(\sigma_{ij}) = \sigma_{yc} \\ \dot\varepsilon_{ij}^{p} &= 0 \quad \text{if} \quad f(\sigma_{ij}) < \sigma_{yc} \end{aligned} \tag{9}$$

6. For a stress state c on the yield surface and another permissible stress state *, Drucker's principle holds

$$\dot\varepsilon_{ij}^{pc}\left(\sigma_{ij}^{c} - \sigma_{ij}^{*}\right) \ge 0 \tag{10}$$

Drucker's principle entails that plastic flow is normal to the yield surface ($g_{ij} = \lambda \frac{\partial f}{\partial \sigma_{ij}}$ outside singularities), and that the yield surface must be convex. These properties are directly related to the stability of plastic flow, and many useful relationships concerning shakedown can be derived from them.

For the above material model, it can be proven [11] that a steady cyclic state of stress and strain rate will be achieved eventually, though not necessarily after a finite (or even small) number of steps. In other words, if the loading is periodic with period T, then

$$\sigma_{ij}(t+T) = \sigma_{ij}(t) \quad \text{and} \quad \dot\varepsilon_{ij}(t+T) = \dot\varepsilon_{ij}(t) \tag{11}$$

after the transient phase. Moreover, during the transient phase the solution is guaranteed to converge to the stable periodic state. Of course, all this is only true if plastic collapse does not occur.

Note that the strain (in particular, the plastic strain) does not necessarily converge to a periodic state, since otherwise ratcheting would be impossible. The condition for shakedown is then

$$\int_{t}^{t+T} \dot\varepsilon_{ij}^{p}(\tau)\, d\tau = 0 \tag{12}$$

after stabilization. Elastic shakedown takes place if $\dot\varepsilon_{ij}^{p} \equiv 0$ throughout the stable cycle. For plastic shakedown, the strain rates cancel over a full cycle according to (11), no matter at what instance t the starting point of the cycle is defined. For ratcheting, each steady cycle gives rise to a fixed strain increment.

For any assessment that involves the steady cycle, it is especially attractive to make use of the purely elastic cycle that would be obtained with a linear-elastic material. The elastic material attains its steady cycle immediately, and the solution requires much less computational effort than the elastic-plastic solution. The purely elastic stress response will be denoted by $\sigma_{ij}^{(e)}$. By using the fact that in the steady cycle, the stresses and elastic strain rates of the elastic-plastic material are also periodic, the following relationship for the plastic energy per cycle can be established [11]

$$\int_{0}^{T}\left(\int_{V}(\sigma_{ij} - \sigma_{ij}^{(e)})\, \dot\varepsilon_{ij}^{p}\, dV\right)dt = 0 \tag{13}$$

This relationship holds for the steady cyclic condition, regardless whether it is elastic or plastic shakedown or ratcheting. Note that both the elastic-plastic and the purely elastic stress fields are in equilibrium with the same external loads, so that their difference is in equilibrium with zero external loads. In other words, it is a residual stress field. The strain energy calculated from any residual stress field and a kinematically admissible strain rate is zero because the work put into the structure by the (zero) external loads acting on the corresponding displacement field is zero.

SHAKEDOWN THEOREMS

For elastic shakedown, an upper and lower bound theorem have been formulated, and these theorems can be used advantageously to derive approximate solutions for the shakedown problem. The lower bound theorem was derived in full generality by Melan [3]. Consider the time dependent purely elastic response, $\sigma_{ij}^{(e)}$, to the applied thermal and mechanical loads. The structure will shake down if a time-independent residual stress field exists $\sigma_{r\,ij}$ such that the sum with the purely elastic stress lies inside the yield surface at all times within the cycle:

$$f(\sigma^{(e)}_{ij} + \sigma_{r\,ij}) \le \sigma_{Yc} \tag{14}$$

Some important properties follow from eq. (13). Secondary stresses themselves have the properties of a residual stress, i.e. they are self-equilibrating. Therefore, a secondary stress that is constant throughout the cycle has no effect - only cyclic secondary stresses make a contribution. For example, when a uniaxially loaded specimen is subjected to purely secondary (deformation controlled) loading cycles with a sufficiently large range, an initially applied mean stress relaxes completely (Lemaitre et al., [12]). The same is not true for the primary stresses, since a residual stress cannot equilibrate a constant primary stress. A purely reversed loading results in stable plastic cycling, therefore it is often the constant primary stress that induces ratcheting. The same effect sometimes can be achieved with a spatially varying secondary stress, however.

Melan's theorem remains valid in the limiting case where the cyclic load is zero and there is only a stationary (primary) load. The residual stress is the plastic stress redistribution in this case, and Melan's theorem becomes a different way of expressing the lower bound collapse theorem of limit analysis. This means that, as the cyclic stress drops to zero, the elastic shakedown region joins the limit collapse line.

For plastic shakedown, Melan's theorem must be supplemented with a time-varying residual stress that ensures that the elastic-plastic stresses do not fall outside the yield surface. Polizotto [6] proposed a restriction on the time variant residual stress term that might lead to an extension of the lower bound theorem to the plastic shakedown regime

The upper bound shakedown theorem (Koiter [4]) uses an expression for the elastic energy per cycle similar to equation (13). For the plastic strain rate, however, now any field $\dot{\varepsilon}^{p}_{0\,ij}$ may be used for which the accumulated strain over a cycle, $\Delta\varepsilon^{p}_{ij}$, is kinematically admissible, i.e. associated to a kinematically admissible displacement field through

$$\Delta\varepsilon^{p}_{ij} = \frac{1}{2}\left(\frac{\partial \Delta u_i}{\partial x_j} + \frac{\partial \Delta u_j}{\partial x_i}\right) \tag{15}$$

Note that the plastic strain rate need not be kinematically admissible throughout the cycle, just its integral over the whole cycle. The condition (15) means that the elastic strains must be periodic for the considered cycle (the thermal stresses are a load and therefore periodic by the definition of a load cycle). The integral

$$\int_0^T \left(\int_V (\sigma_{0\,ij} - \sigma^{(e)}_{ij})\, \dot{\varepsilon}^{p}_{0\,ij}\, dV \right) dt \tag{16}$$

is then greater than or equal to zero for any admissible strain rate $\dot{\varepsilon}^{p}_{0\,ij}$ if the structure shakes down. In the integral, $\sigma_{0\,ij}$ is the stress on the yield surface that is associated with the strain rate $\dot{\varepsilon}^{p}_{0\,ij}$

The significant property of the integral (16) in conjunction with eq. (15) is that it is invariant if an arbitrary residual stress, $\overline{\sigma}_{r\,ij}$, that is constant over the cycle, is added to the stress difference $\sigma_{0\,ij}$-$\sigma^{(e)}_{ij}$. That means, if a residual stress can be found such that the elastic stress path is completely within the yield surface, then the integral (16) is positive for any pair $\sigma_{0\,ij}$, $\dot{\varepsilon}^{p}_{0\,ij}$ by Drucker's principle (10). It is not even necessary to find the constant residual stress, because the integral is invariant under its application. From Melan's theorem, it can then be concluded that the structure shakes down to elastic action. By eq. (13) (and obviously because for shakedown to elastic cycling, the plastic strain rate ends up being zero throughout), the actual steady cycle leads to the integral (16) being zero. For an arbitrary admissible plastic strain rate field, the upper bound to the elastic shakedown load can be obtained by increasing the loads on the structure (or equivalently, by factoring the elastic stress) until the integral (16) becomes zero. For non-cyclic primary loads, (16) can again be shown to reduce to the corresponding upper bound theorem of limit analysis. When the elastic stress cycle lies partly outside the yield surface no matter what constant residual stress is added, the integral (16) will be less than zero for some strain rate fields.

SHAKEDOWN PROPERTIES OF HARDENING PLASTICITY MODELS

Two hardening models will be considered, namely linear kinematic hardening and linear isotropic hardening. The kinematic hardening model has the hardening rule

$$\dot{\sigma}_{\alpha} = K\dot{\varepsilon}^{p}_{ij} \tag{17}$$

The flow rule becomes

$$\dot{\varepsilon}^{p}_{ij} = \lambda(\sigma_{ij} - \sigma_{\alpha\,ij}) \quad , \quad f(\sigma_{ij} - \sigma_{\alpha\,ij}) \le \sigma_y \tag{18}$$

The hardening rule of the isotropic model is

$$\dot{R} = K\dot{\gamma} \tag{19}$$

where $\dot{\gamma} = \sqrt{\dot{\varepsilon}^{p}_{ij}\dot{\varepsilon}^{p}_{ij}}$, with the flow rule

$$\dot{\varepsilon}^{p}_{ij} = \lambda(\sigma_{ij}) \quad , \quad f(\sigma_{ij}) \le R \tag{20}$$

Note that both types of hardening rules are unbounded, i.e. that unidirectional hardening continues forever. Therefore, a structure made from these two material does not collapse plastically under a primary load. In other words, in the context of the present small deformation theory, any applied unidirectional load can be supported if the structure just deforms enough. It is therefore logical that these material models do not allow any ratcheting to occur after some transient period [13]. The length of that period depends on the linear factor K in the hardening rule and on the amount of plasticity per cycle.

For the kinematic hardening model, any ratcheting strain per cycle $\Delta\varepsilon^{p}_{ij}$ would cause hardening in the same direction by virtue of (17), and the back stress $\sigma_{\alpha\,ij}$ would increase in the same direction until ratcheting ceases. Therefore, only either elastic shakedown or plastic shakedown can occur. The structure experiences elastic shakedown if the yield surface can enclose the purely elastically calculated stress path. An equivalent statement, which resembles Melan's theorem, is that a constant stress exists such that the purely elastic stress plus the constant stress lies within the yield surface (Melan [3]). The difference to Melan's theorem is that this time it is just a constant stress that can be added, not necessarily a constant residual stress.

Since the equivalent strain increment $d\gamma = \dot{\gamma}\, dt$ is positive whenever a finite plastic deformation occurs, the linear-isotropic model continues to harden until all cycling is eventually fully elastic. Therefore, only shakedown to elastic action exists. This situation represents cyclic hardening but is of course not realistic in the sense that for actual materials, the amount of hardening is limited. However, Ponter and Karadeniz [14] pointed out an interesting aspect of this model. Consider the two-bar mechanism depicted in Fig. 1. Both bars are coupled and undergo the same displacements. One bar is thicker than the other (shown with twice the section of the thinner bar in Fig. 1), and the thinner one is subject to plastic thermal cycling while a constant load is suspended from the system. When the bars are perfectly-plastic, the thin bar cycles plastically, and the resulting load can be equilibrated by a like section of the thick bar, while the remaining section of the thick bar carries the primary load. As long as the thick bar does not yield, the primary load is carried safely no matter how high the temperature cycles are. The deformation of the system is limited by the elastic deformation of the thick bar. On the other hand, if both bars consist of an isotropic hardening material, the cyclic plasticity causes the thin bar to harden and the cyclic load on the thick bar to increase commensurately. This may cause the thick bar to become plastic, and although both bars will be elastic eventually, the amount of deformation during the transient phase would be higher that the prediction of the perfectly plastic material model. In the example of Fig. 1, the displacement of the hardening model was ten times that of the perfectly plastic one. This should be kept in mind when assessing the deformations during the transient shakedown phase. The perfectly plastic model may not give conservative estimates of the "transient ratcheting" deformation if the material of the structure exhibits strong cyclic hardening.

A comparison to experimental data [10] shows that both the linear kinematic and the linear isotropic plasticity models are not confirmed in practice. Ratcheting is observed in real materials. However, the prediction of ratcheting requires a plasticity model that predicts the plastic response of a real material to general non-proportional loading. Plasticity models that can do this are not currently available (Hassan [15]), and further research is needed in this area. For the purpose of an ASME Code analysis, such detail is not required because ratcheting is not an acceptable operating mode. A conservative estimate of shakedown can be obtained using a perfectly plastic material. As outlined in the previous paragraph, it may be necessary to investigate the transient deformations with that reproduces the expected amount of cyclic (isotropic) hardening in addition to using the perfectly plastic model.

METHODS OF SHAKEDOWN ANALYSIS

Classification

The key objective in shakedown analysis is to determine the steady cycle. Once the steady cycle is known, checking for strain accumulation is a fairly simple task. If the strains are periodic everywhere in the structure, shakedown occurs. If not, the structure ratchets under the applied loads and is unacceptable. The computational methods can be broadly classified into those that perform a transient cyclic plastic analysis and those that do not. The methods that rely on the transient cyclic plastic analysis simulate the actual path towards the steady cycle that the structure would take, whereas the other methods use a different iterative scheme or essentially a guess to establish the steady cycle. In the following, the cyclic method and the elastic core method will be discussed as examples of transient cyclic plastic methods, while the elastic-based ASME Code "3Sm" method, some methods based on Melan's theorem and Ponter's method represent the other class.

The Elastic Method

The ASME Code, Section III Subsection NB uses the range of primary and secondary stress to assess shakedown, which, for linear stress distributions, can be expressed as

$$f(\Delta\sigma_{ij}^{(e)}) \leq 2\sigma_y' \qquad (21)$$

where $\Delta\sigma_{ij}^{(e)}$ is the difference between any two points of the elastically calculated stress history, and σ_y'=1.5 Sm is the cyclic yield stress defined by the Code. It is remarkable that no real distinction is made between primary and secondary stress, since only the combined range enters eq. (21). This seems at odds with the observation that a constant primary stress can contribute to ratcheting as demonstrated already by the Bree problem. Since eq. (21) can be understood as the requirement that purely elastic stress path must fit inside the yield surface, the Code criterion essentially requires elastic shakedown. Fitting the yield surface around the stress path can be regarded as an effective method of finding the optimum time invariant stress state which, when added to the elastically calculated stress, causes elastic shakedown. If such a stress state can be found, this method will find it. The method cannot assure that the additive constant stress state is a residual stress state as demanded by Melan's theorem for elastic shakedown in a perfectly plastic material. Therefore, it is consistent with the linear kinematic material model, not with the perfectly plastic one. In other words, it is assumed that there is no restriction on the translation of the stress path when trying to fit it into the yield surface, while perfectly plastic plasticity permits only residual stresses to effect the translation by eq. (14). Of course, the primary stress criteria restrict the magnitude of the (constant and cyclic) primary stress in the vessel, which mitigates any potential unconservative effect opposite the perfectly plastic model. However, several examples are given below where an unconservative assessment can occur.

If the stress distribution in a component is nonlinear through the thickness, only the linearized stresses need to be evaluated [1]. This reflects the idea that ratcheting requires a cyclic plastic mechanism in the structure in a similar way as plastic collapse requires a sufficient number of plastic hinges in the structure. Localized plasticity at a notch does not cause collapse, and in the same way it does not cause ratcheting in the steady state sense, although ratcheting-like transient behaviour of the plastic strains is often observed [16]. In other words, ratcheting only happens if, during the steady cycle, each element of an entire section of the structure plastifies at some point in time. The section never plastifies entirely (that would be plastic collapse), but different parts at different times during the cycle. Depending on the redundancy, the structure may ratchet only after several sections have plastified. Stress linearization works well for through-thickness effects, but not for local "hot spots" on shells which are frequently encountered, particularly when 3D analysis is necessary. The analyst can address such hot spots only by a judiciously locating the "stress classification lines" where linearization is done. Correct placement depends on the engineering judgement of the analyst, and even experienced analysts may disagree on a particular case. A potential benefit in introducing plastic analysis is that the results may be more consistent from analyst to analyst and less subject to debate.

The Classical Bree Problem

The classical Bree problem [17] applies to pipes and vessels with linear through-wall temperature gradients and pressure loading. In its abstracted form it is a beam in plane stress subject to a constant (primary) axial force and a variable linear temperature gradient through the thickness (Fig. 2). Without loss of generality one can choose to let the temperature gradient vary between zero and a maximum value.

The finite element model achieves shakedown to elastic or cyclic plastic action within one cycle for the Bree problem (Fig. 2). The results are identical with the theoretically expected ones. This behaviour may have created the expectation that any problem will either shake down very rapidly or continue to ratchet. This impression is incorrect as will be demonstrated in later examples. The theoretical shakedown boundary is described by

$$\Delta\sigma_{SB} = \frac{\sigma_y'^2}{\sigma_{PM}} \quad \text{if } |\sigma_{PM}| \le \frac{1}{2}\sigma_y'$$
$$\Delta\sigma_{SB} = 4(\sigma_y' - \sigma_{PM}) \quad \text{if } |\sigma_{PM}| > \frac{1}{2}\sigma_y' \qquad (22)$$

These are the limits given by the Thermal Stress Ratchet criterion in NB-3222.5 of Section III [1].

Fig. 2 shows that the ASME Code "3 Sm" criterion is not a conservative representation of the shakedown boundary if the constant primary stress exceeds 2/3 Sm (= 0.5 σ_y'). The Design stress criteria would limit the primary stress to Sm = 2/3 σ_y'. The potentially unconservative region would therefore be fairly small. Furthermore, at least in vessel design, pressure is not normally considered as non-fluctuating. The only loads that are constant over time are deadweight loads, and the stresses from these tend to be extremely small in nuclear pressure vessels. The same may not be true for support structures.

The case where the membrane stress is almost entirely caused by a fluctuating pressure is therefore more realistic. The membrane stress is taken to vary between 0 and some tensile value. To investigate the effect of the fluctuating membrane stress, the load histories depicted in Fig. 3 were examined. The Code allowable region changes for this new type of history. Since the primary stress now has a range, the secondary stress range is not independent of the primary stress any more. The allowable secondary stress range is 2 σ_y' at zero primary stress range, but only 4/3 σ_y' if the primary stress range is at its allowable limit of 2/3 σ_y'.

A proportional increase and decrease of the membrane stress with the thermal gradient (Fig. 3 a) results in a shakedown boundary significantly above the classical Bree diagram of Fig. 2. For this condition, the Code limit is quite conservative everywhere (Fig. 3). The condition depicted in Fig. 3b is more representative of components in severe thermal service: the membrane stress (pressure) ramps up to its maximum value, and then several thermal cycles (here two) occur at constant membrane stress. After this the membrane stress ramps down and a new cycle begins. So there are 2 thermal cycles to one pressure cycle. Under this condition, the original Bree diagram is recovered exactly. This means that the Bree diagram applies to cases where the pressure variation occurs rarely compared to the thermal cycling and not only to the case where the pressure is constant.

Due to the change in the boundaries of the Code allowable region, the Code 3 Sm limit is now conservative for every combination of primary and secondary stress. There is only one common point between the Code acceptable region and the shakedown boundary (at a membrane stress of 2/3 σ_y'). For this case which is typical for vessel analysis, the Thermal Stress Ratchet criterion of NB-3222.5 is not needed except to justify a secondary stress range above 3 Sm.

The Primary Bree Problem

Another variation on the Bree problem is the case where both the fluctuating moment and the constant axial force are primary loads (Fig. 4). There is a ratcheting region in this case, which supports the notion that a combination of stress fields with different time dependence is the crucial element that causes ratcheting. On closer inspection, the Bree diagram shown in Fig. 4 can be obtained directly from the diagram in Fig. 2 if the range of the reaction moment on the elastic-plastic beam is calculated for the Fig. 2 case. Fig. 4 shows that there is little difference between the ratcheting boundary and the collapse line for the beam. It appears to be a general rule that the ratcheting region for a combination of primary loads is fairly small [11]. Therefore, the factor of 1.5 that the Code applies on collapse is found sufficient in the present example to avoid ratcheting under a combination of primary loads. The example supports the conclusion by Kalnins [16] that a structure that is subject only to primary loading below the Code Design limit will shake down.

The Inverted Bree Problem

In looking at the Bree problem, one might ask, what happens if the beam is subjected to a constant primary moment and a fluctuating secondary membrane load (Fig. 4). This situation is called the inverted Bree problem here. It represents a simplified scenario akin to the loading on certain thermally loaded tubesheets, and could also resemble piping or nozzle problems where a moment due to deadweight and thermal membrane stresses act simultaneously.

When the thermal membrane stress fluctuates with an amplitude of 2σ_y' or more, it will obviously cause plasticity throughout the beam at every cycle, and the beam will not be able to support any additional moment. Conversely, if the moment reaches the limit moment of the beam, the beam will be at incipient collapse and will not sustain any thermal cycling. It can be shown that the shakedown boundary is a straight line between the described extremes:

$$\Delta\sigma_{SM} = 2\,\sigma_y' - \frac{4}{3}|\sigma_{PB}| \qquad (23)$$

where $\Delta\sigma_{SM}$ is the elastically calculated thermal membrane stress range and σ_{PB} is the elastically calculated bending stress. Unlike the classical Bree problem, the inverse Bree problem does not converge to the cyclic solution within the first cycle. The finite element result shown in Fig. 4 was obtained after 25 full cycles using a relative increase in deflection of 10^{-4} as the convergence criterion. On average, the result is still 10% below the theoretical boundary. This indicates that the convergence of the plastic solution to steady cycling is often quite slow in many practical problem. Kalnins [16] found a similar slow convergence for plastic cycling at a notch.

If the applied moment is strictly constant, the shakedown boundary from the Code 3 Sm criterion is everywhere unconservative, except at the point where the bending moment is zero. The Thermal Stress Ratchet criterion in NB-3222.5 offers no alternative because it is applicable only to shells with Bree-type loading. If applied, it would be unconservative in this case.

The Three-Bar Problem

Each of the bars in the three-bar mechanism shown in Fig. 5 is subjected to a heating-cooling cycle in its turn while the end displacement of the bars is coupled. Only one bar is hot at any time, so the within a cycle each bar is hot once, at which time the elastically calculated stress in that bar is compressive. Twice during the cycle when one of the other bars is hot, the elastic stress in the same bar is tensile, and reaches a magnitude of half of the compressive stress. This example from the book by Gokhfeld and Cherniavski [11] is a simplified representation of the thermal stresses in pipes and vessel shells when a thermal gradient moves spatially (e.g. a fluctuating level between a cold and hot fluid).

.

The example can be discussed using Melan's theorem. Suppose that the bars are made from an elastic-perfectly plastic material and that the higher (compressive) stress just exceeds yield. The elastically calculated stress range in each bar at this point is slightly above 1.5 σ_y'. Elastic shakedown occurs if there exists a time-invariant residual stress that can "pull up" the stress in each bar such that it will not exceed yield any more. Each bar experiences the same stress history than the others, just with a phase shift. Therefore, a time-independent stress that moves the stress level in each bar by the desired amount would have to be the same in all three bars. This stress could not be a residual stress because it must have the same magnitude and sign in all three bars. Therefore, the structure cannot shake down to elastic action once yield is exceeded. Furthermore, since yield is exceeded in only one direction, the plastic strains must accumulate, so ratcheting occurs. The finite element results (Fig. 5) confirm that ratcheting occurs as soon as the yield stress is exceeded.

This example is significant for two reasons: firstly, ratcheting occurs under the application of a purely secondary cyclic load and secondly, the calculated purely elastic stress range is 1.5 σ_y' when ratcheting starts. The Code limit of 2 σ_y' is definitely unconservative in this case.

Conclusion – the 3 Sm Method

The Section III Code approach of evaluating shakedown on an elastic stress range basis results in a method that is easy to use. However, the simplifications lead to three significant shortcomings

- the significant effect of constant primary loads is neglected,
- the method allows any time-independent stress state to be added to the cyclic elastic stress, but in reality only a residual stress state is acceptable,
- the current method of allowing for plastic cycling (exclusion of through-thickness stress concentrations and the use of NB-3222.5 in conjunction with simplified elastic-plastic analysis) is very limited and approximate.

Using plastic methods to demonstrate shakedown addresses the shortcomings of elastic analysis, but requires presently a very much larger computational effort than elastic methods. For nuclear applications (PWRs and BWRs), the elastic analysis has been found to provide adequate protection in practice. Reasons are the level of conservatism in the material properties, the conservatism of evaluating a single section while in reality a mechanism of plastic hinges is required for ratcheting. Also, the thermal gradients were typically not sufficient for three-bar style failure. Some of these problems did surface in the design of Fast Breeder Reactors, however.

Methods Based on Melan's Theorem

Significant development work for the European codes focussed on the application of Melan's theorem. This avoids the first two shortcomings listed for the ASME Code 3 Sm method, but is not effective to address cyclic plasticity. Since Melan's theorem is lower bound, it is not necessary to find the actual stable cyclic solution, any approximation gives acceptable conservative loads. The practical difficulty is finding an efficient residual stress field that avoids a very conservative assessment. At present, typical finite element software is not equipped to solve such a problem, other than by simply following the transient solution. A potentially more efficient method would attempt to determine the residual stress field directly. A typical approach relies essentially on an inspired guess from the analyst [18], which may be cumbersome for complicated problems. It was also suggested to solve a nonlinear optimization problem [19], but the numerical technique is not yet fully developed, and presently the computational effort for real-world size problems is extremely high. Mackenzie and Boyle proposed an approach based on adjusting the local elastic modulus, which may currently be the most effective way to apply Melan's theorem in shakedown analysis.

When performed in conjunction with the actual (total) stress field in the component, Melan's theorem can result in a very conservative assessment of the shakedown limits. Even in the most optimistic case, the stress range anywhere in the structure is limited to 2σ_y'. This includes any local structural discontinuities, a very restrictive criterion, which also makes some common finite element modeling simplifications impossible (e.g modeling fillets as sharp corners). Therefore, the global use of Melan's theorem does not result in a practical design criterion. Zeman [18] proposed that "structural" stresses and strains be used, in essence the linearized stress. This puts the method roughly on the same level of conservatism as the ASME Code elastic (3 Sm) criterion. Since the lower bound criterion is satisfied, there are no unconservative cases like in the ASME Code method. However, the use of Melan's theorem requires the linearized stress for every wall section of the structure. Therefore, the method is really only convenient for beam and shell models.

Generally, Melan's theorem cannot be used to assess plastic shakedown. The methods that use it are therefore too conservative for structures with significant thermal loading.

The Cyclic Method

The simplest way to obtain the steady cyclic solution is of course to integrate over time. Once the solution has converged sufficiently, the absence of incremental collapse can be assessed. The Bree problem of Fig. 2 converges after the first cycle has been applied, and this may have led to the belief that convergence can be generally achieved within the first few cycles. In reality, fast convergence is rather the exception, and even relatively benign looking problems, like the cyclic plasticity at a cylindrical hole that was analyzed by Kalnins [16], may take a long time to shake down to all significant digits, possibly hundreds of cycles.

Since an elastic-plastic analysis for such a large number of cycles is not practical, it must be truncated at an earlier time with a suitable criterion that suggests shakedown. The Japan Pressure Vessel Research Council (JPVRC) proposed to define shakedown when the equivalent plastic strain increment over the previous cycle reaches an increment of less than 10^{-4}, while the history of all plastic strain increments indicates a decreasing trend [16, 20]. The decreasing trend is quite important because in the ratcheting structure, the strain increment becomes eventually linear and any decreasing trend at earlier cycles ends eventually, while for shakedown the strain increment must converge to zero.

In practice, this criterion works quite well. Most of the time it induces a conservatism into the shakedown evaluation because when the analysis is terminated after a set number of cycles, some cases that would eventually converge but have not yet converged far enough are not passed. A concern is, of course, the opposite case where the solution has converged far enough to pass the shakedown criterion, but would in reality continue to ratchet. Consider the worst case scenario where the ratcheting strain continues at 10^{-4} / cycle. It will take 100 cycles to reach a permanent strain of 1%. Code Case N-47-28, Appendix T, T-1310 for elevated temperature design defines 1% as the limit on the accumulated inelastic strain averaged through the thickness. Given that the method would be used for severe conditions where cyclic plasticity is possible, 100 cycles in not unrealistic. To cover cases with more cycles, it has been suggested to use a formula of the type

Limit on Equivalent Plastic Strain Increment = 0.01/N

where N is the number of cycles. A rough upper limit can be calculated as follows. Consider a component made from a typical low alloy steel with 3 Sm = 80 ksi. Assume that the linearized stress range is just above the 3 Sm limit, and that the stress distribution is linear. The alternating stress is then approximately 40 ksi, and the fatigue analysis will limit the life of the component to (approximately) 10000 cycles. This is an upper limit on the number of cycles that might be applied when plastic shakedown analysis is used. The strain limit corresponding to 1% accumulated strain would then be 10^{-6} / cycle. While theoretically satisfying, this bounding limit might require too many cycles in practice before the transient stresses subside sufficiently. Therefore, the JPVRC limit or the cycle-dependent limit would be preferred.

As a simple example, consider a hinged beam with constant uniformly distributed load and thermal cycling that causes an axial stress (Fig. 7). Except for the stress biaxiality, the example resembles an unstayed tubesheet subject to thermal and pressure loading. It is a slightly more complicated version of the inverse Bree problem. The curve labeled "shakedown" represents the ratchet boundary as closely as it could be established (about 0.5%). Note that it takes 40 cycles to reach the JPVRC level of accuracy of 10^{-4}. After the initial phase the decay of the incremental strain is approximately exponential. A level of 10^{-6} would have been reached at about 65 cycles (the curve was not extended further because the loss of significant digits when calculating the increment per cycle from the displayed solution caused too much "graininess").

The ratcheting solutions can be clearly identified in Fig. 7 as they branch off the exponentially decaying curve. The solution obtained with a 0.6% increase in the stress range opposite the shakedown range branches off after about 20 cycles, that with 1% increase branches off after 10 cycles. This indicates a fairly good resolution, i.e. a small increase in load will cause a significant deviation of the solutions. However, one should not expect to locate the shakedown boundary with any accuracy after only a few cycles. The 40 cycles required in the present example are by no means unusual. An assessment done after a small number of cycles could be quite conservative. Overall, the computational effort required for this method can be quite high.

Elastic Core

Kalnins [16] suggested that the elastic core method could be a simple and robust way to assess shakedown. The idea is that a structure cannot experience incremental collapse as long as a continuous elastic core in the structure that exists throughout the load cycle supports the applied loads. The same concept can be used for non-incremental plastic collapse; the structure collapses when regions of the structure plastify completely and thus form a plastic hinge. For a rigid-plastic structure, all hinges required to offset the redundancies form simultaneously at the moment of collapse, because there is no previous deformation that would allow plastic flow to occur. In an elastic-plastic structure, on the other hand, it is possible for plastic hinges to form successively. The structure does not necessarily collapse when the first hinge forms. Therefore, through-thickness plasticity, i.e. absence of an elastic core, in one location does not necessarily signify the onset of collapse or incremental collapse.

The beam-strut system in Fig. 8 is an example of a structure where through-thickness yielding occurs long before the onset of collapse or incremental collapse. A cantilever beam is laterally supported at its end by a strut, and a load is applied in the same direction. The example is an abstraction of a stayed tubesheet problem. Let index 1 denote the cantilever beam and 2 the supporting strut. The characteristic dimensionless parameters of the system in Fig. 8 are then

$$A_2 l_1^2 / I_1 = 3,\ l_2 / l_1 = 1,\ \sigma_{y1} / \sigma_{y2} = 1.5,\ E_2 / E_1 = 1.3,\ l_1 / h_1 = 4.5,$$

where A is the cross sectional area, l is the length, I is the moment of area, h is the height (of the beam = distance between the extreme fibres), E is Young's modulus and σ_y is the (cyclic) yield stress. The strut is considerably stiffer than the beam and yields at a displacement where the beam is still elastic. Fig. 8 shows the elastic-plastic response of the system to a cyclic load with range $F/(A_2\ \sigma_{y2}) = 12.15$. Although the strut is fully yielded during most of the cycle and the outer fibres of the beam yield slightly near the load extremes, the response is stable cycling. Incidentally, the shakedown boundary is found to be coincident with the collapse condition because only a single primary load is applied. Again, this example confirms the notion that a shakedown check is not necessary if the structure is loaded only by primary loads. However, the main point is that finding a place where the elastic core vanishes in a structure does not necessarily indicate that ratcheting is about to commence. In the example, the remaining elastic core in the beam still continues to support the load. For a realistic assessment of a redundant structure (and there are almost always some redundancies), the elastic core method requires some engineering judgement.

Strictly speaking, the elastic core method requires the knowledge of the steady cyclic solution because the plastic regions may grow during the transient phase [16]. Therefore, one might think that it does not offer a real advantage over the cyclic method. However, to date the known examples (e.g.[16]) suggest that elastic core might be more robust than the cyclic method. If the elastic core is small, the decision whether or not the ratcheting boundary has been reached is still difficult. The elastic core method is well suited to quickly decide cases where the plasticity is localized, however. Kalnins [16] found that clearly localized plasticity at a notch does not always imply that the cyclic strains converge quickly to the stable solution. Nonetheless, the elastic core method will clearly show the local character of the plastic regions throughout the load cycle without the need for an extremely long cyclic analysis. Further study of examples might be helpful in establishing a growth criterion for the plastic region to simplify the decision when the cyclic analysis can be terminated and a final assessment made.

Ponter's Method

Ponter [21] used an extended upper bound formulation "in excess of shakedown" to devise a method that will yield the shakedown boundary of the elastic or plastic shakedown problem. The problem is decomposed into two separate problems, which are solved with a linear matching technique that matches the response of a linear material to that of the elastic-plastic material. There appears to be no restriction on the type of problem that this method could be applied to. Iterations must be performed to arrive at the steady periodic solution, but not based on a simple integration of the cyclic solution in time. This is not a cyclic method in the sense outlined in the classification section of this paper. Convergence occurs uniformly from above (since this is an upper bound method), and examples show [5, 21] that the number of steps required is typically not less than what the cyclic method would require to converge to a periodic solution. Note that the problem needs to be run to full convergence because this is an upper bound method. However, at the end of this solution the shakedown boundary is known, not just some steady periodic solution inside or outside it.

At present, Ponter's method is the only way to determine the shakedown boundary directly. Therefore, it is the method of choice for this task. As a means to determine whether a given load history leads to shakedown or ratcheting, it is probably at par with the cyclic and elastic core methods. Unlike these, Ponter's method does presently require some user programming in standard FE software.

CONCLUSIONS

1. The ASME Code does not permit ratcheting in a structure. Shakedown to elastic action and shakedown to cyclic plasticity are both acceptable.
2. For a perfectly plastic material, a final steady periodic cycle of the stresses and strain rates will be reached. The final plastic strain is either zero (elastic shakedown), periodic (plastic shakedown) or linearly increasing (ratcheting).
3. Permanent incremental deformation can occur during the transient phase while the steady state is approached. Depending on the function of the structure, deformation limits may need to be specified, and the deformation from the transient phase may need to be evaluated. Cyclic hardening may have to be considered for this evaluation.
4. Shakedown is an asymptotic state. A perfectly plastic structure may need a significant number of cycles to approach it closely (50 cycles is not uncommon)
5. The elastic ASME Code (3 Sm) method is consistent with elastic shakedown for a material exhibiting unbounded kinematic hardening, not perfect plasticity. Compared to a perfectly plastic material, this criterion is unconservative in some cases and conservative when cyclic plasticity is possible.
6. The classical lower bound shakedown theorem (Melan's theorem) is overly conservative for many structures with high thermal loading when cyclic plasticity is possible.
7. If plastic cycling is investigated using time integration of the structural response, the convergence to stable cycling must be shown using the JPVRC or a similar criterion. Very rarely can the stable cycle be reached exactly.
8. The elastic core method is especially suited to quickly and efficiently assess situations where plastic cycling occurs in a localized region.
9. Ponter's method is efficient in looking for the shakedown boundary, as opposed to a yes / no assessment of a specific load set. However, it requires user programming and must be run to complete convergence since it is an upper bound method.

REFERENCES

1. ASME, 2001, *Boiler and Pressure Vessel Code*, American Society of Mechanical Engineers, New York.
2. ASME, 2001, *B31.1 Power Piping Code*, American Society of Mechanical Engineers, New York.
3. Melan, E., 1938, "Zur Plastizität des räumlichen Kontinuums," Ingenieur-Archiv, Vol. 9, pp. 116-123.
4. Koiter, W.T., 1960, "General Theorems of Elastic-Plastic Solids," in: Sneddon, J.N, and Hill, R. (eds.), *Progress in Solid Mechanics*, Vol. 1, pp. 167-221.
5. Ponter, A.R.S., Chen, H., and Habibullah, M., 2002, "Computation of Ratchet Limits for Structures Subjected to Cyclic Loading and Temperature," PVP Vol. 441, ASME PVP Conference, Vancouver, BC, Canada, pp. 143-152.
6. Polizzotto, C., 1993, "A Study on Plastic Shakedown of Structures: Part II – Theorems," ASME J. Appl. Mech., Vol. 60, pp. 324-330.
7. ANSYS, 2001, Ver. 6.0 Online User Guide, SAS IP Inc., Houston, PA.
8. CEN European Committee for Standardization, 2002, European Standard PrEN13445-3, *Unfired Pressure Vessels - Part 3: Design*, Final Draft.
9. Rider, R.J., Harvey, S.J., and Chandler, H.D., 1995, "Fatigue and Ratcheting Interactions," Int. J. Fatigue, Vol. 17 (7), pp. 507-511.
10. Corona, E., Hassan, T., and Kyriakides, 1996, "On the Performance of Kinematic Hardening Rules Predicting a Class of Biaxial Ratcheting Histories," Int. J. Plasticity, Vol. 12 (1), pp. 117-145.
11. Gokhfeld, D.A., and Cherniavsky, O.F., 1980, *Limit Analysis of Structures at Thermal Cycling*, Suthoff & Nordhoff, Alphen aan den Rijn, The Netherlands.

12. Lemaitre, J., and Chaboche, J.-L., 1990, *Mechanics of Solid Materials*, Cambridge University Press, Cambridge, UK.
13. Stein, E., Zhang, G., and Mahnken, R., 1993, "Shake-Down Analysis for Perfectly Plastic and Kinematic Hardening Materials," in: Stein, E. (ed.) *Progress in Computational Analysis of Inelastic Structures*, CISM Courses and Lectures, Vol. 321, Springer-Verlag, Wien-New York, pp. 175-244.
14. Ponter, A.R.S., and Karadeniz, S., 1985, "An Extended Shakedown Theory for Structures that Suffer Cyclic Thermal Loading. Part I: Theory," ASME J. Appl. Mech., Vol. 52, pp. 877-882.
15. Bari, S., and Hassan, T., 2001, "Constitutive Models with Formative Hardening and Non-Linear Kinematic Hardening for Simulation of Ratcheting," Transactions SMiRT 16, Washington, DC.
16. Kalnins, A., 2002, "Shakedown and Ratcheting Directives of ASME B&PV Code and their Execution," PVP Vol. 439, ASME PVP Conference, Vancouver, BC, Canada, pp. 47-55.
17. Bree, J., 1967, "Elastic-Plastic Behaviour of Thin Tubes Subjected to Internal Pressure and Intermittent High-Heat Fluxes with Application to Fast-Nuclear Reactor Fuel Elements," J. Strain Analysis, Vol. 2 (3), pp. 226-238.
18. Zeman, J.L.,2002, "The European Approach to Design by Analysis," PVP Vol. 439, ASME PVP Conference, Vancouver, BC, Canada, pp. 31-37.
19. Staat, M., 2002, "Some Achievements of the European Project LISA for FEM Based Limit and Shakedown Analysis," PVP Vol. 441, ASME PVP Conference, Vancouver, BC, Canada, pp. 177-185.
20. Okamoto, A., Nishiguchi, I., and Aoki, M., 2000, "New Secondary Stress Evaluation Criteria Suitable for Finite Element Analyses," ICPVT-9, Sidney, Vol. 2, pp. 613-620.
21. Ponter, A.R.S., and Chen, H., 2001, "A Minimum Theorem for Cyclic Load in Excess of Shakedown with Application to the Evaluation of a Ratchet Limit," Eur. J. Mech. A / Solids, Vol. 20, pp. 53

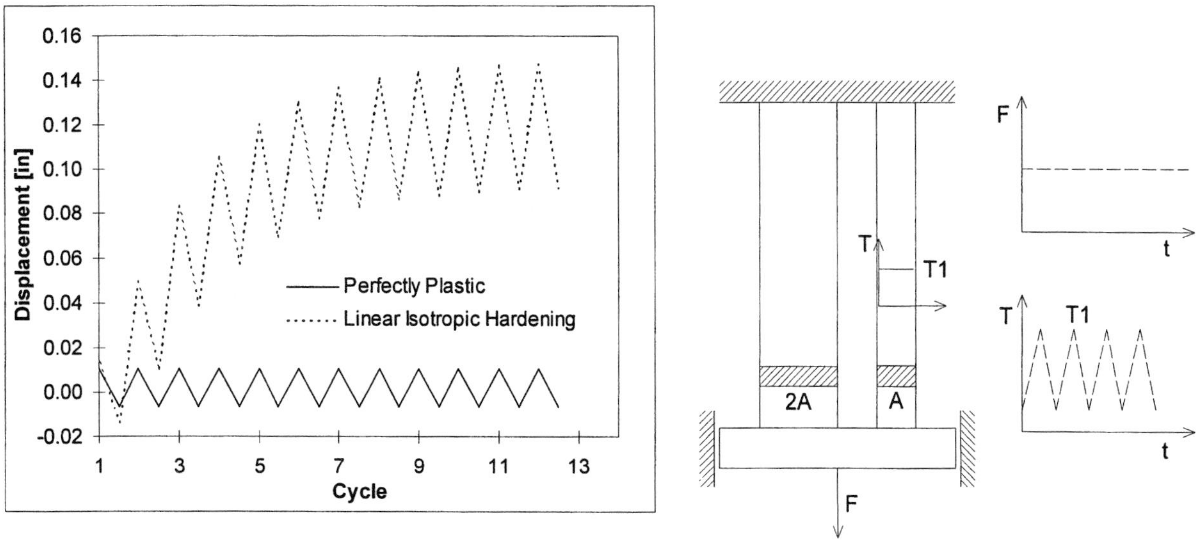

Figure 1: Transient cyclic behaviour of two bar mechanism. Linear isotropic hardening results in increased permanent deflection.

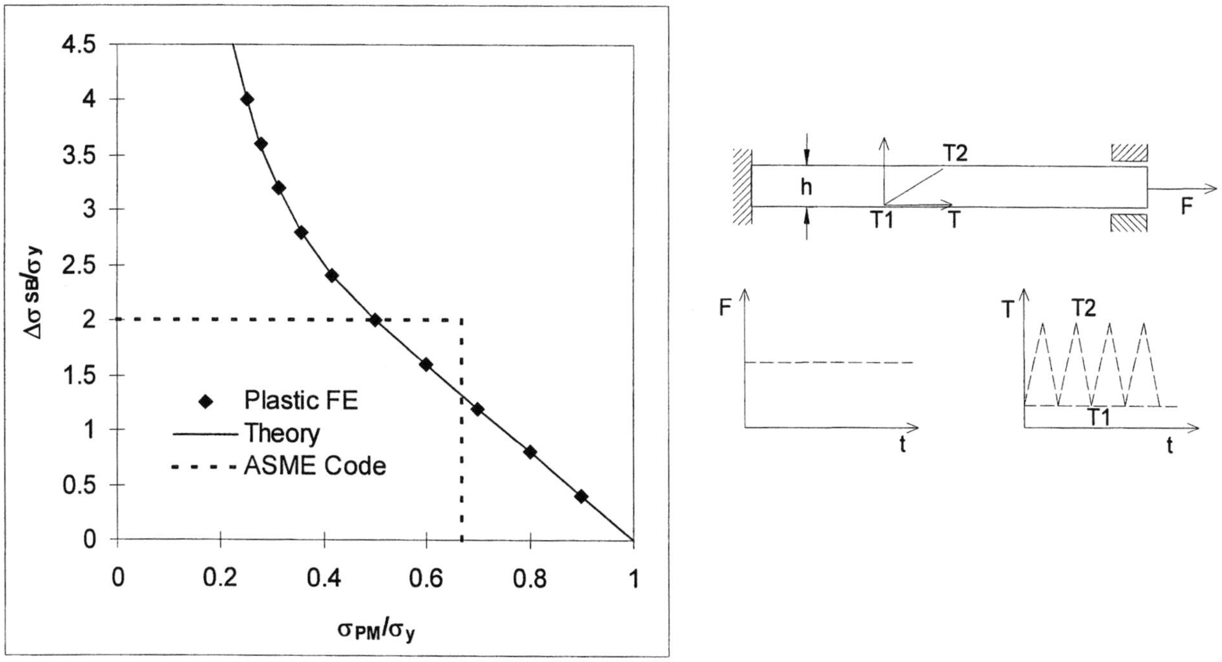

Figure 2: Shakedown boundary of the classical Bree problem (beam with constant primary axial load and cyclic secondary mending).

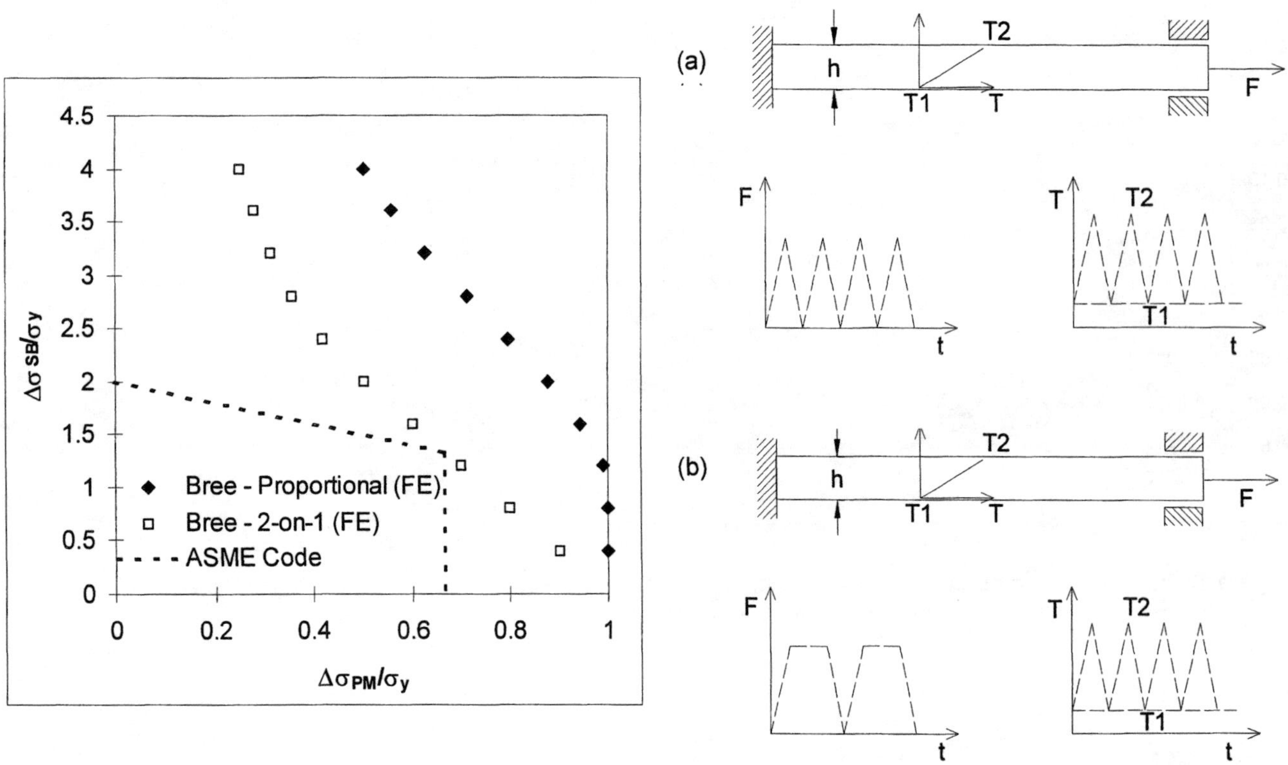

Figure 3: Shakedown boundary of the Bree problem with cyclic primary axial and cyclic secondary bending load – (a) proportional cycling, (b) 2 secondary cycles per primary cycle

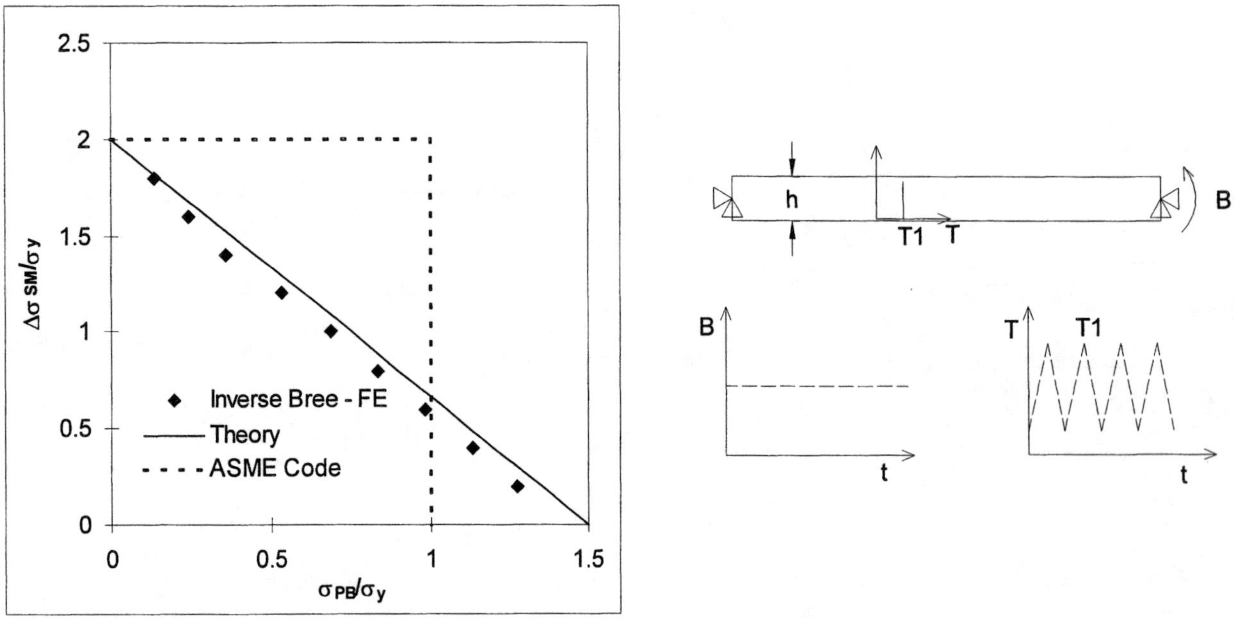

Figure 4: Shakedown boundary of the Inverse Bree problem (constant primary bending and cyclic secondary axial load). FE results were obtained after 25 cyclic load applications.

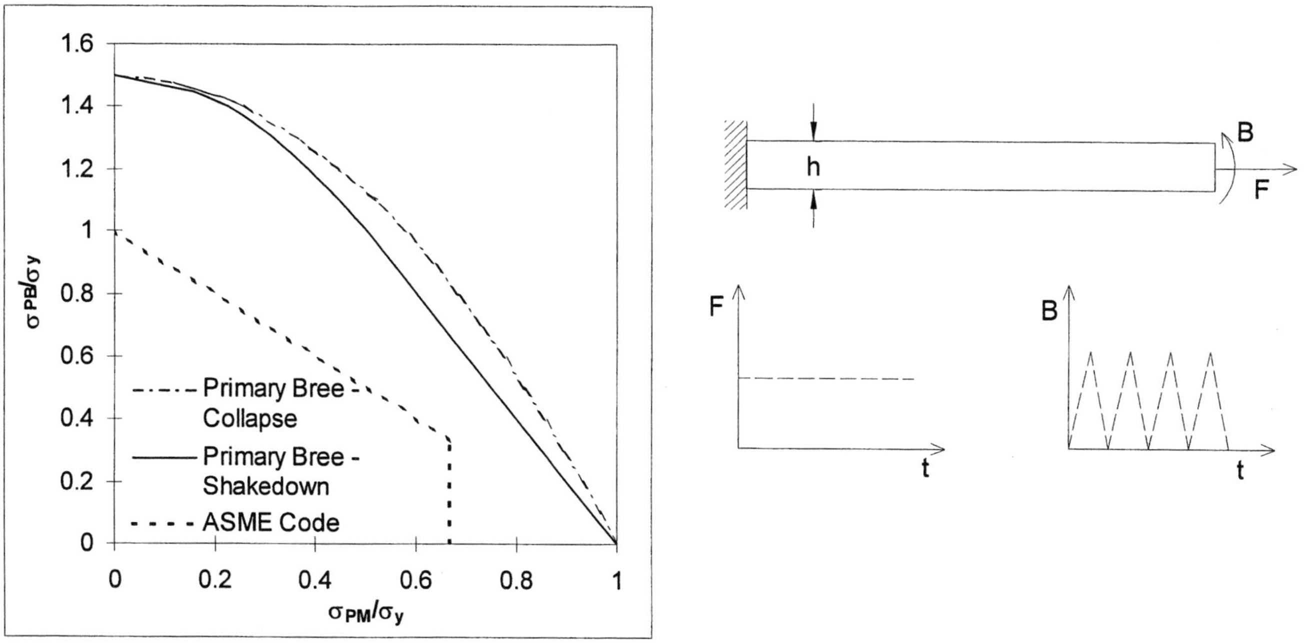

Figure 5: Shakedown boundary of the Primary Bree problem (constant primary axial and cyclic primary bending load).

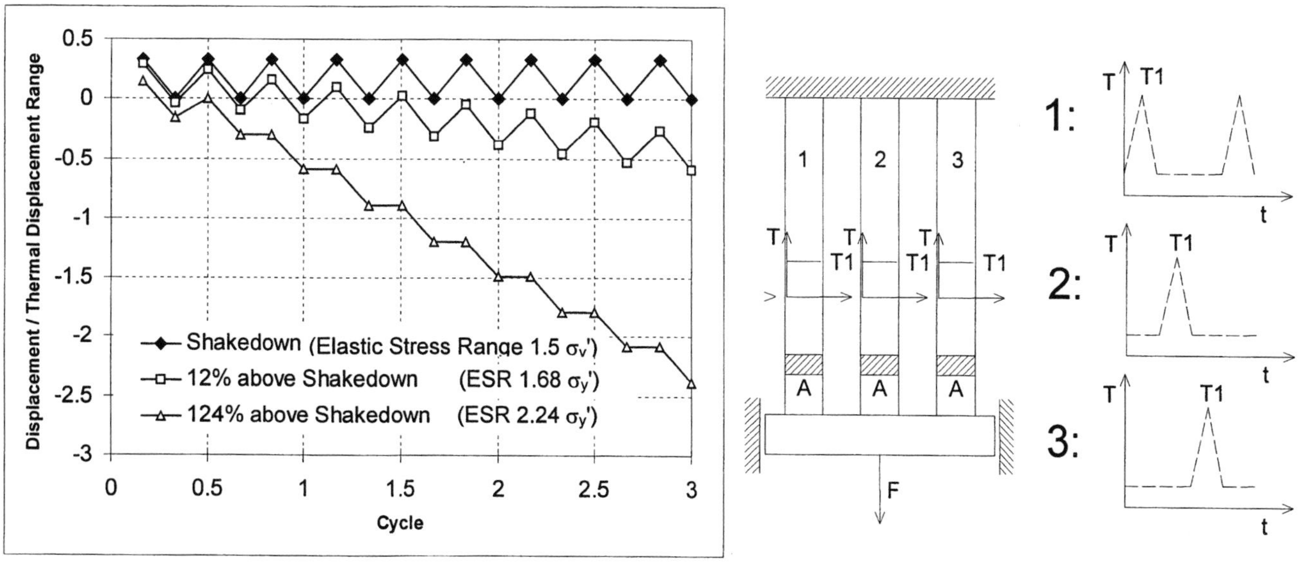

Figure 6: Shakedown of the three bar mechanism with sequential heating of each bar (part 1 to 3 of each cycle). Ratcheting occurs as soon as the yield stress is exceeded.

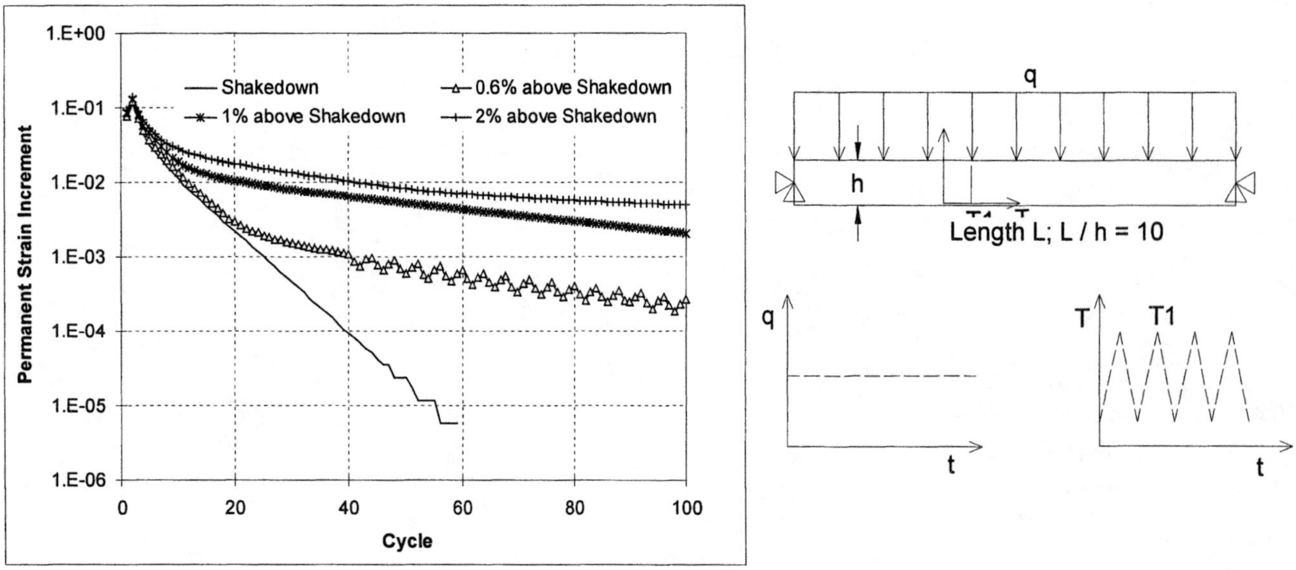

Figure 7: Transient behaviour of cyclic plasticity in a beam near the shakedown boundary (constant primary uniformly distributed shear load and cyclic secondary axial load). Percentages refer to the secondary stress range.

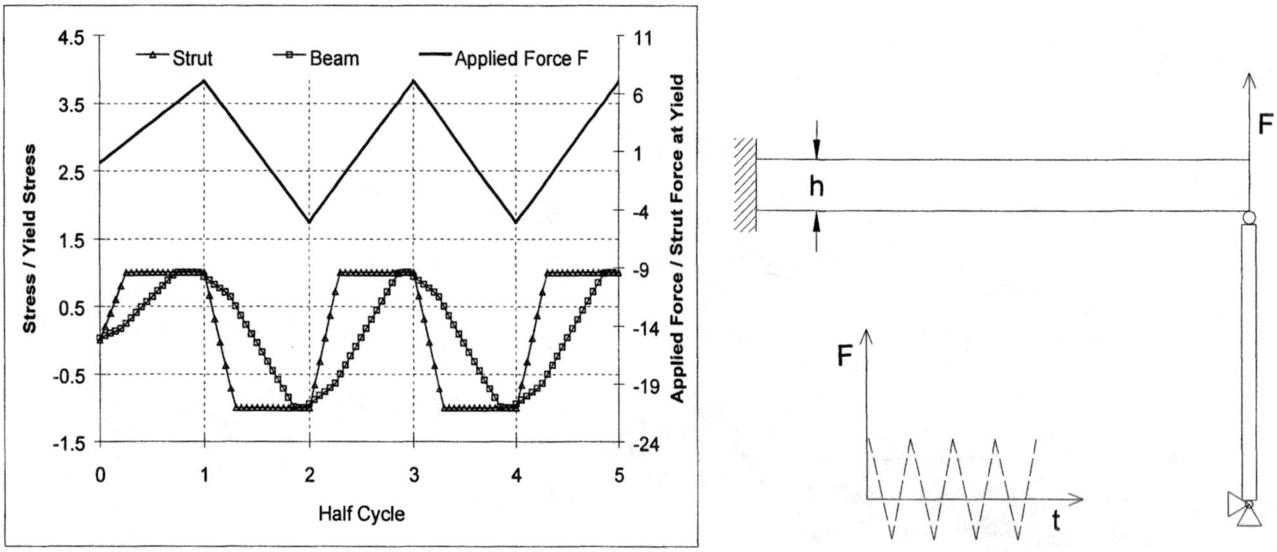

Figure 8: Possible cyclic response of cantilever beam – strut system with cyclic end load. Strut is fully plastic at each cycle.

PVP-Vol. 458, Computer Technology and Applications

PVP2003-1886

BELLOWS DESIGN EQUATIONS SUPPORTED BY LIMIT ANALYSIS

Robert K. Broyles
Senior Flexonics Inc. Pathway Division
Oak Ridge, TN. New Braunfels, TX.
865-483-7444
broylesb@pathwayb.com

ABSTRACT

Limit analysis can be used to design metal bellows for pressure capacity. The objective of limit analysis is to prevent gross plastic deformation with an appropriate design margin. The primary advantage of limit analysis over other methods is that it can find the max. allowable limit load for the structure as a whole and not just for the individual parts. In the case of bellows, an allowable limit pressure can be found which assures a specified margin on plastic collapse. In this paper, a parametric FEA study is performed on a series of two-dimensional axisymmetric models of un-reinforced U-shaped bellows with wide ranging dimensions. The non-linear analysis gives the max. allowable limit pressure for each bellows based on limit analysis and a closed form equation is confirmed to accurately describe the FEA results. Combined with existing equations for column instability and external buckling, limiting design pressure equations are presented for bellows design.

INTRODUCTION

Bellows are dynamic pressure vessels that must safely accommodate both system pressure and thermal expansion movements. This is possible because the convolutions of a bellows have a U-shaped profile. Experience and testing has shown that gross plastic deformation of this profile can lead to catastrophic failure. Therefore, it is imperative that bellows have an adequate design margin to avoid gross plastic deformation due to pressure.

The state of stress in a bellows is complicated. Primary stresses due to pressure are both circumferential and meridional in the convolutions. The membrane and bending stresses combine at the component level. Bellows stress analysis is usually performed with a series of closed form equations that address stresses at separate locations. The calculated stresses are then compared with Code allowable limits. By using limit analysis, it is possible to address the true state of stress in the bellows as a whole and assure an adequate design margin against gross plastic deformation.

LIMIT ANALYSIS

Limit analysis is used to find the maximum load a structure made of ideally plastic material can carry. The deformations of an ideally plastic structure increase without bound at this load, which is termed the collapse load. The analysis can be performed using a closed form equation or finite element analysis (FEA). With FEA, a computer model of the structure is created with appropriate boundary conditions. The model is then exposed to stepwise loading. Component stresses are calculated at each load step. A yield criterion is selected which specifies the state of multiaxial stress corresponding to the onset of plastic flow. The yield criterion can be represented geometrically by a fixed limit surface in the principal stresses space. Commonly used limit surfaces are Tresca and von Mises. The stresses at every point in the structure must lie in or on the limit surface. Since limit analysis is normally based on a yield criterion, it is only applicable for materials operating at temperatures below the creep range.

LIMIT LOADS

The limit load of a structure is defined by Kalnins [1] as the load level at and above which equilibrium of the model in its undeformed configuration cannot be maintained if the stress points in the stress component space are restricted to remain within or on the limit surface. It is calculated for a specified limit surface and a specified value of the limit stress. Usually the limit stress is equated to the yield strength of the material. At this stress, the limit load can be termed the yield point limit load. Below the yield point limit load, the deformation of the structure is acceptable. It is impossible to have loads greater than the limit load for perfectly plastic material.

FINITE ELEMENT ANALYSIS

The software used in this study is Cosmos/M Version 2.7 non-linear. The theory of failure currently used by the ASME Code is the max. shear stress theory; therefore, the Tresca limit surface criterion has been selected with the small displacement formulation The material response is elastic-perfectly plastic and the loading is applied in steps. The program multiplies the specified initial load by a load factor to produce the load increment. The step size is automatically controlled to converge on a lower bound solution. The analysis ends when either the load increment exceeds the initial load or the step size is less than $1x10^{-8}$. Provided the load increment does not exceed the initial load, the lower bound solution is the final load increment.

VERIFICATION PROBLEM

To confirm the accuracy of the software, a verification problem is performed. The problem is taken from WRC Bulletin 464 [1]. It consists of a simply supported beam that is 100 inches long between supports. The beam is 10 inches high and 1 inch wide. It is subjected to an initial uniform pressure load of 1000 psi against the top surface. The material properties have a modulus of elasticity of $30x10^6$ psi, Poisson's ratio of 0.28, and yield strength of 38000 psi. The model uses 4 node plane stress 2D elements. There are 10 elements across the height of the model. The nodes at the centerline support locations on either end are restrained from vertical displacement. The analysis was completed and an exaggerated deformed stress plot for the final load step is shown in Figure 1. The max stress intensity is 38000 psi as expected. The final load factor is 0.75542 so the lower bound solution is 755.42 psi.

The exact yield point limit load for the simply supported beam is determined as follows:

$$P_y = \frac{8S_yZ_p}{L^2} = \frac{8(38000)(25)}{(100)^2} = 760\,\text{psi} \tag{1}$$

The error of 0.59% is considered acceptable for this analysis.

BELLOWS LIMIT ANALYSIS

For bellows limit analysis, a finite element model is created for single ply U-shaped convolutions (See Figure 2). The bellows model has a 48 inch inside diameter (ID), 52 inch outside diameter (OD), 0.05 inch thickness (tp), and 2 inch pitch (q). The material properties have a modulus of elasticity of $30x10^6$ psi, Poisson's ratio of 0.28, and yield strength of 20000 psi. Because of symmetry, only half of one convolution is modeled. The model uses 4 node 2D axisymmetric elements and there are 4 elements through the thickness. The nodes at each end are restrained from horizontal displacement. The elements are subjected to an initial uniform internal pressure of 1000 psig. The analysis was completed and an exaggerated deformed stress plot for the final load step is shown in Figure 3. The analysis was repeated with external pressure on the elements and the results were identical as expected.

To check the accuracy of the solution, the same model was re-meshed with 8 elements through the thickness. The result was 64.06 psi, which is a difference of only 0.14%. Clearly, the 4-element mesh can provide accurate results.

To assure that progressive plastic deformation or ratcheting does not occur after repeated loading, the model was exposed to cyclic loading to 99.7% of the limit pressure or 63.8 psi. After each load cycle, the deformation was checked. The results indicate that shakedown occurs and the deformation does not increase after multiple load cycles.

Changes in geometry of the structure during load application can sometimes affect the results. To assure that geometric weakening does not occur, the analysis was repeated after selecting the updated lagrangian large displacement formulation. The max. stress intensity is still 20,000 psi as expected. The final load factor is 0.06363 so the lower bound solution is 63.63 psi. Since this value is within 0.5% of the small displacement value, the geometric changes during the load application do not significantly weaken the model.

PARAMETRIC BELLOWS LIMIT ANALYSIS

Additional parametric FEA models were generated for a series of bellows with different dimensions. Except for the dimensions, all other aspects of the models are the same. The analysis was completed and the results are shown in Table 1. The FEA limit pressure is the lower bound solution for the final load step.

Table 1 – Bellows Limit Pressure

Model No.	Inside Dia. (in.)	Outside Dia. (in.)	Thk. (in.)	Pitch (in.)	FEA Limit Pressure (psi)
1	48	52	.05	2	64.0
2	48	52	.03	2	26.3
3	48	52	.075	2	117.6
4	48	52	.05	1	55.2
5	48	52	.05	.5	59.5
6	48	50	.05	1	91.8
7	48	50	.03	1	52.4
8	48	50	.075	1	139.5
9	48	53	.05	2	40.0
10	48	53	.03	2	15.8
11	48	53	.075	2	90.8
12	96	100	.05	2	42.5
13	96	100	.03	2	21.7
14	96	100	.075	2	67.3
15	96	100	.05	1	54.9
16	96	100	.05	1.5	50.1
17	96	101	.05	2	37.8
18	96	101	.03	2	13.9
19	96	101	.075	2	71.4
20	24	28	.05	2	77.2
21	24	28	.03	2	32.2
22	24	28	.075	2	157.9
23	24	28	.05	1	57.8
24	24	28	.05	1.5	65.9
25	24	26	.05	1	165.3
26	24	26	.03	1	87.1
27	24	26	.075	1	258.5
28	12.75	15	.05	1.125	215.9
29	12.75	15	.03	1.125	87.8
30	12.75	15	.075	1.125	405.6
31	12.75	15	.018	1.125	36.4
32	12.75	15	.05	.875	210.0
33	6.63	8	.05	.688	506.1
34	6.63	8	.03	.688	219.3
35	6.63	8	.075	.688	815.9
36	6.63	8	.018	.688	88.8
37	18	19.652	.034	.683	154.3
38	17.875	21.849	.048	1.996	79.0

EXPERIMENTAL RESULTS

A test program was completed in 1983 that found the limiting pressures for a series of bellows with U-shaped convolutions. The results are summarized in an ASME paper by Thomas [2]. Models 37 and 38 in Table 1 represent the actual dimensions for test program bellows 1RA and 1HA, respectively. When corrected for the actual yield strength of the test bellows material, the FEA Limit Pressure is 266.2 psi for 1RA and 148.1 psi for 1HA. During the test program, significant plastic deformation was evidenced by inability to maintain pressure due to the increasing internal volume of the bellows. A review of the test program log sheets indicates that the pressure for the onset of significant plastic deformation was 250 psi for 1RA and 150 psi for 1HA. The test results compare favorably with the calculated FEA Limit Pressures.

LIMIT PRESSURE

Bellows design is complex because it involves concurrently satisfying requirements for pressure stress, deflection stress, fatigue life, spring force, stability, and size. The determination of an acceptable design is further complicated by the numerous variables involved such as diameter, thickness, pitch, convolution height, number of plies, material type, and heat treatment. In most cases, the design will involve a compromise of competing requirements. Being an iterative process, it is impractical to create an FEA model for each variation. Therefore, it is advantageous to have a single closed form equation for finding the limit pressure.

Using the results in Table 1, it is possible to find an equation that describes the FEA results. Equation C-28c from the EJMA Standards, 7th Edition [3] was developed to predict the development of plastic hinges in the bellows. The design equation is given as follows:

$$P_{si} = \frac{0.57S_y}{K_2\sqrt{\alpha}} \qquad (2)$$

where

$$K_2 = \frac{D_m}{2nt_p}\,\frac{K_r}{.571+2w/q} \qquad (3)$$

$$A_c = (.571q+2w)nt_p \qquad (4)$$

In accordance with changes planned for the 8th Edition of the EJMA Standards, the factor K_r is included in Eq. (3) to account for the effect of extension movements on circumferential stress. By substitution,

$$P_{si} = \frac{0.57S_y}{\frac{D_m}{2nt_p}\frac{K_r}{.571+2w/q}\sqrt{\alpha}} = \frac{0.57S_y}{\frac{D_m}{2nt_p}\frac{qK_r}{.571q+2w}\sqrt{\alpha}}$$

$$= \frac{1.14S_y(.571q+2w)nt_p}{K_rD_mq\sqrt{\alpha}} = \frac{1.14S_yA_c}{K_rD_mq\sqrt{\alpha}} \qquad (5)$$

From the EJMA Standards [3]

$$\alpha = 1 + 2\,\delta^2 + (1 - 2\,\delta^2 + 4\,\delta^4)^{0.5} \qquad (6)$$

where

$$\delta = \frac{K_4}{3K_2} \qquad (7)$$

$$K_4 = \frac{C_p}{2n}(w/t_p)^2 \qquad (8)$$

By substitution,

$$\delta = \frac{\frac{C_p}{2n}(w/t_p)^2}{\frac{D_m}{2nt_p}\frac{K_r}{.571+2w/q}} = \frac{w^2C_p(.571+2w/q)}{3K_rD_mt_p} \qquad (10)$$

In summary,

$$P_{si} = \frac{1.14S_yA_c}{K_rD_mq\sqrt{\alpha}} \qquad (11)$$

According to Koves [4], the design margin for Eq. (11) is 2.0. Therefore, the equation for the theoretical limit pressure is

$$P\text{limit} = 2P_{si} = \frac{2.28A_cS_y}{K_rD_mq\sqrt{\alpha}} \qquad (12)$$

To evaluate the Eq. (12) for accuracy, the limit pressure is calculated for each of the bellows in Table 1 beginning with Model No. 1. The values for Model No. 1 used in the equation are as follows:

$S_y = 20000$
$t = .05$
$t_p = .05$
$n = 1$
$q = 2$
$w = 1.95$
$D_m = 50$
$C_p = .64257$
$K_r = 1$

The value of the shape factor C_p is found by interpolation from Table I-1 in the EJMA Standards [3]. Since the model is for a U-shaped convolution without any applied movement, the value of K_r is 1.

Table 2 – Bellows Limit Pressure Comparison

Model No.	FEA Limit Pressure (psi)	Plimit (psi)
1	64.0	60.9
2	26.3	24.9
3	117.6	109.3
4	55.2	53.4
5	59.5	58.6
6	91.8	80.5
7	52.4	47.8
8	139.5	119.0
9	40.0	40.0
10	15.8	15.5
11	90.8	86.8
12	42.5	39.2
13	21.7	20.2
14	67.3	60.3
15	54.9	49.2
16	50.1	45.5
17	37.8	35.5
18	13.9	13.9
19	71.4	65.42
20	77.2	72.8
21	32.2	28.9
22	157.9	152.3
23	57.8	55.0
24	65.9	63.1
25	165.3	152.3
26	87.1	80.9
27	258.5	230.6
28	215.9	207.7
29	87.8	82.4
30	405.6	381.8
31	36.4	32.7
32	210.0	198.3
33	506.1	477.6
34	219.3	211.6
35	815.9	756.2
36	88.8	82.7
37	154.3	141.6
38	79.0	72.1

Using these values, the limit pressure is calculated as follows:

$$\delta = \frac{(1.95)^2(.64257)(.571 + 2(1.95)/2)}{3(1)(50)(.05)} = .8213 \quad (13)$$

$$\alpha = 1 + 2\,(.8213)^2 + (1 - 2\,(.8213)^2 + 4\,(.8213)^4)^{0.5} = 3.5619 \quad (14)$$

$$A_c = (.571(2) + 2(1.95))\,(.05)(1) = .2521 \quad (15)$$

$$P\text{limit} = \frac{2.28(.2521)(20000)}{(1)(50)(2)\sqrt{3.5619}} = 60.9 \text{ psi} \quad (16)$$

The calculations are repeated for all the models in Table 1 and the results are presented in Table 2.

The average error is 6.56% and the max. error is 14.68%. This is a conservative result since none of the calculated values for Plimit are higher than the FEA values. Considering the wide range of bellows dimensions, the correlation is remarkably good between the FEA results and the results of Eq. (12).

ALLOWABLE LIMIT PRESSURE

According to the ASME Criteria Document [5], the bellows max design pressure must include a design margin of 1.5 on the limit load. Therefore, an allowable limit pressure can be calculated using the following equation:

$$Pl = \frac{P\text{limit}}{1.5} = \frac{1.52 A_c S_y}{K_r D_m q \sqrt{\alpha}} \quad (17)$$

Only the number of plies actually resisting the pressure should be included in the calculation for Pl.

In the EJMA Standards [3], $C_m S_{ab}$ is used to calculate the yield strength of the bellows material. The value of C_m is 1.5 for bellows in the annealed condition and 3.0 for bellows is the as-formed condition. Therefore, the allowable limit pressure equation can also be written as

$$Pl = \frac{1.52 A_c C_m S_{ab}}{K_r D_m q \sqrt{\alpha}} \quad (18)$$

COLUMN INSTABILITY PRESSURE

Besides failure from gross plastic deformation, a bellows can also fail due to column instability. Column instability is characterized by a gross lateral shift of the bellows centerline. Bellows become unstable when the lateral force, which is proportional to the internal pressure, exceeds the restoring force. It is analogous to a long slender column with an excessive axial compressive load. Equation C-28a from the EJMA Standards [3] gives the limiting internal design pressure based on column instability for an unreinforced bellows with ends fixed. The max. allowable internal pressure based on column instability is as follows:

$$P_{sc} = \frac{0.34\pi C_\theta f_{iu}}{N^2 q} \quad (19)$$

The design margin for Eq. (19) is 2.25 in accordance with the EJMA Standards, Section C-4.1.6 [3].

EXTERNAL BUCKLING PRESSURE

With external pressure or vacuum conditions, bellows can fail due to external buckling. External buckling is characterized by a sudden inward collapse of the bellows with attendant warping of material. The EJMA Standards Section C-5 [2] provides a method for determining the external buckling pressure of a bellows. The first step is to calculate the bellows moment of inertia as follows:

$$I_{11} = N\left\langle \frac{nt(2w-q)^3}{48} + 0.4qnt(w-0.2q)^2 \right\rangle \tag{20}$$

The second step is to find an equivalent pipe thickness with the same moment of inertia as follows:

$$I_{22} = \frac{L_b (t_{pipe})^3}{12(1-v^2)} = I_{11} \tag{21}$$

or

$$t_{pipe} = \left\langle \frac{12(1-v^2)I_{11}}{L_b} \right\rangle^{1/3} \tag{22}$$

Substitution of Eq. (20) into Eq. (22) and letting $q = L_b/N$ gives

$$t_{pipe} = \left\langle \frac{12(1-v^2)}{q} \left\langle \frac{nt(2w-q)^3}{48} + 0.4qnt(w-0.2q)^2 \right\rangle \right\rangle^{1/3} \tag{23}$$

Only the number of plies that actually resist the external pressure should be included in the calculation for t_{pipe}.

The max allowable external working pressure for the equivalent pipe is found using ASME Section VIII, Div 1, par. UG-28 [6]. First, the factor A is determined from ASME Section II, Part D, Subpart 3, Figure G [7] using the ratios L_b/D_m and D_m/t_{pipe} to enter the chart. Using the factor A, the factor B is obtained from the applicable chart in Subpart 3 that matches the bellows material and the curve for the bellows design temperature. Based on par. UG-28, the maximum allowable external pressure based on external buckling for the equivalent pipe is as follows:

$$Pa = \frac{4B}{3(D_m / t_{pipe})} \tag{24}$$

The design margin for Eq. (24) is 3.0 in accordance with ASME Section II, Part D, Appendix 3 [7].

The above method assumes that both ends of the bellows are rigidly supported. If that is not the case, further evaluation of the overall system is required.

LIMITING INTERNAL DESIGN PRESSURE

The limiting design pressure for a bellows with internal pressure is the lesser of the allowable limit pressure and the max. allowable internal pressure based on column instability. Therefore, the equation is

$$P = \frac{1.52A_c C_m S_{ab}}{K_r D_m q\sqrt{\alpha}} \text{ or } \frac{0.34\pi C_\theta f_{iu}}{N^2 q} \quad \text{whichever is less.} \tag{25}$$

Equation (25) applies when both ends of the bellows are rigidly supported and the allowable stress is obtained from time-independent properties.

LIMITING EXTERNAL DESIGN PRESSURE

The limiting design pressure for a bellows with external pressure is the lesser of the allowable limit pressure and the max. allowable external pressure based on external buckling.

Therefore, the equation is

$$P = \frac{1.52A_c C_m S_{ab}}{K_r D_m q\sqrt{\alpha}} \text{ or } \frac{4B}{3(D_m / t_{pipe})} \quad \text{whichever is less.} \tag{26}$$

Equation (26) applies when both ends of the bellows are rigidly supported and the allowable stress is obtained from time-independent properties. Only the number of plies that actually resist the external pressure should be included in the calculation.

NOMENCLATURE

A, B = Factors from ASME Code
A_c = Cross sectional area of one convolution (in^2)
C_m = Material strength factor
C_p = Factor used to relate U-shaped bellows convolution segment behavior to a simple strip beam
C_θ = Column instability pressure reduction factor based on initial angular rotation
D_m = Mean diameter of bellows (in.)
K_2, K_4 = Inplane instability factor
K_r = Circumferential stress factor
L = Beam length (in.)
L_b = Bellows length (in.)
N = Number of convolutions on one bellows
S_{ab} = Allowable stress at the bellows design temperature (psi)
S_y = Yield strength at design temperature of bellows material in the as formed (with cold work) or annealed (without cold work) condition (psi)
Z_p = Plastic modulus (in^3)
f_{iu} = Unreinforced bellows theoretical initial axial elastic spring rate per convolution (lbs/in of movement/conv.)
n = Number of bellows material plies of thickness t
q = Convolution pitch, the distance between corresponding points of any two adjacent convolutions (in.)
t = Bellows nominal material thickness of one ply (in.)

t_p = Bellows material thickness of one ply, corrected for thinning during forming (in.)

t_{pipe} = Equivalent pipe thickness (in)

w = Convolution height less bellows material thickness, nt (in.)

α = Inplane instability stress interaction factor

δ = Inplane instability stress ratio

ν = Poisson's ratio

REFERENCES

1. Kalnins, A., 2001, Guidelines for Sizing of Vessels by Limit Analysis, WRC Bulletin 464, NY.
2. Thomas, R.E., 1984, "Validation of Bellows Design Criteria by Testing", ASME PVP – Vol. 83, Metallic Bellows and Expansion Joints: Part II, pp. 55-64.
3. EJMA, 2000, Standards of the Expansion Joint Manf. Assoc., 7th Edition, 2000 Addenda, Tarrytown, NY.
4. Koves, W. J., 1995, "The Initiation of In-plane Squirm in Expansion Joint Bellows", ASME PVP- Vol. 301, Development on Pressure Vessels and Piping, pp. 109-119.
5. ASME, 1969, Criteria of the ASME Boiler and Pressure Vessel Code for Design by Analysis in Section III and Section VIII, Division 2, NY.
6. ASME, 2001, ASME Boiler and Pressure Vessel Code, Rules for Construction of Pressure Vessels, Section VIII, Division 1, 2001 Edition, NY.
7. ASME, 2001, ASME Boiler and Pressure Vessel Code, Section II, Part D, Properties, 2001 Edition, NY.

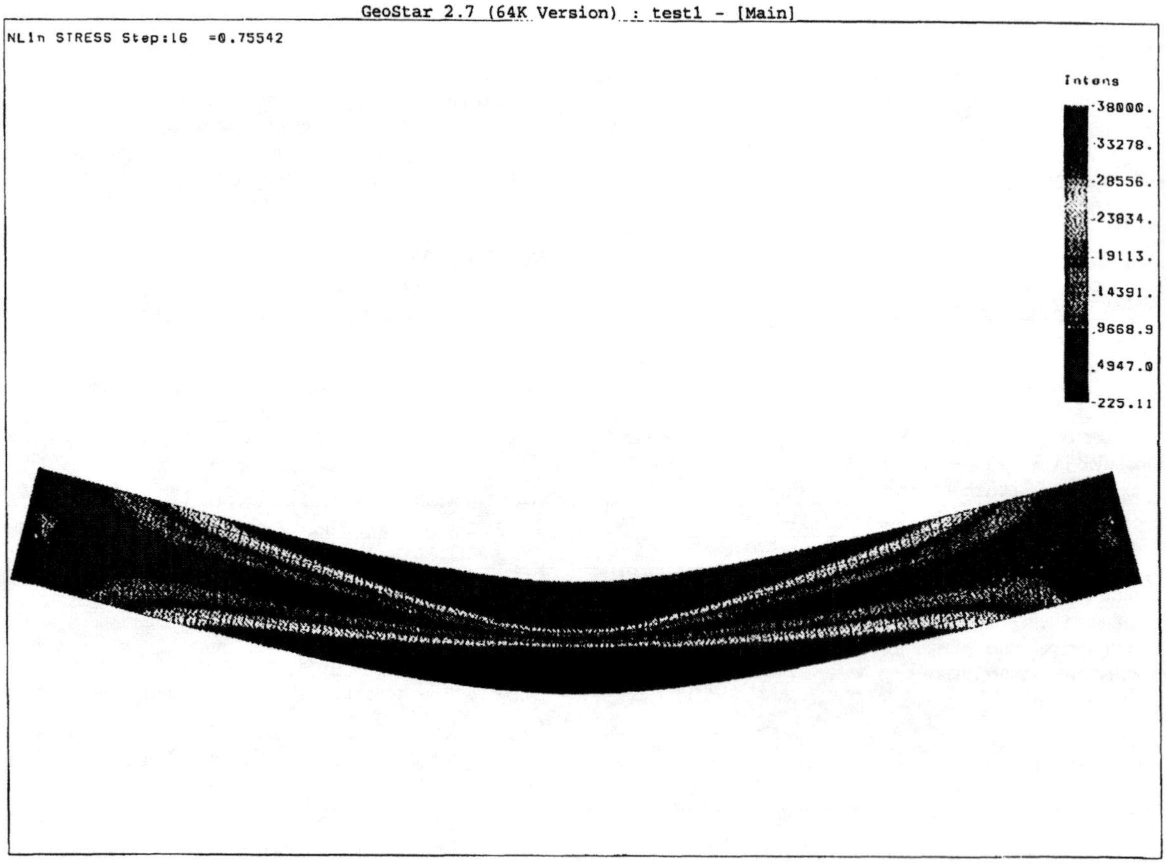

Figure 1 – Beam Deformed Stress Plot

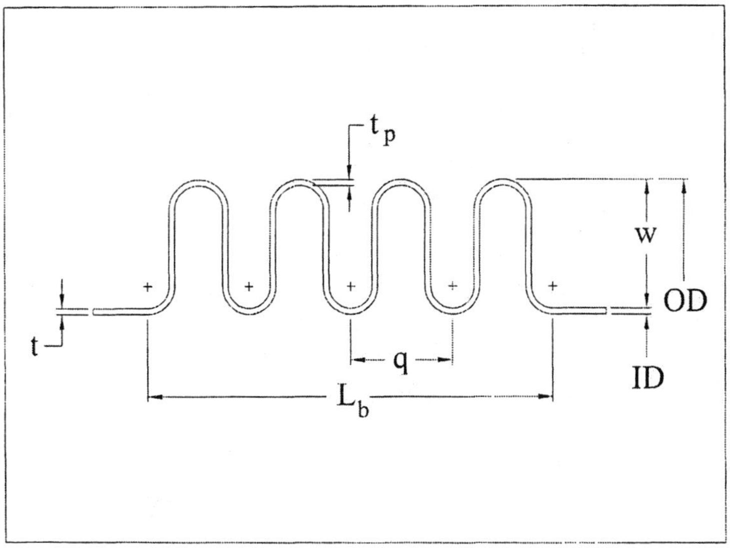

Figure 2 – Bellows Geometry

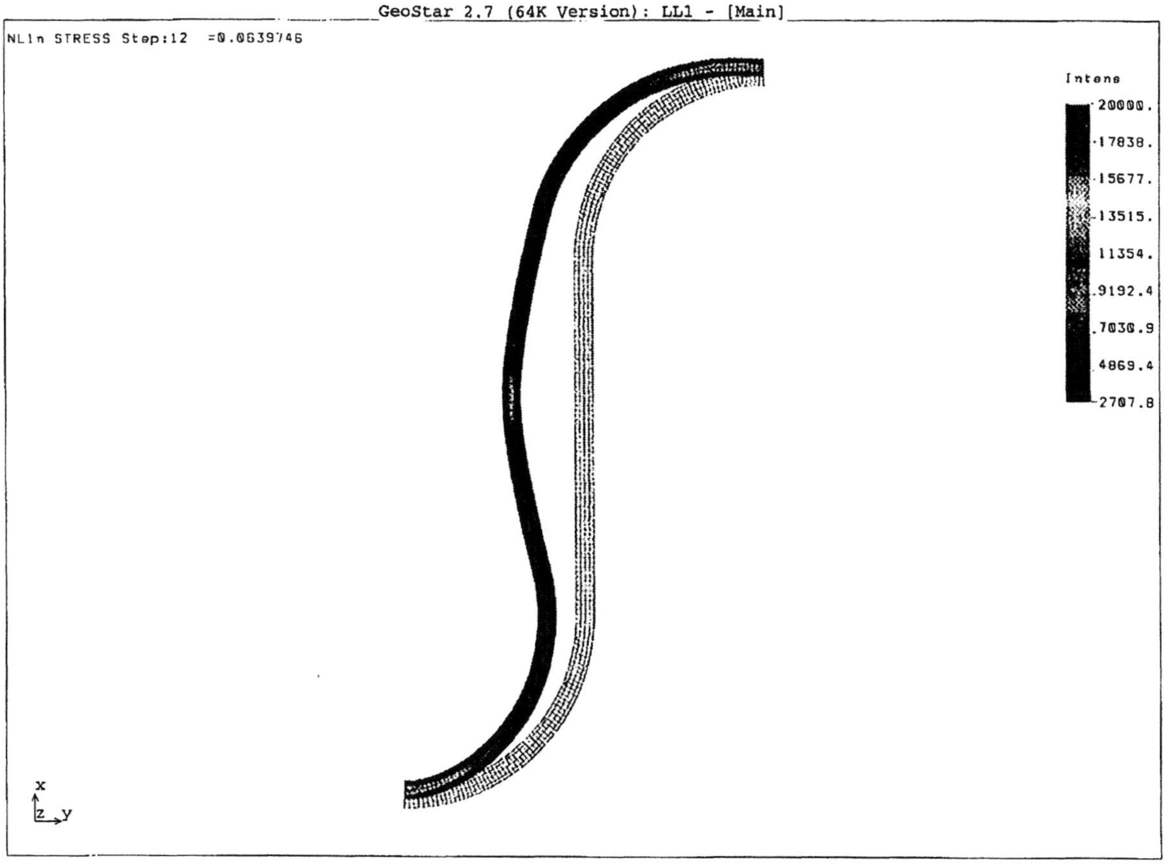

Figure 3 – Bellows Deformed Stress Plot

PVP-Vol. 458, Computer Technology and Applications

PVP2003-1887

LOWER BOUND LIMIT LOAD DETERMINATION: THE m_β - MULTIPLIER METHOD

R Seshadri and H Indermohan

Faculty of Engineering and Applied Science
Memorial University
St. John's, Canada

ABSTRACT

The existing lower bound limit load determination methods that are based on linear elastic analysis such as the classical and m_α-multiplier methods have a dependence on the maximum equivalent stress. These methods are therefore sensitive to localized plastic action, which occurs in components with thin or slender construction, or those containing notches and cracks. Sensitivity manifests itself as relatively poor lower bounds during the initial elastic iterations of the elastic modulus adjustment procedures, or oscillatory behavior of the multiplier during successive elastic iterations leading to limited accuracy.

The m_β-multiplier method proposed in this paper starts out with Mura's inequality that relates the upper bound to the exact multiplier by making use of the "integral mean of yield." The formulation relies on a "reference parameter" that is obtained by considering a distribution of stress rather than a single maximum equivalent stress. As a result, good limit load estimates have been obtained for several pressure component configurations.

INTRODUCTION

The assessment of load carrying capacity under applied loads is an important goal in pressure component design. Limit analysis is carried out to determine the load at which uncontained plastic flow occurs in the component under consideration [1,2]. The concept of reference stress [3], used extensively in the United Kingdom in high temperature integrity assessment procedures and inelastic fracture evaluations [4,5], is related to the limit load.

On the basis of linear elastic finite element analyses with non-homogeneous elastic properties, i.e., spatial variations in elastic moduli, numerous sets of statically admissible and kinematically admissible distributions can be generated [6 - 8], and lower and upper bounds on limit loads can be obtained. The theoretical basis for these methods has been discussed by Ponter et al. [9].

Limit load determination methods, based on a variational principles, was first developed by Mura and coworkers [10] who provided an alternate procedure to the determination of classical lower bound multipliers. Seshadri et al. [11, 12] extended Mura's variational formulation to account for pseudo-elastic stresses that exceed yield stress by making use of the "integral mean of yield" criterion. It is recognized that the classical and the m_α-multipliers exhibit dependence on the maximum elastic stress in a component or a structure. For components with cracks and notches, these multipliers can provide unsatisfactory lower bound limit load values. For slender or shape-sensitive components and structures, the maximum stress locations could "move around" leading to possible oscillatory behavior in the values of m_L and m_α with successive elastic iterations.

In this paper, a multiplier (m_β) is determined, that relies on the entire stress distribution rather than the maximum stress, leading to good lower bound values for components and structures especially with notches and cracks. The method is applied to a number of practical component configurations.

THE m_β - MULTIPLIER METHOD

In limit analysis, a statically admissible stress distribution cannot lie outside the hypersurface of the yield criterion. Mura et al. [10], circumvented such a requirement by making use of the concept of the "integral mean of yield", which can be expressed as

$$\int_{V_T} \mu^o[f(s_{ij}{}^o) + (\phi^o)^2]dV = 0 \qquad (1)$$

The subscript "o" corresponds to a statically admissible state, $s_{ij}{}^o$ is the deviatoric stress corresponding to impending limit state, whereby $s_{ij}{}^o = m^o \tilde{s}_{ij}{}^o$. The deviatoric stresses $\tilde{s}_{ij}{}^o$ equilibriates the applied set of loads. ϕ^o is a point function that takes on a value zero if $s_{ij}{}^o$ is at yield and remains positive below yield.

The von-Mises yield criterion can be expressed as:

$$f(s_{ij})=\frac{1}{2}s_{ij}\,s_{ij}-k^2 \tag{2}$$

The value of k is taken as $\sigma_y/\sqrt{3}$ where σ_y is the yield stress.

The associated flow rule can be expressed as,

$$\dot{\varepsilon}_{ij}=\mu\left(\frac{\partial f}{\partial s_{ij}}\right) \tag{3}$$

$\dot{\varepsilon}_{ij}$ is the strain rate and μ is the flow parameter.

Furthermore, Pan and Seshadri [12] have derived an expression for m^o in order to account for a variable flow parameter, μ, namely

$$m^o=\frac{\sigma_y\sqrt{\int_{V_T}\frac{dV}{E_s}}}{\sqrt{\int_{V_T}\frac{\tilde{\sigma}_e^{\,2}\,dV}{E_s}}} \tag{4}$$

where $\tilde{\sigma}_e$ is the equivalent stress and E_s is the secant modulus at a given location in the component.

Next we consider the inequality derived by Mura [10], which can be expressed as

$$m^o\leq m+\int_{V_T}\mu\left\{f(s_{ij}^{\,o})+(\phi^o)^2\right\}dV \tag{5}$$

Equation (5) can be rewritten as

$$m^o\leq m+\int_{V_T}\mu f^o\,dV \tag{6}$$

A lower bound multiplier (m'') can be obtained from Eq.(6) as:

$$m''=\frac{m^o}{1+Z}\leq m \tag{7}$$

where $Z=\frac{1}{2k^2}\left[\frac{\int_{V_T}\mu f^o\,dV}{\int_{V_T}\mu\,dV}\right]$

In equation (7), μ corresponds to the flow parameter at the limit state, which is not known a priori. In order to decouple the effect of μ, we make use of the Schwarz inequality, according to which the inner product of linear operators of a fairly general class of satisfies

$$(x,y)\leq\|x\|\|y\| \tag{8}$$

where (x, y) is the inner product of x and y, and $\|x\|$ is the norm of x. Integrals for which the integrand is bounded are operators suitable for the application of the Schwarz inequality, and

$$(x,y)=\int x\,y\,dz,\qquad \|x\|=\sqrt{\int x^2\,dz}$$

Consider next the expression for Z in equation (7), i.e.,

$$Z=\frac{1}{2k^2}\left[\frac{\int_{V_T}\mu f^o\,dV}{\int_{V_T}\mu\,dV}\right] \tag{9}$$

Applying the Schwarz inequality identity to both numerator and denominator, we have

$$\int_{V_T}\mu f^o\,dV\leq\sqrt{\int_{V_T}\mu^2\,dV}\sqrt{\int_{V_T}(f^o)^2\,dV} \tag{10a}$$

$$\int_{V_T}\mu\,dV\leq\sqrt{\int_{V_T}\mu^2\,dV}\sqrt{\int_{V_T}1^2\,dV} \tag{10b}$$

Equations (10a) and (10b) are substituted into equation (7), and a parameter β is introduced so that

$$Z_\beta = \frac{\beta}{2k^2}\left[\sqrt{\frac{\int_{V_T}(f^o)^2 dV}{V_T}}\right] \tag{11}$$

A reference parameter $\beta = \beta_R$ is chosen so that $m_\beta \le m$, (see figure 1); i.e.,

$$m_\beta = \frac{m^o}{1+Z_{\beta R}} \le m \tag{12}$$

where $Z_{\beta R}$ corresponds to $\beta = \beta_R$ in equation (11).

Evaluating equation (12) for different β's and iteration variable ζ lead to a set of $m_\beta(\zeta)$ trajectories. The "reference trajectory" corresponding to $\beta = \beta_R$ concurrently satisfies the following requirements:

$$\begin{array}{lll} (1)\ m^o \ge m \ge m_\beta & for\ \zeta \ge 0 & \\ (2)\ \dfrac{dm_\beta}{d\zeta} \ge 0 & for\ \zeta \ge 0 & \\ (3)\ \dfrac{dm^o}{d\zeta} \le 0 & for\ \zeta \ge 0 & \\ and,\ (4)\ m^o = m_\beta = m & as\ \zeta \to \zeta_L & \text{(limit state)} \end{array} \tag{13}$$

In effect, $\beta = \beta_R$ would be the lowest possible value of β that would generate a m_β trajectory that satisfies equations (13). In the above equations, ζ is the iteration variable.

Recognizing that $f^o = \frac{1}{3}\left[(m^o \tilde{\sigma}_e)^2 - \sigma_y{}^2\right]$ and $k^2 = \sigma_y{}^2/3$, equation (12) can be expressed as

$$m_\beta = \frac{m^o}{1+\frac{\beta_R}{2}\sqrt{\frac{\sum_{p=1}^{N}\left[(m^o \bar{\sigma}_{ep})^2 - 1\right]^2 \Delta V_p}{V_T}}} \le m \tag{14}$$

where $\bar{\sigma}_{ep} = \dfrac{\tilde{\sigma}_{ep}}{\sigma_y}$ and ΔV_p is the element volume for element p. As well, $\tilde{\sigma}_{ep}$ corresponds to the applied loads.

When β_R is chosen to satisfy all the conditions set out in equations (13), then $m_\beta \le m$ for all ζ. That is, m_β would be a lower bound for all linear elastic finite element iterations leading to a lower bound limit load given by $P_L = m_\beta\, P$.

MULTIPLIERS – LOWER AND UPPER BOUNDS

The various bounds, as they pertain to the limit load multipliers, have been discussed by Reinhardt and Seshadri [13]. It has been shown that m^o and m_u are upper bounds, where $m^o \ge m_u$. The upper bound m_u is defined as:

$$m_u = \frac{\sigma_y \int_{V_T} \tilde{\varepsilon}_{eq}\, dV}{\int_{V_T} \tilde{\sigma}_e \tilde{\varepsilon}_{eq}\, dV} \Leftrightarrow \frac{\sigma_y \int_{V_T} \frac{1}{E_s}\tilde{\sigma}_{eq}\, dV}{\int_{V_T} \frac{1}{E_s}\tilde{\sigma}^2{}_{eq}\, dV} \tag{15}$$

where $\tilde{\varepsilon}_{eq}$ is the equivalent strain. m_u leads to upper bounds that are closer to the exact multiplier, m, than m^o.

The classical lower bound (m_L) is given by

$$m_L = \frac{\sigma_y}{(\tilde{\sigma}_e)_{\max}} \tag{16}$$

where $(\tilde{\sigma}_e)_{\max}$ is the maximum equivalent stress in the component under consideration. Clearly, m_L would be sensitive to the changes in the magnitude of $(\tilde{\sigma}_e)_{\max}$.

The multiplier (m_α) is given by the expression

$$m_\alpha = 2m^o \frac{2\left(\frac{m^o}{m_L}\right)^2 + \sqrt{\frac{m^o}{m_L}\left(\frac{m^o}{m_L}-1\right)^2\left(1+\sqrt{2}-\frac{m^o}{m_L}\right)\left(\frac{m^o}{m_L}-1+\sqrt{2}\right)}}{\left(\left(\frac{m^o}{m_L}\right)^2 + 2-\sqrt{5}\right)\left(\left(\frac{m^o}{m_L}\right)^2 + 2+\sqrt{5}\right)} \tag{17}$$

In equation (17), m_α also depends on $(\tilde{\sigma}_e)_{\max}$ and $m_\alpha > m_L$ for all iterations except at the converged state. The issue of lower boundedness of m_α is discussed in detail [13].

The multiplier m_β can be evaluated using equation (12), where a value $\beta = \beta_R$ is sought that satisfies the conditions described in equation (13). Clearly, $m_u \geq m \geq m_\beta$ provides the narrowest band on the upper and lower bound multipliers for every iteration.

NUMERICAL EXAMPLES

Successive linear elastic finite element iterations are carried out by modifying the elastic modulus of the various elements as follows:

$$E_{s,i+1} = \left(\frac{\sigma_{ref}}{\tilde{\sigma}_e} \right)_{,i}^{q} E_{s,i} \quad (18)$$

where σ_{ref} is a reference stress, q is the elastic modulus adjustment parameter, and "i" is the iteration index. In order to simulate incompressibility conditions a Poisson's ratio of 0.47 is assumed.

The choice of $\beta = \beta_R$ in equation (14) is problem dependent. It is a simple matter to generate the $m_\beta(\zeta)$ trajectories for various β 's (in the neighborhood of β =1), and then picking the $\beta = \beta_R$ trajectory that satisfies the conditions set out in equations (13). Subject to the usual consequences of discretization, the key conditions to assess the steady state are: (1) $\frac{dm_\beta}{d\zeta} = 0$ and (2) $m^o \geq m_\beta$. Ponter et al. [9] have shown that upper bound multipliers converge to the lowest upper bound. Likewise, m_β subjected to conditions (equation (13)) will converge to the highest lower bound. In this paper $\beta = \beta_R$ has been obtained by inspection. However, equations (13) can be incorporated into a postprocessor enabling a systematic determination of β_R. Once a set of linear elastic iterations has been generated for $\beta = 1$ (say), the $m_\beta(\zeta)$ trajectories for other β values can be readily obtained.

The m_β-multiplier method described in this paper is applied to a variety of pressure component configurations – thick walled cylinder, thick unwelded flat head, thin unwelded flat head and welded-in flat head all under internal pressure and a compact tension specimen subjected to a tensile load. The problems are modeled using ANSYS software (educational version) [14].

1. **Thick Walled Cylinder**: The inside radius is 60 mm, outside radius is 180 mm, modulus of elasticity is 200,000 N/mm^2 and the yield strength is 300 N/mm^2. An internal pressure of 50 N/mm^2 is applied. The cylinder is modeled using four noded isoparametric quadrilateral elements with 30 elements. The $m_\beta(\zeta)$ trajectory corresponding to β_R = 1.0 represents the best lower bound estimate at any given iteration. The variation of limit load multipliers with successive iterations is plotted in figure 2. This corresponds to q = 1.0. The lower and upper bound multipliers converge to exact solution within three iterations.

2. **Thick Unwelded Flat Head**: A flat head configuration with the cylinder thickness 101.6 mm, flat head thickness 101.6 mm, overall length 406.4 mm, fillet radius 101.6 mm is considered. The modulus of elasticity is 207,000 N/mm^2 and the yield strength is 207 N/mm^2. An internal pressure of 50 N/mm^2 is applied. The flat head is modeled using four noded isoparametric quadrilateral elements with 840 elements. The $m_\beta(\zeta)$ trajectory corresponding to β_R = 1.05 represents the best lower bound estimate for any given iteration with q =1.0. The variation of limit load multipliers with successive iterations is plotted in figure 3. The m_β converges at the third iteration.

3. **Welded-In Flat Head**: The Cylinder thickness of 21.5 mm, Flat Head thickness 43 mm, Overall length 243 mm and with the weld groove of 18 mm is considered. The yield strength is 300 N/mm^2; and the modulus of elasticity is 200,000 N/mm^2. An internal pressure of 10 N/mm^2 is applied. The geometry is modeled using four noded isoparametric quadrilateral elements with 936 elements. The $m_\beta(\zeta)$ trajectory corresponding to β_R = 1.2 represents the best lower bound estimate at any given iteration for q =0.5.The variation of limit load multipliers with iteration is plotted in figure 4. The m_β converges at the fourth iteration.

4. **Thin Unwelded Flat Head**: The Cylinder thickness is 101.6 mm, Flat head thickness is 25.4 mm, Overall length is 254 mm, Fillet radius is 25.4 mm. The yield strength is 300 N/mm^2 and the modulus of elasticity is 200,000 N/mm^2. An internal pressure of 0.69 N/mm^2 is applied. The flat head is modeled using four noded isoparametric quadrilateral elements with 940 elements. The $m_\beta(\zeta)$ trajectory corresponding to β_R = 1.82 represents the best lower bound estimate at any given iteration for q =0.25. The variation of limit load multipliers with successive iterations is plotted in figure 5. The m_β converges at the fourth iteration.

5. **Compact Tension Specimen**: A tension specimen of width 100 mm, height 60 mm, thickness 3 mm and the crack length of 46.6 mm is considered. The yield strength is 488.43 MPa and the modulus of elasticity is 211,000 MPa. A tensile load of 100 N is applied. The specimen is modeled using 232 six noded isoparametric triangular elements and crack tip elements with radius of 2.91mm around the crack front. The crack-tip singularity is achieved by moving the mid-side nodes to the quarter points. The $m_\beta(\zeta)$ trajectory corresponding to β_R = 1.9 represents the best lower bound estimate at any given iteration for q =0.5. The variation of limit load multipliers with successive iterations is plotted in figure 6. The m_β converges at the fifth iteration.

CONCLUSIONS

The m_β-multiplier method of limit load determination is based on variational principles that makes use of the "integral mean of yield" condition. It takes into consideration the entire statically admissible stress distributions into Mura's lower bound formulation rather than simply the maximum stress, leading to improved lower bounds. Since the entire stress distribution is accounted for in the formulation of the m_β-multiplier method, components with cracks and sharp notches can now be analyzed and good lower bounds obtained even during early iterations. As well, the fluctuations in the magnitude of the multipliers (m_L and m_α) for slender and shape-sensitive components can be avoided.

The procedures based on elastic modulus adjustments used in conjunction with linear elastic FEA represent "independent verification" methods of inelastic FEA results.

For any given iteration, the (m_β, m_u) pair provides the minimum spread in terms of bounds on the limit load multipliers. The m_β-multiplier is a better lower bound than the m_L and m_α multipliers for the component configurations considered in this paper, and convergence to the inelastic FEA solution has been obtained within five iterations.

REFERENCES

1. ASME Boiler and Pressure Vessel Code, 2001, Section III.
2. ASME Boiler and Pressure Vessel Code, 2001, Section VIII.
3. Webster, G., and Ainsworth, R. A., 1994, "High Temperature Component Life Assessment, Chapman and Hall, London," UK.
4. Ainsworth, R. A., Dean, D. W., and Budden, P. J., 2000, "Development in Creep Fracture Assessments within the R5 Procedure," IUTAM Symposium on Creep in Structures, Nagoya, Japan, pp. 321-330.
5. PD6539:1994, 1994, Guide to Methods for the Assessment of the Influence of Crack Growth on the Significance of Design in Component Operating at High Temperature, BSI, London, UK.
6. Seshadri, R., 1991, "The Generalized Local Stress Srain (GLOSS) Analsysis – Theory and Applications," ASME Journal of Pressure Vessel Technology, **113**, pp. 219-227.
7. Marriott, D. L., 1998, "Evaluation of Deformation or Load Control of Stress under Inelastic Condtions using Elastic Finite Element Stress Analysis," ASME PVP-Vol. 136, pp. 3-9.
8. Mackenzie, D., and Boyle, J. T., 1993, "A Method of Estimating Limit Loads Using Elastic Analysis, I: Simple Examples," International Journal of Pressure Vessels and Piping, **53**, pp. 77-85.
9. Ponter, A. R. S., Fuschi, P., and Engelhardt, M., 2000, "Limit Analysis for a General Class of Yield Conditions," European Journal of Mechanics A/Solids, **19**, pp. 401-421.
10. Mura, T., Rimawi, W. H., and Lee, S. L., 1965, "Extended Theorems of Limit Analysis," Quarterly of Applied Mathematics, **23**, pp. 171-179.
11. Seshadri, R., and Mangalaramanan, S. P., 1997, "Lower Bound Limit Loads Using Variational Concepts: The m_α-Method," International Journal of Pressure Vessels and Piping, **71**, pp. 93-106.
12. Pan, L., and Seshadri, R., 2001, "Limit Load Estimation using Plastic Flow Parameter in Repeated Elastic Finite Element Analysis," ASME PVP-Vol. 430, pp. 145-150.
13. Reinhardt, W. D., and Seshadri, R., 2003, "Limit Load Bounds for the m_α-multiplier," ASME Journal of Pressure Vessel Technology, **125**, pp.34-56.
14. ANSYS Engineering Analysis System User's Manual, Rev. 6.0, Swanson Analysis System, PA, 2001.

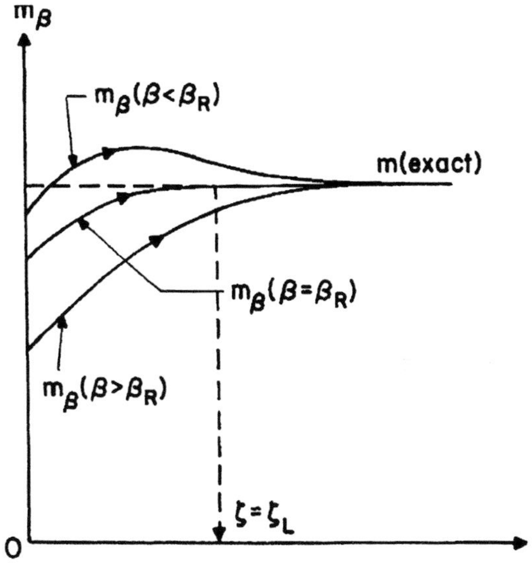

Figure 1. m_β**- multiplier : Reference curve** $(\beta=\beta_R)$

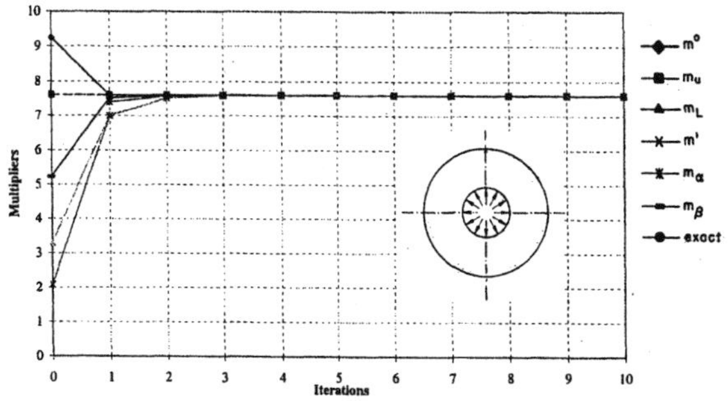

Figure 2. Variation of limit load multipliers for thick walled cylinder

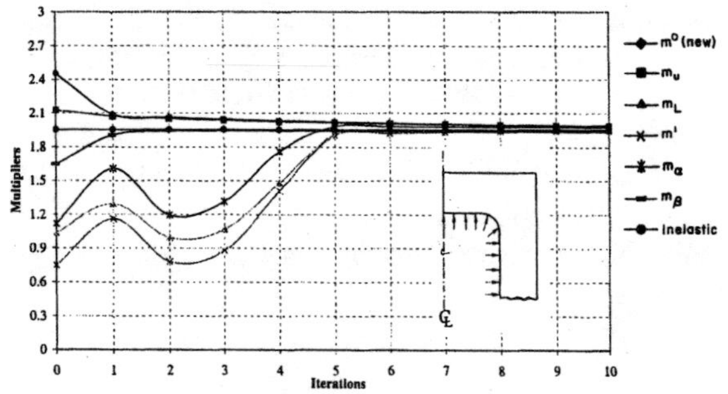

Figure 3. Variation of limit load multipliers for thick flat head

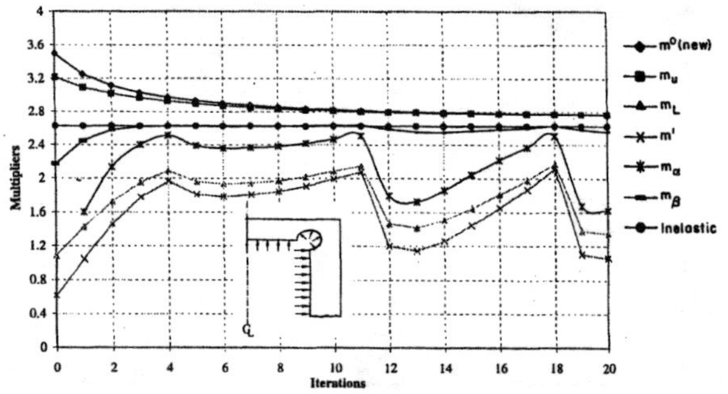

Figure 4. Variation of limit load multipliers for welded-in flat head

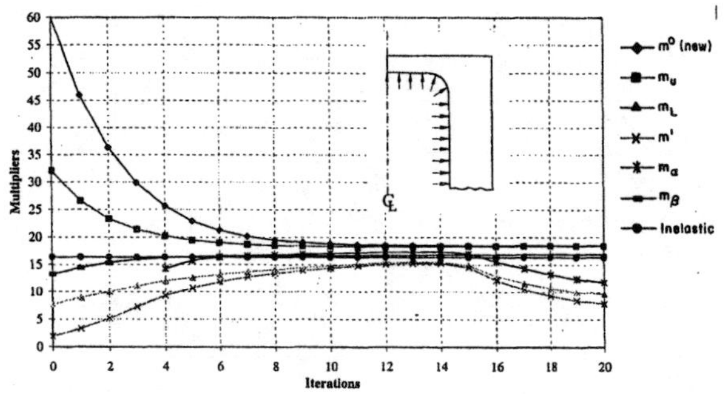

Figure 5. Variation of limit load multipliers for thin flat head

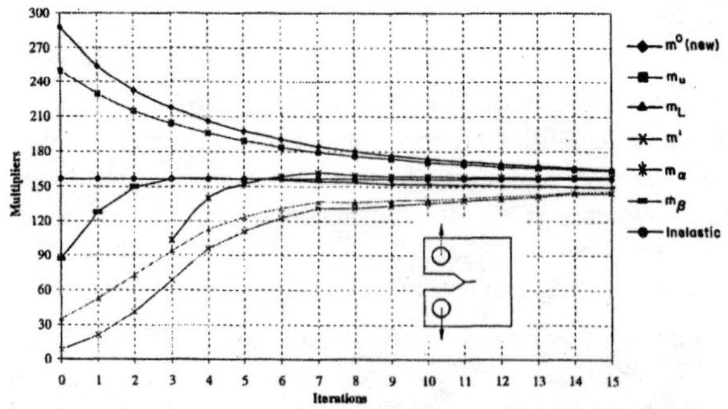

Figure 6. Variation of limit load multipliers for compact tension specimen

PVP-Vol. 458, Computer Technology and Applications

PVP2003-1888

APPROXIMATE ANALYSIS OF CREEP STRAINS AND STRESSES AT NOTCHES

J. E. Nuñez G. Glinka

University of Waterloo, Dept. of Mechanical Engineering
Waterloo, Ontario, N2L 3G1, Canada

ABSTRACT

A method for the estimation of creep induced strains and stresses at notches has been developed. The purpose of the method is to generate a solution for the time-dependent strain and stress at the notch root based on the linear-elastic stress state, the constitutive law, and the material creep model. The proposed solution is an extension of Neuber's rule used for the case of time-independent plasticity. The method was derived for both localized and non-localized creep in a notched body. Predictions were compared with finite element data and good agreement was obtained for various geometrical and material configurations in plane stress conditions.

Keywords: notches, non-localized creep, plane stress, stress-strain analysis

NOMENCLATURE

A	Norton's power law coefficient
C_p	Plastic zone correction factor
E	Young's Modulus
K_t	Theoretical stress concentration factor
K_ε	Actual strain concentration factor
K_σ	Actual stress concentration factor
K_Ω	Total strain energy density concentration factor
r_p	Plastic zone size
Δr_p	Plastic zone size increment
t	Time
Δt_n	nth time increment
α	Stress exponent for creep power law
β	Time exponent for creep power law
ε^0	Total strain component from elastic-plastic solution
ε^{ct}	Time-dependent creep strain component at the notch tip
ε^e	Elastic strain at the notch tip from linear elastic solution
ε^{et}	Time-dependent elastic strain component at the notch tip
ε^f	Elastic strain component in a far field point
ε^{p0}	Mechanically induced plastic strain at $t = 0.0$
σ^0	Total stress from elastic-plastic solution
σ^e	Elastic stress at the notch tip from linear elastic solution
σ^f	Elastic stress component in a far-field point
σ^t	Time-dependent stress component
Ω^e	Total strain energy density at the notch tip from linear-elastic analysis
Ω^f	Total strain energy density in a far field point

1. INTRODUCTION

Modern machines and engineering structures usually contain components with complex geometries. Discontinuities such as holes, grooves, and notches are normally among these complexities. It is well known that these geometrical discontinuities induce a localized stress concentration zone where stresses may exceed the yield limit of the material. Mechanical components work often in very demanding conditions such as a combination of high cyclic stresses with a high operating temperature. In the case of time- and temperature-dependent deformations (creep) the durability analysis requires the detailed knowledge of strains and stresses in the stress concentration zone.

An accurate estimation of these local time-dependent stresses and strains can be generated by using a finite element analysis. However, the visco-plastic finite element programs are expensive, and such analyses are time consuming, particularly when simulating lengthy cyclic loading histories. Therefore, alternative more efficient methods are desirable to perform an estimation of the stresses and strains at the notch root with less effort but without compromising the accuracy.

To date, only a few authors have tried to develop alternative solutions for localized time-dependent creep-plasticity problems. In 1978, Chaudonneret [1] suggested that Neuber's rule [2] could be expressed in its differential form as long as the applied stress increases monotonically and is a continuous function of time. Chaudonneret and Culie [3] demonstrated that the differential form of Neuber's rule could be successfully used to predict notch root stresses and strains in creeping bodies under static and cyclic loading. However a very complex integration procedure had to be incorporated to solve the proposed differential equation, especially when creep strain was present in the net section of the specimen.

In 1984 Kurath [4] utilized a modified form of Neuber's rule to simulate the notch root stress-strain time-dependent behaviour in components made of titanium alloy, and acceptable fatigue life predictions were obtained in comparison with experimental data. Kubo and Ohji [5] extended their "small scale creep" concept, originally proposed for cracks under elastic-creep conditions, to notches under plane strain and in axi-symmetric bodies. The results obtained using the method by Kubo and Ohji are limited to cases where the creep is strictly localized and does not consider the effect of creep away from the notch tip.

More recently, Moftakhar, et al. [6] proposed an extension of Neuber's rule [2] to estimate creep induced multiaxial strains and stresses at the notch tip. Their approach was developed assuming that the total strain energy density at the notch tip does not change when going from a linear elastic solution to a time-dependent elastic-plastic-creep solution. Nevertheless, this assumption may be valid only for cases where the deformation is strictly localized around the point of stress concentration. In 1998, Harkegard and Sorbo [7] reformulated Chaudonneret's model to predict creep stresses and strains in notches under constant nominal stress and constant nominal strain. The authors reduced the integration complexity of Chaudonneret's model [1] by making extensive use of normalized forms of stress, strain and time. Good agreement was reported against generated finite element data.

The method presented below is also based on the Neuber concept but the formulation and the numerical procedure exhibit a relatively high degree of time efficiency and accuracy.

2. APPLICATION OF NEUBER'S RULE TO CREEP ANALYSIS

Neuber's rule [2] was initially proposed for elastic-plastic notched bodies in a state of pure shear stress. The rule is most often presented in terms relating the theoretical elastic stress concentration factor, K_t, and the actual stress, K_σ, and, K_ε, strain concentration factors,

$$K_t^2 = K_\sigma K_\varepsilon \tag{1}$$

Neuber's rule can also be written in the form relating the hypothetical stress and the corresponding elastic strain components from a purely elastic solution to the actual elastic-plastic stress and strain components:

$$\sigma_{ij}^e \varepsilon_{ij}^e = \sigma_{ij}^0 \varepsilon_{ij}^0 \tag{2}$$

In such a case, Eq. (2) for notched bodies in plane stress reduces to:

$$\sigma_{22}^e \varepsilon_{22}^e = \sigma_{22}^0 \varepsilon_{22}^0 \tag{3}$$

Equation (3) results directly from the original Neuber's rule (1) and it can be interpreted as an equivalence of the total strain energy density.

Moftakhar, et al. [6] showed that the total strain energy density at the notch tip remains almost constant as long as the plastic deformation remains localized, even when stresses relax due to creep. In other words, Neuber's rule can be directly extended to time-dependent stress and strain analysis in the following way,

$$\Omega = \sigma_{22}^e \varepsilon_{22}^e = \sigma_{22}^0 \varepsilon_{22}^0 = \sigma_{22}^t \varepsilon_{22}^t \tag{4}$$

where the term on the right hand side represents the total strain energy density at a specific time, t, under constant nominal stress and constant temperature. It has been shown [8] that as long as the plastic yielding and creep remain limited to the zone near the notch tip, Neuber's rule will always overestimate the actual elastic-plastic strains and stresses. Therefore, some authors have called this the upper bound solution as opposed to the ESED method by Molski and Glinka [9] which is believed to give a lower bound solution.

The graphical interpretation of the creep Neuber rule is illustrated in Fig. 1. The equivalence of the strain energy density means that the strain energy density obtained from a linear-elastic solution and represented by area $OA^eB^eC^e$ is equal to the area $OA^aB^aC^a$ representing the actual strain energy density at the notch tip.

3. LOCALIZED CREEP FORMULATION

Consider a notched body at constant temperature and subjected to constant load for an extended period of time. As long as the creep is localized it is reasonable to assume after Neuber that:

$$\sigma_{22}^0 \varepsilon_{22}^0 = \sigma_{22}^t \varepsilon_{22}^t \tag{5}$$

The time-dependent strain can be decomposed into its elastic, mechanically induced plastic, and creep contributions,

$$\varepsilon_{22}^t = \varepsilon_{22}^{et} + \varepsilon_{22}^{p0} + \varepsilon_{22}^{ct} \tag{6}$$

The mechanically induced plastic strain, ε^{p0}, is considered to be constant during the hold time period, since it represents the non-recoverable plastic deformation at time *t=0.0*. This assumption suggests that during the hold time there is only a trade off between the elastic unloading and creep deformation. Substituting Eq. (6) into Eq. (5) results in:

$$\sigma_{22}^0 \varepsilon_{22}^0 = \sigma_{22}^t \varepsilon_{22}^{et} + \sigma_{22}^t \varepsilon_{22}^{p0} + \sigma_{22}^t \varepsilon_{22}^{ct} \tag{7}$$

Differentiating Eq. (7) with respect to time leads to the following equation:

$$0.0 = \sigma_{22}^t \dot{\varepsilon}_{22}^{et} + \dot{\sigma}_{22}^t \varepsilon_{22}^{et} + \dot{\sigma}_{22}^t \varepsilon_{22}^{p0} + \sigma_{22}^t \dot{\varepsilon}_{22}^{ct} + \dot{\sigma}_{22}^t \varepsilon_{22}^{ct} \tag{8}$$

For a uniaxial stress state at the notch tip, the elastic strain can always be obtained as $\varepsilon^{et}=\sigma^t/E$. The elastic strain rate, on the other hand, is defined as $\dot{\varepsilon}^{et} = \dot{\sigma}^t / E$. Therefore, Eq. (8) can be written in the following form:

$$0.0 = \frac{2}{E} \sigma_{22}^t \dot{\sigma}_{22}^t + \dot{\sigma}_{22}^t \varepsilon_{22}^{p0} + \sigma_{22}^t \dot{\varepsilon}_{22}^{ct} + \dot{\sigma}_{22}^t \varepsilon_{22}^{ct} \tag{9}$$

Equation (9) is the general differential equation which together with the material creep law forms the set of two equations needed for the mathematical formulation of the problem. A closed form solution to the two equations is seldom feasible particularly when creep models take complex mathematical forms. Moftakhar, et al. [6] proposed a special time integration method to generate a suitable numerical solution. It was suggested to divide the integration time period into a finite number of discrete steps, Δt_n, and then to generate solutions for subsequent time steps. Based on the proposed approach, the time derivatives can be written in the incremental form as:

$$\dot{\sigma}_{22}^{t} = \frac{d\sigma_{22}^{t}}{dt} \cong \frac{\Delta\sigma_{22}^{t_n}}{\Delta t_n} = \frac{\sigma_{22}^{t_n} - \sigma_{22}^{t_{n-1}}}{t_n - t_{n-1}} \tag{10}$$

and

$$\dot{\varepsilon}_{22}^{ct} = \frac{d\varepsilon_{22}^{ct}}{dt} \cong \frac{\Delta\varepsilon_{22}^{C_n}}{\Delta t_n} = \frac{\varepsilon_{22}^{C_n} - \varepsilon_{22}^{C_{n-1}}}{t_n - t_{n-1}} \tag{11}$$

The substitution of Eq. (10) and Eq. (11) into Eq. (9) results in:

$$0.0 = \frac{2}{E}\frac{\Delta\sigma_{22}^{t_n}}{\Delta t_n}\sigma_{22}^{t_{n-1}} + \frac{\Delta\sigma_{22}^{t_n}}{\Delta t_n}\varepsilon_{22}^{p0} + \frac{\Delta\sigma_{22}^{t_n}}{\Delta t_n}\varepsilon_{22}^{C_n} + \sigma_{22}^{t_{n-1}}\frac{\Delta\varepsilon_{22}^{C_n}}{\Delta t_n} \tag{12}$$

The incremental equation (12) can be solved for the stress increment, $\Delta\sigma_{22}^{t_n}$, occurring during the time step, Δt_n.

$$\Delta\sigma_{22}^{t_n} = \frac{-\sigma_{22}^{t_{n-1}}\Delta\varepsilon_{22}^{C_n}}{\frac{2}{E}\sigma_{22}^{t_{n-1}} + \varepsilon^{p0} + \varepsilon_{22}^{C_n}} \tag{13}$$

As expected, the stress increment is negative; hence, it represents the stress relaxation at the notch tip induced by creep. The increment of creep strain can be directly obtained from the creep law $\dot{\varepsilon}^{c} = f(\sigma)g(t)$, which is often written in a power law form.

$$\dot{\varepsilon}_{22}^{c}(\sigma;t) = A\sigma^{\alpha}t^{\beta} \tag{14}$$

where A, α and β are material constants for a given temperature. The incremental form of the creep law used in the model is subsequently written as:

$$\Delta\varepsilon_{22}^{C_n} = \Delta t_n \cdot \dot{\varepsilon}_{22}^{c}(\sigma;t) \tag{15}$$

4. NON-LOCALIZED CREEP FORMULATION

In most real-life components working in creep conditions, the creep strains in the far field albeit relatively low might be important as well. Analogously to the notch tip, the material in the rest of the component may creep as well. As the stress at the notch tip relaxes, so does the constraint provided by the material from the far field region. Therefore, with the constraint being relaxed, additional creep deformation at the notch tip is possible. This situation is the so-called *non-localized*, or *gross creep* conditions.

Extending further the idea of the equivalence of the total strain energy density, an extra amount of energy is needed to account for the constraint loss imposed on the notch tip by the far field. Moftakhar [8] suggested that the total strain energy density changes occurring in the far field, Ω^{fc}, produce similar but magnified effects at the notch tip. A good measure of such a magnification is the so-called *total strain energy density concentration factor*, K_{Ω}.

The strain energy concentration factor proposed by Moftakhar [6] is defined as

$$K_{\Omega} = \frac{\Omega^{e}}{\Omega^{f}} = \frac{\sigma_{22}^{e}\varepsilon_{22}^{e}}{\sigma_{22}^{f}\varepsilon_{22}^{f}} \tag{16}$$

where Ω^{e} is the total strain energy density at the notch tip obtained from a linear elastic analysis, and Ω^{f} is the total strain energy density at a predefined point in the far field also obtained from the linear elastic solution.

The definition of *far field* used in our analysis is shown in Fig 2. The far field stress in the case of a body subjected to pure axial load is assumed to be equal to the elastic stress found at a distance of three times the notch radius from the notch root. In the case of a pure bending load, the far field stress is defined as one-half of the nominal simple bending stress.

The effect the creep in the far field under constant load has on the notch tip behaviour can be interpreted as an additional input of strain energy density. This additional increment of strain energy density must be included into Eq. (5),

$$\sigma_{22}^{0}\varepsilon_{22}^{0} + K_{\Omega}\Omega^{cf} = \sigma_{22}^{t}\varepsilon_{22}^{t} \tag{17}$$

where,

$$K_{\Omega}\Omega^{cf} = K_{\Omega}\sigma_{22}^{f}\varepsilon_{22}^{cf} \tag{18}$$

Expression (17) represents the total strain energy density contributed as a result of the local and the far field creep. The product shown in Eq. (18) is represented by the area $A^{a}A^{t}B^{t}B^{a}$ in Fig. 1b. Decomposing the time-dependent strain into its elastic, initial plastic and creep contributions, the strain energy density equation at the notch tip can be written as:

$$\sigma_{22}^{0}\varepsilon_{22}^{0} + K_{\Omega}\sigma_{22}^{f}\varepsilon_{22}^{cf} = \sigma_{22}^{t}\varepsilon_{22}^{et} + \sigma_{22}^{t}\varepsilon_{22}^{p0} + \sigma_{22}^{t}\varepsilon_{22}^{ct} \tag{19}$$

Differentiating Eq. (19) with respect to time, results in Eq. (20).

$$K_{\Omega}(\sigma_{22}^{f}\dot{\varepsilon}_{22}^{cf} + \dot{\sigma}_{22}^{f}\varepsilon_{22}^{cf}) = \sigma_{22}^{t}\dot{\varepsilon}_{22}^{et} + \dot{\sigma}_{22}^{t}\varepsilon_{22}^{et} + \dot{\sigma}_{22}^{t}\varepsilon_{22}^{p0} + \sigma_{22}^{t}\dot{\varepsilon}_{22}^{ct} + \dot{\sigma}_{22}^{t}\varepsilon_{22}^{ct} \tag{20}$$

It can be shown that the far field stress changes due to creep are very small compared to the stress changes at the notch tip or none if the external load is constant. Therefore the strain energy density changes in the far field are mainly due to the change in strain. As a result, it can be assumed that the far field stress σ_{22}^{f} remains constant during the hold time, and it takes the value of the elastic stress in the far field calculated at time *t=0.0*. This assumption is drawn from the fact that under constant load the stress due to equilibrium must balance out the load and therefore the far stress field cannot drastically change.

$$\sigma_{22}^{f} = \sigma_{22}^{f}\big|_{t=0.0} = \sigma_{22}^{f0} \tag{21}$$

Furthermore, the elastic strains can be replaced according to Hooke's law by stress. After subsequent substitutions Eq. (20) can be reduced to:

$$K_{\Omega}\sigma_{22}^{f0}\dot{\varepsilon}_{22}^{cf} = \frac{2}{E}\sigma_{22}^{t}\dot{\sigma}_{22}^{t} + \dot{\sigma}_{22}^{t}\varepsilon_{22}^{p0} + \sigma_{22}^{t}\dot{\varepsilon}_{22}^{ct} + \dot{\sigma}_{22}^{t}\varepsilon_{22}^{ct} \tag{22}$$

In order to obtain the recursive form of the general differential equation, Eq. (10) and Eq. (11) need to be substituted into Eq. (22).

$$K_{\Omega}\sigma_{22}^{f0}\frac{\Delta\varepsilon_{22}^{cf_n}}{\Delta t_n}=\frac{2}{E}\frac{\Delta\sigma_{22}^{t_n}}{\Delta t_n}\sigma_{22}^{t_{n-1}}+\frac{\Delta\sigma_{22}^{t_n}}{\Delta t_n}\varepsilon_{22}^{p0}+\frac{\Delta\sigma_{22}^{t_n}}{\Delta t_n}\varepsilon_{22}^{c_n}+\sigma_{22}^{t_{n-1}}\frac{\Delta\varepsilon_{22}^{c_n}}{\Delta t_n} \quad (23)$$

Solving for the actual stress increment, $\Delta\sigma_{22}^{t_n}$, at the notch tip, the following expression is obtained:

$$\Delta\sigma_{22}^{t_n}=\frac{K_{\Omega}\sigma_{22}^{f0}\Delta\varepsilon_{22}^{cf_n}-\sigma_{22}^{t_{n-1}}\Delta\varepsilon_{22}^{c_n}}{\frac{2}{E}\sigma_{22}^{t_{n-1}}+\varepsilon^{p0}+\varepsilon_{22}^{c_n}} \quad (24)$$

where,

$$\left|K_{\Omega}\sigma_{22}^{f0}\Delta\varepsilon_{22}^{cf_n}\right|<\left|\sigma_{22}^{t_{n-1}}\Delta\varepsilon_{22}^{c_n}\right| \quad (25)$$

Eq. (24) is valid as long as inequality (25) is true. Some combinations of load conditions and component geometry may cause the left hand side term of expression (25) to be greater than the term on the right hand side, particularly after a long hold time period, and when the creep rates approach the tertiary creep zone. However, such situations are seldom permitted in engineering practice.

5. PLASTIC ZONE ADJUSTMENT

Moftakhar [8] has shown that in case of the presence of gross creep plasticity, the Neuber's rule approach tends to underestimate the real stresses and strains at the notch tip. Therefore, a correction needs to be introduced for the calculation of the initial total strain energy density obtained from the linear elastic analysis. Glinka [10] introduced the plastic zone correction factor in order to compensate for the stress redistribution occurring in the notch tip region caused by the localized plastic yielding. This correction factor was derived analogously to Irwin's [11] plastic zone correction factor for sharp notches and cracks.

Glinka's plastic zone correction factor, C_p, has been defined as,

$$C_p=1+\frac{\Delta r_p}{r_p} \quad (26)$$

A complete procedure of how the plastic zone size, r_p, and the plastic zone increment, Δr_p, can be estimated for components subjected to tension and pure bending is given in reference [10].

Once the plastic zone correction factor is estimated, the total strain energy density equation (19) can be written as:

$$\sigma_{22}^{0}\varepsilon_{22}^{0}+(K_{\Omega}C_p)\sigma_{22}^{f}\varepsilon_{22}^{cf}=\sigma_{22}^{t}\varepsilon_{22}^{et}+\sigma_{22}^{t}\varepsilon_{22}^{p0}+\sigma_{22}^{t}\varepsilon_{22}^{ct} \quad (27)$$

The resulting final expression for the calculation of the stress decrement at the notch tip is:

$$\Delta\sigma_{22}^{t_n}=\frac{(K_{\Omega}C_p)\sigma_{22}^{f0}\Delta\varepsilon_{22}^{cf_n}-\sigma_{22}^{t_{n-1}}\Delta\varepsilon_{22}^{c_n}}{\frac{2}{E}\sigma_{22}^{t_{n-1}}+\varepsilon^{p0}+\varepsilon_{22}^{c_n}} \quad (28)$$

The actual strain increment can then be calculated as:

$$\Delta\varepsilon_{22}^{t_n}=\Delta\varepsilon_{22}^{c_n}-\frac{\Delta\sigma_{22}^{t_n}}{E} \quad (29)$$

6. APPLICATION OF THE PROPOSED MODEL

In order to assess the accuracy of the proposed model, finite element analyses of creep in notched components were carried out using the ABAQUS 6.1 [12] finite element software. The elastic-plastic stress-strain curve was given in the form of a series of linear segments shown in Table 1. A kinematic-hardening plasticity model was used in the analysis. The calculations from both the finite element and the proposed method were carried out using the *strain hardening* approach, which for a power law creep model of the form of Eq. (14) becomes the following expression:

$$\dot{\varepsilon}_{eq}^{c}=\left(A(\sigma_{eq})^{\alpha}\left[(\beta+1)\varepsilon_{eq}^{c}\right]^{\beta}\right)^{\frac{1}{\beta+1}} \quad (30)$$

where ε_{eq}^{c} is the uniaxial equivalent creep strain, and σ_{eq} is the uniaxial equivalent deviatoric stress. The particular form of the flow rule used in the analyses can be expressed as:

$$\dot{\varepsilon}_{ij}^{c}=\dot{\varepsilon}_{eq}^{c}\cdot n_{ij} \quad (31)$$

where n_{ij} represents the gradient of the deviatoric stress potential.

Three different cases were analyzed including various geometrical, load, and material properties data. Two of them (*case B* and *case C*) accounted for the primary creep as well by introducing a value for parameter β different than zero. Elements at the notch tip had an average size of 1/10 of the notch radius in order to account for the high strain gradient near the notch tip. All components were subjected to a constant external load held for a pre-defined period of time (see Figs. 3c, 4c and 5c). The analyzed configurations were assumed to be in plane stress conditions resulting in a uniaxial stress-strain state at the notch tip. Only the maximum strain component and corresponding stress were generated using the proposed method.

The applied loading histories consisted of two steps, namely: instantaneous loading from zero to maximum load, and hold time under constant maximum load. No creep occurred during the first loading period; however, a significant amount of mechanically induced plastic strain was generated by the load excursion from zero to max (see Figs. 3b, 4b and 5b). The Neuber-type analysis of creep commenced from the moment after reaching the maximum load which was also the reference point for time measurement $(t=0.0)$.

The material constants, time step size, stress concentration factors, as well as the calculated plastic zone correction factors (using Ref. [10]) for all three cases are listed in Table 2. General details of the proposed algorithm are shown in the Appendix. The geometrical configurations of the analyzed components and results obtained from both the proposed method and finite element analyses are shown in Figs. 3a, 4a and 5a.

It is necessary to point out that in order to make a strict evaluation of the method, the initial elastic-plastic solution used in the proposed algorithm was that obtained from the finite element analysis. Therefore, a direct comparison between both methods using the same starting point was possible.

Due to the high strain gradients near the notch root, the simplified method is very sensitive to changes in the theoretical stress concentration factor (K_t). In order to perform a valid comparison between methods, the value of K_t chosen for the simplified approach was obtained directly from a linear-elastic finite element analysis using the same meshes and loads as in the visco-plastic analysis.

7. CONCLUSIONS

It can be observed from Figs. 3 to 5 that the proposed methodology shows a considerably good agreement with the FE results on both the stress and strain concentration factors at the notch tip. The solution obtained from Eqs. (28) and (29) matches closely that obtained from the visco-plastic finite element analysis. The predictions in all cases were within a range of 2% to 7% from the FE solution for both K_σ and K_ε (Fig. 6).

It has been proven that the proposed simplified solution based on Neuber's rule can be successfully extended to cases where creep plasticity is not a local issue. The proposed simplified solution can be easily programmed and is not restricted by the use of specific creep laws. Computational time was found to be 99.7% smaller than that needed for a finite element analysis; although roughly the same amount of time increments were needed to generate a solution.

The proposed technique was found to be sensitive to the location of the far-field point, especially in the bending cases where the most appropriate results were obtained at one-half of the nominal stress. For the uniaxial tension specimens, no difference in the results was observed as long as the far-field point location was greater than 3ρ.

Since the proposed simplified method works accurately for constant load cases, it is reasonable to assume that the same formulation can be used in cases where the specimen is subjected to a variable external load. This new challenge will be approached in future research work.

APPENDIX

Final Set of Equations to Be Solved for Each Increment

Increment of creep strain

$$\Delta\varepsilon_{22}^{c_n} = \Delta t_n \cdot \dot{\varepsilon}_{22}^{c}(\sigma;t) \tag{1A}$$

Decrement of stress due to creep

$$\Delta\sigma_{22}^{t_n} = \frac{(K_\Omega C_p)\sigma_{22}^{f0}\Delta\varepsilon_{22}^{cf_n} - \sigma_{22}^{t_{n-1}}\Delta\varepsilon_{22}^{c_n}}{\frac{2}{E}\sigma_{22}^{t_{n-1}} + \varepsilon^{p0} + \varepsilon_{22}^{c_n}} \tag{2A}$$

Increment of total strain

$$\Delta\varepsilon_{22}^{t_n} = \Delta\varepsilon_{22}^{c_n} - \frac{\Delta\sigma_{22}^{t_n}}{E} \tag{3A}$$

General Stepwise Procedure to Generate a Solution

1. Determine the notch tip stress, σ_{22}^{e}, and strain, ε_{22}^{e}, using the linear-elastic analysis.
2. Determine the elastic-plastic stress, σ_{22}^{0}, and strain, ε_{22}^{0}, using the original Neuber's rule [2], the ESED method [9], or finite element analysis.
3. Begin the creep analysis by calculating the increment of creep strain, $\Delta\varepsilon_{22}^{c_n}$, for a given time increment, Δt_n. A specific creep hardening rule has to be followed.
4. Determine the decrement of stress, $\Delta\sigma_{22}^{t_n}$, due to the previously calculated increment of creep strain, $\Delta\varepsilon_{22}^{c_n}$.
5. Determine the increment of total strain in the notch tip, $\Delta\varepsilon_{22}^{t_n}$, for a given time increment, Δt_n.
6. Repeat steps from 3 to 5 until the total time is completed.

REFERENCES

1. Chaudonneret, M., 1978, "Calcul de Concentrations de Contrainte en Elastoviscoplasticité," *Ph.D. Thesis Dissertation,* Office Natl d'Etudes et de Recherches Aerospatiales (ONERA), Chatillon, France.
2. Neuber, H., 1961, "Theory of Stress Concentration for Shear-Strained Prismatic Bodies with Arbitrary Nonlinear Stress-Strain Law," *Journal of Applied Mechanics, Transactions of the ASME,* **28**, p. 544-551.
3. Chaudonneret, M. and Culie, J. P., 1985, "Adaptation of Neuber's Theory to Stress Concentration in Viscoplasticity," *Recherche Aerospatiale (English Edition)*, n 4, p. 33-40.
4. Kurath, P., 1984, "Extension of the Local Strain Fatigue Analysis Concepts to Incorporate Time Dependent Deformation in Ti-6Al-4V at Room Temperature," T.&A.M. Report No. 464, The University of Illinois, Urbana, IL, USA.
5. Kubo, S. and Ohji, K., 1986, "Development of Simple Methods for Predicting Plane-Strain and Axi-Symmetric Stress Relaxation at Notches in Elastic-Creep Bodies" in *Proceedings of the International Conference on Creep*. Tokyo, Jpn: published by JSME, p. 417-422.
6. Moftakhar, A., Glinka, G., Scarth, D., and Kawa, D., 1994, "Multiaxial Stress-Strain Creep Analysis for Notches" in *ASTM Special Technical Publication 1184*. Philadelphia, PA, USA: published by ASTM, p. 230-243.
7. Harkegard, G. and Sorbo, S., 1998, "Applicability of Neuber's Rule to the Analysis of Stress and Strain Concentration under Creep Conditions," *Journal of Engineering Materials and Technology, Transactions of the ASME*, **120**, n 3, p. 224-229.
8. Moftakhar, A., Buczynski, A., and Glinka, G., 1995, "Calculation of Elasto-Plastic Strains and Stresses in Notches under Multiaxial Loading," *International Journal of Fracture*, **70**, n 4, p. 357-373.
9. Molski, K. and Glinka, G., 1981, "A Method of Elastic-Plastic Stress and Strain Calculation at a Notch Root," *Materials Science and Engineering*, **50**, n 1, p. 93-100.
10. Glinka, G., 1985, "Calculation of Inelastic Notch-tip Strain-stress Histories under Cyclic Loading," *Engineering Fracture Mechanics*, **22**, n 5, p. 839-854.
11. Irwin, G. R., 1968, "Linear Fracture Mechanics, Fracture Transition and Fracture Control," *Engineering Fracture Mechanics*, **1**, p. 241-257.
12. Hibbitt, Karlsson & Sorensen, Inc., 2000, "ABAQUS 6.1," *HKS*: Pawtucket, RI, USA.

Specimen 'A'		Specimen 'B'		Specimen 'C'	
STRESS (MPa)	STRAIN (mm/mm)	STRESS (MPa)	STRAIN (mm/mm)	STRESS (MPa)	STRAIN (mm/mm)
0.000	0.000E+00	0.00	0.000E+00	0.00	0.000E+00
34.400	5.850E-04	230.00	3.067E-03	185.00	2.721E-03
39.400	7.730E-04	240.00	6.933E-03	210.95	6.760E-03
44.313	1.291E-03	250.00	1.080E-02	236.90	1.080E-02
48.000	1.936E-03	261.74	1.967E-02	270.70	2.590E-02
50.600	2.489E-03	290.00	5.380E-02	294.00	5.470E-02
53.500	3.182E-03	308.00	8.530E-02	308.00	8.530E-02
55.525	3.732E-03	326.00	1.168E-01	322.00	1.159E-01

Table 1. Elastic-plastic stress-strain curve data.

	Elastic-Plastic Properties		Creep Constants			Time Increment	Calculated Factors	
	σ_y	E	A	α	β	Δt	K_t	C_p
Case A – Plate with edge notches	34.4 MPa	58800 MPa	$9.484\text{x}10^{-15}$ 1/h	7.7	0.0	$1\text{x}10^{-5}$ h.	2.94	1.17
Case B – Notched cantilever beam	230 MPa	75000 MPa	$6.960\text{x}10^{-14}$ 1/h	4.7	-0.4	$1\text{x}10^{-6}$ h.	3.00	1.22
Case C – Three-point bending beam	185 MPa	68000 MPa	$6.960\text{x}10^{-14}$ 1/h	4.7	-0.4	$1\text{x}10^{-6}$ h.	5.38	1.59

Table 2. Constants and material properties for validation of proposed solution.

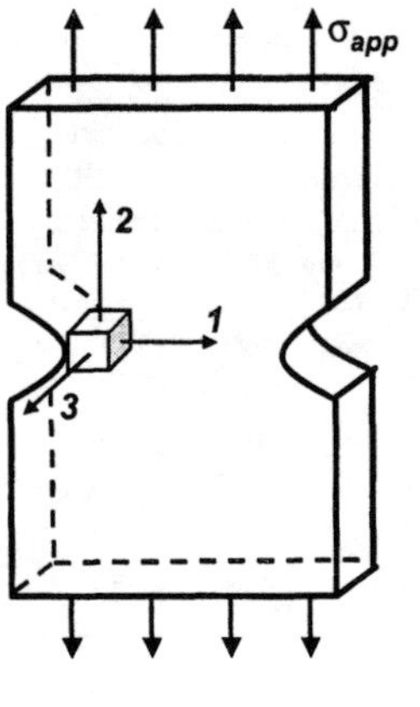

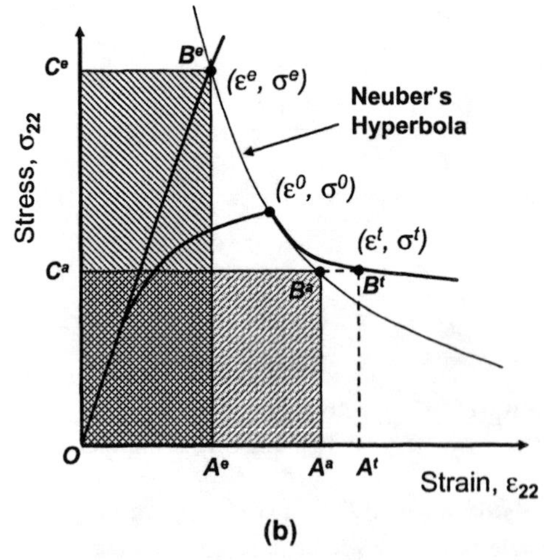

Figure 1. a) Schematic representation of notched specimen under tension loading. b) Graphical interpretation of the Neuber's rule approach to creep analysis.

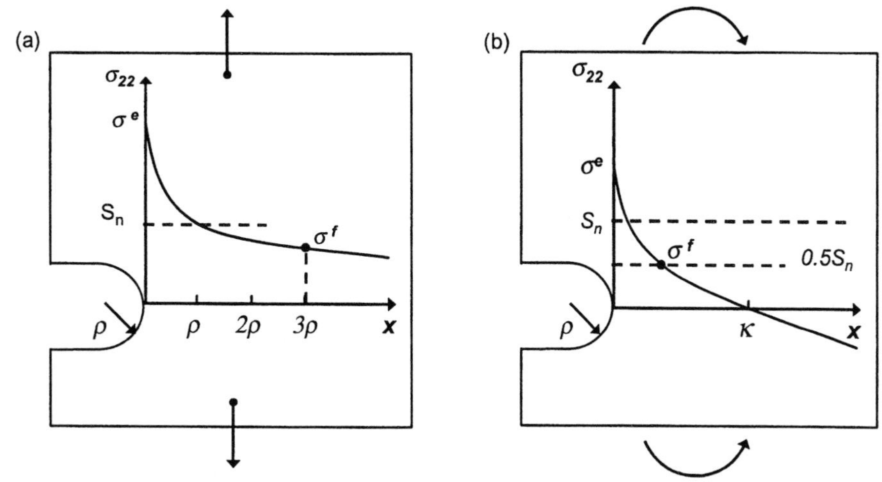

Figure 2. Graphical representation of the far field stress definition in a notched component: a) uniaxial tension, b) pure bending.

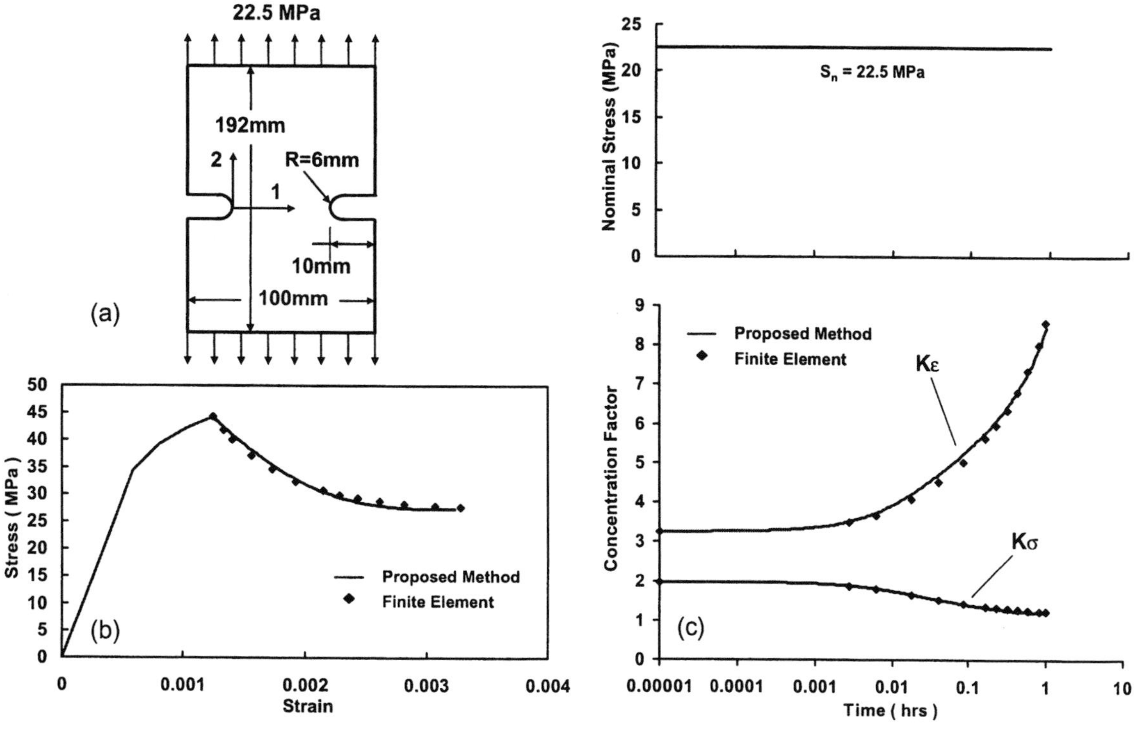

Figure 3. Results for case 'A': (a) Geometrical configuration, (b) stress-strain response for a maximum nominal stress of 22.5 MPa and a hold time of 1 hour, (c) stress and strain concentration factors vs. time.

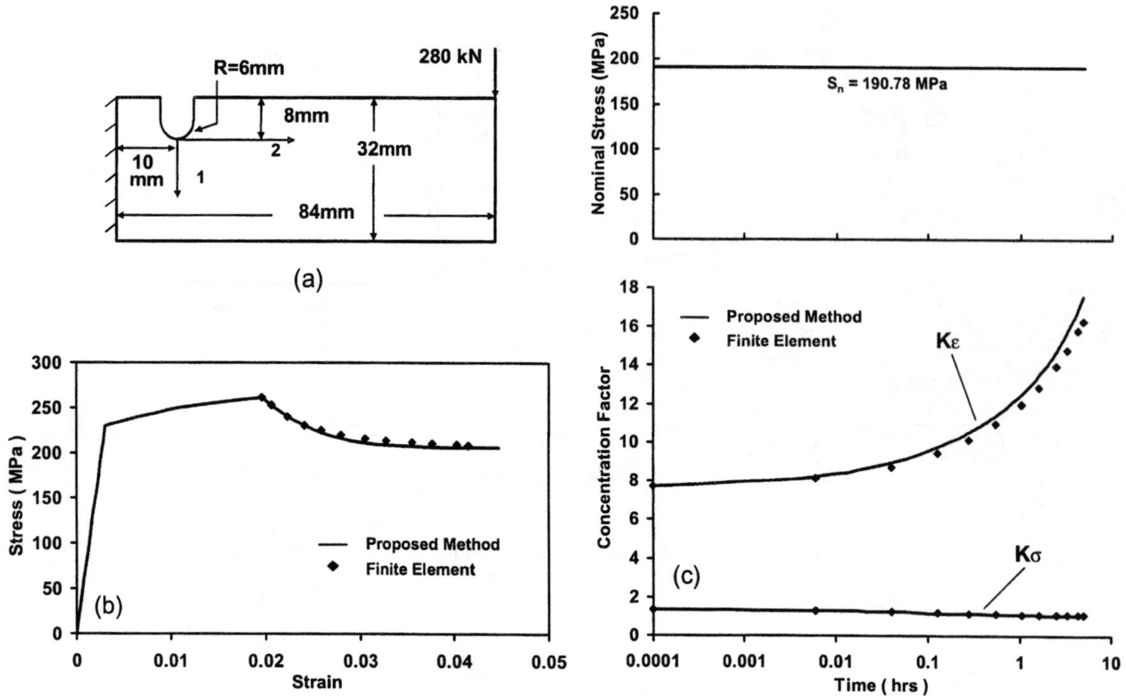

Figure 4. Results for case 'B': (a) Geometrical configuration, (b) stress-strain response for a maximum nominal stress of 190.78 MPa and a hold time of 5 hours, (c) stress and strain concentration factors vs. time.

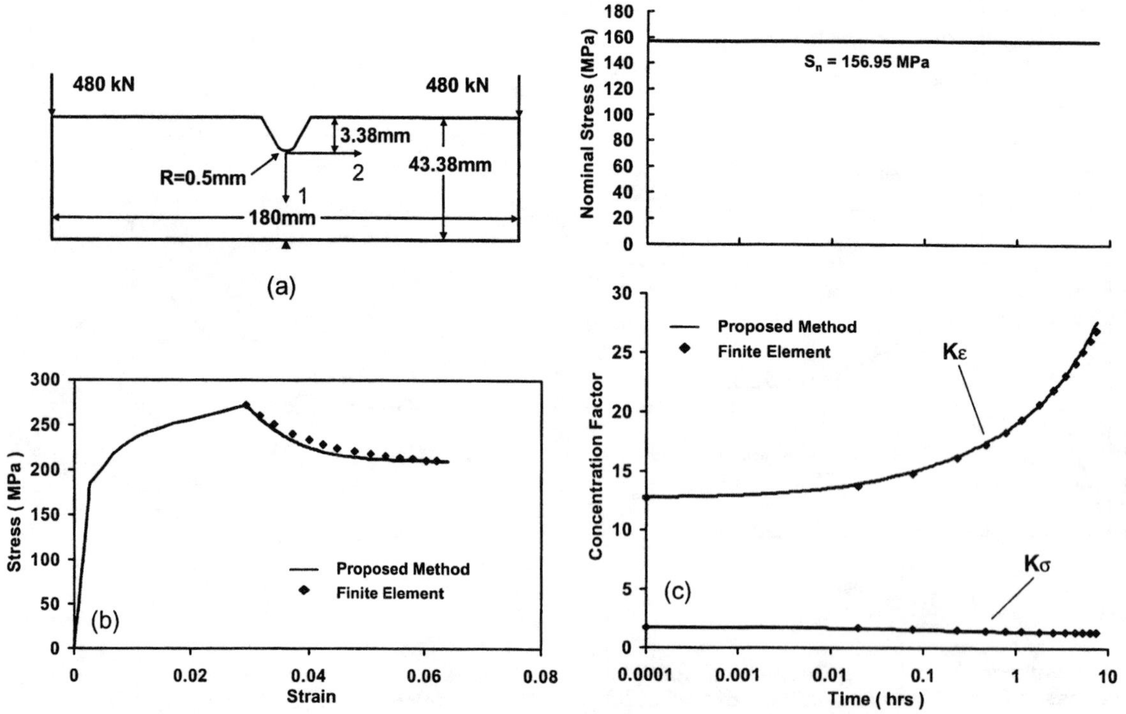

Figure 5. Results for case 'C': (a) Geometrical configuration, (b) stress-strain response for a maximum nominal stress of 156.95 MPa and a hold time of 9 hours, (c) stress and strain concentration factors vs. time.

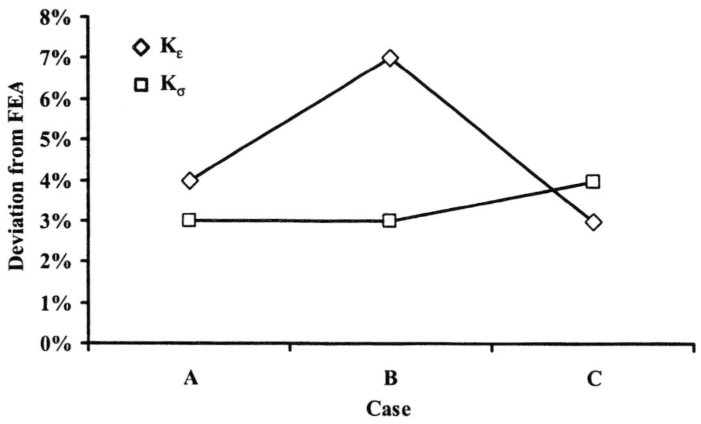

Figure 6. Maximum observed deviation in stress and strain concentration factors for specimens under non-localized creep

PVP-Vol. 458, Computer Technology and Applications

PVP2003-1889

A NON-CYCLIC METHOD FOR PLASTIC SHAKEDOWN ANALYSIS

W. Reinhardt
Babcock & Wilcox Canada
581 Coronation Boulevard
Cambridge ON N1R 5V3
CANADA

ABSTRACT

Shakedown is a cyclic phenomenon, and for its analysis it seems natural to employ a cyclic analysis method. Two problems are associated when this direct approach is used in finite element analysis. Firstly, the analysis typically needs to be stabilized over several cycles, and the analysis of each individual cycle may need a considerable amount of computing time. Secondly, even in cases where a stable cycle is known to exist, the finite element analysis can show a small continuing amount of strain accumulation. For elastic shakedown, non-cyclic analysis methods that use Melan's theorem have been proposed. The present paper extends non-cyclic lower bound methods to the analysis of plastic shakedown. The proposed method is demonstrated with several example problems.

NOMENCLATURE

$f(\sigma_{ij})$	von Mises yield function, eq. (1)
S_I	surface traction
t	time
Δt	period of the applied cyclic loads
u_i	applied displacement; displacement
$d\gamma_p$	equivalent strain increment, eq. (11)
ε^P_{ij}	(infinitesimal) plastic strain
ε^T_{ij}	(infinitesimal) imposed thermal strain
σ_{ij}	stress components
σ'_{ij}	deviatoric stress components
$\sigma_{ij}^{(e)}$	stress obtained with purely elastic material
$\sigma_{r\,ij}$	time-invariant residual stress
$\sigma_{v\,ij}$	time-dependent residual stress
σ_y	yield stress
uniaxial stress =	stress state with only one non-zero component
proportional stress =	two stress states are proportional if they can be related by a scalar factor $\sigma_{ij}^{(1)} = k\,\sigma_{ij}^{(2)}$

INTRODUCTION

For a pressurized component to be acceptable to Section III, Subsection NB of the ASME Boiler and Pressure Vessel Code [1], the absence of plastic growth (ratcheting) under cyclic loads must be demonstrated. The Code rules imply that the material model used in this kind of analysis should be perfectly-plastic. If such a material obeys the usual conventions of classical plasticity and a cyclic load is applied, one can show that the stresses and strain rates will undergo steady cycling after a transient period [2]. The steady response can be either linearly increasing strains per cycle, or ratcheting, which is prohibited by the Code, or periodic straining with or without plasticity, either of which is allowed.

The Code calls the condition where the strains are periodic during the steady cycle "shakedown". In contrast, the classical definition of shakedown extends only to the case where the steady case is purely elastic ("elastic shakedown"). This paper adopts the wider definition of the Code, where shakedown can be either elastic or plastic, where plastic shakedown means that plasticity occurs during the steady cycle. The classical literature has established powerful theorems to address elastic shakedown. These theorems allow the calculation of upper and lower bounds to the loads for which elastic shakedown can be guaranteed. No such general theorems exist presently for plastic shakedown.

The check whether a specific load history gives rise to shakedown proceeds essentially in two steps. In the first step, the steady cycle is obtained and in the second, the strains are evaluated against the shakedown criterion. The shakedown theorems open up a variety of ways to investigate elastic shakedown. Much fewer possibilities exist to establish plastic shakedown.

The most elementary method to establish the steady cycle under a given group of loads is the integration of the response over as many cycles as required, e.g. by an elastic-plastic finite element simulation. There are types of problems for which this method establishes the steady cycle within two or three cycles, for example the Bree problem [3]. Unfortunately, this is not the case for most realistic problems that include local stress concentrations or develop other forms of localized plasticity. For these, the number of cycles to approach the steady solution can be very significant. For problems of realistic size that involve a complicated geometry and loading history, the required number of cycles can make the analysis very time consuming or impractical.

Convergence definitions, such as the criterion proposed by the JPVRC [4,5] and the elastic core concept described by Kalnins [5], can limit the number of cycles that need to be executed, which may however still be significant. Furthermore, when the approximate solution is evaluated for shakedown, there remains some small amount of doubt whether the solution has truly shaken down or a small amount of ratcheting remains. The more design cycles exist, the smaller the tolerable amount of ratcheting would have to be certain to avoid a deformation that could render the structure unserviceable. This may be a theoretical objection, but a valid one as long as convergence cannot be proven in general for the method that is being used.

It seems therefore fruitful to ask whether it is possible to find the boundary between the shakedown and ratcheting domains directly, without having to evaluate the cyclic history. Ponter and Chen [6,7] created an upper bound method that approaches the cyclic solution at the shakedown boundary by matching the response of a linear material to the plastic solution. This method represents a very significant advance in shakedown analysis.

In terms of structural design, the drawback of an upper bound method is that only the fully converged solution is acceptable. Generally, lower bound methods are preferred because they yield conservative approximations to the true solution. The sections that follow outline a proposal for such a method, along with some theoretical justifications and a number of simple example applications. The method is based on the finite element method, and has not yet been established for the most general load histories. It is hoped that suitable extensions to the method can be made in the future.

LOWER BOUND ESTIMATION OF SHAKEDOWN

The response of the model material is elastic-perfectly plastic and has the properties outlined in [8]. The von Mises yield criterion applies

$$f(\sigma_{ij}) = \sqrt{\frac{3}{2}\sigma'_{ij}\,\sigma'_{ij}} \quad \leq \sigma_y \qquad (1)$$

where f() is the yield function and σ_{ij}' is the deviatoric stress. For cyclic application in the context of the ASME B&PV Code [1], σ_y is equal to 1.5 Sm. The associated flow rule is

$$\dot{\varepsilon}^p_{ij} = \lambda\,\sigma'_{ij} \qquad (2)$$

where λ denotes the plastic multiplier.

The loading history consists of time functions of tractions $S_i(t)$ and displacements $u_i(t)$ that are applied on parts of the boundary. Furthermore, a history of the temperature distribution inside the body gives rise to a thermal strain $\varepsilon_{ij}^T(t)$. All of these loads are periodic with the common period Δt

$$\begin{aligned} S_i(t+\Delta t) &= S_i \\ u_i(t+\Delta t) &= u_i \\ \varepsilon^T_{ij}(t+\Delta t) &= \varepsilon^T_{ij}(t) \end{aligned} \qquad (3)$$

The applied loads $S_i(t)$ are primary, whereas the applied strains and displacements are secondary loads.

It is assumed that plastic collapse of the structure due to the primary loads has been precluded with a suitable analysis. The response of the structure to the applied loads is then guaranteed to approach a state where the stresses and strain rates in the structure are periodic with period Δt, just like the loads [2]. The stable cyclic response can be of two types, namely shakedown or ratcheting. Shakedown occurs if the strains in the structure will eventually be periodic as well, i.e.

$$\Delta\varepsilon^*_{ij} = \varepsilon^*_{ij}(t+\Delta t) - \varepsilon^*_{ij}(t) = \int_t^{t+\Delta t} \dot{\varepsilon}^*_{ij}(\tau)\,d\tau \quad = 0 \qquad (4)$$

where ε_{ij}^* is the (total) strain during the steady cycle. Ratcheting implies that there is a fixed strain increment over the steady cycle

The strain increment $\Delta\varepsilon_{ij}^*$ in eq. (4) is independent of the starting time t. The elastic strains are periodic because of the periodicity of the loads (including the thermal strains ε_{ij}^T), and therefore the total strain increment $\Delta\varepsilon_{ij}^*$ equals the plastic strain increment $\Delta\varepsilon_{ij}^p$. For the same reason, the shakedown condition, eq. (4), can be written as

$$\int_t^{t+\Delta t} \dot{\varepsilon}^*_{ij}(\tau)\,d\tau = \Delta\varepsilon^*_{ij} = \int_t^{t+\Delta t} \dot{\varepsilon}^p_{ij}(\tau)\,d\tau \qquad (5)$$

$$\int_t^{t+\Delta t} \dot{\varepsilon}^p_{ij}(\tau)\,d\tau \quad = 0 \qquad (6)$$

Clearly, it may be possible to satisfy the condition (6) with some history of non-zero plastic strain, in which case plastic shakedown occurs. Otherwise, elastic shakedown cases where the plastic strain is zero throughout the cycle will certainly satisfy eq. (6) as well.

For elastic shakedown, a lower bound theorem has been formulated, and this theorem can be used advantageously to derive approximate solutions for the loads which let the structural response reach the elastic shakedown boundary. The lower bound theorem is due to Melan [9]. Consider the time dependent purely elastic response, $\sigma_{ij}^{(e)}$, to the applied thermal and mechanical loads. The structure will shake down to elastic action if a time-independent residual stress field $\sigma_{r\,ij}$ exists, such that the sum with the purely elastic stress lies inside the yield surface at all times within the cycle:

$$f(\sigma^{(e)}_{ij} + \sigma_{r\,ij}) \leq \sigma_y \qquad (7)$$

From eq. (7) follows that adding a time-invariant (constant) residual stress to the purely elastic stress will not be sufficient to keep the stress history inside the yield surface if plastic shakedown occurs. A time-dependent (variable) residual stress, $\sigma_{v\,ij}(t)$ is now needed to satisfy the yield condition. It must be a residual stress because the purely elastic response is already in equilibrium with the applied loads. Thus,

$$f(\sigma_{ij}^{(e)}(t) + \sigma_{r\,ij} + \sigma_{v\,ij}(t)) \leq \sigma_y \tag{8}$$

Since the time dependence of $\sigma_{v\,ij}$ is arbitrary, eq. (8) does not specify the constant residual stress $\sigma_{r\,ij}$ uniquely. In other words, time independent terms can be exchanged arbitrarily between the two residual stress terms. A unique definition of $\sigma_{v\,ij}$ for a given $\sigma_{r\,ij}$ follows from

$$\begin{aligned} f(\sigma_{ij}^{(e)}(t) + \sigma_{r\,ij} + \sigma_{v\,ij}(t)) = \sigma_y \quad & \text{if } \dot{\varepsilon}_{ij}^{p} \neq 0 \\ f(\sigma_{ij}^{(e)}(t) + \sigma_{r\,ij}) \leq \sigma_y \quad & \text{if } \dot{\varepsilon}_{ij}^{p} = 0 \end{aligned} \tag{9}$$

In other words, $\sigma_{v\,ij}$ is non-zero only when plastic flow occurs. This way, eq. (9) reduces automatically to eq. (7) when the cycle is fully elastic. Equation (9) also clearly illustrates a statement of Ponter and Karadeniz [10] that a structure that shakes down must shake down to elastic action outside those regions that exhibit cyclic plasticity.

By itself, eq. (9) cannot ensure shakedown, since it is applicable for any admissible elastic-plastic state of the structure, including ratcheting. An additional constraint that will ensure shakedown derives from eq, (6). The latter can be rewritten as

$$\int_{t}^{t+\Delta t} \lambda\, \sigma'_{ij}\, d\tau = 0 \tag{10}$$

From eq. (2), the multiplier λ can be shown to equal

$$\lambda = \frac{3}{2}\frac{\dot{\gamma}_p}{\sigma_y} = \frac{3}{2}\frac{\sqrt{\frac{2}{3}\dot{\varepsilon}_{ij}^{\,p}\,\dot{\varepsilon}_{ij}^{\,p}}}{\sqrt{\frac{3}{2}\sigma'_{ij}\,\sigma'_{ij}}} \tag{11}$$

whenever plastic flow occurs. Substituting (11) into (10) yields

$$\int_{t}^{t+\Delta t} \sigma'_{ij}\, \dot{\gamma}_p\, d\tau = \int_{\gamma_p(t)}^{\gamma_p(t+\Delta t)} \sigma'_{ij}\, d\gamma_p \tag{12}$$

with

$$\sigma'_{ij} = \sigma_{ij}^{(e)} + \sigma_{r\,ij} + \sigma_{v\,ij} \tag{13}$$

In eq. (12), the stress is integrated with the equivalent von Mises strain increment. This increment is positive whenever plastic flow occurs, and zero otherwise. The stress must change sign to enable the integral (12) to equal zero.

Also, the stress history outside those times when plastic flow occurs is irrelevant because the von Mises strain increment $d\gamma_p$ is zero. This increment represents a plastic strain increment on the equivalent stress-strain curve, and can be understood as an internal time scale of plastic flow. Such a scale must exist because this is rate independent plasticity, and it must be possible to eliminate the time variable from all equations. Note that the von Mises strain increment $d\gamma_p$ can never be negative, it will therefore continue to increase with time whenever plastic flow is present. When there is no plasticity, the internal clock of the material stands still.

There are two important special cases of eq. (12). The first case is time independent loading. The elastic stress is therefore time independent, and so is the residual stress distribution. The analysis reduces to a time independent version of eq. (7), i.e. essentially to a limit analysis.

The second special case is symmetric loading. This does not necessarily mean that the time function of loading must be symmetric, the symmetry is defined through the use of eq. (12). Only those parts of loading where plastic flow occurs are of interest. Symmetry means that from some time point onwards the stresses of the previous history will recur with the opposite sign. The recurring inverted part of the history may be slowed down or sped up compared to the first because the actual time needed to complete a certain part of the history is immaterial (Figure 1). The stress distribution in the entire structure at a time t_0 must correspond to that at t_0+Dt(t_0) through

$$\text{everywhere} \quad \sigma'_{ij}(t_o + Dt(t_o)) = -\sigma'_{ij}(t_o) \tag{14}$$

The time interval Dt(t_0) between the corresponding stress points can be different for every instance of the first part of the history, but the original sequence has to be preserved in the second part. In other words, the transformation from the old to the new time variable must be monotonically increasing. For proportional stressing, such a transformation is possible simply if $(\sigma_{ij}^{(e)})_{Max} = -(\sigma_{ij}^{(e)})_{Min}$. The actual stress history can be equivalently replaced by one that is symmetric in time such that

$$\sigma_{ij}(t_0 + \Delta t/2) = -\sigma_{ij}(t_0) \tag{15}$$

The transformation from the history (14) to (15) must be identical at every point of the structure. If this is possible, the history has zero mean stress, and will therefore not cause ratcheting

OUTLINE OF THE METHOD

Consider a loading history that can be split into a constant part and a number, n, of proportional cyclic parts with zero mean in the sense of the previous section:

$$\begin{aligned} P(t) &= P_0 + \sum_{k=0}^{n} P_{s\,k}(t) \\ u(t) &= u_0 + \sum_{k=0}^{n} u_{s\,k}(t) \\ \varepsilon^{t}(t) &= \varepsilon_0^{T} + \sum_{k=0}^{n} \varepsilon_{s\,k}^{T}(t) \end{aligned} \tag{16}$$

The steady cyclic response of the structure is then given by

$$\begin{aligned} \sigma'_{ij} &= \sigma_{ij}^{(e)} + \sigma_{r\,ij} + \sigma_{v\,ij} \\ \sigma_{ij}^{(e)} &= \sigma_{ij\;0}^{(e)} + \sum_{k=1}^{n} \sigma_{ij\;s\,0}^{(e)}(t) \\ \sigma_{r\,ij} &= \sigma_{r\,ij\,0} \\ \sigma_{v\,ij} &= \sum_{k=1}^{n} \sigma_{v\,ij\;s\,0}(t) \end{aligned} \tag{17}$$

if elastic or plastic shakedown occurs. The elastic response to a fully reversed and proportional load will be fully reversed and proportional. The constant residual stress $\sigma_{r\,ij}$ will be able to redistribute but not fully balance the applied load P_0. It will be able to balance stresses due to the constant part of the imposed displacement, u_0, which can therefore be ignored. Often, but not always, it can compensate the constant thermal stress as well. The constant loads will not contribute to the cyclic residual stress $s_{v\,ij}$. Conversely, the fully reversed symmetric loads will cause a fully reversed elastic stress history, and thus not contribute to the constant stress terms. In this sense, the stress and loading histories have been split into distinct components.

The yield condition (8) must be satisfied for the combined history in eq. (17). Since the yield criterion $f(\sigma_{ij}) = f(\sigma_{ij}')$ is nonlinear by eq. (1), there is no exact additive decomposition of the yield condition. However, there is a decomposition in the sense of a lower bound because

$$f(\sigma_{ij}^{(1)} + \sigma_{ij}^{(2)}) \le f(\sigma_{ij}^{(1)}) + f(\sigma_{ij}^{(2)}) \qquad (18)$$

Essentially, this is the triangle inequality of linear operator theory [11]. Therefore,

$$f(\sigma_{ij}^{(1)}) + f(\sigma_{ij}^{(2)}) \le \sigma_y \Rightarrow f(\sigma_{ij}^{(1)} + \sigma_{ij}^{(2)}) \le \sigma_y \qquad (19)$$

but not the converse. In other words, the method would in general result in a higher equivalent stress level than the exact solution, as would be expected for a lower bound. Equality is reached if the stress states are uniaxial and of the same sign. Therefore, for a "near uniaxial" stress state, the right side of the inequality (19) is expected to give a good lower bound of the left side. The same is true if the stress states are near proportional.

With the preliminaries discussed up to this point, a lower bound method can be devised to analyze elastic or plastic shakedown. The methods involves the following steps:

1. Decompose the loading into constant and fully reversed proportional components according to eq. (16).
2. Create a finite element model of the structure under consideration. Use elastic-perfectly plastic properties with the cyclic yield stress σ_y. This is the initial current yield stress σ_{yc}. Set the counter k to 1.
3. Apply the load range of the kth of n cyclic loading components. Obtain the stress distribution.
4. For each location (element), subtract one half of the von Mises equivalent stress range defined by eq. (1) from the current yield stress $\sigma_{yc}^{(k)}$. The difference is the new local current yield stress $\sigma_{yc}^{(k+1)}$. This step makes use of eq. (19).
5. Repeat steps 3 and 4 for all n fully reversed proportional components according to (16). If plastic collapse occurs at any time, the loading combination is inadmissible (at least according to the lower bound estimate).
6. Using the yield stress distribution $\sigma_{yc}^{(n+1)}$, perform a limit analysis using the constant components of loading. The limit load indicates a lower bound to the allowable constant load that will ensure shakedown. Alternatively, if plastic collapse is shown not to occur for the actual cyclic load, then the structure will shake down to steady strain cycles under the given loads.

For specific histories, the decomposition (16) may not indicate any constant load components. In this case, applying a small constant load can prevent the situation where the cyclic loads have exhausted the yield in a significant part of the structure such that "collapse" (ratcheting) occurs "under the weight of a mouse" on the structure. The analyst would have to choose the location of the additional constant load appropriately.

The example section that follows will show that the outlined procedure is easy to apply in many cases, such as the classical Bree example [3]. For very general loading histories, particularly thermal ones, condition (12) may allow a radical simplification of the history, but it remains to be investigated how this can be done easily before the actual elastic-plastic solution is known. It is also important that the actual loading history can be decomposed into a small number of fully reversed proportional components because otherwise the conservatism of the method may become excessive.

The application of constant loads in step 6 poses another potential difficulty. Since a limit analysis is performed, only applied loads P_0 will make a contribution, not applied constant displacements or applied constant thermal strains. This is entirely appropriate for the displacements because their effect can always be cancelled with a constant residual stress in eqs. (17). There are however examples of loads due to thermal strains that cannot be equilibrated with a residual stress. This case will be discussed further in the examples section.

A strong point of the present method is its ability to consider loads for which the exact time history is not known. Examples are seismic loads and nozzle loads due to the attached piping. The cyclic contribution of such loads can be included conservatively through the use of a separate cyclic component k in (16).

Since the present method uses an elastic-perfectly plastic material model, any parts of the structure that reach yield under the cyclic component of eq. (16) that is being applied do not carry load under any subsequently applied components. Neither do any of the plastified sections carry the final limit load. Therefore, the proposed method can be understood as an implementation and extension of the principle enunciated by Ponter and Karadeniz [10, 12] that shakedown occurs if the plastic parts of the structure are cut away and the remaining parts shake down to elastic action.

EXAMPLE APPLICATIONS

The Classical Bree Problem

One form of the problem described by Bree [3] is a constant axial load, F, is applied to a rectangular beam, which is at the same time subjected to a cyclic temperature distribution that is uniform along the beam but varies linearly through the thickness. The beam is constrained against bending deflection, Fig. 2.

The loading is decomposed into of one cyclic load due to temperature and one constant load due to the applied force. The cyclic load is applied first, and gives rise to a linear (bending) stress distribution that fluctuates between $+\sigma_b$ and $-\sigma_b$ at the extreme fibres, and decreases linearly towards the neutral axis for the case where σ_b is less than the yield stress σ_y. In the other case, the outer fibres become plastic and fluctuate between $+\sigma_y$ and $-\sigma_y$ while the central region has a linear stress distribution. The two cases are indicated in Fig. 3. This corresponds to step 3 of the outline in the previous section.

In step 4, half of the calculated stress range is subtracted from the (initially uniform) yield stress σ_y. The distribution of the yield stress through the beam section after step 4 of the method is also shown in Fig. 3. A linear distribution is obtained. In the case where the outer fibres of the beam reach yield, these fibres will have zero yield stress when subsequent load components are applied and will thus not contribute to the capacity of the beam.

There are no further cyclic load components, and a limit analysis with the constant axial load is performed next. The beam will fail when the applied load equals the area below the yield stress distribution shown in Figure 3b. Therefore, for the first case where the cyclic stress was within yield, the axial stress at collapse, σ_{ac}, in terms of the bending stress range, $\sigma_{br} = 2\,\sigma_b$ will be

$$\sigma_{ac} = \frac{(\sigma_y - \frac{\sigma_{br}}{2})\,h + \frac{\sigma_{br}}{2}\frac{h}{2}}{h} = \sigma_y - \frac{\sigma_{br}}{4} \tag{20}$$

In the other case, yielding occurs in the outer fibres of the beam and the yield stress after step 4 is non-zero only over the height h1 (Fig. 3). By similar triangles, $h1/h = 2\sigma_y/\sigma_{br}$, and the collapse stress becomes

$$\sigma_{ac} = \frac{\sigma_y}{2}\frac{h1}{h} = \frac{\sigma_y^2}{\sigma_{br}} \tag{21}$$

These precisely are the boundaries between the shakedown and ratcheting region found by Bree [14]. A pictorial representation is shown in Fig. 2.

The Inverse Bree Problem

The inverse Bree problem considers a rectangular beam that is constrained against axial motion. It is subject to a constant applied bending moment, B, and a cyclic thermal load that induces thermal membrane stresses, Figure 4.

The load decomposition yields a reversed axial membrane stress with range σ_{ar} to be applied first, and a constant bending moment (maximum elastic bending stress σ_{bc}) for the subsequent limit analysis. Any thermal membrane stresses in the axial direction that do not vary in time can obviously be balanced with a constant residual stress.

The cyclic thermal load that is applied first will reduce the yield strength for the subsequent limit analysis to $\overline{\sigma}_y$:

$$\overline{\sigma}_y = \sigma_y - \frac{\sigma_{ar}}{2} \tag{22}$$

Plastic collapse occurs then under the moment that is applied in the next step if

$$\sigma_{bc} = \frac{3}{2}\overline{\sigma}_y = \frac{3}{2}\sigma_y - \frac{3}{4}\sigma_{ar} \tag{23}$$

A comparison of this shakedown boundary against the results of cyclic finite element analysis is shown in Fig. 4. The cyclic FE analysis defines shakedown if a relative displacement increment of 10^{-4} or less is reached after 50 cycles. In contrast, the theoretical shakedown boundary requires an increment of 0 after infinitely many cycles, which can lead to a slightly different boundary. However, the linear character of the boundary (23) is clearly reproduced by the FE results. It can also be shown theoretically that eq. (23) is correct.

The Three-Bar Problem

Each of the bars in the three-bar mechanism shown in Fig. 5 is subjected to a heating-cooling cycle in its turn while the end displacement of all bars is coupled. Only one bar is hot at any time, so the within a cycle each bar is hot once, at which time the elastically calculated stress in that bar is compressive. Twice during the cycle when one of the other bars is hot, the elastic stress in the same bar is tensile, and reaches a magnitude of half of the compressive stress. This example comes from the book by Gokhfeld and Cherniavski [14]. It is special in that only secondary stresses occur in the structure, and yet ratcheting is found as soon as the compressive stress exceeds yield (i.e. when the elastic stress range in the bars is 1.5 σ_y).

Denote the peak thermally induced elastic stress by $-\sigma_{th}^{(e)}$. This is the compressive stress in the hot bar. The stress in the other two bars at the same time is $0.5\,\sigma_{th}^{(e)}$. The stress cycles can be decomposed as follows. If a constant stress of $-0.25\,\sigma_{th}^{(e)}$ in each bar is assumed, then the cyclic stress component in each bar will fluctuate between $0.75\,\sigma_{th}^{(e)}$ and $-0.75\,\sigma_{th}^{(e)}$. Therefore, the cyclic component in each bar is fully reversed.

Although the stresses are all secondary, the constant stress of $0.25\,\sigma_{th}^{(e)}$ cannot be equilibrated with a residual stress because it occurs in all three bars simultaneously. The only way to establish equilibrium would be to apply a force at the juncture of the three bars, which is obviously not a residual stress. One has to conclude that the constant stress must be applied to the structure in the final "limit analysis", although its secondary origin would preclude it from causing a true limit failure. It can, however, give rise to ratcheting. Therefore, the steps of the method proceed as follows:

First, apply the fully reversed stress. Over the complete cycle, this causes a stress range of $1.5\,\sigma_{th}^{(e)}$ assuming that yield is not exceeded. The updated fictitious yield stress after this cycle is then

$$\overline{\sigma}_y = \sigma_y - \frac{3}{4}\sigma_{th}^{(e)} \tag{24}$$

The "limit analysis" of the subsequent step applies an equal membrane stress in all 3 bars. "Collapse" occurs if

$$\frac{1}{4}\sigma_{th}^{(e)} = \overline{\sigma}_y \tag{25}$$

Combining eq. (24) and eq. (25) yields as the condition for the shakedown boundary

$$\sigma_{th}^{(e)} = \sigma_y \tag{26}$$

This is the analytical result [14].

CONCLUSIONS

The proposed method allows a elastic-plastic shakedown analysis to be performed with a series of non-cyclic plastic analyses. Shakedown is here defined as the absence of ratcheting [1]. It is assumed that the applied loads (including the thermal loads) can be equivalently decomposed into proportional, fully reversed cycles and constant components, such that the sum of all components equals the original system of applied loads. The analysis is elastic-perfectly plastic and applies each cyclic component successively. At each load application, the current yield stress is reduced by the von Mises equivalent stress amplitude. The constant load component is applied last. Any self-equilibrating stresses that do not vary, such as those due to any time-invariant temperature distribution, may be eliminated from the constant load. The remaining load is applied, and the final limit analysis indicates ratcheting if the structure cannot sustain the applied load ("collapses") and shakedown otherwise.

In three simple examples involving uniaxial stress states, the method was shown to yield the exact boundary between the shakedown and ratcheting regions. In multiaxial applications, the method is expected to result in a lower bound estimation.

In the development of the method, it has been assumed that the yield stress is temperature independent. To account for the temperature dependence in real materials, it is suggested to use the yield stress at the average temperature of the complete cycle for each of the component analyses. The yield stress used for the component analyses should be kept consistent.

The method is suitable for an implementation into a commercial Finite Element code. This would require a small amount of data manipulation between each of the successive analyses, since the yield stress needs to be adjusted at each element. Such an adjustment can be achieved by a post-processing step after each solution phase. Some work remains to be done to ensure the numerical stability and accuracy of the obtained solution (the yield stress may be reduced to zero in parts of the structure). More work is planned to demonstrate the application of the method to actual pressure vessel design problems.

REFERENCES

1. ASME, 2001, *Boiler and Pressure Vessel Code*, American Society of Mechanical Engineers, New York.
2. Gokhfeld, D.A., and Cherniavsky, O.F., 1980, *Limit Analysis of Structures at Thermal Cycling*, Suthoff & Nordhoff, Alphen aan den Rijn, The Netherlands.
3. Bree, J., 1967, "Elastic-Plastic Behaviour of Thin Tubes Subjected to Internal Pressure and Intermittent High-Heat Fluxes with Application to Fast-Nuclear Reactor Fuel Elements," J. Strain Analysis, Vol. 2 (3), pp. 226-238.
4. Okamoto, A., Nishiguchi, I., and Aoki, M., 2000, "New Secondary Stress Evaluation Criteria Suitable for Finite Element Analyses," ICPVT-9, Sidney, Vol. 2, pp. 613-620.
5. Kalnins, A., 2002, "Shakedown and Ratcheting Directives of ASME B&PV Code and their Execution," PVP Vol. 439, ASME PVP Conference, Vancouver, BC, Canada, pp. 47-55.
6. Ponter, A.R.S., and Chen, H., 2001, "A Minimum Theorem for Cyclic Load in Excess of Shakedown with Application to the Evaluation of a Ratchet Limit," Eur. J. Mech. A / Solids, Vol. 20, pp. 53.
7. Ponter, A.R.S., Chen, H, and Habibullah, M., 2002, "Computation of Ratchet Limits for Structures Subjected to Cyclic Loading and Temperature", PVP Vol. 441, ASME PVP Conference, Vancouver, BC, Canada, pp. 143-152.
8. Reinhardt, W., 2003, "Distinguishing Ratcheting and Shakedown Conditions in Pressure Vessels", ASME PVP Conference, Cleveland, OH.
9. Melan, E., 1938, "Zur Plastizität des räumlichen Kontinuums," Ingenieur-Archiv, Vol. 9, pp. 116-123.
10. Ponter, A.R.S., and Karadeniz, S., 1985, "An Extended Shakedown Theory for Structures that Suffer Cyclic Thermal Loading. Part I: Theory," ASME J. Appl. Mech., Vol. 52, pp. 877-882.
11. Naylor, A.W., and Sell, G.R., 1982, *Linear Operator Theory in Engineering and Science*, Applied Mathematical Sciences Vol. 40, Springer Verlag, New York.
12. Ponter, A.R.S., and Karadeniz, S., 1985, "An Extended Shakedown Theory for Structures that Suffer Cyclic Thermal Loading. Part II: Applications," ASME J. Appl. Mech., Vol. 52, pp. 883-889.

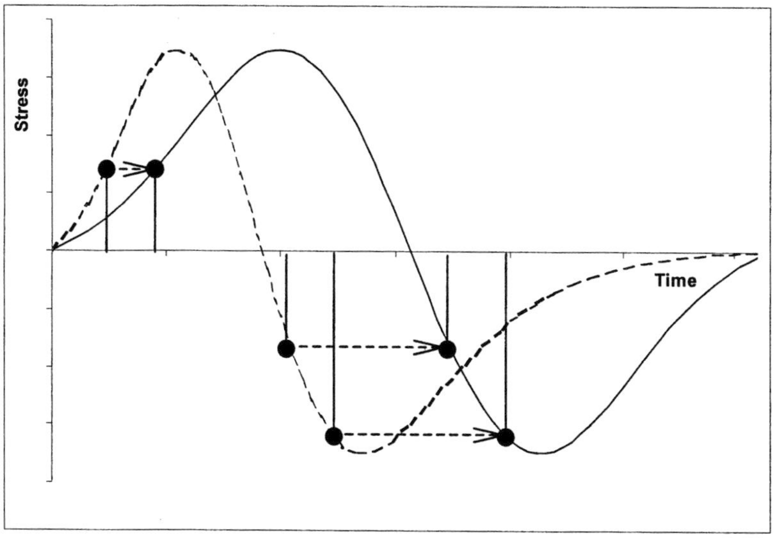

Figure 1: Equivalent load (stress) histories: A monotonic transformation of the time scale transforms one history into the other.

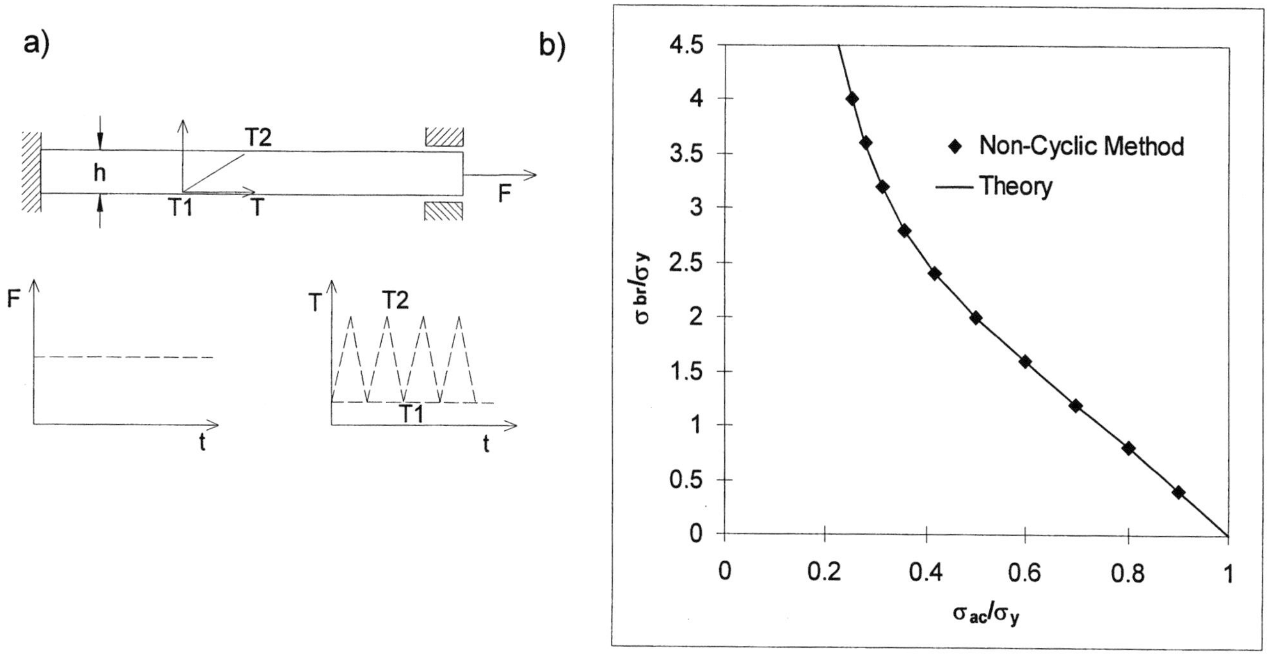

Figure 2: a) Schematic representation of the Bree problem, and b) shakedown region predicted by Bree [3] and by the present method

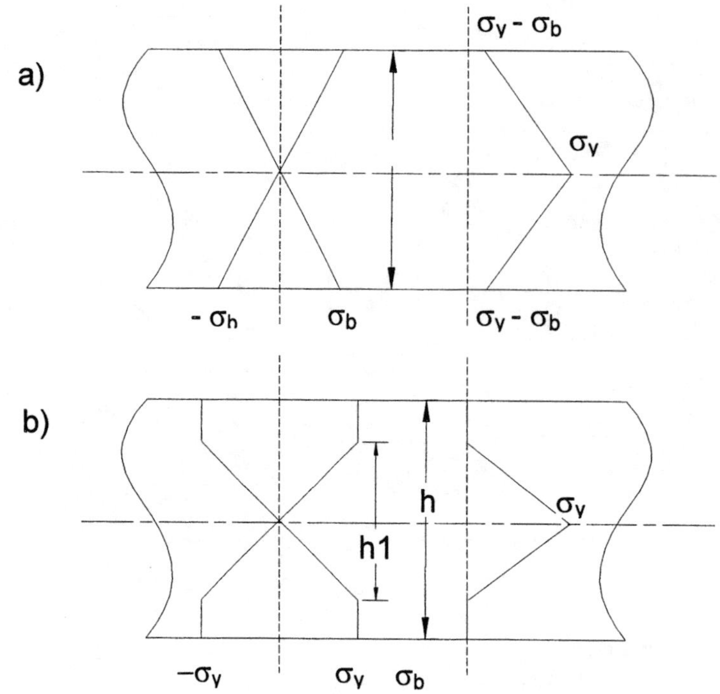

Figure 3: Application of the non-cyclic method to the Bree problem: Stress profile due to fully reversed bending load and subsequent yield stress distribution a) if bending stress remains below yield, b) if bending stress exceeds yield

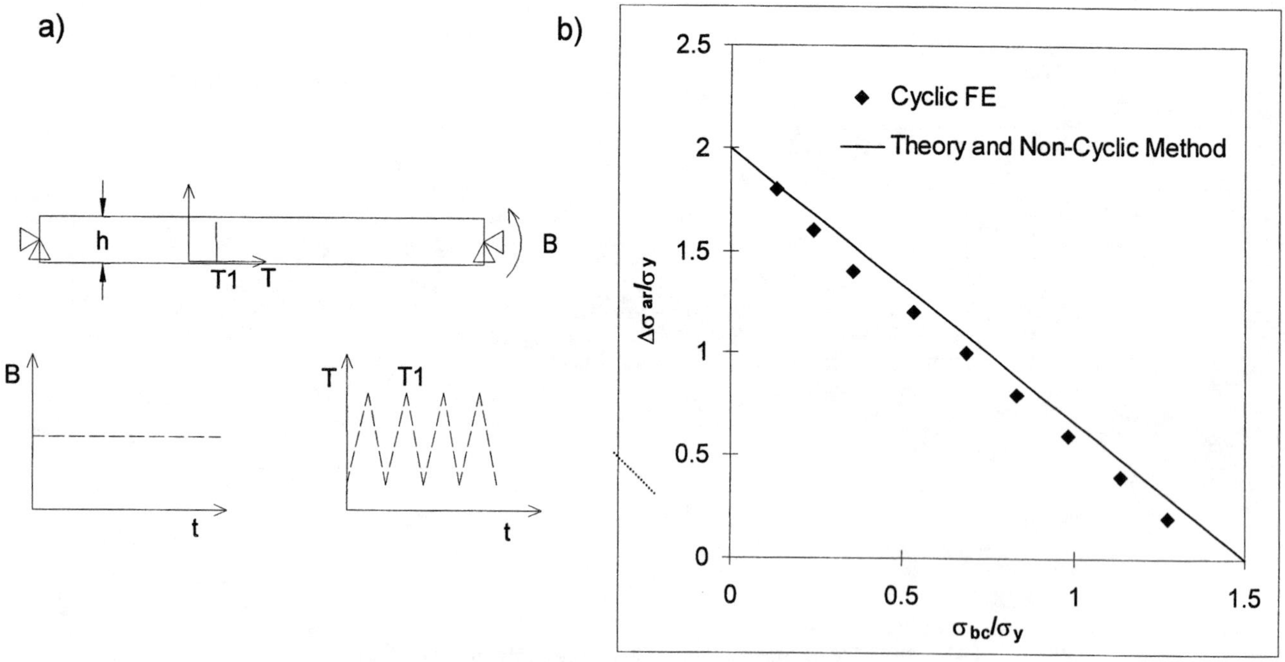

Figure 4: a) Schematic representation of the inverse Bree problem, and b) shakedown region predicted by cyclic FE, theory and by the present method

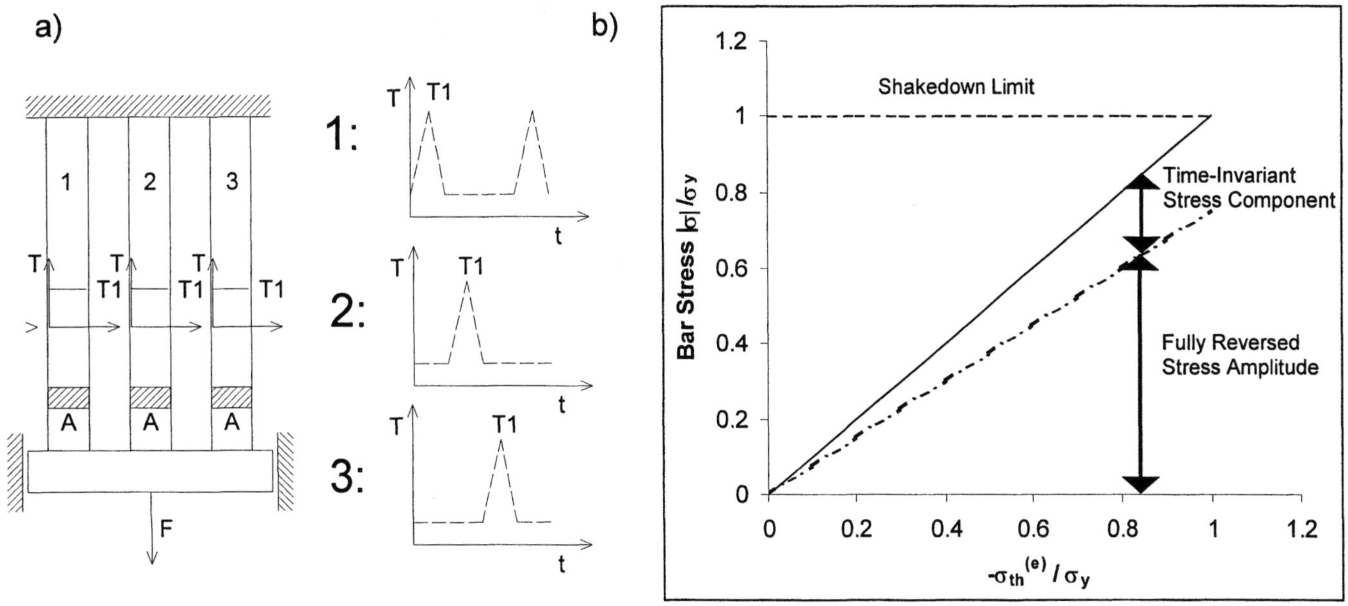

Figure 5: a) Schematic representation of the Three Bar problem, and b) constant and fully reversed stress components used by the non-cyclic method

PVP-Vol. 458, Computer Technology and Applications

PVP2003-1890

ACCELERATED LIMIT LOADS USING REPEATED ELASTIC FINITE ELEMENT ANALYSES

Prasad Mangalaramanan[1]
Dana Corporation, P.O. BOX 4097, Kalamazoo, MI 49003-4097, USA. Prasad.Mangalaramanan@dana.com

ABSTRACT

This paper demonstrates the limitations of repeated elastic finite element analyses (REFEA) based limit load determination that uses the classical lower bound theorem. The r-node method is prescribed as an alternative for obtaining better limit load estimates. Lower bound aspects pertaining to r-nodes are also discussed.

NOMENCLATURE

B	Constant in power-law stress-strain relation
E	Young's modulus
$\overline{E}$	Recoverable energy
E_o	Original Young's modulus
E_s	Modified Young's modulus
ECM	Elastic Compensation Method
N	Elastic iteration number
P	Applied load
P_L	Limit load
$P_{L\_r\_node}$	Limit load by r-node method
$\dot{U}$	Velocity
V	Volume
$\overline{V}$	Partial volume
$V\downarrow$	Volume where REFEA stresses relax
$V\uparrow$	Volume where REFEA stresses increase
V_R	Proposed reference volume
V_{RP}	Reference volume [5]
V_T	Total volume
V_δ	Volume of element having maximum stress
i	A given finite element; initial
f	Final
j	Element sequence number
m_o	Upper bound multiplier [17]
$\overline{m}_o$	Upper bound multiplier [18]
n	Power-law index
$\overline{n}$	Total number of finite elements
q	Attenuation index
r_i	Inside radius of cylinder
r_o	Outside radius of cylinder
r_R	Reference radius of cylinder
ε	Strain
ε_{ij}	Strain tensor
$\lambda = r_o / r_i$	Ratio of inside and outside radii of cylinder
σ	Stress
σ_{arb}	Arbitrary stress
σ_e, σ_{eqv}	von Mises equivalent stress
σ_{ij}	Stress tensor
$\sigma_{\max\_r\_node}$	Maximum r-node stress
σ_R	Reference stress
σ_y	Material yield strength

1 INTRODUCTION

Lower bound limit analysis is an acceptable method for designing pressure components [1]. While analytical methods for calculating limit loads are applicable only for simple structures, non-linear finite element analysis (FEA) is generally considered as a reliable and generic alternative. Since pressure vessel industries have traditionally relied on stress classification using linear elasticity, there was a motivation to develop limit analysis techniques based on linear elastic FEA. Over the last fifteen years, a number of such methods have emerged due to the pioneering efforts of Marriott [2], Seshadri [3], Mackenzie et al. [4] and Ponter et al. [5]. Although addressed by different names, all the methods developed by the above researchers invariably use a set of linear elastic finite element analyses, with each analysis aimed at producing a stress field that is closer to limit than the one before. While Marriott [2] and Mackenzie et al. [4] used the classical lower bound theorem to determine limit loads, Seshadri [3] invoked the concept of creep reference stress using the so

[1] Also Adjunct Assistant Professor, Department of Mechanical and Aeronautical Engineering, Western Michigan University, Kalamazoo, MI 49008, USA

called r-nodes, which are near equilibrium locations least sensitive to inelastic stress redistribution. Ponter et al. [5] proposed an upper bound method and proved the upper bound limit load would converge to an asymptotic value and the lower bound (loosely defined) would converge to the least upper bound. The present paper proposes an accelerated method for determining limit loads using r-nodes and demonstrates that it would be beneficial to use the maximum r-node stress rather than the maximum von Mises stress for determining lower bounds.

2 CLASSICAL LOWER BOUNDS USING REPEATED ELASTIC ANALYSES

Marriott [2] showed that performing a sequence of elastic FEA while systematically changing the elastic moduli of selected elements after each analysis would lead to statically admissible stress fields that are capable of producing progressively better lower bound limit loads. Mackenzie and Boyle [4] developed the elastic compensation method, which, unlike Marriott's method, modified the elastic moduli of every element by using the expression [3]:

$$E_s = E_o\left[\frac{\sigma_{arb}}{\sigma_{eqv}}\right] \quad (1)$$

A number of papers written by Mackenzie and his coworkers [6] demonstrated the simplicity and power of their method. The expression given by equation (1) was, however, found to be too aggressive for several components and was therefore attenuated by Mangalaramanan and Seshadri [7] as:

$$E_s = E_o\left[\frac{\sigma_{arb}}{\sigma_{eqv}}\right]^q \quad (2)$$

$$0 \le q \le 1$$

The factor q is called the attenuation index. For beam bending and thick cylinders under internal pressure, Mangalaramanan and Reinhardt [8] showed that "q" exactly corresponded to the index "n" in the power-law stress strain relationship given by

$$\varepsilon = B\sigma^n \quad (3)$$

such that

$$q = 1 - \frac{1}{n} \quad (4)$$

It has been shown [9] that $q = 1$ is likely to cause convergence difficulties in problems with dominant shear, and however tempting it may seem, $q > 1$ does not carry any physical meaning and may actually have a detrimental effect on limit loads.

3 CONCEPT OF R-NODES

Marriott and Leckie [10] observed that there are points in components undergoing transient creep at which the stress does not change with time. These locations were named skeletal points. Somewhat similar to skeletal points, Seshadri [3] introduced the concept of redistribution nodes more than a decade ago as a simple method for determining limit loads using two linear elastic finite element analyses. While not much further research was done on skeletal points, Seshadri [3] established a connection between the concepts of creep reference stress, primary stress and load control, and also prescribed a method for determining r-nodes using two linear elastic finite element analyses. Subsequent work [11,12] involving multi-bar collapse models offered a number of numerical examples to demonstrate that the method can be used to obtain quick estimates of limit loads for redundant structures, assuming elastic-perfectly plastic conditions. Each bar was identified with a single r-node peak that could act as an independent plasticity nucleation center and thereby develop into a plastic hinge. Almost all examples solved were either two-dimensional or simple three-dimensional cases. Mangalaramanan [13] proposed systematic procedures for proper identification of r-nodes peaks in order to obtain good limit load estimates. Methods for recognizing valid redistributed elastic analyses stresses were also suggested in the same reference. This paper is an attempt to show that determination of limit loads using a structure's maximum equivalent stress is counter-productive in some cases and r-nodes present a reliable and rational alternative for determining lower bounds of generic two and three-dimensional mechanical components and structures.

R-Nodes can be defined as approximate load-controlled locations in a structure. When inelastic stress redistribution under constant load occurs due to a change in the assumed constitutive behavior, r-nodes are the most resistant locations to stress changes. Therefore, increase or decrease of r-node stresses does not depend as much on the inelastic effect occurring in the structure as on the external load(s) that the component is subjected to. As the applied load is increased to its limit value, the r-node stresses increase in direct proportion by virtue of near equilibrium condition, regardless of stress redistribution elsewhere. In practice, r-nodes can be determined as locations that have constant equivalent stress from two linear elastic finite element analyses. The first elastic analysis is a regular (homogeneous) one and the second one is carried out after modifying the elastic modulus of every finite element using equation (2). It should be noted that r-nodes are not locations in pure equilibrium with the external loads. Analytical and numerical studies reveal that as inelasticity progresses from linear elastic to limit state, r-node stresses do undergo changes, albeit not as much as the rest of the structure.

4 A METHOD FOR ACCELERATED LIMIT LOADS

To identify r-nodes with kinematically admissible collapse mechanisms, Seshadri and Fernando [11] proposed multi-bar models that would relate r-node peaks with plastic hinge locations in a structure. Although this method consistently produced limit loads lower than the exact, there is no formal proof, yet, it is a lower bound. More often than not, it is impossible to know the presence of plastic hinges beforehand. Another drawback is that it is not always possible to obtain distinct plastic hinges that would lead to a collapse mechanism. Plastic hinges in elastic-plastic structures may offer a convenient means for theoretically calculating upper bound limit loads, but such distinct hinges may not always occur in reality. Inelastic analysis of structures such as torispherical shells reveal a plastic zone smudge before collapse rather than three clear-cut hinges. Mangalaramanan [13] identified this problem and proposed methods for locating valid r-node peaks. Conceptual models for understanding the role of r-nodes in plastic collapse were also proposed [14]. While all these efforts have led to a better insight on the nature of r-nodes and their effect on collapse loads, a systematic proof that r-node based limit loads are lower bounds remains elusive to date.

It is imperative that designers accept a procedure only based on assurance of consistent lower bound limit loads. With this in mind, an accelerated method for determining lower bound limit loads using repeated elastic moduli modification method is proposed. For this purpose, the current paper renounces multi-bar collapse models and concentrates only on the maximum r-node stress. It is also shown that use of maximum r-node stress provides a more rational means of determining limit loads as compared to the maximum von Mises equivalent stress. Finally, the lower bound nature of maximum r-node stress is explained.

4.1 Generic Procedure for Determining Limit Loads using R-Nodes

A simple procedure for determining r-nodes in two and three-dimensional components is explained below:

1. Perform the first linear elastic finite element analysis
2. Modify the elastic moduli of every element using equation (2)
3. Carry-out modified moduli analyses using small values of "q" for smooth convergence
4. Determine r-nodes by interpolating the stresses as shown in Figure 1
5. Obtain the maximum r-node stress

Since r-node stresses are essentially determined from linear elastic analyses, these can be scaled linearly such that the

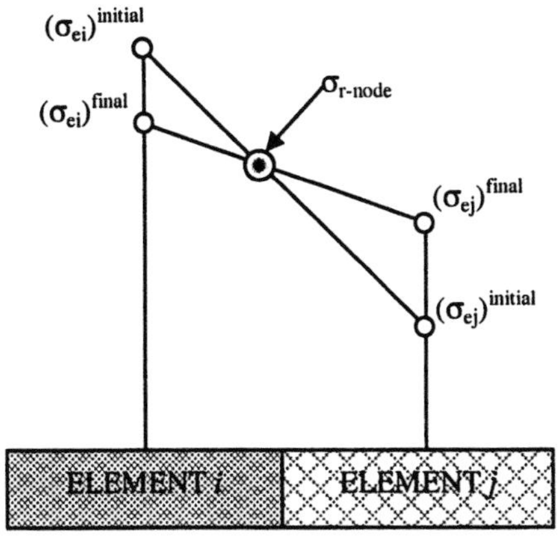

Figure 1: Determination of R-Node

maximum r-node stress reaches the yield value. The corresponding applied load is the limit load. For an arbitrary applied load, P, the limit load is given by:

$$P_{L\_r-node} = \frac{\sigma_y}{\sigma_{\max\_r-node}} P \tag{5}$$

Assuming P to be the exact limit load, P_{L_r-node} would be a lower bound if $\sigma_{\max\_r-node}$ asymptotically reduces to the yield stress value or above, when repeated elastic analyses are carried out. Another stipulation is that the maximum r-node stress should lie within a volume that would most likely become plastic during impending collapse. This volume is referred to as the reference volume. Asymptotic change in the maximum r-node stress is straightforward to monitor in a computer run. The concept of reference volume and the procedure for ascertaining the same is introduced next.

4.2 Reference Volume Concept

Ponter et al. [15] used the reference stress concept for determining the force exerted by a moving ice sheet on a stationary structure. Ice sheets are usually assumed to be semi-infinite slabs. Since not the entire vast expanse of a moving ice sheet has an effect when it crosses path with an offshore structure, it becomes necessary to consider a sub volume in which appreciable inelastic activities are supposed to take place. Inelasticity is assumed to rapidly become negligible beyond this region. This active volume is defined as the reference volume. The authors [15] determined the reference volume of a cylinder en route to validating the accuracy of the reference stress concept. Since the original paper has some minor typographical errors, the correct derivation for a thick cylinder's reference volume is presented in this paper and the result is compared with the proposed simplified method. According to reference [15], the external work and the internal energy rates are equated as:

$$P\dot{U} = \int \sigma_{ij}\dot{\varepsilon}_{ij} dV \tag{6}$$

In terms of reference stress and reference strain, the right side of equation (6) can be expressed as:

$$P\dot{U} = \sigma_R \dot{F}_\varepsilon[\sigma_R] V_{RP} \tag{7}$$

where σ_R is the reference stress and $\dot{F}_\varepsilon[\sigma_R]$ is the equivalent strain that corresponds to the equivalent reference stress (through the material stress-strain relation). V_{RP} is the reference volume that can *loosely be thought of* as that part of the total volume over which inelastic deformation occurs.

The reference stress is chosen so that:

$$\frac{\sigma_R}{P} = \frac{\sigma_y}{P_L} \tag{8}$$

where P_L is the limit load that corresponds to plastic collapse of a structure with the same geometry but made of a perfectly-plastic material with yield stress, σ_y.

For a thick-walled cylindrical tube of unit length with internal radius r_i and external radius λr_i, subjected to internal pressure, p, equation (6) can be expressed as:

$$2\pi r_i p\dot{U} = \sigma_R \dot{F}_\varepsilon[\sigma_R]V_{RP} \tag{9}$$

Assuming the von Mises yield criterion, the limit pressure for a thick-walled tube is [16]:

$$P_L = \frac{2}{\sqrt{3}}\sigma_y \ln\lambda \tag{10}$$

From equations (8) and (10), the reference stress can be calculated as:

$$\sigma_R = \frac{p}{\left(2/\sqrt{3}\right)\ln\lambda} \tag{11}$$

An elementary elastic solution assuming a linear elastic incompressible material shows that the radial displacement U induced by pressure p is

$$U = \frac{3pr_i}{2E(1-\lambda^{-2})} \tag{12}$$

Here U and E are to be interpreted as the radial displacement and the Young's modulus for an elastic material, and as the velocity and the ratio between uniaxial stress and strain rate for a linear viscous material. The strain rate $\dot{F}_\varepsilon[\sigma_R]$ corresponding to the reference stress σ_R is then σ_R/E. Substituting into (9),

$$2\pi r_i p\frac{(3/2)pr_i}{(1-\lambda^{-2})E} = \frac{p}{\left(2/\sqrt{3}\right)\ln\lambda}\frac{p}{\left(2/\sqrt{3}\right)E\ln\lambda}V_{RP} \tag{13}$$

Therefore, the reference volume becomes:

$$V_{RP} = 4\pi(\ln\lambda)^2 r_i^2/(1-\lambda^{-2}) \tag{14}$$

This sub volume of the cylinder is assumed to be the region of dominant inelastic deformation. It is interesting to note that in reality the entire volume of the cylinder undergoes active inelasticity at the limit state and the reference volume, in this case, happens to be a conservative estimate.

Since analytical solutions are readily available for a cylinder, a closed form expression for the reference volume became possible. However, this is not the case for a generic structure. A simpler method for determining the reference volume is described next.

4.3 Reference Volume using Repeated Elastic Analyses

The idea that it is possible to have a single expression for both lower and upper bound limit loads by considering local and global regions of a structure forms the basis for the proposed method of reference volume determination. Seshadri and Mangalaramanan [17] provided an expression for upper bound multiplier as:

$$m_o = \sqrt{\frac{\sigma_y^2 \int dV}{\int \sigma_{ei}^2 dV}} \tag{15}$$

Pan and Seshadri [18] allowed the flow parameter to vary, thereby paving way for an improved upper bound multiplier [19]:

$$\overline{m}_o = \sqrt{\frac{\sigma_y^2 \int dV/E_s(V)}{\int \sigma_e^2 dV/E_s(V)}} \tag{16}$$

It is interesting to note that, if the highly stressed finite element alone is considered, the factors m_o and $\overline{m}_o$ actually become the classical lower bound limit load multiplier. Assuming $V_\delta \le \overline{V} \le V_T$, where the infinitesimal volume V_δ is that of the most highly stressed element and, $\overline{V}$ progressively increases to the total volume in the order of decreasing equivalent stress, the factor $\overline{m}_o$ is bounded by the classical lower bound multiplier for $\overline{V} = V_\delta$ and upper bound for $\overline{V} = V_T$. As $\overline{V}$ is increased from V_δ to V_T, a transition point exists from lower to upper bound where $\overline{m}_o$ is least sensitive to the elastic iterations. This value of $\overline{V}$ is called as the reference volume and can be assumed as the volume that is most likely to become plastic during collapse.

In the next section, the aforementioned concept is applied for finding the reference volume of a thick cylinder.

<u>Reference Volume of a Thick Cylinder</u>

In a thick cylinder subjected to internal pressure, it has been shown [8] that performing modified moduli analysis with $q=1$ and a Poisson's ratio of 0.5 would produce a limit type stress field. In other words, two elastic analyses are sufficient for determining collapse load of the cylinder. Substituting $E_s(V)$ from equation (2) and carrying out the necessary simplifications, equation (16) becomes:

$$\overline{m}_o = \sqrt{\frac{\sigma_y^2 \int_V \overline{\sigma}_e dV}{\int_V \overline{\sigma}_e \sigma_e^2 dV}} \tag{17}$$

In the above expression, σ_e is the von Mises equivalent stress of a second elastic analysis and $\overline{\sigma}_e$ is obtained from the first analysis.

The limit von Mises equivalent stress (called as the reference stress) is given by [16] as:

$$\sigma_e = \frac{\sqrt{3}p}{2\ln\frac{r_o}{r_i}}; \quad \text{(for } q=1\text{)} \tag{18}$$

The stress corresponding to first linear elastic analyses [16] is:

$$\overline{\sigma}_e = \frac{\sqrt{3}p}{r^2}\frac{r_o^2 r_i^2}{r_o^2 - r_i^2}; \quad \text{(for } q=0\text{)} \tag{19}$$

Equation (17) recast in terms of the reference radius, r_R, is:

$$(\overline{m}_o)_{q=0} = \sigma_y \sqrt{\frac{\int_{r_i}^{r_R} r\,dr}{\int_{r_i}^{r_R} (\overline{\sigma}_e)^2\, r\,dr}} \tag{20}$$

$$(\overline{m}_o)_{q=1} = \sigma_y \sqrt{\frac{\int_{r_i}^{r_R} \overline{\sigma}_e r\,dr}{\int_{r_i}^{r_R} \overline{\sigma}_e \sigma_e^2\, r\,dr}} \tag{21}$$

The reference volume corresponds to the one calculated using the reference radius such that $(\overline{m}_o)_{q=0} = (\overline{m}_o)_{q=1}$.

Equations (18) and (19) can be substituted in equations (20) and (21), and the reference volume can be calculated as:

$$V_R = \frac{V_{RP}}{1-\lambda^{-2}} - \pi r_i^2 \tag{22}$$

Figure 2 shows the comparison of the reference stress values between equations (14) and (22). The two equations, albeit derived using different assumptions, show a remarkable agreement in results.

V_R is always less than V_{RP} thus underestimating the plastic volume for all values of λ. Since the function $f(\lambda)$ in Figure 2 does not depend on r_i, the maximum difference between V_R and V_{PR} for any thick cylinder subjected to internal pressure is no more than 6%.

<u>Reference Volume of Generic Components</u>

For a cylinder, it is known that $q=1$ would produce limit state [8]. In generic components, especially with redundancies, this is not the case.

Taking cognizance of equation (2), the following sets of equations can be written:

$$E_S^I = E_0;\; E_S^{II} = E_S^I\left[\frac{\sigma_{arb}^I}{\sigma_e^I}\right]^q; \quad \cdots;\; E_{SN} = E_S^{N-1}\left[\frac{\sigma_{arb}^{N-1}}{\sigma_e^{N-1}}\right]^q \tag{23}$$

$$r_i := 7 \qquad \mathrm{Vrp}(\lambda) := 4\cdot\pi\cdot r_i^2\cdot\frac{(\ln(\lambda))^2}{1-\lambda^{-2}} \qquad \mathrm{Vr}(\lambda) := \left(\frac{\mathrm{Vrp}(\lambda)}{1-\lambda^{-2}}\right) - \pi\cdot r_i^2 \qquad f(\lambda) := \frac{\mathrm{Vrp}(\lambda) - \mathrm{Vr}(\lambda)}{\mathrm{Vrp}(\lambda)}$$

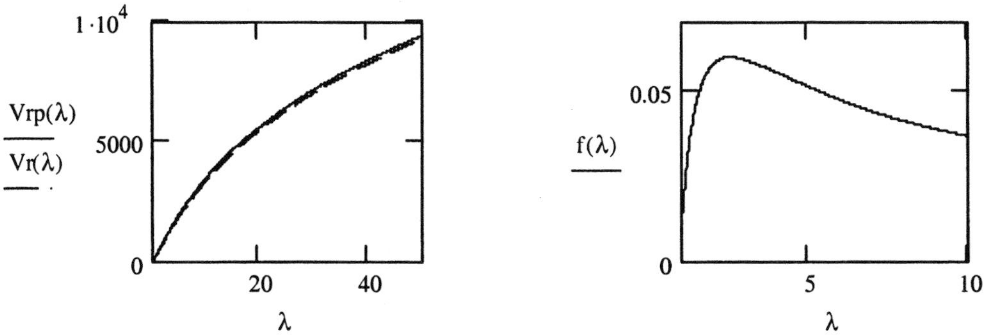

Figure 2: Comparison of reference volumes in a thick cylinder

Equation (23), when substituted into equation (16), leads to

$$\overline{m}_o^N = \sigma_y \sqrt{\frac{\int \left[\sigma_e^I \cdot \sigma_e^{II} \cdots \sigma_e^{N-1}\right]^q dV}{\int \left[\sigma_e^I \cdot \sigma_e^{II} \cdots \sigma_e^{N-1}\right]^q \left(\sigma_e^N\right)^2 dV}} \tag{24}$$

In terms of finite elements, equation (24) can be rewritten as:

$$\overline{m}_o^N = \sigma_y \sqrt{\frac{\sum_{i=1}^{j} \left[\sigma_{ei}^I \cdot \sigma_{ei}^{II} \cdots \sigma_{ei}^{N-1}\right]^q \Delta V_i}{\sum_{i=1}^{j} \left[\sigma_{ei}^I \cdot \sigma_{ei}^{II} \cdots \sigma_{ei}^{N-1}\right]^q \left(\sigma_{ei}^N\right)^2 \Delta V_i}} \tag{25}$$

In equations (24) and (25), $I, II, \cdots$ are the iteration numbers and N is the total number of iterations including the initial linear elastic finite element analysis.

The following steps are proposed for determining the reference volume of generic structures:

1. Perform first linear elastic analysis
2. Sort elements in the descending order of equivalent stress
3. Determine values of $(\overline{m}_o)^i$ as shown in Table 1
4. Repeat elastic analyses with Young's modulus modification until satisfactory convergence
5. Sort the resulting equivalent stresses in the same order corresponding to step 2. This method of sorting is valid since, at a constant load and for increment of q in small values, the stress at any location would by and large keep either monotonically increasing or decreasing (stress redistribution), or remain the same (r-nodes). The order of the stress sequence as determined in step 2 is maintained to a significant extent.
6. Determine $(\overline{m}_o)^f$ as shown in Table 1
7. The reference volume is one that corresponds to $\text{Min}\left[abs\left((\overline{m}_o)^i - (\overline{m}_o)^f\right)\right]$.

Table 1: Determination of Reference Volume, V_R

*Element Sequence Number, j	Von Mises stress		Partial Volume	$\overline{m}_o$ equation (25)			Comments
	Initial	Converged	$\sum_{i=1}^{j} V_i$	Initial, i	Inequality	Final, f	
1	$\overline{\sigma}_{eqv1}$	σ_{eqv1}	V_1	$(\overline{m}_o)_1^i$	<	$(\overline{m}_o)_1^f$	Lower Bound Regime
2	$\overline{\sigma}_{eqv2}$	σ_{eqv2}	V_2	$(\overline{m}_o)_2^i$	<	$(\overline{m}_o)_2^f$	
3	$\overline{\sigma}_{eqv3}$	σ_{eqv3}	V_3	$(\overline{m}_o)_3^i$	<	$(\overline{m}_o)_3^f$	
4	$\overline{\sigma}_{eqv4}$	σ_{eqv4}	V_4	$(\overline{m}_o)_4^i$	<	$(\overline{m}_o)_4^f$	
⋮	⋮	⋮	⋮	⋮	<	⋮	
$R-1$	$\overline{\sigma}_{eqv\,R-1}$	$\sigma_{eqv\,R-1}$	V_{R-1}	$(\overline{m}_o)_{R-1}^i$	<	$(\overline{m}_o)_{R-1}^f$	
R	$\overline{\sigma}_{eqv\,R}$	$\sigma_{eqv\,R}$	V_R	$(\overline{m}_o)_R^i$	≈	$(\overline{m}_o)_R^f$	Fuzzy
$R+1$	$\overline{\sigma}_{eqv\,R+1}$	$\sigma_{eqv\,R+1}$	V_{R+1}	$(\overline{m}_o)_{R+1}^i$	>	$(\overline{m}_o)_{R+1}^f$	Upper Bound Regime
⋮	⋮	⋮	⋮	⋮	>	⋮	
$\overline{n}-3$	$\overline{\sigma}_{eqv(\overline{n}-3)}$	$\sigma_{eqv(\overline{n}-3)}$	$V_{\overline{n}-3}$	$(\overline{m}_o)_{\overline{n}-3}^i$	>	$(\overline{m}_o)_{\overline{n}-3}^f$	
$\overline{n}-2$	$\overline{\sigma}_{eqv(\overline{n}-2)}$	$\sigma_{eqv(\overline{n}-2)}$	$V_{\overline{n}-2}$	$(\overline{m}_o)_{\overline{n}-2}^i$	>	$(\overline{m}_o)_{\overline{n}-2}^f$	
$\overline{n}-1$	$\overline{\sigma}_{eqv(\overline{n}-1)}$	$\sigma_{eqv(\overline{n}-1)}$	$V_{\overline{n}-1}$	$(\overline{m}_o)_{\overline{n}-1}^i$	>	$(\overline{m}_o)_{\overline{n}-1}^f$	
$\overline{n}$	$\overline{\sigma}_{eqv\overline{n}}$	$\sigma_{eqv\overline{n}}$	$V_{\overline{n}}$	$(\overline{m}_o)_{\overline{n}}^i$	>	$(\overline{m}_o)_{\overline{n}}^f$	

*Not to be confused with element number. Element sequence number corresponds to the decreasing order of von Mises equivalent stress for the first linear elastic analysis, such that $\overline{\sigma}_{eqv1} \geq \overline{\sigma}_{eqv2} \geq \cdots \geq \overline{\sigma}_{eqv\overline{n}}$.

4.4 The Lower Bound Nature of Maximum R-Node Stress

In order to understand the lower bound nature of r-nodes, it would be of interest to understand the broad differences between inelastic and repeated elastic analyses. These, assuming an elastic-perfectly plastic material, are listed in Table 2.

Table2: Inelastic versus Repeated Elastic Analysis

	Inelastic Analysis	Repeated Elastic Analysis
1	Actual plastic flow	No plastic flow, per se
2	Work performed not fully recoverable due to plastic flow	Work is fully recoverable
3	Stress and strains are true	Initial elastic and limit distributions are scaled equivalent of plastic analysis
4	Maximum stress increases and remains constant after yield	Maximum stress relaxes as elastic iterations progress
5	Plasticity progresses with monotonic load increase	External load remains constant with elastic iterations
6	Infinite work due at collapse due to uncontained plasticity	Finite work even at limit state
7	Distinct elastic and plastic regimes	No such distinction. Power-law kind of stress distribution.
8	Strains are meaningful and real	Strains are not real since these depend on an arbitrary Young's moduli value.

It is possible to show that the maximum r-node stress would lead to lower bound limit loads by considering the recoverable (elastic) energy. Excluding time dependence and assuming plastic yield, the recoverable energy, $\bar{E}$, can be expressed as:

$$\bar{E} = \int_{V\_plastic} \frac{\sigma_y^2}{2E} dV + \int_{V\_elastic} \frac{\sigma_{eqv}^2}{2E} dV \qquad (26)$$

Since the energy increases as plasticity progresses, the rate of change of energy with respect to the applied load is

$$\frac{d\bar{E}}{dP} = \int_{V\_plastic} \frac{1}{E}\frac{d(\sigma_y^2)}{dP} + \int_{V\_elastic} \frac{1}{E}\frac{d\sigma_{eqv}}{dP} dV \qquad (27)$$

Assuming elastic-perfectly plastic conditions, no additional load is required to create plastic strains once the material has yielded. Therefore, equation (27) becomes:

$$\frac{d\bar{E}}{dP} = 0 + \frac{1}{E}\int_{V\_elastic} \frac{d\sigma_{eqv}}{dP} dV \qquad (28)$$

In other words $d\bar{E}/dP$ is zero in the plastic region. This can be easily imagined by way of a two bar mechanism subjected to tension with one bar yielding before the other. The yielded bar stretches under a constant load, while all additional load is taken by the elastic bar. Increasing the total load does not alter the recoverable energy in the yielded bar.

In case of repeated elastic analysis under constant external load, it is the elastic iterations and Young's moduli change that simulate inelasticity. Equation (26) can be recast as:

$$\bar{E} = \int_{V\downarrow} \sigma_{eqv}\varepsilon_{eqv} dV + \int_{V\uparrow} \sigma_{eqv}\varepsilon_{eqv} dV = \bar{E}_{V\downarrow} + \bar{E}_{V\uparrow} \qquad (29)$$

In equation (29), $V\downarrow$ refers to the portion of the structure where the stresses relax with iterations, and $V\uparrow$ represents the remaining region where the stresses increase. The transition point between $V\downarrow$ and $V\uparrow$ is the r-node. From equation (29), the rate of change of energy is given by

$$\frac{d\bar{E}}{dN} = \frac{d\bar{E}_{V\downarrow}}{dN} + \frac{d\bar{E}_{V\uparrow}}{dN} \qquad (30)$$

Monotonic increase in external load during inelastic analysis is tantamount to performing a number of repeated elastic iterations at constant load in modified moduli methods. Therefore, in equation (30), the number of iterations, N has been used instead of P.

Recoverable energy is considered in equation (26) due to the fact that strains due to repeated elastic analyses are fully recoverable upon unloading, at all stages. Mangalaramanan [9] showed that there exists a one to one correspondence between stress fields due to repeated elastic and inelastic analyses in a thick cylinder subjected to internal pressure. Therefore, the overall behavior of equation (28) should correspond to equation (30). Equation (28) shows that it is the nature of the component to accumulate recoverable energy at a higher rate in the elastic region. It should also be noted that the external load is constant in case of equation (29). Any energy gain in the outer region of the cylinder comes at the expense of energy loss in the inner region. Therefore, it can

$p := 10 \qquad r_i := 3 \qquad r_o := 9 \qquad q := 0.1$

$$\sigma_e(r,N) := \left[\frac{\sqrt{3}\,p\cdot(1-q)^{(N-1)}}{r^{2\cdot(1-q)^{(N-1)}}}\right]\cdot\frac{r_o^{2\cdot(1-q)^{(N-1)}}\cdot r_i^{2\cdot(1-q)^{(N-1)}}}{r_o^{2\cdot(1-q)^{(N-1)}} - r_i^{2\cdot(1-q)^{(N-1)}}}$$

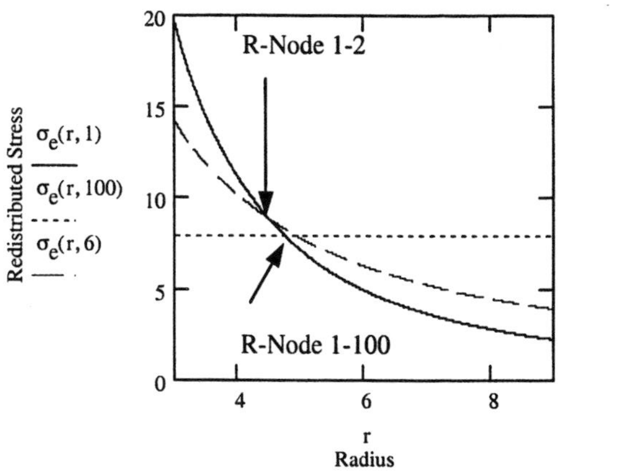

Figure 3: Movement of R-Nodes in a Thick Cylinder

be concluded that in order to compensate for higher rate of energy increase in $V\uparrow$, $V\downarrow$ needs to significantly relax as iterations progress. This is evident from both analytical and numerical examples. But even this relaxation in the inside radius of a cylinder (for incremental small values of q) is not sufficient to balance the energy level since $d\overline{E}_{V\uparrow}/dN$ is greater than $d\overline{E}_{v\downarrow}/dN$ due to the nature of equation (28). As stress relaxation rate decreases with iterations (tantamount to availability of less elastic region for stress redistribution in the plastic case), the only way to ensure energy balance between either sides of the r-node is by shifting the r-node towards the outer radius of the cylinder so that an even larger volume is available for stress relaxation. In other words, the r-node stresses would decrease as the elastic iterations increase until limit state is reached. For a given value of q, the stress distribution in the

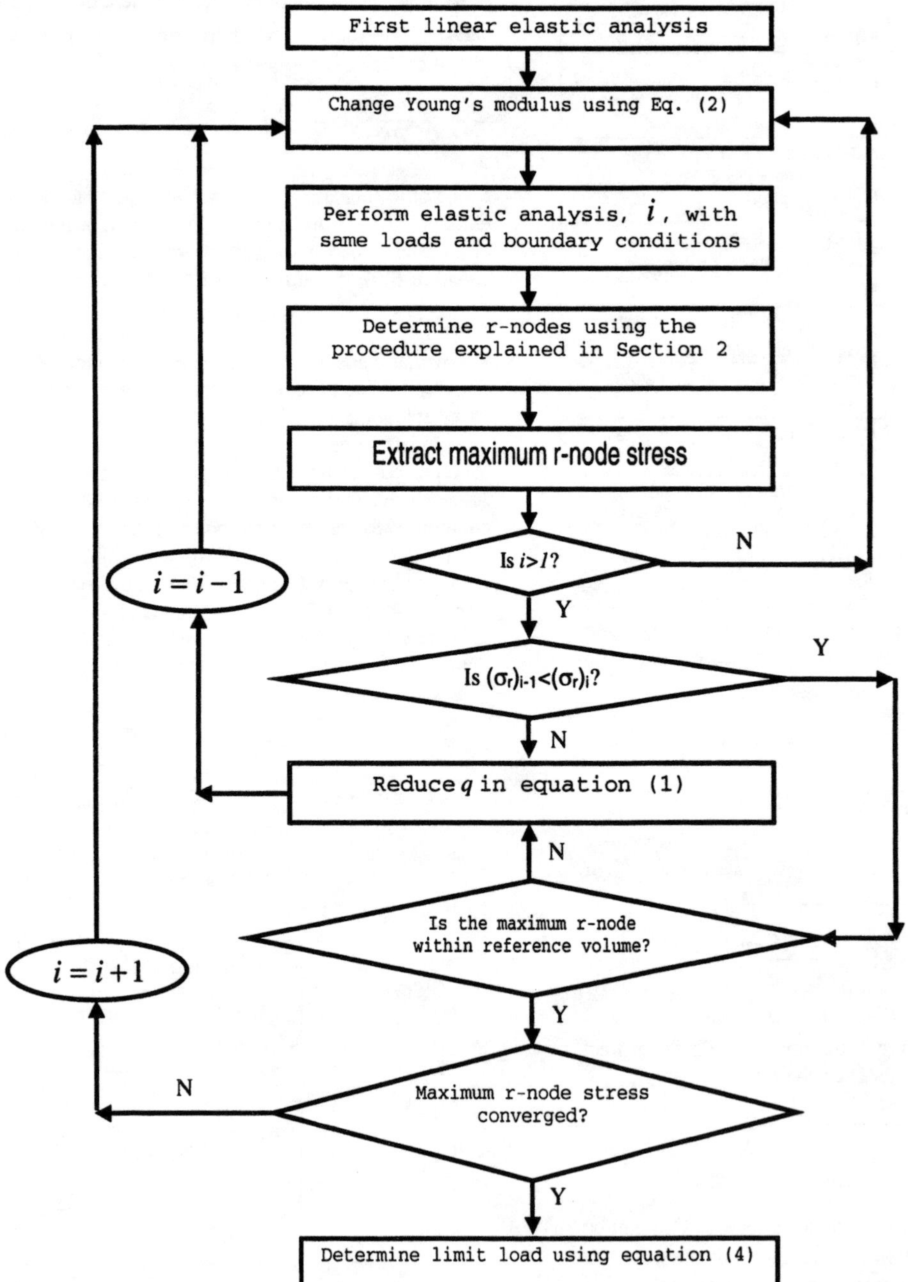

Figure 4: Determination of R-Nodes

cylinder can be obtained from reference [8] as:

$$\sigma_e(N) = \frac{\sqrt{3}p(1-q)^{N-1}}{r^{2(1-q)^{N-1}}} \times \frac{r_i^{2(1-q)^{N-1}} r_o^{2(1-q)^{N-1}}}{r_o^{2(1-q)^{N-1}} - r_i^{2(1-q)^{N-1}}} \quad (31)$$

Figure 3 shows the movement of r-nodes using equation (31). A simple real life example using an amusement seesaw would help understand the aforementioned argument. Consider a seesaw having two persons on either ends and in perfect balance. Assume that an additional person sits on the right hand side. The only way the seesaw can remain in balance is by shifting the fulcrum point to the right so that the person sitting on the left gets additional moment arm. The fulcrum can be considered as the r-node. The increase in load on the right side of the seesaw is the additional energy sought by the elastic region closer to the outside radius during repeated elastic analysis. The demand for this energy requirement is met by steeper stress gradients and by movement of r-nodes towards the outer radius thereby increasing $V \downarrow$.

Although the lower bound nature of r-node has been explained using a thick cylinder example, the theoretical underpinnings for generic problems are exactly the same. The nature of energy balance favors movement of r-nodes from conservative to limit state.

4.5 Conditions for Lower Bound Limit Load using the R-Node Method

From the above discussions, the following conditions are to be met in order to obtain accelerated lower bound limit loads:

1. R-Nodes are determined between any linear elastic finite element analysis and the first one.
2. The maximum r-node stress should monotonically decrease with increasing iterations.
3. The maximum r-node must be located within the reference volume.

5 NUMERICAL EXAMPLES

Two numerical examples are presented in this section. The first example is a L-shaped cantilever beam with a sharp crack. The purpose of this example is to illustrate some of the pitfalls of classical lower bound limit load estimates and show that the r-node method provides more rational limit loads. The second problem is a cylindrical pressure vessel with an oblique nozzle to demonstrate that r-nodes can be determined with ease even for a complex three-dimensional problem. The proposed method has been completely automated for both two and three-dimensional components. In both cases, the material is elastic-perfectly plastic with the following material properties:

Young's modulus = 206.85×10^6 Pa
Poisson's ratio = 0.3
Yield Strength = 206.85×10^3 Pa

5.1 L Shaped Cantilever Beam with a Notch

Figure 5 shows the cantilever beam (all dimensions in m) subjected to an end load of 1000 N. The beam has a vertical member designed to fail by plastic hinge formation and a sharp crack that would only lead to local plasticity. The purpose of this exercise is to understand as to how many number of iterations it would take for repeated elastic analysis to recognize the collapse mode. The crack-tip was modeled using isoparametric six-noded triangular elements with the mid-side nodes moved to quarter point. It can be seen that while the peak r-node is located in the plastic hinge region, the maximum stress is always located near the crack-tip. The limit load results are shown in Figure 6. While the r-node limit load rapidly converges to that obtained from non-linear analysis, it takes many iterations for the classical lower bound limit load based solution. The r-node plot of the component is shown in Figure 7, and the reference volume and plastic region plots are illustrated in Figures 8 and 9. This example shows that the accelerated method provides quicker estimates of limit loads. This becomes particularly important in situations where too many number of elastic iterations can cause the difference between the maximum and minimum Young's moduli to increase, thereby ill-conditioning the stiffness matrix.

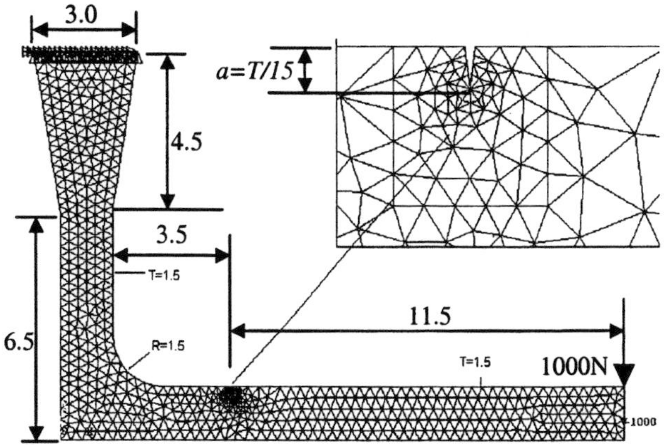

Figure 5: Dimensions of the L-Shaped Cantilever Beam

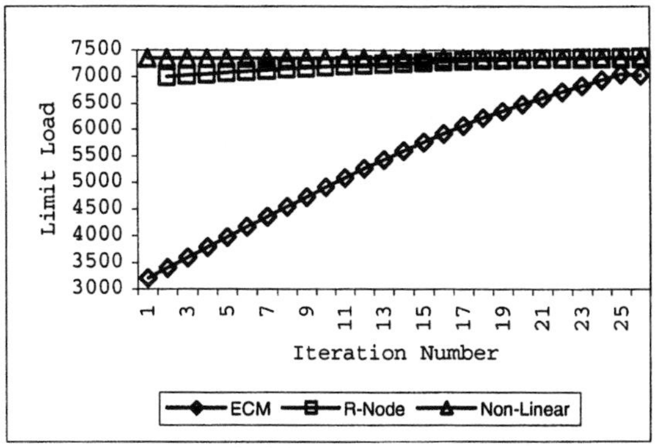

Figure 6: Limit Loads of the L-Shaped Cantilever

5.2 Cylindrical Pressure Vessel with an Oblique Nozzle

The cylindrical pressure vessel with nozzle (all dimensions in m) is shown in Figure 10. The ANSYS [20] FE model contains around fifty thousand tetrahedral elements with mid-side nodes. Uniform internal pressure is applied on the inside walls of the vessel while ensuring that the end faces are only constrained from moving along the respective axes of the nozzle and the vessel. For performing repeated elastic analysis, the pressure applied is 1000 kPa. The finite element mesh is shown in Figure 11. The limit load estimates are presented in Figure 12.

Inelastic analysis of this component was anything but easy. The problem was solved in a Pentium 4, 1.7 GHz, dual processor PC with 1GB memory. The SPARSE matrix solver, which was tried at first, showed extremely slow convergence. Subsequent attempt using PCG solver exhibited rapid convergence until the point where load-step bisection became inevitable. The PCG solver, however, wrongly attributed the consequent lack of convergence to PCG iterations rather than bisecting the load. The solver, as a result, went into a mode of performing endless PCG iterations and the job had to be forcefully terminated. Thankfully, knowledge of limit load due to repeated elastic analyses enabled careful control of load steps to achieve desired convergence. This challenge offered by non-linear analysis was in sharp contrast to the easy convergence characteristics of repeated elastic analyses methods with relatively moderate computer resource requirements.

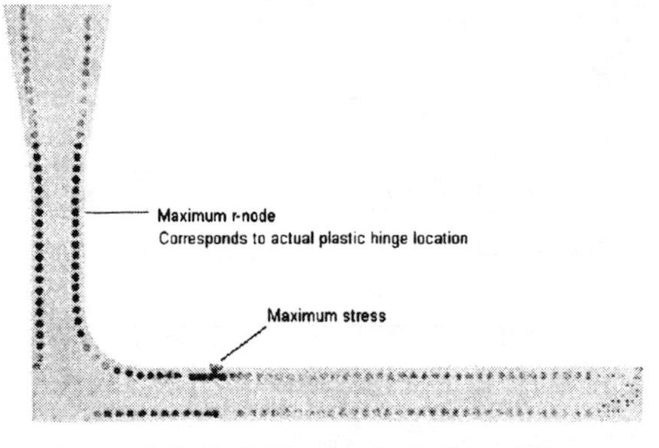

Figure 7: R-Node Plot for the L-Shaped Beam

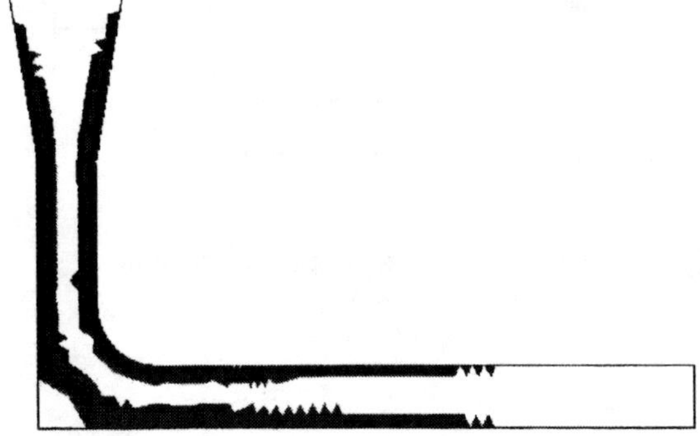

Figure 8: Reference Volume in the L-Shaped Beam

As in the previous example, the accelerated method converges faster than elastic compensation method, as shown in Figure 12.

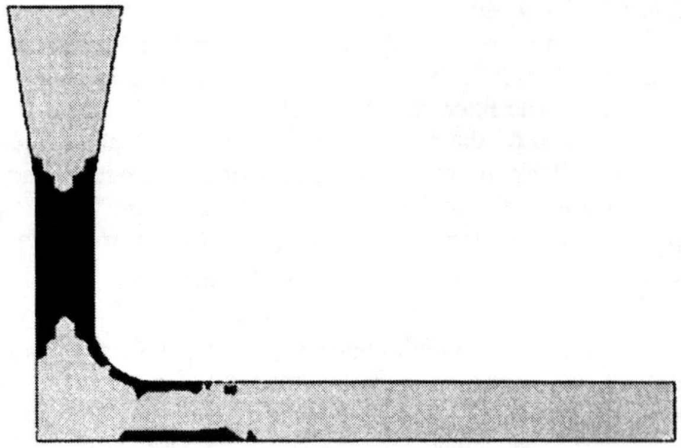

Figure 9: Plastic Volume in the L-Shaped Beam

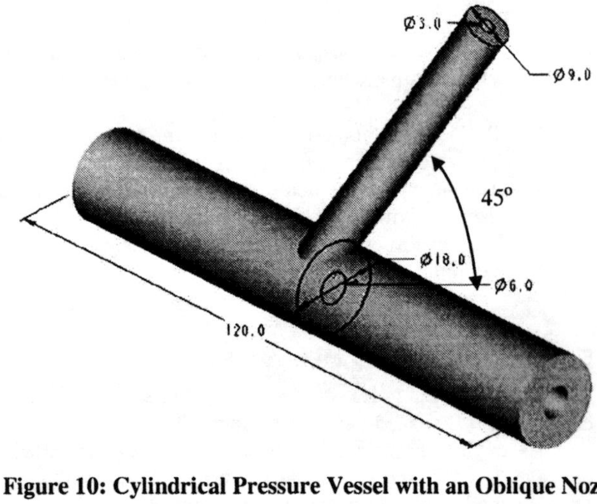

Figure 10: Cylindrical Pressure Vessel with an Oblique Nozzle

6 CONCLUSION

This paper shows the limitations of lower bound stress analysis methods such as the elastic compensation method, which makes use of the maximum stress for determining limit loads. The r-node method is presented as an alternate method in conjunction with repeated elastic analyses for determining lower bound limit loads. The r-node method is found to provide rational limit load estimate. A beam with crack has been considered to verify this claim. While the elastic compensation method after many iterations indicated failure at the crack tip, the maximum r-node showed the location of the plastic hinge more accurately. At least during the initial iterations, the elastic compensation method can be easily deceived by sharp stress gradients. On the other hand, r-

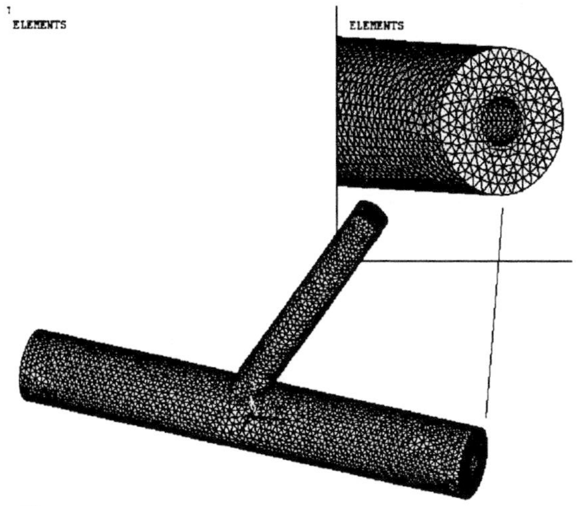

Figure 11: Finite Element Mesh for the Oblique Nozzle

nodes seem to be reasonably immune to steep stress gradients. Maximum stress location in a component only shows the initiation of plasticity and not necessarily a plastic hinge. R-Nodes, on the other hand, exhibit sectional properties, and therefore a plastic hinge. It is also shown that the accelerated

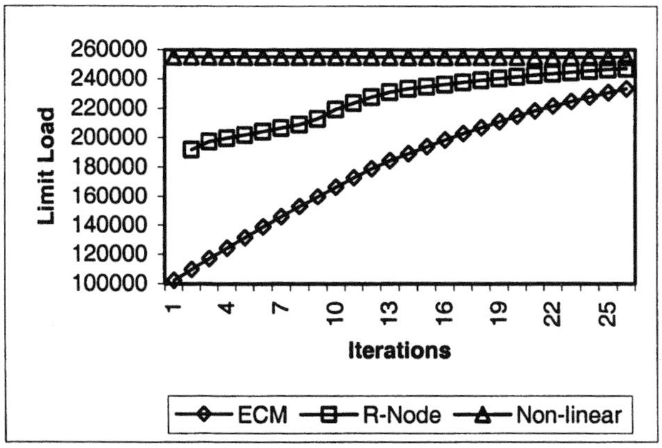

Figure 12: Limit Load estimates for the Oblique Nozzle

method using r-nodes is a lower bound. The concept of reference volume is explained in the context of repeated elastic analyses. Comparison of reference volumes between the proposed method and reference [15] is quite good.

In designing complex pressure vessels for primary stress, defeaturing of the CAD model becomes an essential step to simplify the geometry. Primary stresses, as a rule, should not be affected by details. Absence of details such as fillets and rounds do not affect the results of non-linear finite element analysis. On the other hand, elastic compensation method can be misguided by the presence of sharp edges, which inherently "capture" the peak elastic stresses. A very small value of q and a large number of elastic iterations becomes necessary in such cases. However, carrying out a large number of iterations has the problem of ill-conditioning the finite element stiffness matrix, let alone consumption of additional computational time. It would be therefore advantageous to achieve convergence as early as possible. The accelerated method for determining limit loads aims at accomplishing such an objective.

7 ACKNOWLEDGMENT

The author would like to acknowledge the encouragement and support provided by Mr. Paul Pollock, Chief Engineer, Dana Corporation, Kalamazoo and Mr. Jim Ridge, Senior Designer, Dana Corporation, for the CAD model of the oblique nozzle. Many thanks are due to Dr. Wolf Reinhardt for his kind and constructive criticisms. The author also wishes to appreciate his wife Bama and infant son Narayanan Aiyer for enduring neglect during the long hours that were spent in preparing this paper.

8 REFERENCES

[1] *ASME Boiler and Pressure Vessels Code,* 1989, Section III, Division 1, Subsection NB-3000.

[2] Marriott, D.L., 1988, "Evaluation of Deformation and Load Control of Stresses under Inelastic Conditions using Elastic Finite Element Stress Analysis," Proceedings of the ASME-PVP Conference. Pittsburgh, vol. 136, pp. 3-9.

[3] Seshadri, R., 1991, "The Generalized Local Stress Strain (GLOSS) Analysis – Theory and Applications," ASME Journal of Pressure Vessel Technology, vol. 113, pp. 219-227.

[4] Mackenzie, D., and Boyle, J.T., 1993, "A Method for Estimating Limit Loads by Iterative Elastic Analysis. I-Simple Examples," International Journal of Pressure Vessels and Piping, vol. 53, pp. 77-95.

[5] Ponter, A.R.S., Carter, K.F., 1997, "Limit state solutions based on linear elastic solutions with spatially varying elastic modulus," Computer Methods in Applied Mechanics and Engineering, vol. 140, pp. 249-279.

[6] Mackenzie, D., Boyle, J.T., and Hamilton, R., 2000, "The elastic compensation method for limit and shakedown analysis: a review," Journal of Strain Analysis, vol. 35(3), pp. 171-188.

[7] Mangalaramanan, S.P., and Seshadri, R., 1997, "Limit Loads of Layered Beams and Layered Cylindrical Shells using the R-Node Method," Proceedings of the ASME Pressure Vessels and Piping Conference, Orlando, PVP vol. 353, pp. 201-215.

[8] Mangalaramanan, P., and Reinhardt, W.D., 2001, "On relating inelastic and redistributed elastic analyses stress fields," International Journal of Pressure Vessels and Piping, vol. 78, pp. 283-293.

[9] Mangalaramanan, P, 2002, "Can repeated elastic analyses always lead to exact limit load estimates?" Proceedings of the ASME Pressure Vessels and Piping Conference, Vancouver, vol. 441, pp. 215-223.

[10] Kraus, H., *Creep Analysis,* 1980, Wiley-Inter Science Publication, John Wiley & Sons.

[11] Seshadri, R., and Fernando, C.P.D., 1992, "Limit Loads of Mechanical Components and Structures using the

GLOSS R-Node Method," ASME Journal of Pressure Vessels Technology, vol. 114, pp. 201-208.

[12] Mangalaramanan, S.P., and Seshadri, R, 1994, "Robust Limit Loads of Plate Structures," Transactions of the CSME, vol.1, pp. 108-129.

[13] Mangalaramanan, S.P., 1997, PhD thesis, *Robust Limit Loads using Elastic Modulus Adjustment Techniques, Chapter: 5 - In search of redistribution nodes*, pp. 114-149.

[14] Mangalaramanan, S.P., 1997, "Conceptual models for Understanding the Role of R-Nodes in Plastic Collapse," ASME Journal of Pressure Vessel Technology, vol. 119, pp. 374-378.

[15] Ponter, A.R.S., Palmer, A.C., Goodman, D.J., Ashby, M.F., Evans, A.G., and Hutchinson, J.W., 1983, "The Force exerted by a moving Ice Sheet on an Offshore Structure. Part 1. The Creep Mode," Cold Regions Science and Technology, vol. 8, pp. 109-118.

[16] Calladine, C.R., 1969, Engineering Plasticity, Pergamon Press, Oxford.

[17] Seshadri, R., and Mangalaramanan, S.P., 1997, "Lower Bound Limit Loads using Variational Concepts: The m_α method," International Journal of Pressure Vessels and Piping, vol. 71, pp. 93-106.

[18] Pan, L., and Seshadri, R., 2001, "Limit Load Estimation using Plastic Flow Parameter in Repeated Elastic Finite Element Analysis", Proceedings of the ASME Pressure Vessels and Piping Conference, Atlanta, vol. 430, pp. 145-150.

[19] Reinhardt, W.D., and Seshadri, R., 2003, "Limit Load Bounds for the m_α multiplier", ASME Journal of Pressure Vessel Technology, Vol. 125, pp. 1-8, February.

[20] *ANSYS Users Manual*, 2000, Canonsburg, PA.

9 APPENDIX

The variation of $\overline{m}_o$ as a function of volume over a number of elastic iterations is shown below as a MathCAD™ worksheet for a thick cylinder subjected to uniform internal pressure. The region in which $\overline{m}_o$ is least sensitive to elastic iterations is identified as the reference volume. This worksheet also demonstrates as to how a single expression provided by equation (16) can be used for both lower and upper bound limit loads as expressed below:

Lower bound limit load: $$P_{LL} = \lim_{\substack{V \to 0 \\ \text{or } r \to r_i}} (\overline{m}_o) \times P \qquad (32)$$

Upper bound limit load: $$P_{LU} = \lim_{\substack{V \to V_T \\ \text{or } r \to r_o}} (\overline{m}_o) \times P \qquad (33)$$

for an arbitrary applied load, P.

Inside radius of the cylinder $\quad r_i := 3$

Outside radius of the cylinder $\quad r_o := 9$

Total number of elastic iterations $\quad N := 1, 2 .. 100$

Reference volume per reference [15]

$$vrp := 4 \cdot \pi \cdot r_i^2 \cdot \frac{\left(\ln\left(\frac{r_o}{r_i}\right)\right)^2}{1 - \left(\frac{r_i}{r_o}\right)^2}$$

$vrp = 153.57$

Proposed reference volume

$$vr := \left[\frac{vrp}{1 - \left(\frac{r_i}{r_o}\right)^2}\right] - \pi \cdot r_i^2$$

$vr = 144.49$

Reference radius from proposed reference volume

$$r_r := \sqrt{\frac{vr}{\pi} + r_i^2}$$

$r_r = 7.42$

Von Mises eq. stress in cylinder due to elastic moduli modification

$$s(r, N, q) := \left[\frac{(1-q)^{N-1}}{r^{2 \cdot (1-q)^{(N-1)}}}\right] \cdot \left[\frac{r_i^{2 \cdot (1-q)^{N-1}} \cdot r_o^{2 \cdot (1-q)^{N-1}}}{r_o^{2 \cdot (1-q)^{N-1}} - r_i^{2 \cdot (1-q)^{N-1}}}\right]$$

Upper bound multiplier

$$mo(N, r_R, q) := \sqrt{\frac{\int_{r_i}^{r_R} r\,dr}{\int_{r_i}^{r_R} s(r, N, q)^2 \cdot r\,dr}}$$

$mo(1, r_r, 0) = 2.2$

$mo(100, r_r, 0.1) = 2.2$

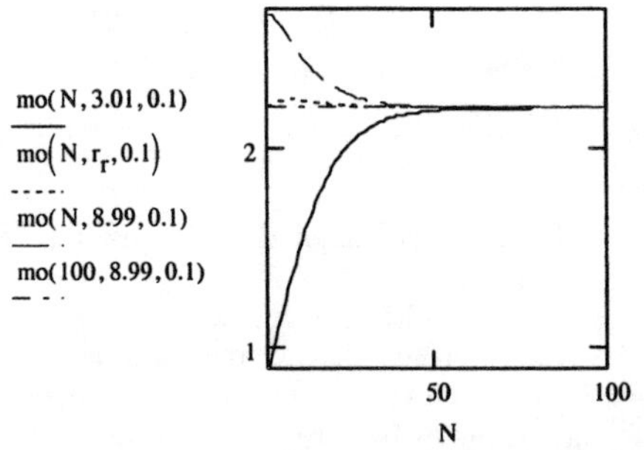

Determination of reference volume

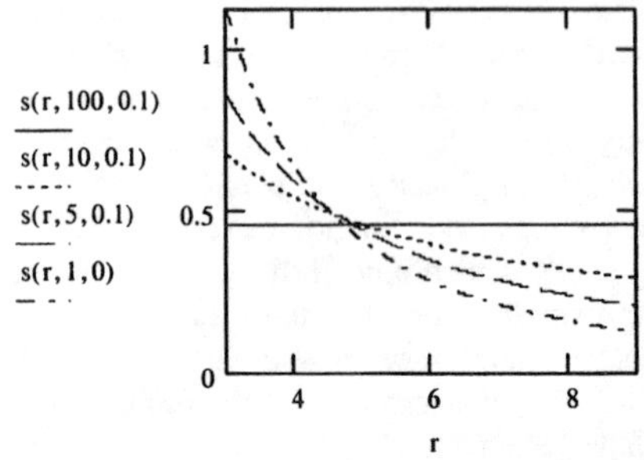

Variation of stress due to elastic iter.

PVP-Vol. 458, Computer Technology and Applications

PVP2003-1891

ON THE CONSTRAINT-BASED FAILURE ASSESSMENT FOR SURFACE CRACKED WELDED PLATES UNDER TENSION

X. Wang
Department of Mechanical
and Aerospace Engineering
Carleton University
Ottawa, ON K1S 5B6
CANADA

W. Reinhardt
Nuclear Engineering
Babcock and Wilcox Canada
Cambridge, ON N1R 5V3
CANADA

ABSTRACT

Conventional failure assessment schemes (CEGB-R6, BS-7910) use a lower bound toughness obtained from highly constrained test specimens. A lower crack tip constraint leads to enhanced resistance to both cleavage and ductile tearing. The cracks in many real engineering structures are not highly constrained, which makes failure predictions using conventional failure assessment schemes based on the lower bound fracture toughness overly conservative. Too much conservatism in the structural assessment can lead to unwarranted repair or decommissioning of structures, and thus cause unnecessary cost and inconvenience. Due to recent developments in constraint-based fracture mechanics, it is important to include the constraint effect in the practical assessment of defective components. For example, the recent revision of R6 and the newly developed structural integrity assessment procedures for European industry (SINTAP) have suggested a framework for failure assessments that can take the constraint effect into account. In this paper, the constraint-based failure assessment of a surface cracked welded plate under uniaxial tension load is presented. A constraint-based failure assessment diagram and a method for combining primary and the secondary loads are discussed. Finite element based correlations are used to calculate the stress intensity factors, and constraint parameters, while the limit loads are derived from existing closed form approximations. It is demonstrated that when the lower constraint effect is properly accounted for, the maximum allowable tensile stress level can increase 15% or more.

1. INTRODUCTION

Fracture assessments of cracked structures with the failure assessment diagram approach are well established [1, 2]. However, all these conventional failure assessment schemes use a lower bound measure of fracture toughness that was obtained from highly constrained test specimens. Over-conservative assessments can occur in situations where the crack in a component is under conditions of low constraint. Constraint effects on fracture have, therefore, received considerable attention recently in an effort to reduce the conservatism of the analysis. Two parameter, (J, T) or (J, Q), fracture mechanics has been developed to describe the stress and strain fields ahead of a crack tip under various loading conditions corresponding to different constraint levels [3, 4]. Here J stands for the J-integral, which is the fracture parameter. In linear-elastic fracture mechanics, J is directly related to K, the factor for the singular $1/\sqrt{r}$ term of the near tip stress distribution. The parameter T is the elastic T-stress, and Q is the hydrostatic Q-stress which are possible constraint parameters. The T stress is the constant term of the near tip stress distribution. Based on two-parameter fracture mechanics, a framework for including the constraint effect in the failure assessment diagram approach was proposed in [5]. The recent version of R6 [1], and the newly developed structural integrity assessment procedure for the European industry (SINTAP) [6], include procedures for a failure assessment that accounts for the constraint effect.

A surface cracked plate under tensile load represents a typical model for welded pressure vessel components with surface defects. Therefore, the assessment of such plates is of great importance for the life extension of the pressure vessels. It has been observed that a tensile load causes a relatively low crack tip constraint compared to a plate under bending load [7]. The lower constraint effects can be advantageously included when conducting the failure assessment of the plate. In doing so, excessive conservatism in the assessment can be removed and unwarranted repair or decommissioning of the cracked structure can be avoided.

This paper presents a constraint-based failure assessment of a surface cracked welded plate under tensile load. The constraint-based failure assessment diagram approach and the approach to the combination of primary and secondary loads are described in section 2. Section 3 presents the calculation of the stress intensity factors and the T-stress from finite element based correlations. Plastic limit loads for the surface cracked plate are taken from the literature. In section 4, an example failure assessment is conducted for the case where a primary load is applied alone, and for the case of a combined primary and secondary load. When the lower constraint effect is properly accounted for, the maximum allowable tensile stress level increases by as much as 15 – 20% in this particular example.

2. CONSTRAINT-BASED FAILURE ASSESSMENT DIAGRAM

This section outlines the construction of the constraint-based failure assessment diagram and discusses the procedure for incorporating secondary loads in the constraint-based evaluation.

2.1 Constraint-Based Approach

The R6 Code [1] first included constraint effects into the failure assessment diagram approach. More recently, the SINTAP procedures were proposed [6]. The procedure that is outlined here follows the R6 recommendations [1].

The conventional failure assessment diagram methods embedded in R6 [1] or PD7910 [2] use only a single fracture parameter. That is, the single value of toughness or crack opening displacement is assumed to govern fracture. The diagram derives the assessment of failure from two calculated parameters, (K_r, L_r) which are defined by

$$K_r = \frac{K_I}{K_{IC}} \tag{1}$$

and

$$L_r = \frac{\sigma}{\sigma_L} \tag{2}$$

In eq. (1) and (2), K_r is the stress intensity ratio. K_I is the stress intensity factor for the crack and K_{IC} is the toughness of the material. The parameter L_r is the stress ratio, defined as the ratio of the applied stress to the limit stress. The plastic limit solution of the cracked structure that gives rise to the limit stress is calculated based a perfectly plastic material model where the onset of plasticity occurs at the yield stress of the material. Failure is avoided if the point (K_r, L_r) lies within a failure assessment diagram, represented by a curve,

$$K_r = f(L_r) \tag{3}$$

and the cut-off value of L_r at

$$L_r = L_r^{max} \tag{4}$$

The two parameter fracture mechanics version of R6 [1] (Appendix 14) recognizes that the actual fracture resistance K_{mat}^C in a structure subject to conditions of low constraint may exceed the conventional fracture resistance K_{IC} that is always measured under conditions of high constraint. To include this effect, the failure assessment curve (3) is modified to:

$$K_r = f(L_r)\left(\frac{K_{mat}^C}{K_{IC}}\right) \tag{5}$$

Following a detailed theoretical and experimental analysis, [3-5] represented this increase in fracture toughness by:

$$K_{mat}^C = K_{IC}\left[1+\alpha(-\beta L_r)^m\right] \tag{6}$$

where α and m are material dependent constants, which define the dependence of fracture toughness on constraint. The parameter β is the normalized constraint parameter, which will be defined in terms of the T-stress in equation (8). Substituting (6) into (5), we have

$$K_r = f(L_r)\left[1+\alpha(-\beta L_r)^m\right] \tag{7}$$

Equation (7) is the constraint-based failure assessment curve that replaces (3) in the failure assessment diagram. Failure is avoided if the assessment point (K_r, L_r), as calculated from eqs. (1) and (2), lies below eq. (7) and L_r is less than the material dependent cut-off value L_r^{max} (4).

The normalized constraint parameter β is defined from the hydrostatic Q-stress or the elastic T-stress:

$$\beta = \frac{Q}{L_r} = \frac{T}{\sigma_Y L_r} \quad \text{if } T, Q \le 0 \tag{8}$$
$$\beta = 0 \quad \text{otherwise}$$

where σ_Y is the yield stress of the material. Note that here the following estimation of Q stress in terms of T-stress is used [5]:

$$Q = \frac{T}{\sigma_Y} \tag{9}$$

The reason for eq. (8) is that a positive T-stress corresponds to a condition of high constraint, as would be present in a plane strain fracture specimen. The K_{mat}^C then reduces to K_{IC}. For negative T-stress, a negative β will be substituted into eq. (7). In linear elastic fracture mechanics, the T-stress is also linear. The T-stress from different load components, like membrane and bending or different biaxial components is therefore additive. This T-stress combination must occur before the resultant T-stress is substituted into eq. (8) and (7). The stress intensity factor and the limit load for the cracked component as well as the T-stress or the Q stress must be determined to conduct the constraint-based assessment. As an example, section 3 will obtain these for a plate with a semi-elliptical surface crack.

2.2 Effect of Secondary Stress

Secondary stresses are often present in the components that are being evaluated, for example, thermal stresses or the residual stress distribution in an as-welded component. The approach to combining primary and secondary loading in the FAD methodology that was developed in [8] can be used in conjunction with the constraint-based approach. When the assessment includes the secondary stress effect, the equations for the FAD curves remain the same while the assessment point (K_r, L_r) will be adjusted. Since secondary stresses do not contribute to plastic collapse, L_r is unchanged (i.e. eq. (2) that was derived under primary loading only is used). The fracture parameter K_r is adjusted as follows:

$$K_r = \frac{K_I^P + K_I^S}{K_{IC}} + \rho \tag{10}$$

where K_I^P and K_I^S are the respective stress intensity factors for primary and secondary stresses. The ρ-factor accounts for the interaction of primary and secondary stresses in the elastic-plastic region of the FAD. Tables of ρ-factor have been developed in the literature [1] for the conventional FAD approach. To include the constraint effect on the interaction factor eq. (10) is replaced by [9]:

$$K_r = \frac{K_I^P + K_I^S}{K_{IC}} + \rho\left[1+\alpha(-\beta^{P+S} L_r)^m\right] \tag{11}$$

where β^{P+S} is the normalized constraint parameter for combined loading. This is the adjustment of the assessment point. The failure assessment curve of eq. (11) becomes:

$$K_r = f(L_r)\left[1+\alpha(-\beta^{P+S} L_r)^m\right] \tag{12}$$

Following Eq. (8), the constraint parameter β^{P+S} can be estimated as follows:

$$\beta^{P+S} = \frac{T^{P+S}}{\sigma_Y L_r} \tag{13}$$

where T^{P+S} is the T-stress under combined loading.

In summary, if only primary load is considered, the constraint-based failure assessment is conducted using eqs. (1), (2) and (7). Under combined primary and secondary loading, the assessment is conducted using eqs. (11), (2) and (12).

3. FRACTURE PARAMETER AND LIMIT LOAD SOLUTIONS FOR SURFACE CRACKED PLATE

This section presents the calculation of the fracture parameters and the limit load for a plate with a semi-elliptical surface crack. The stress intensity factor and elastic T-stress are calculated from the finite element based correlations presented in Appendix A and B Figure 1 shows the plate geometry. Since this is a 3D problem, the stress intensity factor and the T-stress vary along the crack front. However, it has been observed that the failure of a surface cracked plate under tension occurs generally at the deepest point. That means, only the deepest point needs to be considered in the current analysis.

3.1 Stress Intensity Factor Solution

Reference [10, 11] obtained the stress intensity factor correlations for semi-elliptical surface cracks of varying aspect ratios (a/c) in finite thickness plates (crack depth to thickness ratio (a/t)) using finite element methods. For the current analysis, the stress intensity factors were taken from [11]. Reference [11] presents the stress intensity factors for semi-elliptical surface cracks within the range of $0 \le a/c \le 1$ and for a plate geometry within $0 \le a/t \le 0.8$ under general linear and nonlinear loads. The stress intensity factor is normalized as follows,

$$F = \frac{K}{\sigma_0\sqrt{\pi a / Q_e}} \tag{14}$$

where F is the so-called boundary correction factor, σ0 is the nominal tensile stress and Q_e is shape factor of an ellipse. The shape factor depends on the crack aspect ratio a/c, and is given by the following equation:

$$Q_e = 1.0 + 1.464\left(\frac{a}{c}\right)^{1.65} \tag{15}$$

The correlations for the boundary factor F for different crack a/c and plate a/t ratios are presented in Appendix A of the present paper.

3.2 Elastic T-Stress Solution

Elastic T-stress solutions for the plate with semi-elliptical surface crack under tension and bending loads were obtained in [12] using the finite element method. In [12], correlations of the T-stress solution are derived for a crack aspect ratio between $0 \le a/c \le 1$ and a relative plate thickness between $0 \le a/t \le 0.8$. The T-stress is normalized with σ_0, the nominal tensile stress, as

$$V = \frac{T}{\sigma_0} \tag{16}$$

where V is the normalized T-stress. For convenience, the solutions from [12] are presented in Appendix B of the present paper.

3.3 Limit Load Solution

Different limit load solutions for surface cracked plate under tensile loads exist in the literature. For example, Mattheck et al. [13] proposed an expression for ligament yielding based on the Dugdale strip yield model. More recently, Kim et al [14] suggested expressions based on a detailed finite element analysis. In the current work, a very simple formula that is based on the effective load carrying area is used [15]:

$$\sigma_L = \sigma_F\left(1 - \frac{A_c}{A}\right) \tag{17}$$

where Ac is the area of the semi-elliptical crack:

$$A_c = \pi a c / 2 \tag{18}$$

A is the effective load-bearing area:

$$A = t(t + 2c) \tag{19}$$

and σ_F in the stress at failure, here set equal to the yield stress, σ_y).

Equation (17) is fairly conservative if the plate dimensions are substantially larger than the crack dimensions. The reason is that the surrounding plate section reinforces the crack ligament to a much larger degree than the effective load bearing area of eq. (19) would suggest. An exploratory finite element calculation performed for this paper confirmed that failure in tension can occur at a noticeably higher load than that predicted by eq. (17).

4. FAILURE ASSESSMENT OF SURFACE CRACKED PLATE

Based on the stress intensity factor, T-stress and limit load solutions presented in the last section, an example failure assessment is now conducted. Section 4.1 demonstrates the assessment of a surface cracked plate under a purely primary load A plate under combined primary and secondary load is assessed in section 4.2.

4.1 Assessment for Primary Load

The analyzed surface cracked plate has a thickness of 3.00 inch and is made from a High Nickel Alloy material. The tensile yield stress at room temperature, is 45 ksi, and the ultimate strength is 114 ksi. The Young's modulus of the material is 30.3×10^{3} ksi. The initiation toughness of the material J_{IC} is taken to be 1300 in-lb/in2 based on a conservative estimation. The fracture toughness K_{IC} is obtained by converting the J_{IC} value. The constraint related material constants α, and m, which define the dependence of fracture toughness on constraint, are 1.5 and 1, respectively. Following ASME Section III Appendix G, a semi-elliptical surface crack was assumed, the crack length (2c) being 1.5 times of the thickness of the plate, or 4.5 in. For the assessment, a range of stresses is applied which corresponds to different operating conditions. The maximum crack depth, a (represented by a/t), for each loading condition is obtained from the failure assessment diagram (FAD). The surface cracked plate is initially assessed using the conventional FAD procedure, and then reassessed using the constraint based procedure.

The failure assessment diagram equation comes from Level 3 of PD6493 [2]. The lower bound FAD curve in [2] is independent of the geometry and the material:

$$K_r = (1 - 0.14\, L_r^2)\left[0.3 + 0.7\exp(-0.65\, L_r^6)\right] \tag{20}$$

The load ratio L_r is defined in terms of the yield strength and can therefore be greater than 1. The typical cut-off is at 1.2 for C-Mn steel and at 1.8 for austentic stainless steel [2]. Since the flow stress is known for the material of the plate in the current analysis, the ratio of flow stress to yield stress, σf/σy, is used as the cut-off. Figure 2 shows a plot of the failure assessment diagrams for the given crack length c.

The applied tensile stress is normalized as follows:

$$P_n = \frac{\sigma}{\sigma_f} \tag{21}$$

where σ_f is the flow stress of the material equals $0.5(\sigma_y + \sigma_u)$, with σ_y and σ_u being the yield and ultimate stress, respectively. Figure 3 shows the resulting maximum normalized axial stresses P_n (stress at predicted failure) for various a/t ratios.

The constraint-based assessment uses the failure assessment curve given by eq. (7):

$$K_r = (1 - 0.14\,L_r^2)\left[0.3 + 0.7\exp(-0.65\,L_r^6)\right]\left[1 + 1.5(-\beta L_r)\right] \tag{22}$$

Several of the resulting FAD curves for the current analysis are plotted in Figure 2. Based on eq. (22), the maximum normalized axial stress P_n for different a/t ratios are obtained, Figure 3. Again, excluding the effect of low constraint is found to be overly conservative. By accounting for this effect, the increase in the allowable tensile stress can be 17% to 30% in the present example.

4.2 Assessment for Combined Primary and Secondary Load

In section 4.1, the effect of potential weld residual stresses was not included. Since a post-weld heat treatment is not required for austenitic materials, welded residual stresses may be present and should be accounted for. Since the residual stresses in an as-welded component are typically unknown, the residual stress can be conservatively approximated as a membrane stress field with a magnitude of the yield stress. In the current analysis for the welded plate, this assumption is made:

$$\sigma^s = \sigma_y \tag{23}$$

Following the procedure in section 2.2, the failure assessment can be conducted using Eqs. (2), (11) and (12). Note that in the current analysis the interaction factor ρ is assumed to be zero, because the assumption of Eq. (23) is considered conservative enough. With the secondary stress effect included, the plate was assessed using the conventional FAD procedure first, and then reassessed using the constraint based procedure.

The resulting maximum normalized axial stress P_n for different a/t ratios is presented in Figure 4. As expected, a comparison between Figures 3 and 4 indicates that for the same a/t ratio, the maximum allowable stress levels are lower when the effect of secondary residual stress is included. Figure 4 demonstrates that by accounting for the effect of low constraint, the increase in the allowable tensile stress is again as much as 15% to 25% compared to the conventional one-parameter fracture assessment. Interestingly, the constraint based assessment that includes the residual stress effect predicts failure loads fairly close to those of the conventional assessment without residual stress effect.

From the present analysis, it is demonstrated that for the current example case, when the lower constraint effect is properly accounted for, the maximum allowable tensile stress level can increase 15% or more. It is consistent with numerical/experimental observations obtained in Ref. [7]. Note that the assessment points for the current analysis are in the region of "ductile failure" as demonstrated in Figure 2, that is where overly conservatism occurs if the constraint effects are not included in the failure assessment.

5. CONCLUSIONS

In this paper, the constraint-based failure assessment of a welded plate with semi-elliptical surface crack under tensile load is performed. The constraint-based failure assessment diagram approach and a procedure to include both primary and secondary loads are described. Correlations for calculating the fracture parameters, i.e. stress intensity factors and T-stress are presented. A conservative relationship for the limit load of the surface cracked plate was obtained from the literature. An example failure assessment has been conducted for primary load only and for a combined primary and secondary load. For both the primary load and the combined primary and secondary load cases, the maximum allowable tensile stress level can increase by as much as 15 – 30% when the lower constraint effect is properly accounted for. Since a constraint based failure assessment has already been included in the R6 [1] and SINTAP [6] guidelines, incorporating a similar procedure into the FAD method described in Appendix H of Section XI of the ASME Code [16] should be taken into consideration.

ACKNOWLEDGMENTS

The authors gratefully acknowledge the financial support from Materials and Manufacturing Ontario (MMO).

REFERENCES

1. R6, "Assessment of the Integrity of Structures Containing Defects, Procedure R6, Revision 3," 1997, Nuclear Electric Ltd. Gloucester, U.K.
2. PD 6493:1991, "Guidance on Some Methods for the Derivation of Acceptance Levels for Defects in Fusion Welded Joints", August 1991, British Standards Institution.
3. O'Dowd, N.P. and Shih, C.F., "Family of Crack Tip Fields Characterized by a Triaxiality Parameter," Journal of Mechanics and Physics of Solids, Vol. 39, 1991, pp. 989-1015.
4. Betegon, C., Hancock, J.W., "Two-Parameter Characterization of Elastic-Plastic Crack Tip Fields," ASME Journal of Applied Mechanics, Vol. 58, 1991, pp. 104-110.
5. Ainsworth, R.A. and O'Dowd, N.P., "Constraint in the Failure Assessment Diagram Approach for Fracture Assessment," ASME Journal of Pressure Vessel Technology, Vol. 117, 1995, pp. 260-267.
6. Ainsworth, R.A., Sattaru-Far, I., Sherry, A.H., Hooton, D.G. and Hadley, I., "Methods for Including Constraint Effects within the SINTAP Procedures," Engineering Fracture Mechanics, Vol. 67, 2000, pp. 563-571.
7. Dodds, R.H. Jr., Shih, C.F. and Anderson, T.L, 1993, "Continuum and Micromechanics Treatment of Constraint in Fracture," International Journal of Fracture, Vol. 64, 1993, pp. 101-133.
8. Ainsworth, R.A., "The Treatment of Thermal and Residual Stresses in Failure Assessments," Engineering Fracture Mechanics, Vol. 24, 1986, pp. 15-21.
9. Ainsworth, R.A., Sanderson, D.J., Hooton, D.G. and Sherry, A.H., "Constraint Effects in R6 for Combined Primary and Secondary Stresses," ASME PVP, Vol. 324, Fatigue and Fracture, Vol. 2, 1996, pp. 117-131.
10. Raju, I.S. and Newman, J.C., "Stress Intensity Factors for a Wide Range of Semi-Elliptical Surface Cracks in Finite Thickness Plates," Engineering Fracture Mechanics, Vol. 24, 1979, pp. 817-829.
11. Wang, X. and Lambert, S.B., "Stress Intensity Factors for Low Aspect Ratio Semi-Elliptical Surface Cracks in Finite Thickness Plates Subjected to Nonuniform Stresses," Engineering Fracture Mechanics, Vol. 51, 1995, pp. 517-532.
12. Wang, X., "Elastic T-Stress Solutions for Semi-Elliptical Surface Cracks in Finite Thickness Plates," Engineering Fracture Mechanics, Vol. 70, 2003, pp. 731-756.

13. Mattheck, C., Morawietz, P., Munz D. and Wolf B., "Ligament Yielding of a Plate with Semi-elliptical Surface Cracks under Uniform Tension," International Journal of Pressure Vessel and Piping, Vol. 16, 1984, pp. 131-143.
14. Kim, Y.J., Shim, D.J., Choi, J.B. and Kim, Y.J., "Elastic Plastic Analyses for Surface Cracked Plates under Combined Bending and Tension," Journal of Strain Analysis, Vol. 37, 2002, pp. 33-45.
15. Miller, A.G., "Review of Limit Loads of Structures Containing Defects," International Journal of Pressure Vessel and Piping, Vol. 32, 1988, pp. 197-327.
16. The American Society of Mechanical Engineers, ASME Boiler and Pressure Vessel Code, Section XI, Rules for Inservice Inspection of Nuclear Power Plant Components, 2001.

APPENDIX A

Stress Intensity Factor Solutions for Semi-Elliptical Surface Cracks

From Wang and Lambert [11] the stress intensity factor for the problem described in Figure 1 under tensile load is normalized as follows,

$$F = \frac{K}{\sigma_0 \sqrt{\pi a / Q}}$$

where σ0 is the nominal tensile stress and Q is shape factor of an ellipse given in Eq. (15). At the deepest point, F is found to be:

$$F = B_0 + B_1(\frac{a}{t})^2 + B_2(\frac{a}{t})^4 + B_3(\frac{a}{t})^6$$

where

$$B_0 = 1.0929 + 0.2581(\frac{a}{c}) - 0.7703(\frac{a}{c})^2 + 0.4394(\frac{a}{c})^3$$

$$B_1 = 0.456 - 3.045(\frac{a}{c}) + 2.007(\frac{a}{c})^2 + \frac{1.0}{0.147 + (\frac{a}{c})^{0.688}}$$

$$B_2 = 0.995 - \frac{1.0}{0.027 + (\frac{a}{c})} + 22.000(1.0 - \frac{a}{c})^{9.953}$$

$$B_3 = -1.459 + \frac{1.0}{0.014 + (\frac{a}{c})} - 24.211(1.0 - \frac{a}{c})^{8.071}$$

Note that this solution is valid for surface cracks of $0 \le a/c \le 1$ and $0 \le a/t \le 0.8$.

APPENDIX B

T-Stress Solutions for Semi-Elliptical Surface Cracks

From Wang [12] the elastic T-stress for problem described in Figure 1 under tensile load is normalized as follows,

$$V = \frac{T}{\sigma_0}$$

where σ0 is the nominal tensile stress. At the deepest point, V is found to be:

$$V = A_0 + A_1(\frac{a}{t})^2 + A_2(\frac{a}{t})^4 + A_3(\frac{a}{t})^6$$

where

$$A_0 = -0.581176 + 0.429125(\frac{a}{c}) - 0.670213(\frac{a}{c})^2 + 0.290259(\frac{a}{c})$$

$$A_1 = 0.795958 - 6.53119(\frac{a}{c}) + 12.1073(\frac{a}{c})^2 - 6.47294(\frac{a}{c})^3$$

$$A_2 = -4.70569 + 28.4706(\frac{a}{c}) - 51.1588(\frac{a}{c})^2 + 27.3436(\frac{a}{c})^3$$

$$A_3 = 6.16031 - 40.1692(\frac{a}{c}) + 68.7053(\frac{a}{c})^2 - 35.4684(\frac{a}{c})^3$$

Note that this solution is valid for surface cracks of $0.2 \le a/c \le 1$ and $0 \le a/t \le 0.8$.

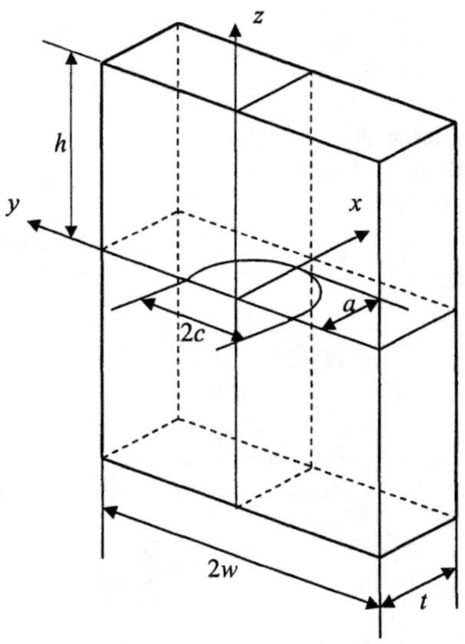

Figure 1: Surface Cracked Plate Geometry

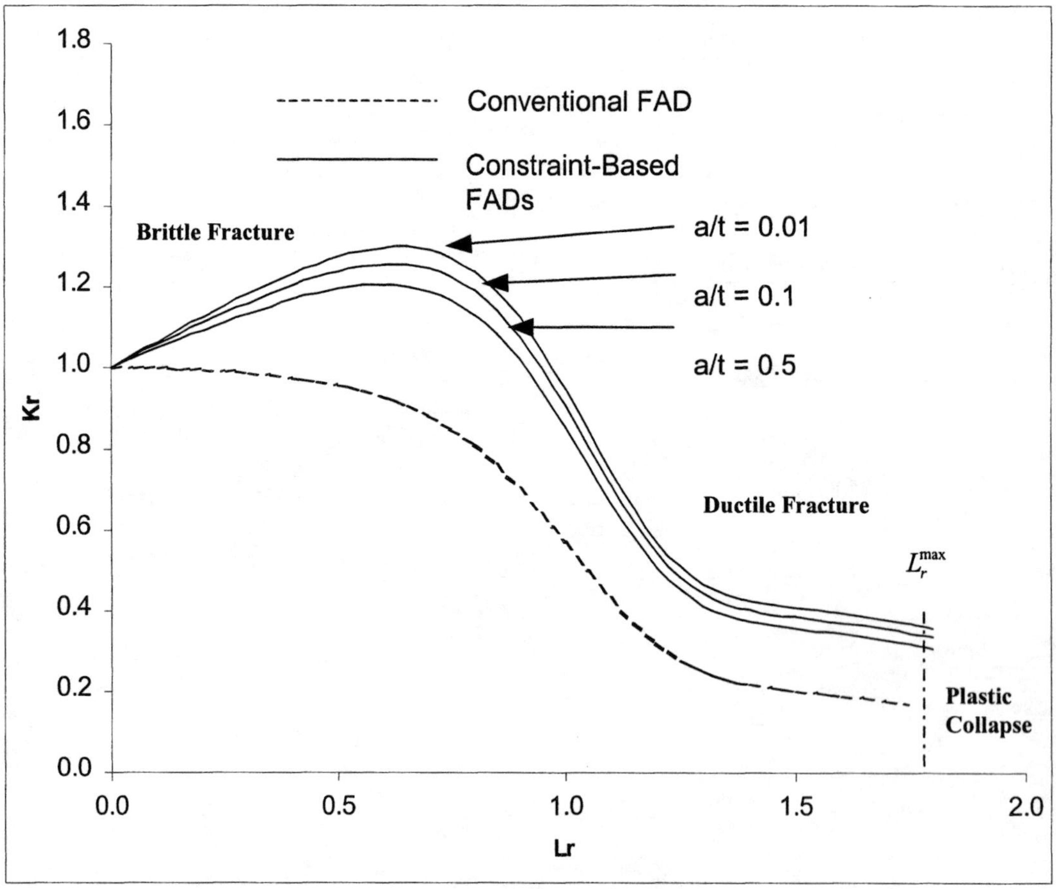

Figure 2: Example Failure Assessment Diagrams for Primary Load (Crack Length c = const.)

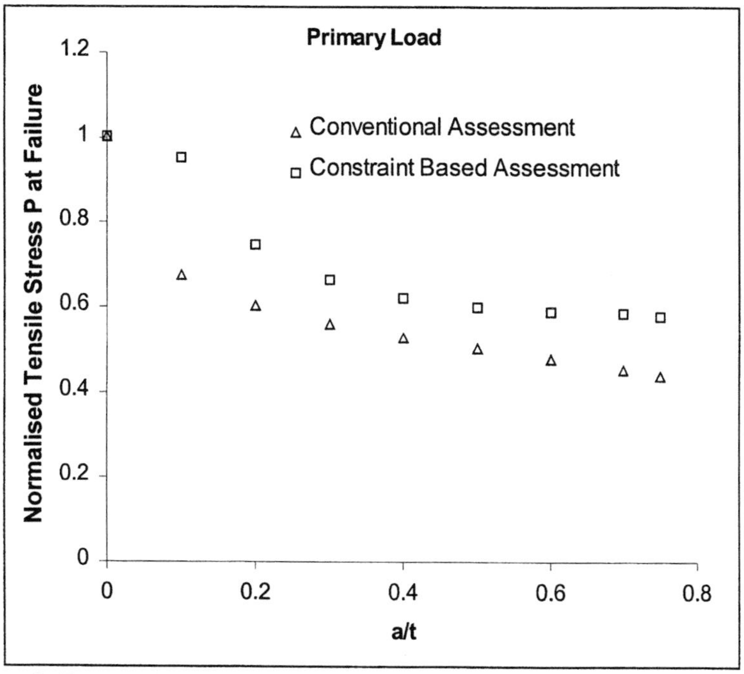

Figure 3: Results from Assessment of Primary Load (Crack Length c = const.)

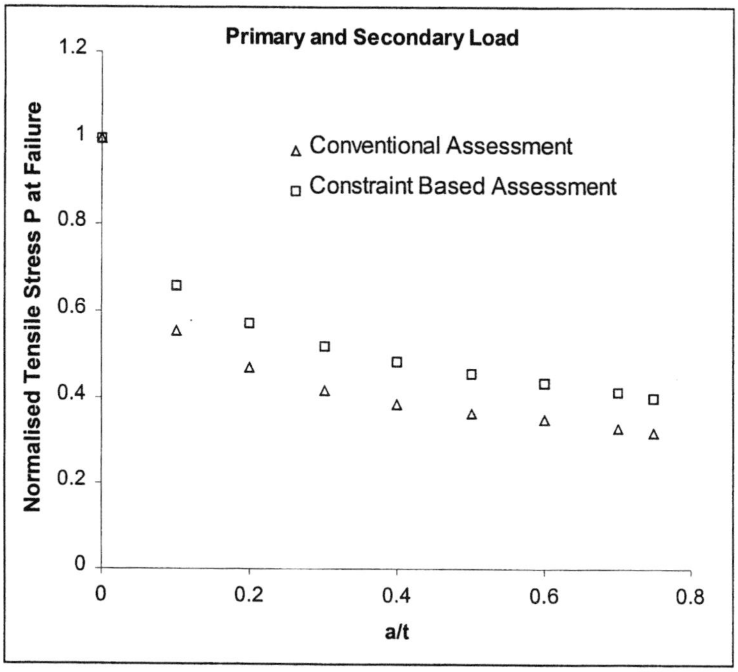

Figure 4: Results from Assessment of Combined Primary and Secondary Loads (Crack Length c = const.)

Finite Element Manufacturing Simulations and Field Applications Introduction

John Martin

Lockheed Martin, Inc
Albany, NY

This section includes four technical papers presented in the Computer Technology sponsored session "Finite Element Manufacturing Simulations and Field Applications" session. Three of these papers discuss correlations between finite element computational results and experimental data based on test specimens and/or prototypical component testing. The fourth paper presents finite element based methods, models, and assumptions used in the simulation of a cold-working tube forming process.

The paper "Stress Analysis of a Pin and Clevis Assembly" examines an instrumented pin and clevis test hardware loaded statically to failure. Test data is compared with computational results to assess the adequacy of boundary condition assumptions for modelling purposes. Comparisons are also made with standard strength of materials shear failure formulations.

The second paper titled "Modelling Studies of the Fatigue Behavior of a Nozzle-to-Vessel Intersection" investigates the fatigue behavior of a nozzle/vessel intersection for a typical geometry using computational tools, fatigue test specimens, and full scale prototypical components. Examination of the results of these computational models and experimental results are reported along with a comparison of predictions based on current design code procedures.

A third paper, "A Finite Element Study of Friction Effects in Four-Point Bending Creep Test" examines the differences between assumed finite element boundary conditions and actual test fixtures when using modelling tools to simulate creep test specimen testing. The paper identifies potential errors in predicting material properties when using modelling tools with inappropriately defined friction boundary conditions.

The last paper, "Methods, Models, and Assumptions Used in Finite Element Simulation of a Pilgering Metal Forming Process" proposes a finite element modelling approach to simulate a complex three-dimensional cold-working rolling/extrusion tube forming process. This modelling is undertaken to gain a fundamental understanding of the stresses and strains generated by this manufacturing process. Sensitivity studies are performed and correlations with qualitative trends observed in commercial tubing are reported.

I wish to thank all authors, reviewers, and presenters for contributing to this session. Without your efforts, neither this session nor the ASME 2003 PVP conference would have been successful.

PVP-Vol. 458, Computer Technology and Applications

PVP2003-1892

STRESS ANALYSIS OF A PIN AND CLEVIS ASSEMBLY

Brian Pechatsko, James Stambolia,
Elvin B. Shields, Jack B. Davenport, and Arnold D. Orning

Department of Mechanical and Industrial Engineering
Youngstown State University
Youngstown, Ohio

ABSTRACT

The shear yield strength of the material is one of the controlling factors in determining the pin diameters for a pin and clevis connection. Typically, for steels used in machine fabrication, an accepted method is to approximate the shear yield strength as 75% of the material's ultimate tensile strength Blodgett [1].

In a pin and clevis connection, the pin can be subjected to static and dynamic loads that can generate stresses on the order of the shear yield strength of the steel. The loads produce bending and shear stresses in the pin. The highest stressed region is the clearance gap between the blade and clevis. Finite Element Analysis (FEA) results produced valuable insight into the problem. The various methods available to constrain the computer models produced different results. Enhanced FEA computer models provided the most realistic insight into the actual pin stresses.

Full-scale steel test hardware was tested with static loads until pin failure occurred. A standardized pin size of 12.7 mm (0.500 in) diameter was used in order to maintain the loads at a manageable testing level. Strain gages placed in the clearance gap recorded bending strains. The pins failed from shear stress in the area of the gap region.

NOMENCLATURE

A - Cross sectional area of pin in mm^2 (in^2)
C - Clearance gap for strain gages = 2.0 mm (0.080 in)
D - Diameter of pin = 12.7 mm (0.500 in)
E - Modulus of elasticity for steel pin = 2.07 x 10^5 MPa (30 x 10^6 psi)
H - Distance from base to center of pin in mm (in)
I - Moment of inertia of the pin cross-section in mm^4 (in^4)
K_t - Stress concentration factor at corners = 1.7
L - Width of blade = 48.8 mm (1.92 in)
M -Bending moment in mm-N (in-lbf)
N - Width of clevis = 12.7 mm (0.500 in)
N_{DF} - Design Factor = 4
P - Load on pin in N (lbf)
R - Reaction force of clevis in N (lbf)
S_a - Allowable stress in MPa (psi)
S_u - Ultimate tensile stress of 1040 cold-drawn steel = 620 MPa (90,000 psi)
S_y - Yield strength of 1040 cold-drawn steel = 414 MPa (60,000 psi)
V and V_{yz} - Shear force in N (lbf)
W - Thickness of blade and clevis in mm (in)
c - Distance from neutral axis to outside edge = 6.4 mm (0.25 in)
t - Nominal thickness of strain gage and adhesive glue = 0.038 mm (0.0015 in)
ε_{exp} - Experimental strain in m/m (in/in)
ε_{raw} - Uncorrected experimental strain in m/m (in/in)
σ and σ_z - Bending stress at surface of pin in MPa (psi)
$\sigma_1>\sigma_2>\sigma_3$ - Principal stresses in MPa (psi)
σ_{exp} - Experimental stress in MPa (psi)
σ_{vm} - von Mises stress in MPa (psi)
τ_{oct} - Octahedral shear stress in MPa (psi)
τ - Shear stress in MPa (psi)

INTRODUCTION

Pin and clevis connections are used in industrial machines to transmit loads while allowing rotational movement of machine parts. For example, a shearing device for a steel plate can contain multiple pin and clevis connections. The connections allow for an efficient way of performing a desired task or operation. A frequent problem encountered is pin failure due to shear stresses caused by the loads while the machine is in operation.

Engineers have a difficult time choosing an appropriate pin size because there is not a sole standard to select a pin diameter based on the applied load. In past machine design problems, pin sizes have been determined largely through trial and error. For example, Greene et al [2] did specific work on the space shuttle tang-clevis joint.

In this study, design codes like the Boiler and Pressure Vessel Code, ASME sections VIII and III were not consulted for pin shear. Some specific formulas from Shields [3] are given in equation (1). Since their origin and validity are not documented, the formulas should not be used for design without confirmation.

$$1.0 \le \frac{L}{D} \le 3.0$$
$$1.25 \le \frac{H}{D} \le 2.0$$
$$2.0 \le \frac{W}{D} \le 3.0 \qquad (1)$$
$$0.62 \le \frac{N}{D} \le 1.2$$
$$\frac{P}{2 \cdot A} \le 24{,}000$$

PROBLEM STATEMENT

It is desired to formulate a general design equation to determine pin diameter sizes to satisfy general machine design problems. Two possible formula types are given as equations (2) and (3) for shear stress and bending stress, respectively (Mott [4]). The choice for the allowable stress is the crux of the problem. In addition, the allowable stress must be modified by a design factor (N_{DF}) to account for uncontrolled variables. For machine elements constructed of ductile materials with some combination of shock load, untried material, and uncontrolled environment the design factor should be 4 or more (Mott[4]). The form shown in equation (2) will be used in this study.

$$\tau := \frac{P}{2 \cdot A} \le \frac{S_{a_1}}{N_{DF}} \qquad (2)$$

$$\sigma := \frac{Mc}{I} \le \frac{S_{a_2}}{N_{DF}} \qquad (3)$$

TABLE 1. PHYSICAL TEST HARDWARE DIMENSIONS

Component	Dimension mm (in)
Nominal Pin Diameter, D	12.7 (0.500)
Clevis Width, N	12.7 (0.500)
Blade Width, L	48.8 (1.92)
Clearance Gap, C	2.0 (0.080)

EXPERIMENTAL ANALYSIS

The test hardware dimensions describing the pin and clevis connection are shown in Table 1 and sketched on Figure 1. Several hundred tests were performed on pin and clevis connections. In reality, some connections are subjected to both dynamic and static loads. Although large impact loads can create massive internal stresses in cyclic form and frictional rotation can produce thermally induced stresses, the effects of such loads were not considered in these experiments.

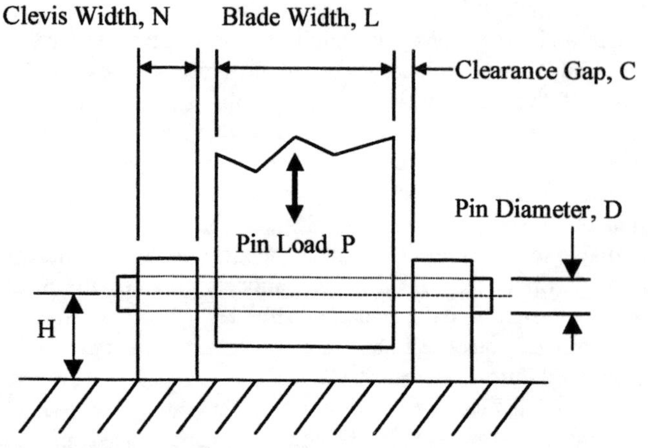

FIGURE 1. FRONT VIEW OF PIN AND CLEVIS ASSEMBLY

A steel pin and clevis assembly was fabricated from mild steel plates to the required dimensions. It was then tested with strain gages under various loads. Individual components of the test hardware and a completed assembly are shown on Figures 2 and 3, respectively. For a push fit, a radial clearance of 0.05 mm (0.002 inch) was used to ensure smooth operation of the pin and clevis connection.

FIGURE 2. CLEVIS, BLADE, AND PIN TEST HARDWARE

FIGURE 3. COMPLETED ASSEMBLY

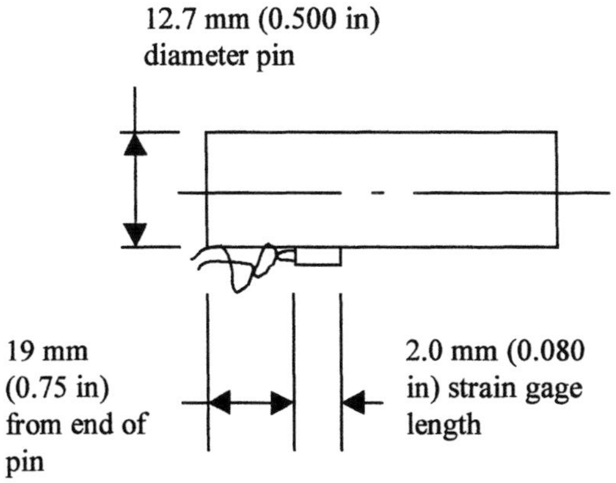

FIGURE 4. STRAIN GAGE PLACEMENT

The test hardware was modified to allow for the positioning of the strain gage in the clearance gap region. The strain gage measured 2.0 mm (0.080 inch) in length, so the clearance gap between the blade and clevis was increased from 0.38 mm (0.015 inch) to 2.0 mm (0.080 inch). This allowed the gage to measure the strain without the blade or clevis overlapping the gage and interfering with the readings.

Originally, only a full size 12.7 mm (0.500 in) diameter pin of 1040 cold-drawn steel pin was to be tested. However, with usage, the sliding clearance between the pin and yoke gradually increases due to wear and deformation. To account for this wear two undersize pins with diameters of 12.45 mm (0.490 in) and 12.57 mm (0.495 in) were tested. The undersize pins (2% and 1% smaller than the full size pin) were tested to show the stresses on a loose-fitting or worn pin. For instance, worn pins can be found in overworked or neglected machinery.

A Measurements Group strain gage type EA-13-060LZ-120 with a 2.085 gage factor was mounted axially on the pin in the clearance gap area (Vishay [5]) as shown on Figure 4. Wires were attached to the strain gage and then connected to a Measurements Group strain indicator (model P-3500). To start the experiment, the pin and clevis assembly was turned upside down, weights were hung from the blade of the assembly and compressive strain readings were recorded for each weight increase. When using the individual weights, a maximum load of 2.4 kN (540 lbf) was reached due to space limitations. Then a Dake 445 kN (50-ton) hydraulic press and a 89 kN (20,000 lbf) capacity load cell (model ALD COMP) were utilized to produce the larger static loads. The pin and clevis assembly was right side up in the hydraulic press as shown on Figure 3. The hydraulic press and load cell allowed the loads to be applied in the required load steps.

Tensile strain readings were recorded at 890 N (200 lbf) intervals and corrected strain values are presented in Table 2. These values are plotted on Figure 5. The plots show that the strain initially increases linearly. The 2%-undersize diameter pin yielded at 3800 $\mu\varepsilon$, the 1%-undersize diameter pin yielded at 3100 $\mu\varepsilon$, and the full size diameter pin yielded at 2500 $\mu\varepsilon$. As Figure 5 shows, the strain in the two undersize pins was greater for the same size load than for the 12.7 mm (0.500 in) full size diameter pin. The higher strain was due to the increased clearance between the pin, clevis, and blade that allowed the components to move more freely. For each load applied by the hydraulic press, the pin and clevis assembly was loaded slowly and then returned to a no load condition. A permanent strain reading indicated yielding. The pins were loaded to shear failure.

The raw strain values were corrected for the nominal thickness of the strain gage and adhesive glue of 0.038 mm (0.0015 in) as shown in equation (4). Then the corrected strain values were converted into stress values by equation (5) and plotted on Figure 6 versus the pin shear stress P/2A. The modulus of elasticity for steel (E) was taken as 2.07×10^5 MPa (30×10^6 psi). Since the surface of the pin is free of shear stress, the experimental bending stress (σ_{exp}) is the only non-zero stress. Hence, at the surface of the pin the experimental bending stress is equal to the first principal stress ($\sigma_1 > \sigma_2 > \sigma_3$) with the other principal stresses equal to zero. This situation will result in the von Mises stress being equal to the experimental stress ($\sigma_{vm} = \sigma_1 = \sigma_{exp}$).

$$\varepsilon_{exp} = \varepsilon_{raw} \cdot \left(\frac{\frac{D}{2} - t}{\frac{D}{2}} \right) \qquad (4)$$

$$\sigma_{exp} = \varepsilon_{exp} E \qquad (5)$$

To determine the allowable stress (S_a), the 2% undersize pin data curve on Figure 6 was examined in further detail as shown on Figure 7. The bending stress values were converted to octahedral shear stress values by equation (6) in accordance with Budynas [6]. The octahedral shear stress theory is based on normal and shear stresses acting on the faces of an eight-sided element. The results are equivalent to the von Mises Distortion Energy Theory for ductile materials.

$$\tau_{oct} \cdot = \cdot \frac{\sqrt{2}}{3} \sigma_{vm} \qquad (6)$$

In Figure 7, the R^2 value for the curve is 0.9776, which is a balanced straight line. Since the ultimate tensile stress of the 1040 cold-drawn steel pin is 620 MPa (90,000 psi), the shear yield stress by the 75% rule (Blodgett [1]) would be 465 MPa (67,500 psi). Then the value for P/2A is calculated by equation (7) to be 137 MPa (19,900 psi). Thus, the value for the allowable stress for 1040 cold-drawn steel was set at Sa = 138 MPa (20,000 psi). For a design factor (N_{DF}) of 4, the ratio of S_a/N_{DF} = 35 MPa (5,000 psi). This approach can be generalized to include other ductile steels by setting up the allowable stress (S_a) as 1/3.4 of 0.75 S_u and the result is shown in equation (8).

$$\tau \cdot = \cdot 3.4 \cdot \frac{P}{2 \cdot A} \qquad (7)$$

$$S_a \cdot = \cdot 0.22 \cdot S_U \qquad (8)$$

From the free body diagram of the pin, equation (9) is the moment at the center of the clearance gap. Ideally, the load would be centered on the assembly and produce symmetric shear and moment diagrams.

$$M = \frac{P}{2}\left(\frac{N+C}{2}\right) \qquad (9)$$

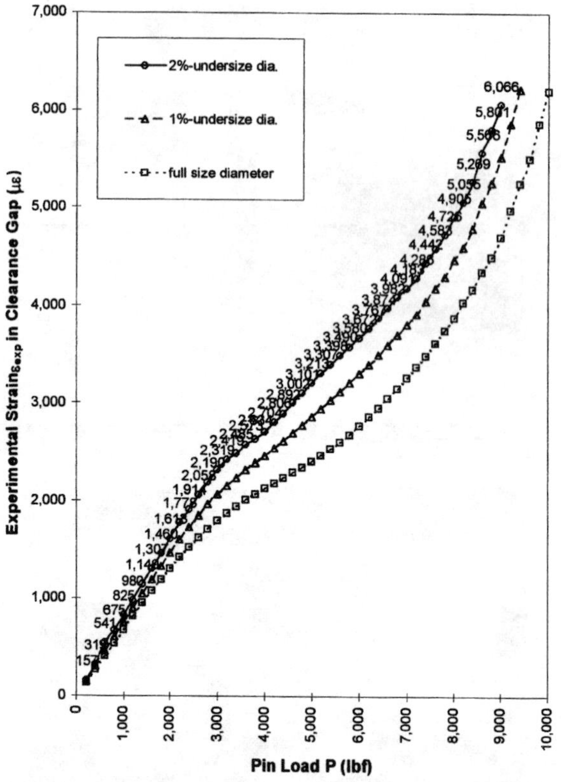

FIGURE 5. EXPERIMENTAL PIN STRAIN COMPARISON

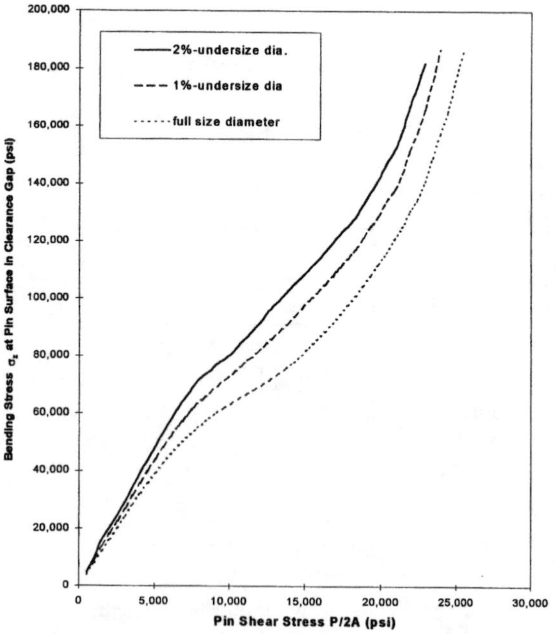

FIGURE 6. PIN SHEAR STRESS VS BENDING STRESS IN CLEARANCE GAP

TABLE 2. CORRECTED EXPERIMENTAL STRAIN DATA

Pin Load (kN)	Pin Load (lbf)	12.45 mm (0.490 in) 2% undersize diameter (με)	12.57 mm (0.495 in) 1% undersize diameter (με)	12.7 mm (0.500 in) full size diameter (με)
0.890	200	157	152	133
1.779	400	319	302	266
2.669	600	541	457	405
3.558	800	675	605	539
4.448	1,000	825	752	675
5.338	1,200	980	901	816
6.227	1,400	1,149	1,049	948
7.117	1,600	1,307	1,192	1,077
8.006	1,800	1,460	1,332	1,196
8.896	2,000	1,613	1,467	1,311
9.786	2,200	1,778	1,605	1,421
10.675	2,400	1,914	1,726	1,525
11.565	2,600	2,058	1,851	1,625
12.454	2,800	2,190	1,963	1,714
13.344	3,000	2,319	2,068	1,797
14.234	3,200	2,419	2,154	1,873
15.123	3,400	2,485	2,234	1,943
16.013	3,600	2,573	2,315	2,008
16.902	3,800	2,634	2,384	2,069
17.792	4,000	2,704	2,458	2,126
18.682	4,200	2,806	2,539	2,181
19.571	4,400	2,892	2,615	2,235
20.461	4,600	3,002	2,694	2,287
21.350	4,800	3,101	2,775	2,342
22.240	5,000	3,213	2,863	2,401
23.130	5,200	3,307	2,944	2,463
24.019	5,400	3,398	3,030	2,531
24.909	5,600	3,490	3,118	2,599
25.798	5,800	3,580	3,210	2,679
26.688	6,000	3,672	3,305	2,765
27.578	6,200	3,767	3,395	2,859
28.467	6,400	3,874	3,493	2,951
29.357	6,600	3,982	3,596	3,049
30.246	6,800	4,091	3,703	3,150
31.136	7,000	4,183	3,804	3,261
32.026	7,200	4,286	3,911	3,369
32.915	7,400	4,442	4,047	3,488
33.805	7,600	4,583	4,182	3,614
34.694	7,800	4,726	4,300	3,747
35.584	8,000	4,905	4,473	3,873
36.474	8,200	5,055	4,597	4,035
37.363	8,400	5,269	4,789	4,169
38.253	8,600	5,566	5,053	4,352
39.142	8,800	5,801	5,255	4,496
40.032	9,000	6,066	5,527	4,699
40.922	9,200	failed	5,863	4,974
41.811	9,400	failed	6,221	5,247
42.701	9,600	failed	failed	5,507
43.590	9,800	failed	failed	5,865
44.480	10,000	failed	failed	6,203

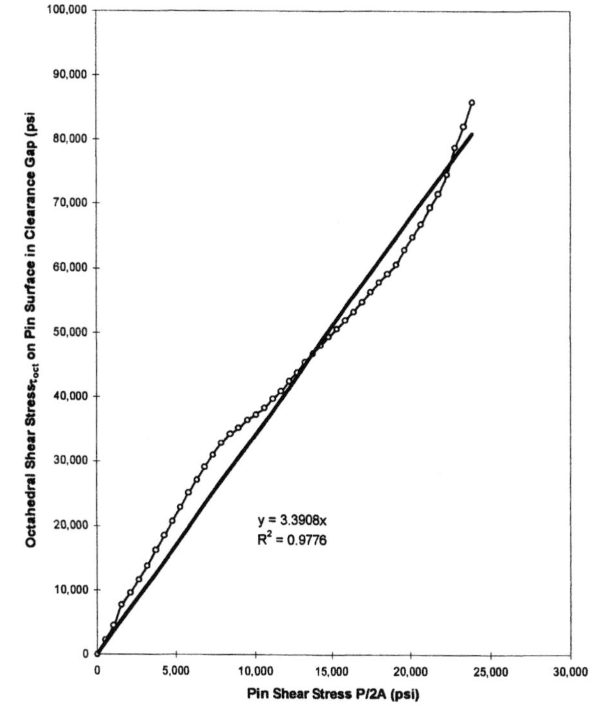

FIGURE 7. PIN SHEAR STRESS VS OCTAHEDRAL SHEAR STRESS FOR 2% UNDERSIZE DIAMETER PIN

MANUAL ANALYSIS

Using Budynas [5], the normal stress is found by equation (10) and converted into the principal stress by equation (11). There is no shear stress on the surface of the pin. Then the von Mises stress is calculated in the usual way by equation (12). Since strain gages read actual stresses, a stress concentration factor (K_t) has been added to equation (12) to account for the corners of the clearance gap. For the situation of a press fitted pin in a disk, Spotts and Shoup [7] recommended a value of approximately 1.7 for the stress concentration factor. The manually calculated von Mises stresses for several loads are shown in Table 3.

$$\sigma = \frac{\mathbf{M} \cdot \mathbf{c}}{\mathbf{I}} \tag{10}$$

$$\sigma_{1,3} = \frac{\sigma_z}{2} \pm \sqrt{\left(\frac{\sigma_z}{2}\right)^2 + \tau_{yz}^2} \cdot \mathbf{and} \cdot \sigma_2 = 0 \tag{11}$$

$$\sigma_{vm} = \frac{K_t}{\sqrt{2}} \sqrt{(\sigma_1 - \sigma_2)^2 + (\sigma_2 - \sigma_3)^2 + (\sigma_3 - \sigma_1)^2} \tag{12}$$

TABLE 3. EXPERIMENTAL VS. THEORETICAL VON MISES

Pin Load kN (lbf)	Exper. Stress MPa (psi)	Manual ($K_t = 1.7$) MPa (psi)	FEA (Gap Elem.) MPa (psi)	FEA (Only BC's) MPa (psi)
4.448 (1,000)	139.6 (20,248)	138.5 (20,087)	134.7 (19,536)	146.0 (21,177)
8.896 (2,000)	271.2 (39,333)	277.0 (40,173)	269.4 (39,072)	286.2 (41,507)
13.344 (3,000)	371.7 (53,915)	415.5 (60,260)	404.1 (58,608)	430.7 (62,472)
17.792 (4,000)	439.8 (63,785)	554.0 (80,347)	not available	569.4 (82,590)
22.240 (5,000)	496.5 (72,015)	692.5 (100,433)	not available	689.5 (100,000)

FINITE ELEMENT ANALYSIS

A static, linear elastic finite element analysis (FEA) was performed on the various pin and clevis models using the Algor[1] computer software. A finite element analysis of the pin and clevis connection was performed on two- and three-dimensional models. Two-dimensional models were made to closely resemble the actual assembly under a uniformly distributed static load and were made with beam-type elements. For all FEA models, a uniform pin and clevis size was used, as listed in Table 1. In addition, the modulus of elasticity used for all models was 2.07×10^5 MPa (30×10^6 psi) for steel.

The pin was uniformly loaded with 4.448 kN (1,000 lbf) distributed over the 48.8 mm (1.92 in) blade. The ends of the pin were constrained to act as the resisting clevis. Boundary constraints that allowed axial translation and rotation about one end of the pin were used for the analyses. The 2.0 mm (0.080 in) clearance gap was left unloaded and without constraints.

The maximum stress produced in the two-dimensional models in the gap region was approximately 18 MPa (2,600 psi). The stress calculated was extremely low compared to the experimental stress; therefore, the two-dimensional models were not analyzed further and are not given as a part of this report.

Standard eight-noded brick elements were used in all three-dimensional models. Brick elements provided more realistic data than beam elements because of the similarity in the model construction and the constraints used. Two different FEA models were processed using only half models due to symmetry.

The first 3-D FEA model was constrained with simple boundary conditions. Models with only boundary conditions (no gap elements) are more rigid than in reality. Therefore, the computed stresses are higher due to the fixed constraints. A dither of the stress distribution is shown on Figure 8. It is obvious that the area of maximum stress is located in the clearance gap region.

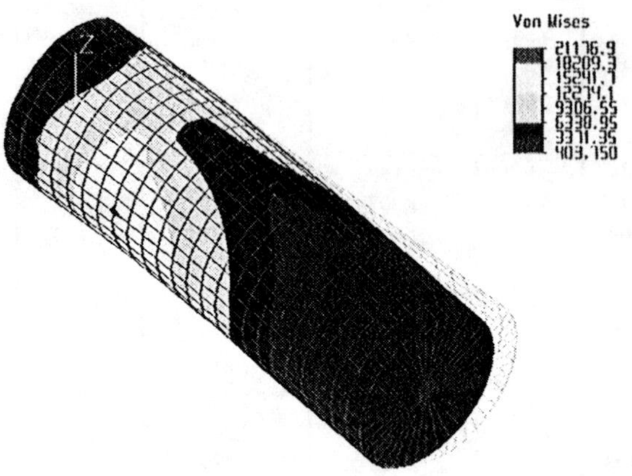

FIGURE 8. FEA VON MISES STRESSES FOR THE HALF PIN MODEL WITH ONLY BOUNDARY CONDITIONS

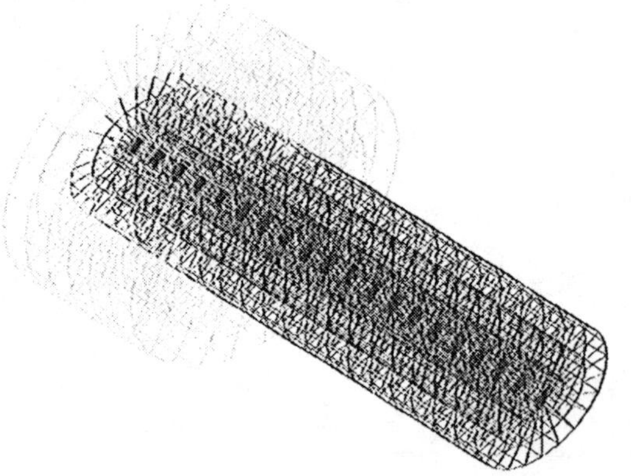

FIGURE 9. FINITE ELEMENT MODEL OF HALF OF THE PIN AND CLEVIS USING GAP ELEMENTS

The second 3-D FEA model used gap elements. Figure 9 shows a half model of the pin and separate clevis. The clevis was rigidly fixed with boundary conditions and connected to the pin using gap elements. The gap elements mimic the actions of a sliding fit, allowing the pin to move as if it were in an actual clevis. In the linear elastic region, the stresses calculated with gap elements are theoretically more accurate. In Table 3, von Mises stress values are shown for both three-dimensional FEA models for selected loads.

[1] Algor, Inc., 150 Beta Drive; Pittsburgh, PA 15238-2932 USA

Results and Conclusions

Figure 10 shows a stress comparison between the experimental, manual, and FEA methods. The experimental stresses and the second FEA model with gap elements are within 10% of each other. In addition, the manual calculations tended to match the first FEA model with only boundary elements. The theoretical methods (manual and FEA models) gave values that were too high at loads greater than approximately 8,900 N (2,000 lbf) because the experimental results became nonlinear after this point.

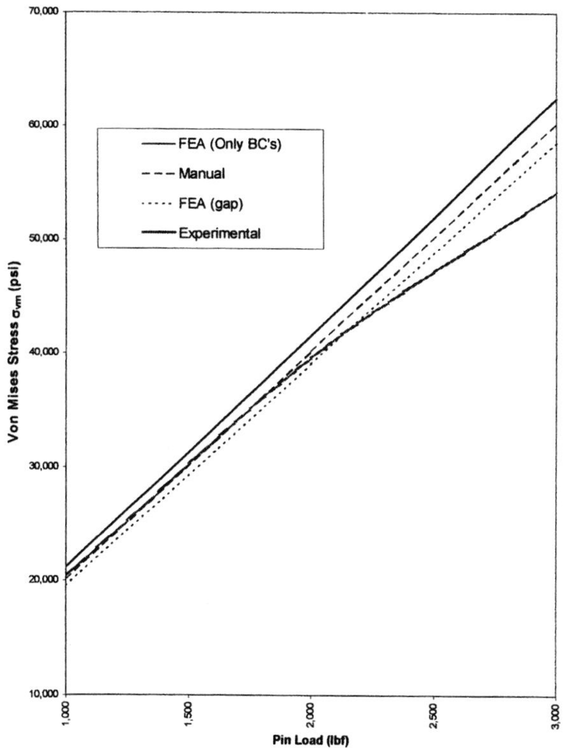

FIGURE 10. EXPERIMENTAL VS. THEORETICAL VON MISES STRESSES

The traditionally accepted method of selecting a pin diameter by using the form of equation (13) has been selected with an allowable value specified by equation (8). This should be a sufficiently conservative approach to the problem as long as an appropriate design factor (N_{DF}) is used. For example, by equation (14) a load of P = 44,500 N (10,000 lbf) for a steel pin with S_U = 620 MPa (90,000 psi) and a design factor of N_{DF} = 4 would require a pin diameter of 29 mm (1.13 in). These conclusions apply to 1040 cold-drawn steel and other ductile steels with a 12% or higher elongation in 51 mm (2 in).

$$\tau := \frac{\mathbf{P}}{2\cdot \mathbf{A}} \leq \frac{\mathbf{S_a}}{\mathbf{N_{DF}}} = \frac{0.22\cdot \mathbf{S_U}}{\mathbf{N_{DF}}} \quad \textbf{(13)}$$

$$\mathbf{D} \geq 1.7\cdot \sqrt{\frac{\mathbf{N_{DF}}\cdot \mathbf{P}}{\mathbf{S_U}}} \quad \textbf{(14)}$$

For equations (13) and (14) to be valid for any diameter other than 12.7 mm (0.500 in), C << N and N/D = 1. Regarding Figure 6, the x-axis value of P/2A changes with the diameter of the pin since $A = \pi D^2/4$. The y-axis value of σ also changes with the diameter of the pin. Using equations (9) and (10), the bending stress can be shown to equal equation (15). In this study, the clearance gap (C) was small compared to the width of the clevis (N) and N/D = 1. Therefore, the slope of the curves depicted on Figure 6 will remain constant at a value of approximately 4 for different pin diameters.

$$\sigma = \frac{\mathbf{P}}{2\cdot \mathbf{A}}\cdot \left[\frac{4\cdot(\mathbf{N}+\mathbf{C})}{\mathbf{D}}\right] \quad \textbf{(15)}$$

This study of a pin and clevis assembly dealt only with static loads. If the effects of impact loading, thermal friction, and torque were included, the conclusions reached may not fully hold true. In addition, nonlinear finite element analysis using nonlinear contact elements instead of linear gap elements may provide a more accurate representation of the actual condition.

BIBLIOGRAPHY

1. Blodgett, O.W., 1982, "Design of Welded Structures", The Cleveland. James F. Lincoln Arc Welding Foundation, Cleveland, OH, section 2.1-4.
2. Greene, W.H., Knight Jr., N.F., and Stockwell, A.E., 1986, "Structural Behavior of the Space Shuttle SRM Tang-Clevis Joint", National Aeronautics and Space Administration, Washington, D.C.
3. Shields, E.B., 1996, Machine Design notes without references, Youngstown State University, Youngstown, OH.
4. Mott, R.L., 1999, *Machine Elements in Mechanical Design*, Prentice Hall, Upper Saddle River, NJ.
5. Vishay Measurements Group, 1997, "Student Manual for Strain Gage Technology", Bulletin 309D, Raleigh, NC.
6. Budynas, R.G. 1999, *Advanced Strength and Applied Stress Analysis*, WCB/McGraw-Hill, p.103.
7. Spotts, M.F., and Shoup, T.E., 1998, *Design of Machine Elements*, Prentice-Hall, Inc., Upper Saddle River, NJ, p. 634.

PVP-Vol. 458, Computer Technology and Applications

PVP2003-1893

MODELLING STUDIES OF THE FATIGUE BEHAVIOUR OF A NOZZLE-TO-VESSEL INTERSECTION

Abílio M.P. De Jesus
University of Trás-os-Montes and Alto Douro
Quinta de Prados, 5000-911 Vila Real, Portugal
Email: ajesus@utad.pt

Alfredo S. Ribeiro
University of Trás-os-Montes and Alto Douro
Quinta de Prados, 5000-911 Vila Real, Portugal
Email: aribeiro@utad.pt

António A. Fernandes
Faculty of Engineering, University of Porto
Rua Dr.Roberto Frias, 4200-465 Porto, Portugal
Email: aaf@fe.up.pt

ABSTRACT

In this paper the fatigue behaviour of a nozzle-to-vessel intersection is studied. This geometry is typical in pressure vessels. Numerical models are investigated in order to predict the fatigue life. These studies are validated by an experimental program including a fatigue test of a real size pressure vessel, fatigue tests of structural details and fatigue tests of small scale and smooth specimens. Comparisons between predictions obtained using proposed models and predictions obtained using procedures included in design codes are carried out.

NOMENCLATURE

d – Medium diameter of the cylindrical shell of a pressure vessel
da/dN – Crack growth rate
E – Young modulus
K' – Cyclic strength coefficient
K_{max}, K_i - Maximum and actual stress intensity factors
K_t – Elastic stress concentration factor
K_σ – Elasto-plastic stress concentration factor
K_ε – Elasto-plastic strain concentration factor
LPG – Liquid Petroleum Gas
N – Number of cycles
N_f – Number of cycles to failure
n' – Cyclic strain hardening exponent
R – Stress ratio
R_ε – Strain ratio
R^2 – Coefficient of determination
t – Wall thickness of the vessel
$2N_f$ – Number of reversals to failure
ΔK – Stress intensity factor range
Δp – Pressure range
$\Delta\varepsilon$ – Strain range
$\Delta\varepsilon^e$ – Elastic strain range
$\Delta\varepsilon_{nom}$ – Nominal strain range
$\Delta\varepsilon^p$ – Plastic strain range
$\Delta\sigma$ – Stress range
$\Delta\sigma_{hoop}$ – Nominal hoop stress acting in cylindrical shell of the pressure vessel
$\Delta\sigma_{long}$ – Nominal longitudinal stress acting in cylindrical shell of the pressure vessel
$\Delta\sigma_{nom}$ – Nominal stress range
δ_{max}, δ_i – Maximum and actual crack point increments
σ_{max} – Maximum stress

INTRODUCTION

The fatigue behaviour of a nozzle-to-vessel intersection is studied. This detail is located in a horizontal pressure vessel which is fatigue tested until failure. This pressure vessel is made of steel plate P355NL1, according to EN 10028-3 standard [1], has a volume of 1000 litres and is used to store LPG. The failure of the pressure vessel is due to a fatigue crack initiated at the crotch corner of the nozzle which propagated until leakage occurred. In order to model the fatigue behaviour of the nozzle-to-vessel intersection, several modelling strategies are investigated, namely the nominal stress approach to model the total failure (leak before break), the local strain approach to model the crack initiation and the linear elastic fracture mechanics to model the crack propagation until critical dimensions are achieved. Experimental fatigue data such as S-N curves based on nominal stresses is obtained using specimens representative of the structural detail. Also ε-N curves are derived in order to evaluate the application of local strain approach. Finally, crack propagation data is obtained in order to evaluate the application of the linear elastic fracture mechanics.

Linear elastic and non-linear elasto-plastic finite element calculations are performed in order to derive notch stresses and local strains, respectively. Finally the predictions made with the above models are compared with the experimental results and with the predictions obtained using the procedures proposed by ASME VIII – Division 2 [2] and the recently approved European standard EN 13445 [3].

Numerical modelling of pressure equipment details has been addressed by some authors [4,5], but only few workers [6,7] have used an experimentally validated approach, as reported in the present paper.

EXPERIMENTAL DETAILS

The current study involves three levels of fatigue tests. The first level consists of a fatigue test of a real size pressure vessel, including two nozzle-to-vessel intersections, as illustrated in Figure 1. The second level of tests consists of fatigue tests of a structural detail, representative of the nozzle-to-vessel intersection included in the vessel, namely a nozzle-to-plate intersection. Its dimensions and geometry are illustrated in Figure 2. Finally, small-scale specimens are tested in order to determine basic cyclic fatigue properties of the material.

Fatigue test of a real size pressure vessel

A horizontal pressure vessel, made of steel plate P355NL1 (EN10028-3) is tested under fluctuating internal pressure. Table 1 summarizes the main characteristics of the pressure vessel, obtained from the manufacturer. The pressure vessel is tested using water as hydraulic fluid. The variation of internal pressure is set between the minimum of 1 Bar and the maximum of 35 Bar. The failure is observed at the vertical nozzle-to-vessel intersection as pointed out on Figure 1. A fatigue crack initiated at the crotch corner of the nozzle, propagated towards the outer surface leading to a fluid leakage. The number of cycles observed at failure by leakage is **12 455** cycles.

Fatigue tests of structural components corresponding to a nozzle-to-plate connection

A structural component, corresponding to a nozzle-to-plate connection, represented in Figure 2, is tested in a servo hydraulic machine under load control and with an approximately zero load ratio (R-ratio). Fifteen specimens are tested for several nominal stress ranges and results are summarized in the graph represented in Figure 3. This graph includes the experimental data points, the mean S-N curve (50% failure probability) and confidence bands corresponding to mean S-N curve plus/minus one, two and three standard deviations. All failures are due to a half-penny shaped crack initiated at the crotch corner (corner crack). This crack extends through the nozzle and plate thickness and the test is terminated as soon as the crack becomes a part-through crack, corresponding to a leak before break failure criterion.

The mean S-N curve can be described through the following equation:

$$\Delta\sigma^{3.203} N_f = 2.679E + 12 \qquad (1)$$

where $\Delta\sigma$ is the nominal stress range, in MPa, and N_f is the number of cycles to failure. The previous equation is obtained by linear regression analysis of the experimental results, with a coefficient of determination, R^2, of about 0.864.

The lower confidence band resulting from the mean S-N curve minus two standard deviations corresponds to a probability of failure of 2.3 %, while the lower confidence band, resulting from the mean S-N curve minus one standard deviation, corresponds to a probability of failure of 15.9 %. Finally, a lower confidence band, resulting from S-N curve minus three standard deviations, corresponds to a probability of failure of about 0.5 % [3,8]. Equations (2), (3) and (4), correspond to the lower confidence bands with probability of failure of 0.5, 2.3 and 15.9 % respectively:

$$\Delta\sigma^{3.203} N_f = 6.9108E + 11 \qquad (2)$$

$$\Delta\sigma^{3.203} N_f = 1.085E + 12 \qquad (3)$$

$$\Delta\sigma^{3.203} N_f = 1.702E + 12 \qquad (4)$$

Fatigue tests of small scale specimens

Small scale specimens of steel P355NL1 are also tested in order to derive strain-life data, stress-life data and crack propagation data.

Two series of smooth specimens are tested under strain control, according to ASTM E606 standard [9]. In the first series, 15 specimens are tested with a strain ratio of -1. In the second series, 20 specimens are tested with a strain ratio equal to 0. The results are presented in Figure 4. It can be observed that the two curves are almost coincident, which means that this material is not sensitive to the strain ratio. A mean stress relaxation is observed for fatigue tests with null strain ratio, leading to test conditions, in terms of stress states, similar to those observed in the series with strain ratio equal to -1. This phenomenon of cyclic stress relaxation, typical on carbon and low alloy steels [10], justifies the little difference between the curves presented in Figure 4.

Using the Coffin-Manson relation [11,12] the strain-life equations are as follows:

$$\frac{\Delta\varepsilon}{2} = \frac{971E6}{200E9}\left(2N_f\right)^{-0.0965} + 0.2872\left(2N_f\right)^{-0.5302}, \; R_\varepsilon = 0 \qquad (5)$$

$$\frac{\Delta\varepsilon}{2} = \frac{1003E6}{200E9}\left(2N_f\right)^{-0.1086} + 0.4218\left(2N_f\right)^{-0.5543}, \; R_\varepsilon = -1 \qquad (6)$$

Since both relations are almost coincident it is derived a single relation that averages the results:

$$\frac{\Delta\varepsilon}{2} = \frac{992E6}{200E9}\left(2N_f\right)^{-0.0990} + 0.3784\left(2N_f\right)^{-0.5511} \qquad (7)$$

The strain-life equations fairly describe the fatigue behaviour of components with high stress concentrations, even if these components are tested under remote load control. The deformation in those highly stressed zones is controlled by the surrounding elastic material. The strain ratio expected for the highly stressed zones depends on the extension of elastic zone but, for a remote null stress ratio, a null or positive strain ratio is expected. The strain-life data obtained can be extended to other strain ratios due to the observed low dependency with the strain ratio.

In order to evaluate the mean stress effect on this steel, fatigue tests under stress control, are performed. Small and smooth specimens are used. Three stress ratios are investigated, namely, $R=-1$, $R=-0.5$ and $R=0$. Results are presented in Figure 5, in terms of maximum stress *versus* number of cycles to rupture. The analysis of stress-strain test records reveals the existence of some ratchetting for higher stresses, manly for stress ratios equal to $R=-0.5$ and $R=0$. Some fatigue tests with null strain ratio are characterised by an elastic shakedown, after an initial ratchetting. Equations (8), (9) and (10) represent the S-N curves for stress ratios $R=-1$, $R=-0.5$ and $R=0$, respectively:

$$\sigma_{\max}^{14.389} N_f = 1.959E+41 \tag{8}$$

$$\sigma_{\max}^{22.573} N_f = 5.704E+63 \tag{9}$$

$$\sigma_{\max}^{16.892} N_f = 1.691E+50 \tag{10}$$

Equations (8), (9) and (10) are obtained by linear regression analysis, with coefficients of determination, R^2, equal to 0.9968, 0.9837 and 0.8915, respectively. The description of these S-N curves through straight lines is only a first approximation. An improved description of these S-N curves can be carried out using three line segments or using high order curves. These alternatives allow, for example, the description of the convergence of S-N curves to the ultimate tensile strength, for the first half cycle [17].

Crack growth rates, for the base material, steel P355NL1, are experimentally determined. Five compact tension specimens are tested according to the ASTM E647 standard [13], with three different stress ratios, namely, $R=0.01$, $R=0.5$ and $R=0.7$. Figure 6 illustrates the crack growth rates correlation with the stress intensity factor range. For the stress intensity ranges considered in the tests (~250-1500 $N.mm^{-1.5}$), the influence of stress ratio is small [10] and so a unique relation relating the crack growth rate with the stress intensity factor range is established. The linear relation between the crack growth rate and the stress intensity rage, proposed by Paris [14], seems to be a good fit, leading to the following relation:

$$\frac{da}{dN} = 1.3092E-14\Delta K^{3.4243} \tag{11}$$

with da/dN expressed in *mm/cycle* and ΔK in $N.mm^{-1.5}$. The previous equation is obtained using a linear regression analysis characterized by a coefficient of determination, R^2, equal to 0.9843.

TWO-PHASE MODEL FOR FATIGUE LIFE ESTIMATION

A model based on the partial estimations of the initiation and propagation of a crack is used and applied to predict the total fatigue life. The initiation phase corresponds to the initiation of a macroscopic crack (a crack with 0.5 mm depth). The propagation phase corresponds to the growth of a macroscopic crack until it achieves critical dimensions, leading to leak before break. The prediction of the initiation phase is usually based on strain-life relations obtained with tests of smooth specimens under strain control, equations (5-7). The propagation phase can be modelled using the linear elastic fracture mechanics.

Modelling the crack initiation phase

The Coffin-Manson equations (5-7) are used to estimate the number of cycles to crack initiation assuming that the total local elasto-plastic strain is known.

Usually the nominal stresses are known. In order to evaluate the local elasto-plastic strains, the rules proposed by Neuber [15] and Ramberg-Osgood [9,16] are used. The general form of Neuber's equation is as follows:

$$K_t^2 = K_\sigma K_\varepsilon = \frac{\Delta\sigma}{\Delta\sigma_{nom}} \frac{\Delta\varepsilon}{\Delta\varepsilon_{nom}} \tag{12}$$

where K_t is the stress concentration factor, $\Delta\sigma_{nom}$ and $\Delta\varepsilon_{nom}$ are the nominal stress and strain range and $\Delta\sigma$ and $\Delta\varepsilon$ are the local stress and strain range.

The Ramberg-Osgood relation expresses the cyclic stress-strain curve of the material and has the following form:

$$\frac{\Delta\varepsilon}{2} = \frac{\Delta\varepsilon^e}{2} + \frac{\Delta\varepsilon^p}{2} = \frac{\Delta\sigma}{2E} + \left(\frac{\sigma}{2K'}\right)^{\frac{1}{n'}} \tag{13}$$

where K' and n' are respectively the cyclic strength coefficient and the strain hardening exponent.

Another method to evaluate the local strains is by means of a continuum plasticity model in conjunction with a non-linear finite element code. In the work reported, a plasticity model, based on von Mises yield criterion is used. In order to describe the stabilised cyclic behaviour of the steel, a non-linear kinematic hardening with saturation of the yield surface is considered [17]. The analysis is performed using the ANSYS program [18].

Table 2 summarizes the properties required to model the material behaviour according to Ramberg-Osgood relation and the continuum plasticity model [19].

Modelling the crack propagation phase

The crack propagation is simulated using the software FRANC3D [20]. The software FRANC3D is a fracture analysis code for simulating arbitrary non-planar 3D crack growth. This software is based on the boundary element method. An initial crack can be propagated automatically in an incremental way. For each crack increment, stress intensity factor values, at a specified set of points of the crack front, are evaluated. The maximum crack extension, δ_{max}, is previously defined (constant) and is assigned to the crack front point with the highest stress intensity factor, K_{max}. The remaining crack front points will suffer an increment, δ_i, proportional to their stress intensity factor values, K_i ($\delta_i = K_i.\delta_{max}/K_{max}$). The directions of these increments are calculated using the maximum shear stress theory.

The stress intensity history is recorded during the crack propagation for subsequent fatigue life prediction. This software allows several crack growth models (e.g. Forman, Paris, etc). The Paris model is used in the current analysis.

In the current work an initial quarter-penny crack with a radius of 0.5 mm is assumed at the crotch corner of the nozzles. A constant maximum increment of 0.5 mm is also considered. The simulation of crack growth is interrupted as soon as the crack front achieved the opposite surface, leading to a part-through crack. Using the stress intensity history generated by the FRANC3D the number of cycles is calculated by integrating the Paris law of the material, equation (11).

APPLICATION OF A TWO-PHASE MODEL FOR FATIGUE LIFE PREDICTION OF THE NOZZLE-TO-PLATE INTERSECTION

The model described previously, based on separate predictions for the crack initiation and propagation phases, are used to generate alternative S-N curves for the nozzle-to-plate intersection. The strain-life relation (7) is used to obtain the number of cycles to crack initiation. The elasto-plastic strains are determined using the Neuber and Ramberg-Osgood equations, (12) and (13). Alternatively, the elasto-plastic analysis is carried out using an inelastic rate independent plasticity model, based on von Mises yield criteria with a non-linear kinematic hardening. In both cases a finite element model of the structural detail is generated, using ANSYS 6.0 software. In the first case the finite element model is used to evaluate the stress concentration factor, K_t. In the second case it is applied to perform a full elasto-plastic analysis in order to evaluate directly the local strain state. Figure 7 illustrates the finite element mesh used in the analysis of the nozzle-to-plate intersection. A stress concentration factor, K_t, equal to 3.8 is assumed. The finite element model is constituted of 20-node structural solid (brick) elements. A total of 4160 elements and 20010 nodes are used. Only ¼ of full geometry is analysed, taking advantage of existing symmetries. A uniform tension is applied to the plate.

Figure 8 represents the S-N curves obtained for the detail using the Neuber and Ramberg-Osgood relations. Figure 9 represents the S-N curves obtained using the continuum plasticity model. The difference between these two graphs is due to different predictions of the number of cycles to crack initiation. The Neuber and Ramberg-Osgood rules lead to a lower number of cycles to crack initiation. Thus, S-N curves resulting from the application of these rules are more conservative. In both cases the number of cycles obtained for the propagation phase is very small when compared with the number of cycles obtained for the initiation phase. The S-N curve obtained with elasto-plastic strains derived from the continuum plasticity model is non-conservative, for lives above $5\text{x}10^5$ cycles. Below this number of cycles this S-N curve is appropriate since it allows safe predictions. However, this second S-N curve requires greater computation effort.

APPLICATION OF DESIGN CODE PROCEDURES IN THE FATIGUE ASSESSMENT OF THE NOZZLE-TO-PLATE INTERSECTION

In this work a comparison between S-N curves obtained experimentally and S-N curves derived using design codes for pressure vessels is carried out for the nozzle-to-plate intersection. The design codes considered are the ASME Boiler and Pressure Vessel Code, namely its section VIII – Division 2 [2], and the recently approved European pressure vessel design code, EN 13445 [3]. The graph represented in Figure 10 illustrates the experimental S-N curves as well as the S-N curves obtained using the design codes. It can be concluded that both design codes are very conservative. The ASME code is more conservative than the European code. The European code takes into account the surface finishing, through the peak-to-valley height, *Rz*. For high number of cycles (about $1\text{x}10^6$) the S-N curves obtained using the European code are close to the lower band of the experimental S-N curve minus two standard deviations. Figure 10 also includes the S-N curve obtained using the two-phase model with elasto-plastic finite element calculations. It is observed that the predicted curve tends to the ASME curve for low cycle fatigue domain. For high cycle fatigue domains this curve is non conservative, in opposition to the code S-N curves.

PREDICTION OF THE NUMBER OF CYCLES TO FAILURE OF THE PRESSURE VESSEL

Several methodologies previously presented are applied in the evaluation of the number of cycles to failure of the pressure vessel nozzle. The model based on strain-life data for prediction of a crack initiation phase is applied. The linear elastic fracture mechanics is also applied to model the propagation of an initial quarter-penny shaped flaw with 0.5 mm of radius. A prediction using total elasto-plastic notch stresses is used based on equations (8), (9) and (10). The ASME VIII–Division 2 and the EN 13445 design codes are also used to obtain the number of cycles to failure of the tested vessel. Finally, equations (1) and (2) representing, respectively, the S-N curves with 50 and 0.5 % probabilities of failure, for the nozzle-to-plate connection, are used. The nominal hoop and longitudinal stresses acting in the cylindrical shell of the vessel, in area where the nozzle connection is located, can be calculated by the equations:

$$\begin{cases} \Delta\sigma_{hoop} = \dfrac{\Delta p \cdot d}{2t} = 332 \text{ MPa} \\ \Delta\sigma_{long} = \dfrac{\Delta p \cdot d}{4t} = 165.82 \text{ MPa} \end{cases} \quad (14)$$

Performing a linear elastic analysis of the nozzle-to-vessel intersection, using the finite element method, a stress concentration factor of 2.4 at the critical location (crotch corner) is calculated. This value is slightly lower than the value obtained using Decock equation [21] that gives a K_t equal to 2.8. The Decock equation does not account for the presence of the weld fillet. The stress concentration is applied to the nominal hoop stress in the shell to determine the maximum stress acting in the crotch corner in the circumferential direction, with respect to the nozzle. Figure 11 illustrates the finite element mesh used in the analysis of the nozzle-to-vessel intersection. This model uses 20-node structural solid (brick) elements. The shell thickness is modelled with one solid element, for remote zones; three solid elements are used to model shell thickness for zones near intersection. The model is composed by a total of 1940 elements and 10624 nodes. Internal pressure acting in the cylindrical shell and nozzle is considered. Symmetry conditions and longitudinal pressure, acting at end of cylindrical shell, are applied.

Non-linear analyses are also carried out in order to derive notch strain and stresses at the crotch corner of the nozzle-to-vessel intersection.

Figure 12 presents the results of the predictions of the pressure vessel fatigue life, using several methodologies. The diagram also indicates, for comparison, the number of cycles to failure observed in the experimental test of a real size vessel. From the analysis of the graph it can be seen that the prediction based on notch stresses overestimates the number of cycles to failure. This can be explained by the fact that for fatigue tests under stress control, the surface finishing is an important parameter, influencing the process of initiating a crack. The stress-life data is obtained using smooth specimens with a very good surface finishing, which does not correspond to surface finishing obtained with the current industrial practice of manufacturing the pressure vessel. The fatigue test of the vessel induced important inelastic strains at the crotch corner of the nozzle-to-vessel intersection. Thus the notch stress analysis is not recommended. It is preferable to use strain-life data since the estimation of inelastic stresses is more prone to error than the estimation of inelastic strains, due to the very low tangent elasto-plastic modulus.

The code based predictions are conservative as expected. The ASME code gives more conservative results. This prediction has the same order of magnitude of the crack propagation prediction. The model that takes into account the crack initiation, using strain-life data, and crack propagation, using linear elastic fracture mechanics, gives the same prediction as the experimental model based on the S-N curve with 0.5 % probability of failure and is close to the prediction obtained using the EN 13445 code. The S-N curve with 2.3% probability of failure gives the best prediction in the side of conservative ones. The S-N curve corresponding to 50% probability of failure makes a very unsafe prediction.

CONCLUSIONS

Several methodologies for fatigue assessment of a pressure vessel are investigated and compared. The required fatigue properties are obtained using an experimental work involving three test levels, including the test of a real size pressure vessel. The well established ASME VIII –Division two design code and the recently approved European design code are also compared. This comparison demonstrates that the American code is more conservative than the European one.

ACKNOWLEDGEMENTS

The authors whish to acknowledge the support of the company A.S. Matos, Lda, to this project, on providing the tested pressure vessel and specimens. The authors also wish to acknowledge the Cornell Fracture Group for making available the software FRANC3D for fracture analysis.

REFERENCES

[1] EN 10028-3, 1996, "EN 10028-3 Steel plates for pressure vessel equipment – Part 3: Weldable normalized fine grain structural steels", European Standard (published in French).

[2] ASME VIII, 1998, "ASME Boiler and Pressure Vessel Code, Section VIII - Rules for construction of Pressures Vessels, Division 2, Alternative rules".

[3] European Standard. Final Draft. PrEN13445-3, March 2002: Unfired Pressure Vessels – Part 3: Design. CEN - European Committee for Standardization.

[4] Nigel Taylor at al, 1999, "The Design by Analysis Manual", European Commission, Joint Research Centre – Patten, Report EUR 19020EN.

[5] Weiβ, E., Rudolph, J., 1995, "Finite element analyses concerning the fatigue strength of nozzle-to-spherical shell intersections", International Journal of Pressure Vessel & Piping, **64**, pp. 101-109.

[6] Zengliang, G., Liangfeng, X., and Kangda, Z., 1990, "Fatigue crack growth in the nozzle corner of a pressure vessel", International Journal of Pressure Vessel & Piping, **42**, pp. 1-13.

[7] Giglio, M., Vergani, L., 1995, "Life prediction of pressure vessel nozzles", International Journal of Pressure Vessel & Piping, **63**, pp. 199-204.

[8] Maddox, S.J., 1991, "Fatigue Strength of Welded Structures", Abington Publishing, Second Edition.

[9] ASTM E606-92, 1998, "Standard practice for strain-controlled fatigue testing", In Annual Book of ASTM Standards, Part 10, American Society for Testing and Materials, pp. 557-571.

[10] Ellyin, F., 1997, "Fatigue Damage Crack Growth and Life Prediction", Chapman & Hall, London.

[11] Coffin, L.F., 1969, "A study of the effect of cyclic thermal stresses on a ductile metal", Translations of the ASME, **76**, pp. 931-950.

[12] Manson, S.S., 1953, "Behavior of Materials under conditions of thermal stress", NACA Report 1170.

[13] ASTM E647-93,1993, "Standard test method for measurement of fatigue crack growth rates", In Annual Book of ASTM Standards, Part 10, American Society for Testing and Materials, pp. 569-596.

[14] Paris, P. C., Gomez, M. P. and Anderson, W. E., 1961, "A rational analytic theory of fatigue", The trend in Engineering, **13(1)**, pp. 9-14.

[15] Neuber, H., 1961, "Theory of stress concentrations for shear strained prismatic bodies with arbitrary non-linear stress-strain law", J. Applied Mechanics, **28**, pp. 544-551.

[16] Miller, J. A., 1946, "Stress-strain and elongation graphs for aluminium alloy R301 sheet", NACA Technical Note nº 1010.

[17] Lemaitre, J., Chaboche, J. L., 1990, "Mechanics of Solid Materials", Cambridge University Press.

[18] ANSYS, Swanson Analysis Systems, Inc., Houston, Version 6.0, October 2001, Run on Pentium IV/Windows XP.

[19] De Jesus, A. M. P., Ribeiro, A. S., Costa J.D., Fernandes, A.A., 2002, "Low cycle fatigue and cyclic elasto-plastic behaviuor of P355NL1 steel", Proceedings of Portuguese Fracture Conference, UTAD, Vila Real.

[20] FRANC3D – 3D Fracture Analysis Code, Cornell Fracture Group, Version 1.15, May 2001, Run on Pentium IV/Windows XP (www.cfg.cornell.edu/software/FRANC3D.html).

[21] Decock, J., 1973, "Determination of stress concentration factors and fatigue assessment of flush and extruded nozzles in welded pressure vessels", 2nd International Conference on Pressure Vessel Technology, Part II, ASME, San Antonio, Texas, Paper II-59, pp 821-834.

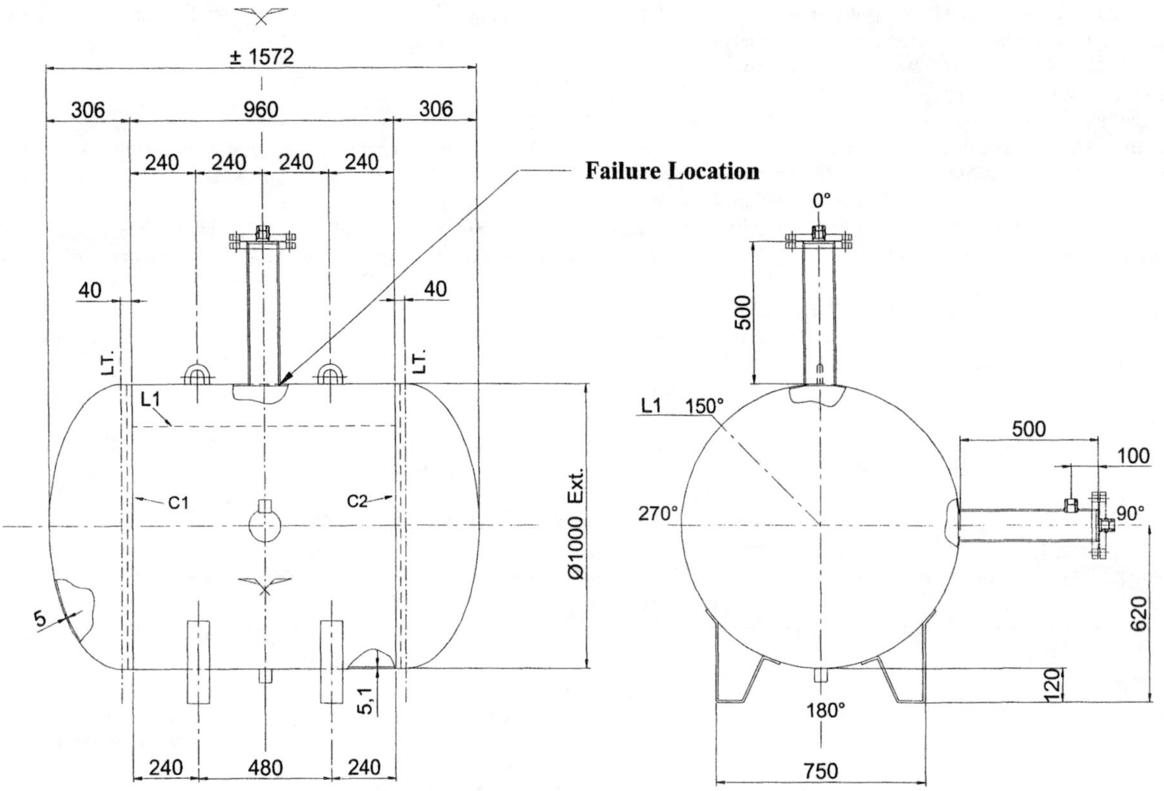

Figure 1. Horizontal pressure vessel tested under fluctuating internal pressure (dimensions in mm).

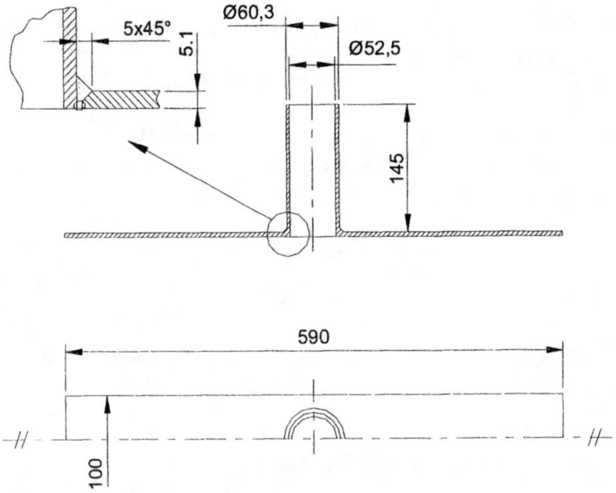

Figure 2. Structural detail corresponding to a nozzle-to--plate connection (dimensions in mm).

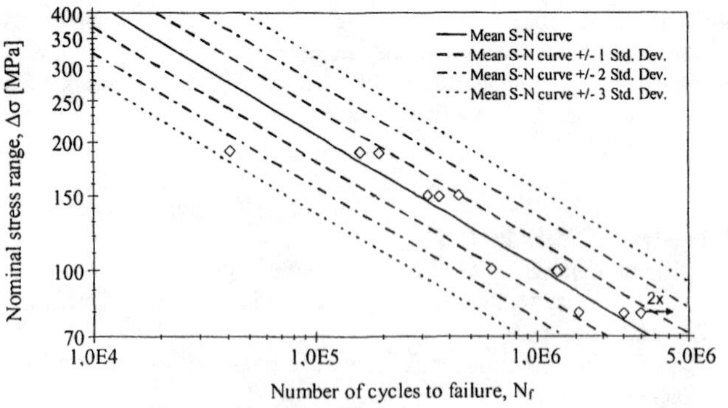

Figure 3. S-N curves obtained for detail corresponding to the plate-to-nozzle intersection.

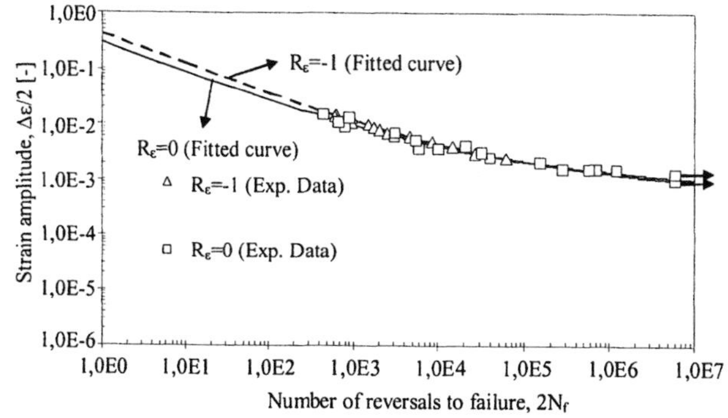

Figure 4. Strain controlled fatigue data: total strain amplitude *versus* reversals to failure of the P355NL1 steel.

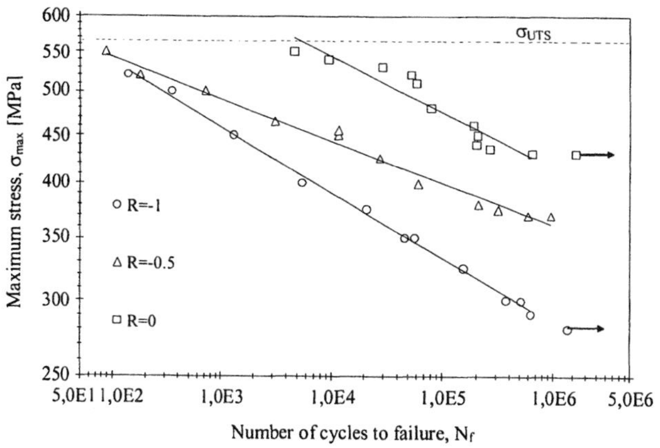

Figure 5. Stress controlled fatigue data: maximum stress *versus* number of cycles to failure of the P355NL1 steel.

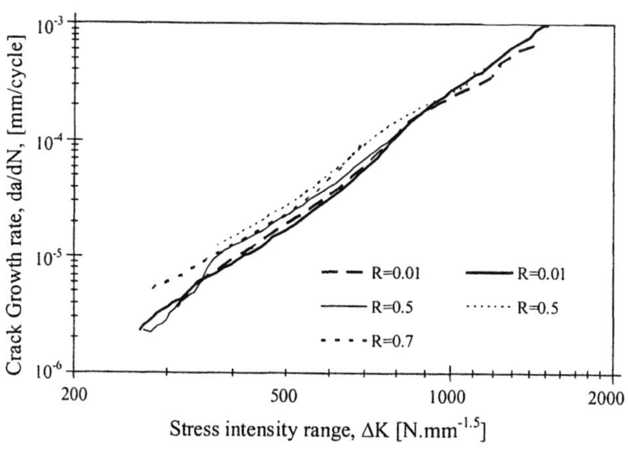

Figure 6. Experimental data for fatigue crack growth rate of the steel P355NL1 .

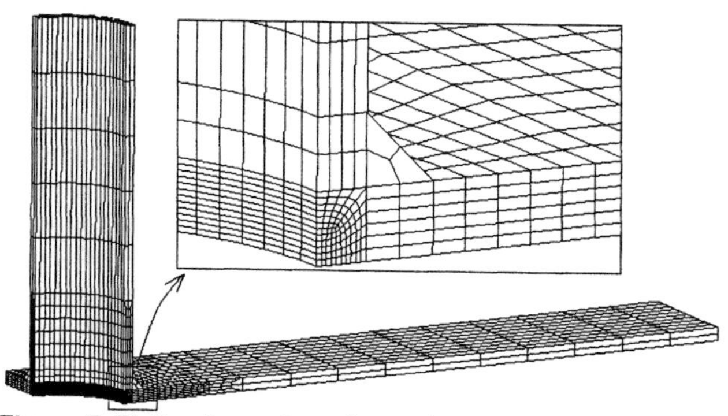

Figure 7. Finite element mesh used to model the nozzle-to--plate intersection.

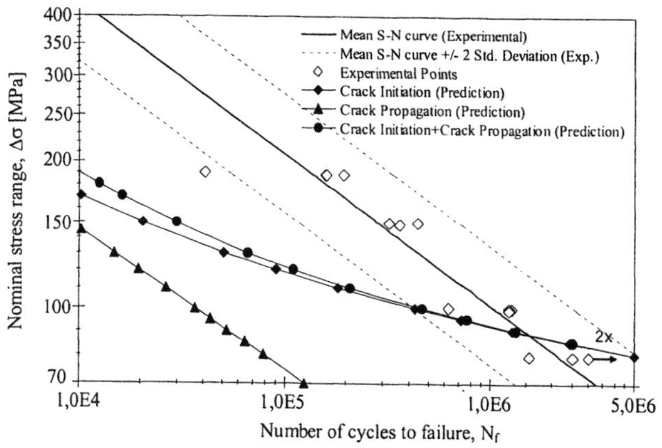

Figure 8. S-N curves for the Nozzle-to-plate intersection obtained using the Neuber and Ramberg-Osgood rule.

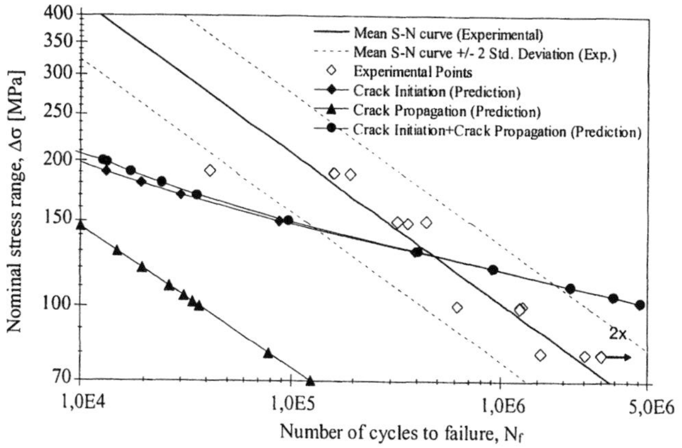

Figure 9. S-N curves for the Nozzle-to-plate intersection obtained using a continuum plasticity model.

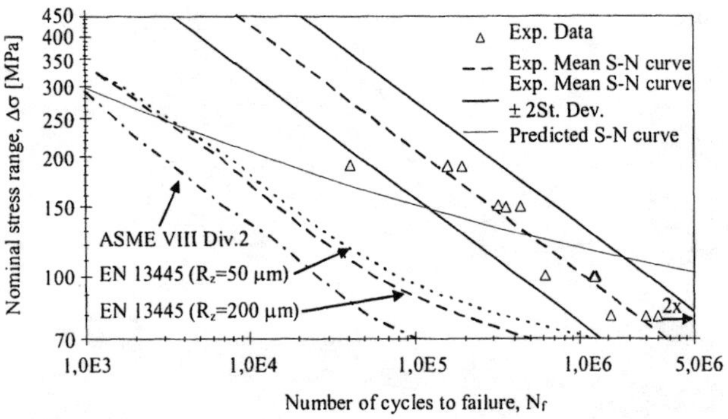

Figure 10. Comparison between S-N curves obtained experimentally and by applying the design codes.

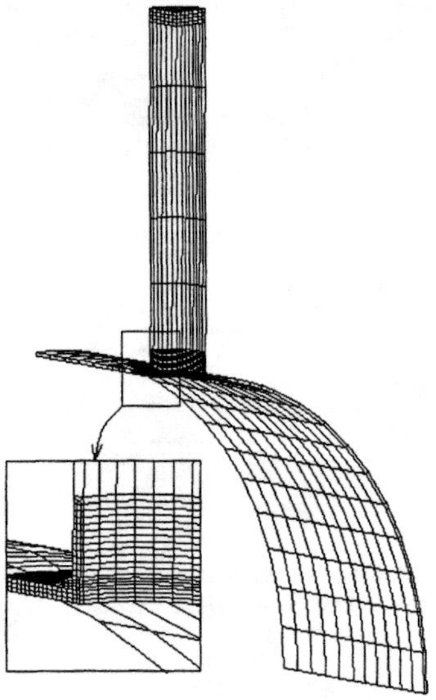

Figure 11. Finite element mesh used in the analysis of the nozzle-to-vessel connection.

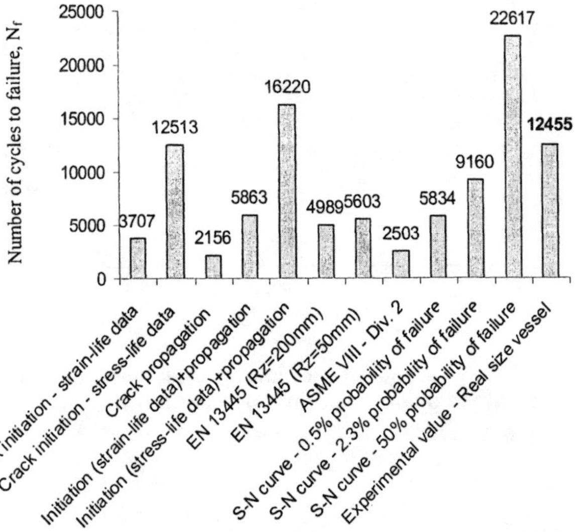

Figure 12. Predictions of the number of cycles to failure of the nozzle-to-vessel intersection using several methods.

Table 1. Main characteristics of the tested pressure vessel.

PROJECT CHARACTERISTICS	
Construction code	CODAP 95 - B
Service Conditions	
Service pressure	17.5 Bar
Service temperature	-20 to +50 °C
Maximum Empty weight	±320 kgf
Contained fluid	G.P.L.
Capacity	1.1 m^3
Calculation conditions	
Design pressure	17.64 Bar
Design temperature	-20 to +50°C
Corrosion allowance	0 mm
Joint efficiency	0.85
Safety factor	2.4
Control conditions	
NDT - RX	100% all crossings
Hydraulic test pressure	26.64 Bar (Water)
Coupons test plates	According to code
Dye penetrating liquids	100% Nozzles

Table 2. Elasto-plastic properties of the steel P355NL1.

Strength properties	
Tensile strength, σ_{UTS} (MPa)	568
Monotonic yield strength, σ_Y (MPa)	418
Elastic Properties	
Young modulus, E (MPa)	200000
Poisson ratio	0.27
Constants for Ramberg-Osgood's relation	
Cyclic strength coefficient, K' (MPa)	479.84
Cyclic hardening exponent, n'	0.195
Constants for Continuum Plasticity model	
Linear kinematic hardening coefficient, C (MPa)	118300
Non-linear kinematic hardening coefficient, γ	350
Cyclic yield stress for continuum plastic model, σ'_Y (MPa)	200

PVP-Vol. 458, Computer Technology and Applications

PVP2003-1894

A FINITE ELEMENT STUDY OF FRICTION EFFECT IN FOUR-POINT BENDING CREEP TEST

Young Suk Kim and Don R. Metzger
Department of Mechanical Engineering
McMaster University

ABSTRACT

Creep tests are often performed in four-point bending and the stress distribution in the bending specimen is nonlinear, so creep properties are estimated from bend creep test data. However, getting creep properties from bending creep tests is often doubted because of uncertainties from contact point shift and frictional effect between loading pin and specimen in four-point bending creep tests. Finite element simulations of the four-point bending creep tests were performed with geometric models which include contact conditions. It was found from simulation studies that friction in the bend tests can cause error in the estimation of creep properties, but when no friction was applied in simulations the creep properties were well predicted from bend test data.

INTRODUCTION

The most common method of creep testing involves applying a constant load directly in tension or compression to a specimen. For many materials including ceramics, tensile specimens and fixturing required to grip the specimens are expensive, and premature failures can occur near the grip. Therefore four-point bending creep tests are often used because the specimens are inexpensive rectangular bars, and the loading fixtures are simple and stable. However, the stress distribution in the bending specimen is nonlinear, so a proper interpretation method is needed to extract the creep parameters from the bend test data.

Hollenberg et al. [1] developed a method to get creep parameters from bend test data where the creep rates in tension and compression are the same (symmetric creep). This method is easy to apply because it has a closed-formed solution, but is restricted to materials which have symmetric creep properties. Chuang [2] generalized the method for creep properties where the creep rate in tension is different from in compression (asymmetric creep). In Chuang's method the equations are highly nonlinear, so an iterative solution scheme is required.

The method of Hollenberg et al. [1] (hereafter called Hollenberg's method) assumes that the neutral axis is at the center of section (symmetric creep) during creep bending, and at steady state the material follows the power-law equation in a form

$$\dot{\varepsilon}_c = A\sigma^n \tag{1}$$

where $\dot{\varepsilon}_c$ is the steady state creep strain rate, A and n are creep parameters. The strain rate is linearly dependant on Y (the distance away from the neutral axis) so that

$$\varepsilon_c = \frac{Y}{\rho} = KY \text{ and } \dot{\varepsilon}_c = \dot{K}Y \tag{2}$$

where ρ is radius of curvature and $\dot{K}$ is curvature rate. Upon combining Equation (1) and Equation (2), the stress distribution over the cross-section is

$$\sigma(Y) = \left(\frac{Y\dot{K}}{A}\right)^{1/n} \tag{3}$$

Moment equilibrium requires that the total bending moment from the stress over the cross-section be equal to the applied bending moment. After some mathematical manipulations, moment equilibrium becomes

$$\dot{K} = \left(\frac{6M}{BH^2}\right)^n \frac{2A}{H}\left[\frac{(2n+1)}{3n}\right]^n \tag{4}$$

where B, H and M are beam width, height and applied moment respectively (see Figure 1). The creep parameters (A and n) can be evaluated by fitting a line to the curvature rate

($\dot{K}$) vs. the applied moment (M) on a log-log plot.

Some materials show higher creep rate in tension than in compression (asymmetric creep). Significant errors can occur if the Hollenberg's method is applied to asymmetric creep cases because the neutral axis migrates towards the compression side of the specimen. To account for this, Chuang [2] developed a generalized method with the following two equations which describe the asymmetric creep behaviors

$$\dot{\varepsilon}_c = A_c \sigma^{n_c} \qquad \text{compression} \qquad (5a)$$

$$\dot{\varepsilon}_c = A_t \sigma^{n_t} \qquad \text{tension} \qquad (5b)$$

The subscripts c and t refer to compression and tension respectively. Equation (3) still applies and two stress equations are obtained for compression and tension. The force equilibrium requires that the force acting on the compression side of the cross-section be equal to the force on the tension side.

$$B\int_0^{H_c} \sigma dY = B\int_0^{H_t} \sigma dY \qquad (6)$$

where $H_c + H_t = H$. Moment equilibrium requires that

$$M = \int_0^{H_c} \sigma YBdY + \int_0^{H_t} \sigma YBdY \qquad (7)$$

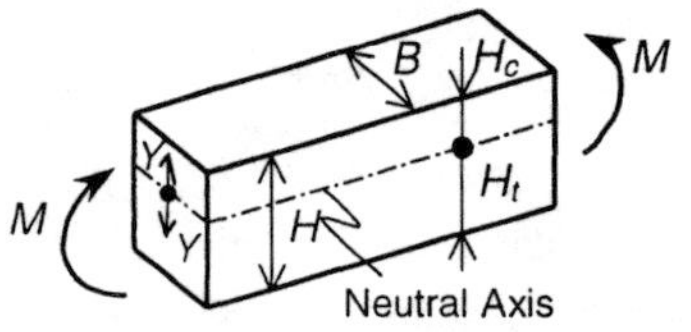

Figure 1. Segment of beam in four-point bending

From these two equilibrium equations (6) and (7), two nonlinear equations

$$f(\dot{K}, H_c, A_t, A_c, n_t, n_c \ldots) = 0 \qquad (8)$$

$$M = f(\dot{K}, H_c, A_t, A_c, n_t, n_c \ldots) \qquad (9)$$

are obtained. Equations (8) and (9) constitute a system of algebraic equations for the two unknowns H_c and $\dot{K}$, while the remaining parameters such as the applied moment M and the creep parameters A and n are considered to be known or given. These two coupled non-linear algebraic equations can be solved numerically. More details and examples are in Chuang's paper [2] [3].

Quinn and Morrell [4] reviewed the problems of using bend test data for design purposes and concluded that "Flexure testing by itself is not recommended for quantitative creep analyses." Uncertainties from frictional constraints of fixed load points and contact point shifts in four-point bending creep tests were among the reasons for their conclusion. Also, Jakus and Wiederhorn [5] showed creep curvature data which indicated nonuniform curvatures and highly concentrated curvatures near load points. Both Hollenberg's and Chuang's analyses assumes that the crept bar behaves like a simple beam in the four-point bending. So, questions arise as to whether Hollenberg's and Chuang's methods are still valid in spite of Jakus and Wiederhorn's observation and how friction affects the estimations of the methods.

The best way to verify the Hollenberg's and Chuang's methods is to perform tension, compression, and bend creep tests together, and to compare creep parameters measured from tension and compression tests with the parameters estimated from bending creep test data by the methods. At this time there is very little literature which has tension, compression and bend creep test data together. Ferber et al. [3] performed tension, compression and bend creep tests together, but the estimation by Chuang's method did not agree well with experimental results because in his bend creep tests the specimens at the two highest applied stresses failed before they reached the steady state. Chen and Chuang [6] performed compression and bend creep tests together and showed that estimated compression creep strain rate is in fair agreement with compression creep test result.

To fully test the methods, finite element analysis can be effectively applied by performing four-point bending creep test simulations with selected creep parameters as inputs. The data collected from simulations (curvature rates) are used to estimate the creep parameters using the methods. Then, the estimated parameters can be compared with original input parameters and the validity of the methods be evaluated.

EXPLICIT FINITE ELEMENT CREEP SIMULATION

To make simulations meaningful for comparison, contact and friction between loading rollers and specimen must be considered in the simulation, and material model for asymmetric creep behavior be applied in the simulation (to verify Chuang's method). To handle material non-linearity (creep) and sliding of material interfaces (contact between loading rollers and specimen), an explicit finite element method is preferred because traditional implicit methods can encounter difficulties solving this problem.

In explicit finite element methods, non-linear static problems are viewed as the steady state condition of critically damped dynamic problem. This approach is often called dynamic relaxation [7] [8]. The equation of dynamic motion at time t is given as

$$M\ddot{u}^t + C\dot{u}^t + F_{int}^t = F_{ext}^t \qquad (10)$$

Using $C = 2\omega M$ in the central difference scheme, this leads to

$$\dot{u}^{t+\Delta t/2} = \frac{\left[\Delta t M^{-1}\left(F_{ext}^t - F_{\text{int}}^t\right) + (1 - \omega\Delta t)\dot{u}^{t-\Delta t/2}\right]}{(1+\omega\Delta t)} \qquad (11a)$$

$$u^{t+\Delta t} = u^t + \Delta t \dot{u}^{t+\Delta t/2} \qquad (11b)$$

In dynamic relaxation algorithm the ω, Δt and M are chosen judiciously so that the transient response disappears fast, resulting in the static solution

$$F_{int} = F_{ext} \quad (12)$$

This algorithm is conditionally stable, and usually M is chosen so that the critical time step is $\Delta t = 1$. Then Equation (11) becomes

$$\dot{u}^{t+1/2} = \frac{\left[M^{-1}\left(F_{ext}^{t} - F_{int}^{t}\right) + (1-\omega)\dot{u}^{t-1/2}\right]}{(1+\omega)} \quad (13a)$$

$$u^{t+1} = u^{t} + \Delta t \dot{u}^{t+1/2} \quad (13b)$$

such that t becomes an iteration counter rather than time. More details of dynamic relaxation are shown in [8].

Creep Material Model

A material model for asymmetric creep was developed for this research by modifying an existing symmetric creep model. The asymmetric creep model is explained here within the context of symmetric creep model.

In the dynamic relaxation algorithm for creep calculation, the creep strain rate at the current time step is given by

$$\dot{\varepsilon}_c^{t} = f(\sigma) = A\left(\sigma^{t}\right)^{n} \quad (14)$$

Then, a creep strain increment is obtained using a creep time increment $\Delta\tau$ as

$$\Delta\varepsilon_c^{t} = \Delta\tau\dot{\varepsilon}_c^{t} \quad (15)$$

As shown above, there are two kinds of time steps (Δt and $\Delta\tau$) in the algorithm. The time step Δt is a dynamic relaxation cycle counter and it is not related to physical time. The creep time step $\Delta\tau$ is a real time interval for creep deformation. Within one creep time step ($\Delta\tau$), dynamic relaxation iterations (Δt) proceed while $\Delta\varepsilon_c$ is continually updated and the converged state is the solution of the given creep time step. An example for creep calculation process is given in Figure 2. Each of the right most points represent equilibrium being reached to within a convergence tolerance. Through iterations within each creep time step, the displacement continually changed and reached a converged solution and based on the converged condition, the iteration goes on with the next creep time step.

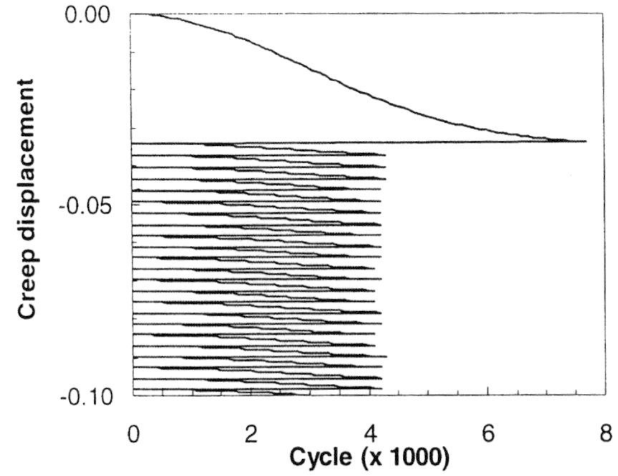

Figure 2. Displacements of a node in creep-dynamic relaxation cycles

The creep time step $\Delta\tau$ affects the solution stability, and as a rule of thumb the creep strain increment should not exceed one half of the total elastic strain [9]. For simulations in this paper, the critical creep time step was calculated from the formula [10] [11]

$$\Delta\tau_{cr} \leq \frac{2\bar{\sigma}}{EAn\bar{\sigma}^{n}} \quad (16)$$

where $\Delta\tau_{cr}$ is critical creep time step, E is modulus of elasticity, $\bar{\sigma}$ is effective stress, and A and n are creep parameters.

To model the asymmetric creep behavior, a simple modification was performed in the symmetric creep material model. Instead of using the single creep equation, the two equations (5a) and (5b) were used selectively to calculate creep strain. The hydrostatic stress determines the compression/tension condition of the element. Instability problems were encountered in the asymmetric modification when current hydrostatic stress was used at each iterations. In bending creep simulation, at elements near the neutral axis, the hydrostatic stress may fluctuate between tension and compression during dynamic relaxation iterations and convergence is slow or never achieved. This problem was solved when a single creep law ($\dot{\varepsilon}_c = A_c\sigma^{nc}$ or $\dot{\varepsilon}_c = A_t\sigma^{nt}$) was applied during iterations within one creep time step. The hydrostatic stress of converged solution from last creep time step was used to decide compression/tension condition of the element. Figure 3 summarizes the procedure.

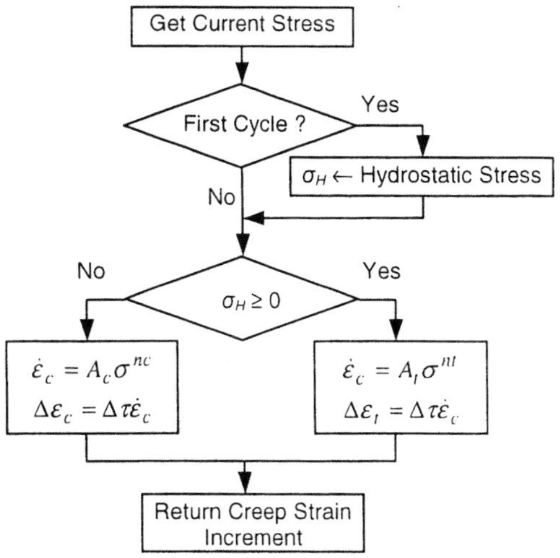

Figure 3. Creep strain calculation module

Verification of Creep Model

The asymmetric creep material model implemented in the explicit code was verified before being used in the bending creep test simulations of this research. For this purpose, a C-ring compression creep test simulation was performed with modified code and the result was compared with the published experimental and simulation data.

Chuang et al. [12] performed compression creep tests with siliconized silicon carbide (Si-SiC) C-ring. In their test, a Si-SiC C-ring specimen was placed between two SiC push rods inside a furnace, a constant compression load was applied at 1300 °C and the separation distance of C-ring was measured as a function of time.

Chuang et al. [13] also performed a finite element simulation of C-Ring Compression creep test. They used a two-dimensional finite element model with plane strain conditions as shown in Figure 4. Quadratic elements were used to model the C-ring and a concentrated load was applied to one node on top. Creep parameters used for their simulation had been established by Wiederhorn et al. [14] from uniaxial tension and compression creep tests. The displacement rate from the simulation was 11.00 $\mu m/h$ at $t = 100$ h which is in good agreement with the experimental data of $\dot{\Delta}_p$ = 11.55 $\mu m/h$.

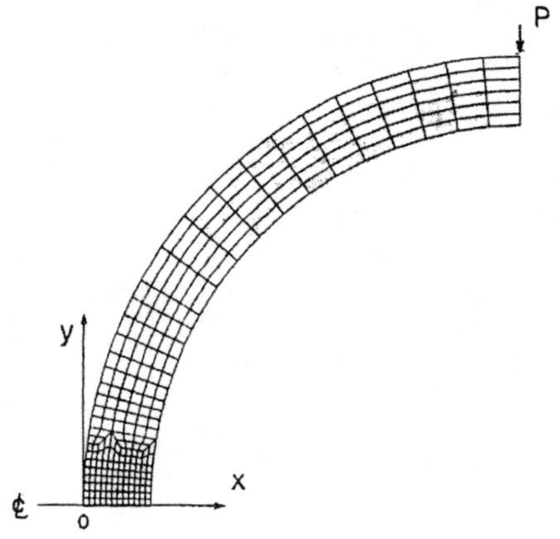

Figure 4. The two-dimensional finite element model of C-ring [13]

The asymmetric creep model was implemented in the general purpose finite element code H3DMAP [15] and the C-ring compression creep test simulation was performed to compare its result with Chuang's experiment and simulation. The same creep parameters, specimen dimensions and load [13] used in Chuang's experiment and simulation were applied. Because of symmetry of geometry and load, only the upper half of the ring and a half of the thickness were modeled. The plane stress condition was applied so that thickness could change as the ring bent. Instead of concentrated nodal loads, a pushing rod was modeled with rigid surface, and contact elements were modeled both on the rigid surface and on the top surface of the ring, as shown in Figure 5. The stable creep time step (50 sec) was calculated from the Equation (16) and 360,000 sec (100 hr) creep was simulated. The friction coefficient between the pushing rod and C-ring was assumed as $\mu = 1$ because the pushing rod and specimen are almost stuck together due to high temperature.

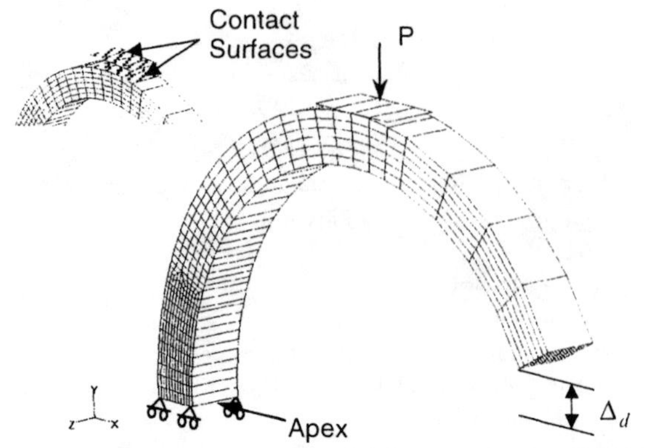

Figure 5. Finite element model of C-ring

Figure 6 shows a comparison between current simulation and published data and Figure 7 shows the stress redistributions as time across the thickness at the apex from the simulation. The simulation is in good agreement with both experimental and simulation data of Chuang et al. [12] [13]. The deflection rate from the current simulation is 11.23 $\mu m/h$ at 100 hr, which is compared with 11.55 $\mu m/h$ from experiment and 11.00 $\mu m/h$ from Chuang's simulation. The discrepancies in transient responses between experiment and simulations are because the transient creep is not considered in the simulation. The steady-state creep rate is the main concern of this research. Therefore the validity of the asymmetric creep model implemented in the explicit code is established from the comparison.

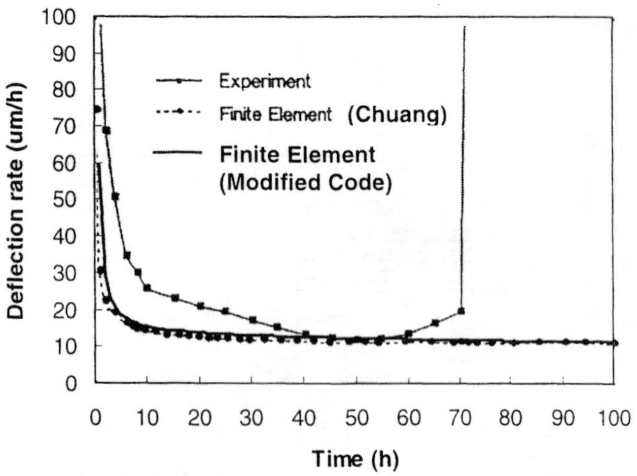

Figure 6. Deflection rate vs. time (comparison)

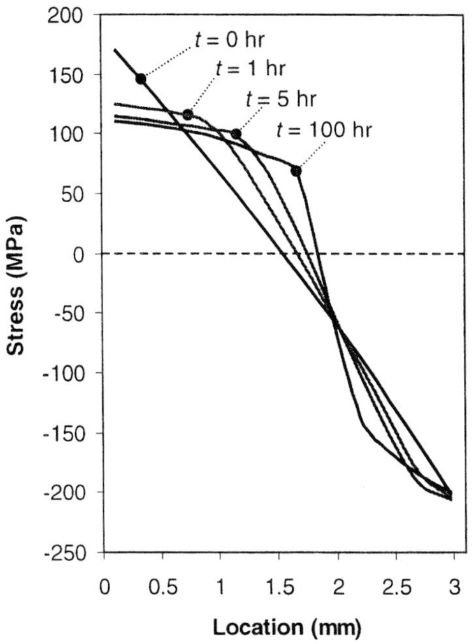

Figure 7. Stress distribution at the ring apex as a function of time

To examine the error of using nodal loads instead of including contact conditions between the pushing rod and C-ring, a simulation was performed with simplified nodal loads. In the C-ring compression test, the contact point migrates toward the inside of C-ring as seen in Figure 8. With nodal loads, this contact point shift is not considered in the simulation and the applied moment increases as deformation proceeds. In the current simulations, the effect of contact shift was minor, but as can be seen in Figure 9 the difference increases as deformation proceeds. Thus, caution is required to use simplified nodal loads in simulations with large deformation.

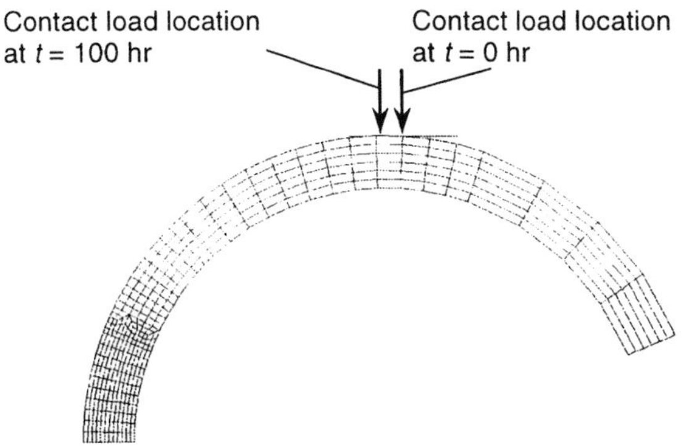

Figure 8. Contact point shift in the C-ring compression test

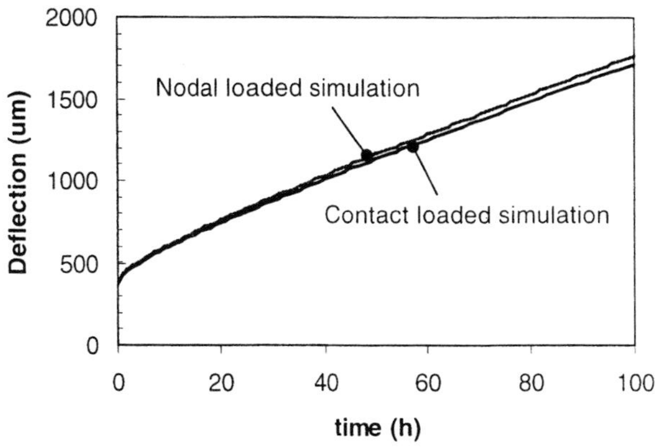

Figure 9. Total deflection vs. time (comparison)

FOUR-POINT BENDING CREEP TEST

In this section, the Jakus and Wiederhorn's experiments are simulated to examine if their observation (nonuniform curvature and highly concentrated curvature near load points) is reproduced in simulations. Then, the methods of Hollenberg and Chuang are evaluated with respect to the effect of load point friction.

To see the friction effects in four-point bending, loading rollers were included in the finite element modeling which interact with a bending specimen through contact surfaces as shown in Figure 10. Linear hexahedral solid elements were used to model the bending specimen and shell elements were used to model the loading rollers. The nodes on the pushing roller were tied to each other so that they work as a rigid surface, and the total force (P) was applied on the pushing roller. A plane stress boundary condition was applied on the model.

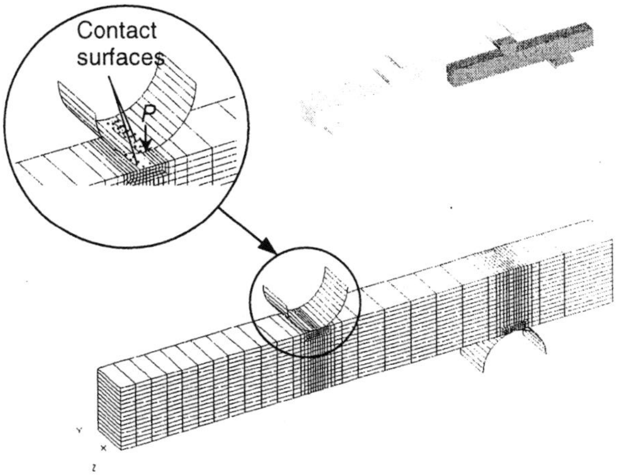

Figure 10. Finite element model

Simulation of Jakus and Wiederhorn's Experiment

In Jakus and Wiederhorn's experiment [5], crept specimens showed nonuniform curvatures and highly concentrated curvatures near load points as can be seen from Figures 11 and 12. Two materials used in their experiment were glass and alumina. The glass is known for its linear viscous creep, which means that its creep is symmetric and its creep exponent is $n = 1$. Alumina is one of the ceramic materials which shows asymmetiric creep behavior. Because the creep parameters were not given in their paper [5], the following creep parameters were assumed based on the limited information given in papers [2] [5].

Glass (symmetric creep) :

$$A = 2.70 \times 10^{-7}\ \mathrm{s}^{-1},\ n = 1 \tag{16}$$

Alumina (asymmetric creep) :

$$A_t = 0.885 \times 10^{-13}\ \mathrm{s}^{-1},\ A_c = 4.25 \times 10^{-9}\ \mathrm{s}^{-1}$$

$$n_t = 4,\ n_c = 0.5 \tag{17}$$

The geometry (3 × 5 × 50 mm, 10 mm inner loading span, 40 mm outer span) and load (30 MPa outer fiber elastic stress) shown in the experiments were used. The localized curve-fitting scheme used by Jakus et al. [5] to determine curvatures from experiments was applied to calculate curvatures from simulations. To get data for this scheme, nodal displacement data from simulations were connected with a smooth curve by cubic spline interpolation.

Results and Discussion

The beam deflection data from the simulation were flipped for the opposite half so that the curvature curves for the whole specimen length could be viewed. The curvature curves from simulations didn't match quantitatively with the curve from the experiment because creep parameters were assumed. However, simulation reproduces the trend of the experiments. Nonuniform curvatures in the mid-span and concentrated curvatures near the load-points are apparent both in the experiments and simulations. Results obtained with no friction and high friction conditions are shown in Figures 11 and 12. From comparison, one can notice that friction lowers the curvature at the loading point. On this basis, the nonuniform occurrence of friction in the experiments can be seen and understood.

Figure 13, the stress (σ_{xx}) distribution at the beam cross section in the mid span was plotted as a function of time for glass from a simulation without friction. It shows an important fact about four-point bending creep tests. The stress distribution curves are linear as expected from creep parameter $n = 1$ (viscous creep). The total bending moment in the section can be calculated as the summation of moments produced by stresses of elements in the cross section. Therefore, this plot shows that the total bending moment applied in the beam section decreases with time in the four-point bending creep tests.

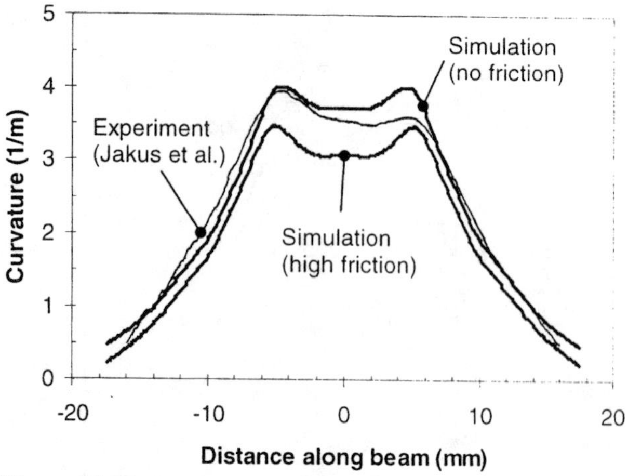

Figure 11. Curvature curves from simulations and experiment for symmetric creep

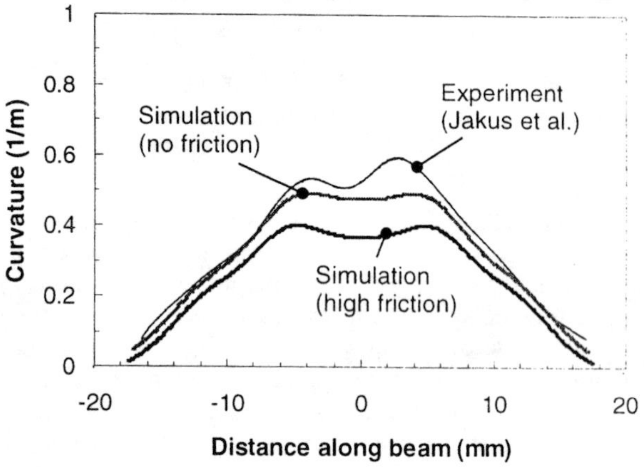

Figure 12. Curvature curves from simulations and experiment for asymmetric creep

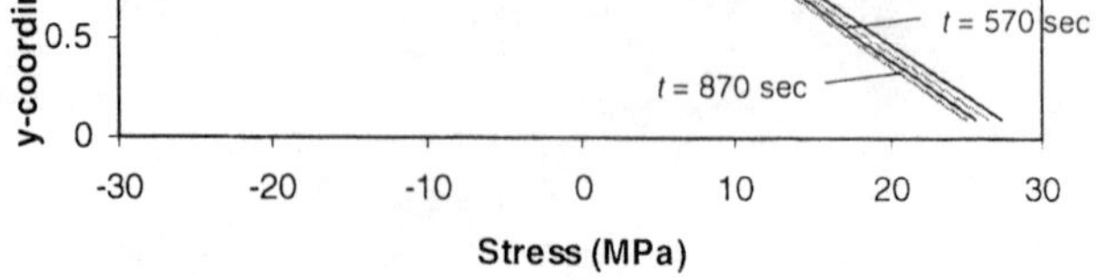

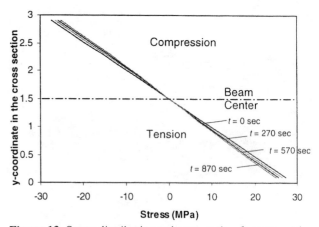

Figure 13. Stress distribution at beam section for symmetric creep ($n = 1$)

Evaluation of Hollenberg's Method

To evaluate Hollenberg's method, simulations were performed at four load levels with the following geometries and creep parameters.

Specimen size : 3 × 5 × 50 mm

Inner/outer spans : 10 mm / 40 mm

$$A = 3.273\times10^{-12}\ \text{s}^{-1},\ n = 2.2 \quad (18)$$

From simulations, mid span displacements (y_R) were calculated by $y_R = y_{pin} - y_{center}$ (14). Then curvature rates were calculated from the mid-span displacement rates by the formula

$$\dot{K} = \frac{8}{a^2}\dot{y}_R \quad (19)$$

where a is inner loading span (10 mm) and $\dot{K}$ is curvature rate of the inner loading span. This process also was repeated for simulations with four load levels. The creep parameters were then estimated by Hollenberg's method using equation (4) by fitting a line to the curvature rate vs. the applied outer-fiber elastic stress on a log-log plot as can be seen in Figure 15. A typical evolution of the stress distribution in the beam section as a function of time is shown in Figure 16.

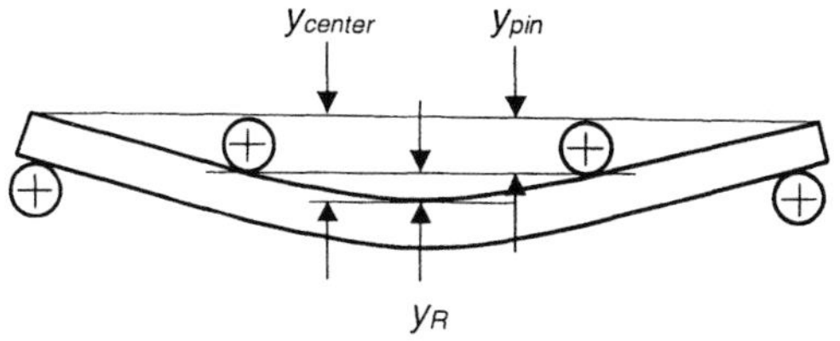

Figure 14. Mid span displacement y_R

To study the frictional effect on bending creep test, another set of simulations were performed with high friction ($\mu = 1$) and creep parameters were estimated. The Table 1 compares the creep parameters estimated from simulations with and without friction.

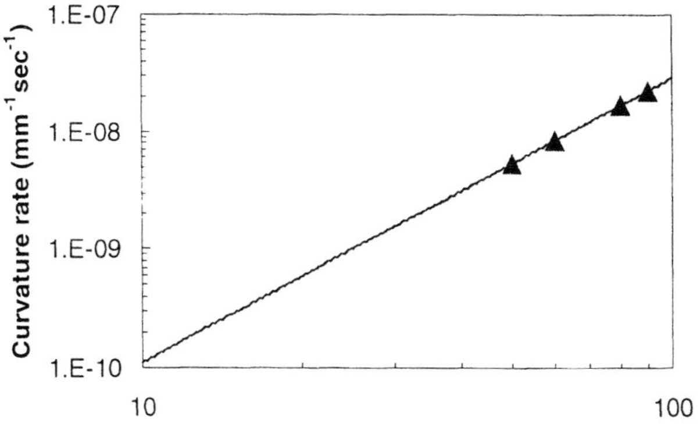

Figure 15. curvature rate vs. the applied outer-fiber elastic stress on a log-log plot

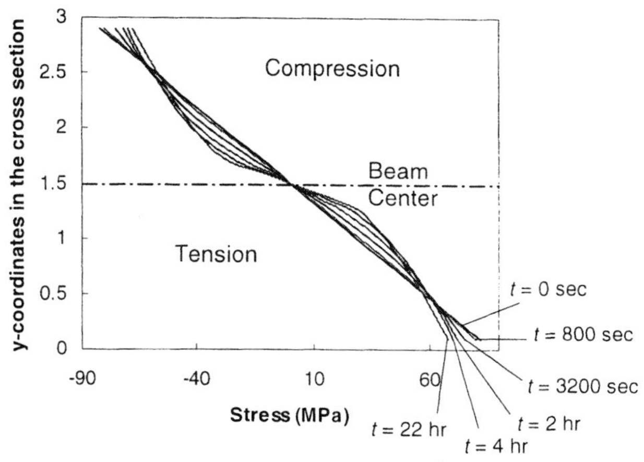

Figure 16. Stress distribution in the beam section as a function of time ($\sigma_e = 90$ MPa)

Table 1. Frictional effect on estimation of creep parameters

Creep Parameters	Original Input	Simulation without Friction ($\mu = 0$)	Simulation with Friction ($\mu = 1$)
n	2.2	2.41	2.47
A	3.273×10^{-12}	1.061×10^{-12}	5.79×10^{-14}

Discussion

In simulations without friction, the estimated creep exponent n was 10% higher than original input n and the estimated pre-exponent A was about one third of original input A. Still, both are within allowable error range in creep experiments. The friction did not affect the creep exponent n estimation much (2.5% higher than no friction case), but the pre-exponent A was seriously underestimated (15 times lower than no friction case) when friction is applied in simulations.

Finite Element Study of Chuang's Method

The Chuang's method to extract creep parameters from bend test data for asymmetric creep is evaluated using finite element simulations. The simulations were performed at five different load levels ($\sigma_e = 90, 80, 60, 50, 39$ MPa) and used the following data.

Specimen size : 3 × 5 × 50 mm

Inner/outer spans : 10 mm / 40 mm

$$A_t = 2.972\times10^{-17}\ \text{s}^{-1},\ A_c = 5.555\times10^{-13}\ \text{s}^{-1},$$

$$n_t = 5.6,\ n_c = 1.7 \quad (20)$$

The steady-state curvature rates were obtained from the simulations at different load levels. Figure 17 shows a typical evolution of the stress distribution in the beam section as a function of time.

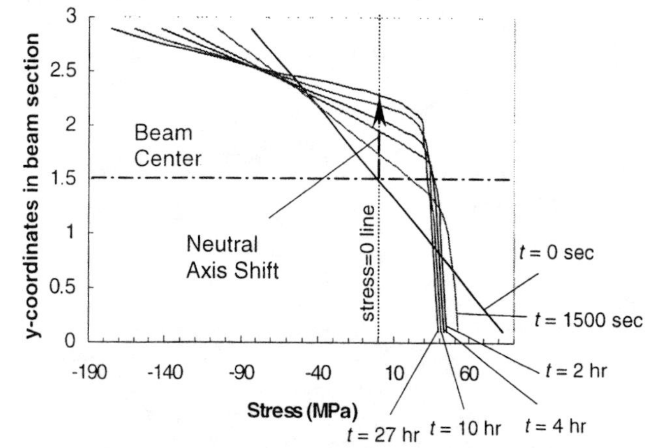

Figure 17. Stress distribution in the beam section as a function of time (σ_e = 90 MPa)

Chuang's method has two highly non-linear coupled equations (8) and (9), so an iterative numerical solution is required to estimate creep parameters. In the current research, efforts were not directed at programming an iterative method such as the whole iteration steps of Chuang's method. Instead, curvature rates were estimated from the given creep parameters (A_t, A_c, n_t, n_c) by using the Equations (8) and (9). Figure 18 compares the curvature rates estimated from Chuang's method and the curvature rates from simulations.

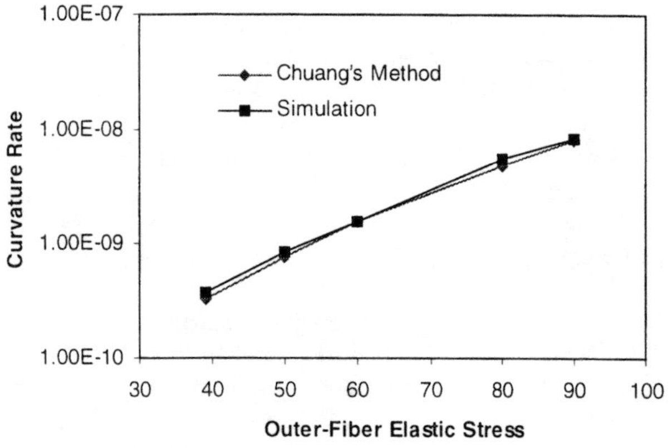

Figure 18. Curvature rates from Chuang's method vs. simulations

Additional simulations were performed with various frictions between loading roller and specimen to see how friction affects the bending creeps. Figure 19 summarize the results.

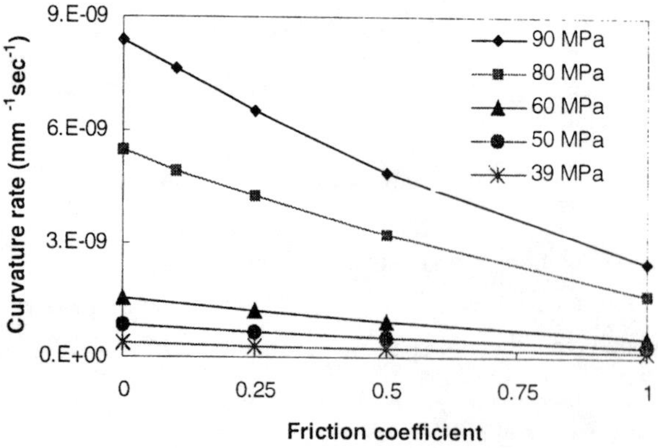

Figure 19. Curvature rate changes depending on frictions

Discussion

The curvature rates from Chuang's method were well matched with the curvature rates from simulations without frictions. Friction lowered the outer-fiber stresses and creep rates as can be seen in Figure 20. From the view of current simulation studies, the creep parameters estimated from bend creep data by Chuang's method are reliable when the friction between loading rollers and specimen is maintained low enough during creep test.

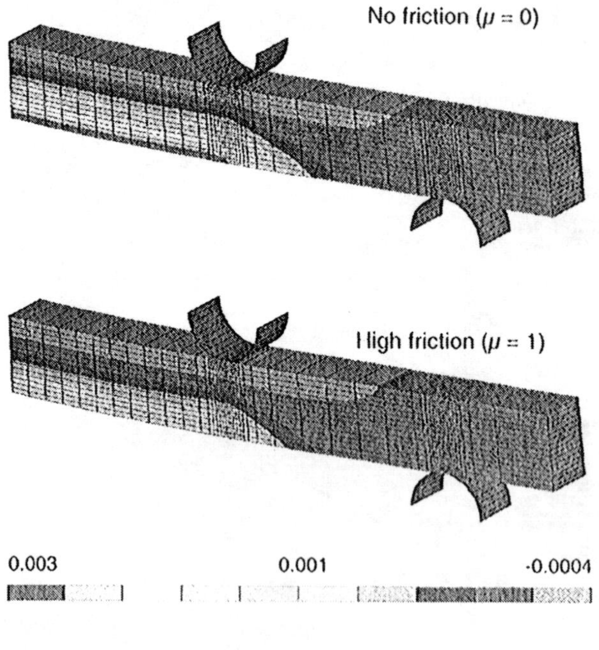

Figure 20. Creep strain (ε_{xx}) distribution at t = 27 hr (σ_e= 90 MPa)

CONCLUSIONS

The methods of Hollenberg and Chuang to extract creep parameters from bend test data were evaluated by finite element analyses where friction effects were included. When friction between loading rollers and specimen was not applied in the simulations, both methods well predicted creep parameters from bend creep simulation data. But, when frictions were high, the parameter (A) was highly underestimated, while prediction of the parameter (n) was not affected much by friction.

REFERENCES

[1] Hollenberg, G. W., G. R. Terwilliger, and R. S. Gordon, 1971, "Calculation of Stresses and Strains in Four-Point Bending Creep Tests," Journal of the American Ceramics Society, Vol. 54, pp. 196-199

[2] Chuang, T.-J., 1986, "Estimation of Power-Law Creep Parameters from Bend Test Data," Journal of Materials Science, Vol. 21, pp. 165-175

[3] Ferber, M. K., M. G. Jenkins, and V. J. Tennery, 1990, "Comparison of Tension, Compression, and Flexure Creep for Alumina and Silicon Nitride Ceramics," Ceram. Eng. Sci. Proc. Vol. 11, pp. 1028-1045

[4] Quinn, G. D., and R. Morrell, 1991, "Design Data for Engineering Ceramics: A Review of the Flexure Test," Journal of the American Ceramics Society, Vol. 74, pp. 2037-2066

[5] Jakus, K., and S. M. Wiederhorn, 1988, "Creep Deformation of Ceramics in Four-Point Bending," Journal of the American Ceramics Society, Vol. 71, pp. 832-836

[6] Chen, C.-F., and T.-J. Chuang, 1990, "Improved Analysis for Flexural Creep with Application to Sialon Ceramics," Journal of the American Ceramics Society, Vol. 73, pp. 2366-2373

[7] Underwood, P., 1983, "Dynamic Relaxation," Computational Methods for Transient Analysis, Vol. 1, pp. 245-265

[8] Sauve, R. G., and D. R. Metzger, 1995, "Advances in Dynamic Relaxation Techniques for Nonlinear Finite Element Analysis," Transactions of the ASME, Vol. 117, pp. 170-176

[9] Zienkiewicz, O. C., 1991, "The Finite Element Method," 4th edition, McGraw-Hill, London

[10] Cormeau, I., 1975, "Numerical Stability in Quasi-Static Elasto/Visco-Plasticity," International Journal of Numerical Methods in Engineering, Vol. 9, pp. 109-127

[11] Sauve, R. G., and N. Badie, 1989, "Creep Response of Fuel Channels in Candu Nuclear Reactors," Ontario Hydro Research Division Report No. 89-2-K

[12] Chuang, T.-J., W.-J. Liu, and S. M. Wiederhorn, 1991, "Steady-State Creep Behavior of Si-SiC C-Rings," Journal of the American Ceramics Society, Vol. 74, pp. 2531-2537

[13] Chuang, T.-J., Z.-D. Wang, and D. Wu, 1992, "Analysis of Creep in a Si-Sic C-Ring by Finite Element Method," Journal of Engineering Materials and Technology, Vol. 114, pp. 311-316

[14] Wiederhorn, S. M., and T.-J. Chuang, 1988, "Damage-Enhanced Creep in a Siliconized Silicon Carbide: Mechanics of Deformation," Journal of the American Ceramics Society, Vol. 71, pp. 595-601

[15] Sauve, R.G., "H3DMAP Version 6.2 – A General Three-Dimensional Finite Element Code for Linear and Nonlinear Analyses of Structures", Ontario Hydro Technologies Report No. 181.1-H3DMAP-1988-RA-0001, 1999

PVP-Vol. 458, Computer Technology and Applications

PVP2003-1895

Methods, Models, and Assumptions Used in Finite Element Simulation of a Pilger Metal Forming Process

Prepared by
John Martin, Principal Engineer
Structural Engineering/Project Structural Design
Lockheed Martin, Inc.
jmartin8@nycap.rr.com

ABSTRACT

The pilger process is a cold-worked mechanical process that combines the elements of extrusion, rolling, and upsetting for the formation of thin-walled tubes. This complex manufacturing process relies on the results of trial and error testing programs, experimental parameter sensitivity studies, and prototypical applications to advance the technology. This finite element modelling effort describes the methods, models, and assumptions used to assess the process parameters used to manufacture thin-walled tubing. The modelling technique breaks down the manufacturing process into smaller computer generated models representing fundamental process functions. Each of these models is linked with the overall process simulation. Simplified assumptions are identified and supporting justifications provided. This work represents proof of principle modelling techniques, using large deformation, large strain, finite element software. These modelling techniques can be extended to more extensive parameter studies evaluating the effects of pilger process parameter changes on final tube stress and strain states and their relationship to defect formation/propagation.

Sensitivity studies on input variables and the process parameters associated with one pass of the pilger process are also included. The modelling techniques have been extended to parameter studies evaluating the effects of pilger process parameter changes on tube stress and strain states and their relationship to defect formation. Eventually a complex qualified 3-D model will provide more accurate results for process evaluation purposes. However, the trends and results reported are judged adequate for examining process trends and parameter variability.

BACKGROUND

Pilgering is a complex, nonlinear, highly plastic cold-working process for forming thin-walled seamless tubing. Currently the pilger process relies on experience and trial and error variations to the process as a means of identifying potential process parameter changes capable of achieving the desired final tubing dimensions without introducing surface defects. The process entails a series of incremental wall thickness reductions, along with diametral deformations and axial elongations. References [a], [b], and [c] reported on a limited 3-D finite element model representation of a pilger process based on simplifying assumptions. Each paper utilized a single, incremental material reduction, coarse element configuration, 3-D model to examine the generated stress and strain. Reference [a] is one of the earliest attempts to use finite element methodology to address the stresses and strains generated in a workpiece during a pilger process. Reference [b] continues the work documented in Reference [a] with a more involved evaluation of localized effects. A more recent work (Reference [c]) uses a 2-D cross sectional model to address the radial and circumferential stress/strain states generated by the process.

The models presented in this paper discuss the modelling methodology needed to simulate the process parameters of a pilger process. The primary objective of this work is to provide proof of principle modelling techniques for undertaking more extensive parameter studies on the effects of pilger process parameter changes on final tube stress and strain states and the potential to generate defects.

DISCUSSION

Pilgering is a cold-worked, large deformation, metal forming process used to manufacture pipes and tubing. The process consists of a rolling/extrusion form of elongation process, while simultaneously producing a diametral and wall thickness reduction through an upsetting type of process. The inner diameter and tube cross-section reduction are facilitated through contact between a gradually reducing mandrel cross-section and the shape of the cyclic oscillating cam rolling die form. Tube material elongation is also an outgrowth of this high reduction rolling action.

Current practice relies on empirically developed manufacturing parameters related to cross-sectional geometry changes. These parameters are referenced to as "Q" and "R". "Q" is defined as the ratio of the wall thickness reduction (initial to final wall thickness) to the outer diameter reduction (initial outer diameter to final outer diameter). "R" is defined as the overall cross-sectional area reduction. The later variable is established as a means of accounting for the tube elongation. Both these parameters are solely dependent on geometry and empirically based observations. Neither parameter is set up to help establish a qualitative, fundamental understanding of the various stress/strain states that occur during the process and their potential relationship to generated defects.

The basic methodology used is to develop simplified model representations of the complex pilger process. This is accomplished by breaking down the actual process into individual parts; then using applied mechanics and metal forming principles to evaluate that section of the process. Initially, simplified models using the large deformation, large strain finite element software code MSC MARC (Reference [d])were developed to represent each phase of the actual pilger process. However, the proof of principle modelling also supports the viability of a complex 3-D model of the entire pilger process.

Pilger Process

A simplified description of the general pilger process is provided to facilitate comparison of the simulation models with their respective parts of the actual process. The process consists of the workpiece, two dies and a mandrel. The workpiece is a material that will be cold-worked, over the course of the multiple bi-directional passes, into the final tubing dimensions. The dies are two special shaped cam rollers. The mandrel is tapered as a means of providing the guide necessary to obtain the required inner diameter reduction during the pilger process. The two cam dies are shaped to accommodate both the elongation of the tubing material and the thinning of the diameter/ wall thickness during the extrusion/rolling process. (see Figure 1.)

Formation of the cold worked tube occurs in a series of rapid die motions intended to incrementally form the tube while controlling the strain levels as a means of preventing voids within the material and surface cracking to occur. The process begins with the workpiece being fed along a flat portion of the mandrel until it reaches the start of the tapered section. This tapered section creates the diametral reduction. The process is continuous, requiring a constant feeding force be maintained on the workpiece.

To create the reduced diametral and thinner wall thickness, the roller cams begin at the wider diameter portion of the mandrel and roll along the length of the mandrel. The cam is designed to coincide with the pitch of the mandrel. This design is such that as the roller proceeds along the length of the mandrel it gradually increases the depth of penetration into the tube thereby forcing an elongation of the workpiece. The pilger process being modelled involves coldworking the tube on both the forward and return strokes.

The cams do not encompass the entire outer diameter of the workpiece. A side relief pocket is provided at the axis of the workpiece diameter parallel to the rolling direction. This side relief prevents the development of a hydrostatic stress state throughout the entire cross-section during the forming process. The side relief pocket dimension is a function of the process parameters including: material hardness, elongation and reduction rate, feed rate, and friction. The side relief value is developed through empirical testing and is generally proprietary information. However, the inclusion of this pocket creates a clearance that results in ovality within the tube during the forming process. Subsequent loadings during the process are designed to eliminate this ovality. The amount of ovality may be detrimental to the final product if it is too severe.

In the pilger process being modelled, the cam is momentarily released from the workpiece after the roller completes the forward and return pass along the workpiece. During this release, the workpiece is rotated incrementally to reposition the ovality formed by the previous pass. The cam rolling process is then repeated with further diametral reduction, workpiece elongation, and circumferential relocation of the ovality.

Some of the process factors that are critical to the manufacturing of acceptable tubing include: (1) the effects of friction on both the cam roller surface and the workpiece interface with the mandrel during material elongation; (2) the frictional restraints that exist between the mandrel and the workpiece impact the force on the inner surface during the rotational portion of the process; (3) the amount of reduction and wall thinning required per pass; and (4) the amount of workpiece elongation that occurs with each pass of the cam.

Assumptions

In generating finite element models to simulate the pilger process, some simplifying assumptions were required. The impact of some of these assumptions on the trends and conclusions are assessed using bounding sensitivity studies.

These bounding sensitivity studies evaluated the variability associated with the input parameters used for the pilgering finite element modelling effort. The results presented are representative of completion of the initial pass of the multipass pilger process and address the rolling portion of the process only. The comparisons are primarily based on the plastic strains calculated during the process simulation. The premise is that the cyclic nature of the strain during the loading and unloading cycles may be a contributor to the formation of defects. The studies undertaken addressed the (a) material hardening behavior law assumptions; (b) effects of material variability on the process produced strain; (c) coefficient of friction values assumed between the tube and the mandrel and the tube and the die; and (d) overall effect of the mandrel geometry on the strain in the work piece.

Material Properties

Figure 2 represents the true stress-true strain nickel-based material curve used for this study.

Material Hardening Law

A separate study examined the difference in results between an assumed isotropic and kinematic material hardening law (graphical results not shown). The total equivalent plastic strain results indicate very little difference between the two hardening laws. However, this is not a completely accurate comparison since the material stress-strain curve used was obtained from a uniaxial tension specimen test results. The cyclic nature of the loading/unloading throughout the cold-working process requires a cyclic strain controlled test to generate a series of true stress- true strain curves to allow the computational model to better account for any hardening that would occur as a result of this cold-working process (i.e. ORNL material model).

Material Variability

A sensitivity study was undertaken to examine the effects material property variability will have on the calculated strain results. The materials used for the comparison were two nickel based alloys (A625 and A690). Their respective true stress- true strain curves are compared in Figure 3.

Using these material properties, a coefficient of friction $\mu = 0.1$ and examining the total equivalent plastic strain at both the inner and outer tube surfaces indicated there was minimal difference in the resulting strain values in spite of the marked difference in the material property curves (Figure 3). This similarity is judged to be attributed to the same applied strain rate and the loading being a displacement driven problem.

The intent of this was to examine the effect a change in material property might have on the residual strain. Preliminary analysis indicates that as a consequence of the large deformations and strains experienced by the material during the process, there is little difference between the final strain for either A625 or A690 materials. This study and conclusion is based solely on the yield strength and the stress-strain curve and does not address other material features such as grain size and material hardening.

Based on the above study it is judged that a change to the material properties should not influence the methods, models, assumptions, observed trends, or preliminary conclusions stated in this report. Given the cyclic nature of the loading, and the apparent loading/unloading of the material during the process, a kinematic hardening law was assumed for this study. However, confirmation of the actual yield surface behavior under cyclic loading would need better material property definitions (i.e. a cyclic true stress-true strain curve developed through variable strain controlled cyclic testing) if quantitative material conditions later prove to be desirable for model qualification testing.

Models

Two orthogonal models were developed to simulate the pilger process and predict the stress and strain states that will occur during the process. Each model assumes a plane strain condition. Using two independent orthogonal plane strain models will evaluate the two phases of the pilger process, e.g. rolling and extrusion. It is recognized that these plane strain assumptions for the transverse and longitudinal models do not reflect the three-dimensional nature of the material flow that occurs during the forming process. However, preliminary results indicate a relatively uniform through thickness strain gradient is developed within the longitudinal model. If this uniform gradient is reflected in all the models, then a generalized plane strain model can be developed for application to the tangential model to better reflect the three-dimensional nature of the material flow. Thus, after each of the two models has successfully completed simulating a representative portion of the pilger process, a generalized plane strain state can be generated linking the results from each of the models in a pseudo-three-dimensional state model. The eventual goal being to generate a 3-D model as a means of qualifying process evolution.

Rolling Cam (Longitudinal) Model

One model represents the cam rolling phase of the process. This model is developed from the viewpoint of representing the loading and unloading of a single location on the workpiece surface during the course of the pilger process. Beginning at the workpiece's first contact with the roller cam, the contacted location will undergo a brief point of contact where it will experience both a reduction and extrusion. The roller will then proceed along the entire length of the workpiece, but this localized region will no longer experience any contact with the cam roller until subsequent passes. After the initial pass the workpiece is rotated thereby reducing the amount of direct contact with the cam on its next pass. This process proceeds numerous times over the length of the working section of the mandrel until the selected localized region has passed through the entire length of the working section of the mandrel.

To simulate this process in the model, the approximate number of strokes experienced by the workpiece and the length of travel were compared to estimate the number of times that the specific localized region would come in direct contact with the cam. Taking into account that there were two cams situated 180° apart and an incremental rotation after each stroke, multiple direct contacts were estimated to occur during the travel of the workpiece. Therefore the finite element model used to represent the rolling process was based on eighty independent rollers (Figure 8). The rollers were located sufficient distance apart to not have any overlapping stress/strain field influence with the adjacent contact areas. In addition, the travel of the workpiece along the mandrel represents a reduction in wall thickness and inner diameter. This was accomplished by incrementally lowering the center of rotation for each of the subsequent rollers to match a constant incremental reduction in diameter. A rigid body pusher is located at the end of the workpiece to simulate the constant feed that the actual workpiece experiences during the process.

Figure 9 is a blow up of a generic time step during the incremental rolling process. The results report the shear stress under a roller during contact of a single roller. A distinct difference is observed between the residual shear stress at the roller contact and the residual stress along the mandrel.

Strain values from this model can be used as input into the cross section model as a means of defining a uniformly applied plane strain condition. This sequence represents a single pass, forward roll contact pilger process.

Cross-Section Model

The cross-section (transverse) model simulates the contact of the cam with the workpiece at any specific time during the pilger process. The major forming events simulated by this phase of the modelling are the reduction of the cross section, the decreased inside diameter that occurs with each pass of the roller, and the circumferential rotation of the workpiece with respect to the dies. The features addressed in this model simulate the extrusion portion of the pilger process.

To simulate these conditions a 3-D model of the dies was generated and a one element thick model of the workpiece were used. The single element workpiece model (considered to be in the XY plane) maintained the plane strain assumption by restraining displacements in the planes perpendicular to the rolling axis (Z-axis). Although this is a conservative assumption, the strain restraint in the Z direction can be replaced with a generalized plane strain formulation based on the rolling direction model. The model included twenty-eight (28) independent surfaces representing the roller cam at incremental locations around the workpiece. The input parameters that define body contact used for this model were structured such that only two opposing surfaces were in contact during any specific loadcase. Thus with each subsequent loadcase the surfaces in contact represented the 15° rotation that occurred during the process. The only difference was that the model has the dies rotating whereas in the actual process the workpiece is rotating. However, the relative motion between the parts is similar.

The last part of the simulation is the reduction of the inner diameter surface that occurs with each ensuing incremental rotational contact. This is accomplished through the use of a 3-D mandrel. The mandrel model is tapered along the Z-axis to represent the actual mandrel. Each loadcase is set up to not only rotate the dies (through contact table callout) but also to incrementally move the mandrel along the Z-axis to gradually reduce the inner diameter. Hence the model will reflect the relative rotation of the workpiece with respect to the roller cam and the reduction of the inner and outer diameters through die displacement and mandrel motions.

For this model a side relief of 0.010 inches was assumed. Sensitivity studies will eventually be performed to reflect the effects of a variable side relief dimension. These studies will assess the affect of ovality on the residual stress/strain states.

The transverse model is run with multiple steps to simulate the rotation of the dies from the 0° position to the 90° position as a means of demonstrating the proof of principle modelling techniques.

These models represent the primary forces and boundary conditions judged to impact the formation of a pilger processing generated tube section. Additional models may be developed as needed to address any process modelling uncertainties or assumptions that may arise during further investigation of the process parameters.

Process Dimension Selection

The models developed for these studies are discussed only in a qualitative manner. However, sensitivity studies are planned to address variability in these parameters along with providing a fundamental understanding of the effect of process parameter changes on tube stress and strain states.

Die Modelling

Die hardness is a factor in the number of tubes that can be manufactured within tolerance prior to die replacement. The hardness of the dies is generally significantly higher than that of the workpiece. Therefore, it is reasonable to assume the dies can be modelled as rigid bodies for calculating an upper bound of the stress and strain introduced into the workpiece during the pilger process and prior to annealing. Once the model has been demonstrated to have the capabilities to reflect the actual stress and strain states, and possibly defect generation, and reduced grain sizes, the impact of die flexibility and wear can be introduced into the model. However, the introduction of deformable dies will significantly increase the complexity of the modelling effort and is judged to not be warranted for this initial investigation.

It is also assumed that the temperature generated by the cold working plastic deformation that occurs over the course of the process is low enough to not effect a shift in material properties. In addition, a constant flow of lubrication/coolant mitigates the development of heat during the process. Hence, heat generation calculations have been omitted.

Modelling Coefficient of Friction (Coulomb versus Shear Friction)

Reference [a] utilized a Coulomb frictional relationship with a limit on the maximum shear stress. The Coulomb friction model is based on the relationship between normal and tangential forces while a shear friction model is generally defined as a function of flow stress. A Coulomb friction model assumes that the shear stress is proportional to the normal load between the two bodies in contact. For the shear friction (constant friction) model the shear stress is proportional to the flow strength of the material.

With the application of the Coulomb friction model it is possible to develop stresses much higher than the material can support. For this condition the potential for failure would be high and thereby result in unrealistically "high" local strains.

Utilizing the same coefficient of friction value but within different friction models, a different value of resisting stress will be developed. In almost all the cases, the shear model coefficient of friction will report a lower resistance stress. A comparative investigation of the results of the Coulomb and shear friction models shows different surface interactions between the roller and tube surface and the tube and mandrel surface. A Coulomb friction model results in a stress distribution closer to the surface with higher peaks but faster dissipation through the thickness while a shear friction model shows a lower surface peak stress but a greater and deeper through thickness stress variation.

Given a condition where the hydrostatic stress in the material is greater than the flow stress the Coulomb model would develop very high shear stress; potentially greater than yield. If Coulomb friction models are utilized in this case then the value of μ would need to be reduced as the shear levels approached the flow stress. However, the shear stress model uses the calculated material equivalent stress as the basis for developing friction formulations. With the shear friction model the material flow stress will never be exceeded. (This is closer to reality and best approximates bulk forming modelling.) For these reasons a shear friction model was chosen.

However, if a quantitative resolution is to be achieved, resolution of the appropriate friction model and coefficient of friction through additional testing needs to be addressed. The preliminary modelling work comparing various pilgering parameters reported in this paper assumes a shear friction model with a coefficient of friction value of 0.1 for calculated stresses and strains at both the interface of the roller cam to the workpiece and the workpiece to the mandrel. A sensitivity study addressing the variability of μ at either the roller or mandrel surfaces is discussed below. This sensitivity study varies the shear friction law coefficient of friction between 0.1 and 0.5 for both the roller and mandrel surfaces independent of the effect on the other's strain state.

Coefficient of Friction Variability Study

The value of the coefficient of friction assumed for the model can affect the calculated results. The pilger process is a cold-working process that relies on large deformation to create the desired final workpiece shape. During the course of this deformation very high strains will be generated throughout the workpiece. The highest of these strains are projected to occur at the inner and outer surfaces of the tube. Each surface is exposed to a different tool and tool motion, hence the likelihood of each interface experiencing a different coefficient of friction. A series of models were run varying the assumed coefficient of friction from $\mu = 0.1$ to 0.5 for the die to tube interface and $\mu = 0.1$ to 0.4 for the tube to mandrel interface. Each run varied the coefficient of friction on only one of the two interfaces at a time.

Figure 5 shows approximately a 14% difference in total equivalent plastic strain (Tepe) between values when assuming a coefficient of friction of 0.1 and 0.5 between the roller and the tube. Based on the overall strain values, it is judged that the rolling friction between the workpiece and the die will not significantly affect the results of the strain that will exist within the material subsequent to the forming process.

Unlike along the rolling surface, the mandrel remains stationary during the rolling portion of the cold-working process. Hence as the material flows due to plastic deformation, a higher strain is likely when the coefficient of friction is increased. As seen in Figure 6 there appears to be an exponentially increasing effect on the total equivalent plastic strain as the coefficient of friction increases from 0.1 to 0.4. The roller-workpiece interface easily obtained a solution at $\mu = 0.5$. However, in running $\mu = 0.5$ on the mandrel, convergence issues arose due to the large change in strain across elements. This non convergence can be overcome by decreasing the time steps but then the comparison based on time (increment) would be skewed- so that extra run has been omitted. The results shown in Figure 6, indicate a 15% increase in Tepe from a value of $\mu = 0.1$ to a value of $\mu = 0.3$. Then an increase of 59% in Tepe from a value of $\mu = 0.3$ to a value of $\mu = 0.4$. This demonstrates a significant impact on the strains calculated for the inner surface as the coefficient of friction increases. Based on this, any improvements that will minimize the coefficient of friction between the mandrel and the tube should be incorporated into the process.

Process Geometry Simplifications

The actual mandrel geometry is a proprietary feature that contributes to the final strain state in the formed tube. A sensitivity study of the tube forming mandrel geometry was performed with the results shown in Figure 7. The study used two mandrel geometries. One mandrel was linearly ramped along an arbitrary length to facilitate forming from the initial to final outer tube diameter. The other mandrel geometry used an elliptical shape to provide 60% of the overall reduction over 30% of the arbitrary length with the remaining length formed from a linear reduction. Using the second geometry two additional cases were examined. One simulated a multiple pass one directional rolling reduction process; the other simulated a multiple pass reduction process where the roller would reverse its angular rotation relative to the workpiece surface during the reverse stroke but provide incremental tube reduction after each pass. The results reported are based on the inner diameter tube surface. This second comparison indicated that alternating the rolling direction had a negligible effect on the strain in the rolling direction along the inner diameter tube surface. It is observed, however, that the curved mandrel will generate higher strains early in the pilger process but eventually results in lower strains when the tube has reached its final state. This supports the change to the curved surface but also identifies the need to have a better definition of the actual mandrel shape if quantitative strains are to be expected.

Typically a mandrel is designed to provide for a majority of the reduction (60-70%) to occur in approximately the first 35% of the overall travel of the rolling cam during each step. This can be accounted for by varying the incremental wall reduction along the length of the cam travel. Initially the proof-of-principle modelling effort assumed a uniform, linear reduction in wall thickness for each incremental modelling step. However, other comparisons address an elliptical representation of the mandrel geometry.

In addition, the rolling modelling simulations do not differentiate among the entry pocket, working section, and exit point geometries along the mandrel. It is judged that these geometries, although representing discontinuities in the process, are not representative of the bulk of the forming process. Each of these geometries can be evaluated for its impact on the development of inside diameter defects should the overall process warrant this evaluation.

Mesh Density

Selection of an appropriate mesh density will affect the accuracy of the results. A mesh density study has been completed using a slightly coarser mesh than all the rolling models run to date. The results indicate sufficient mesh refinement considerations have been made to have confidence in the trends and results presented. In some cases, the use of finite element techniques to simulate manufacturing requires significant mesh adaptation and refinement due to the large deformation incurred during the process. However, the pilger process is more representative of extrusion and rolling processes, as opposed to the type of large localized deformation experienced during thread rolling simulations (References [e],[f]) Thus the pilger process is better related to material flow problems similar to those associated with computational fluid dynamics. Based on this observation, the need to invoke an adaptive meshing solution is not warranted. An adaptive meshing approach was initially employed for the two models used in this study and found to provide similar solutions as the final selected mesh but with a significant increase in computer resources.

RESULTS/OBSERVATIONS

General Observations

The results/observations presented in this section demonstrate qualitative trends applicable to the pilger process. The results/observations reflect workpiece response to the different aspects of the process.

Figure 11 plots the rolling direction stress (direction of workpiece elongation) over the initial pass final increments prior to process completion. The figure demonstrates the relative magnitude of the alternating stress that occurs during the process. Examination of the stress state shows an alternating stress cycle occurs during the rolling portion of the pilger process. Using the accompanying strain values would permit evaluation for surface defect initiation in a qualifying model. Although the stress plots indicate a complete stress reversal exists for the longitudinal stress, the same plots for the longitudinal strain show only a slight decrease in strain values with each unloading. This gradually increasing strain change reflects the elongation that occurs during the course of the rolling process.

Figure 12 shows the through thickness stress state at the same time during the process reflected in Figure 11. Again, a similar stress state reversal is observed to exist for the through thickness state. However, the strain in the through thickness undergoes a constant interval step change increase in the compressive strain state. Each unloading results in a slight relaxation of the compressive state but no strain reversal is shown.

This coincides with the steady reduction in wall thickness and inner diameter.

The total equivalent plastic strain shows a steadily increasing strain state associated with each loading and unloading that occurs during longitudinal simulation of the pilger process. This final value can be used to assess the potential for fatigue induced cracking to occur and as input into developing conditions for determining the generalized plane strain boundaries for use in the cross-section model.

For the cross-section model, the results shown in Figures 13 and 14 represent the initial loading where the roller cam has completed the first increment of loading and unloading prior to rotation. Figure 13 reports the Von Mises stress at the maximum depth of penetration for that specific time during the process. Figure 14 reports the stress state for the unloaded conditions while maintaining the same die positions prior to the simulation of the workpiece rotation. Similar to the rolling direction stress plots, an alternating stress pattern occurs at both the inner and outer diameter surfaces.

The cyclic nature of the loading caused pockets of stress reversal conditions just below the surface and within the material. In addition, there was a significant difference in the final stress state at the points of starting and stopping the process as opposed to those locations representing a continuously applied load. These pockets of flowing material that occur during the forming process representing an alternative material stress state may contribute to the creation of defects on the inside diameter surface.

The results reported provide a proof of principle of the modelling techniques used to generate a finite element model simulation of a complex pilger process.

UNCERTAINTIES

Assumptions and approximations that are made to simplify obtaining a solution may lead to a degree of uncertainty regarding the analysis results/observations/conclusions. Some of these uncertainties and their projected impact are identified below.

The nature of the cyclic results indicates use of a single stress-strain curve depicting the material behavior might not accurately reflect any material hardening that may occur. A cyclic strain controlled test to examine the material behavior history may be warranted as this highly cyclic behavior may contribute to the formation of process generated defects.

The change in reduction generated by a multi-pass process with the subsequent reduction being the greater of the previous reductions, may introduce processing problems similar to those observed in extrusion processes. The extrusion process has the potential to create voids internal to the surfaces if the forces and entrance and exit angles of the die are not compatible with the material (Reference [f]) The increased amount of reduction may contribute to the formation of surface defects. However, until a failure criteria or yield surface can be developed for the selected material under the cold worked conditions, only qualitative comparisons associated with predicting cause and effects can be correlated with process variables.

The coefficient of friction will have a large part on the accuracy of the results. Friction will impact the extrusion process that occurs during rolling. The frictional forces under the roller will be different from those required to flow the material along the mandrel. The frictional forces will have an impact of the strains generated in the material when the workpiece rotates around the mandrel. As the amount of reduction increases the torsional force needed to release the workpiece will also increase. In addition, the potential warping of the workpiece due to the slender cross-section/length and the force to overcome the torsional frictional force may also cause tearing at the inner diameter surface. Hence, friction testing should be considered to accurately reflect the coefficient of friction that exists at the surfaces between the dies and the workpiece.

FUTURE WORK

To enhance the work presented in this report and to eventually be in a position to provide additional guidance to the industry the following steps can be considered. Initially a series of steps can be taken to complete a qualitative assessment of the pilger process. These steps include: (a) model dimensional modification to represent the actual current process parameters including multiple pass material reductions; (b) address both the rolling and cross-section models sufficiently to understand the impact of the input parameters, Q, R, and initial ID to final ID on the results; (c) develop input parameters to include generalized plane strain conditions to better predict final strain and stress states; (d) evaluate sensitivity studies in support of process changes and uncertainty variability; and (e) comparison to current database information as a means of providing model verification/qualification as appropriate. Successful simulation efforts have the potential to provide insight into tubing manufacturing processes which may in turn support pilger process improvements.

Additional technical issues that might be considered should the current modelling techniques prove to be successful include: 1) inside diameter surface defect formation; 2) understanding the mitigation or growth behavior of intentional flaws and experimental results from qualification studies; 3) predicting tooling and pilger machine loads for possible relation to mandrel failure or other equipment problems; 4) understanding and predicting cases of metal pick-up or metal transfer between tubing and tooling (mandrels); and 5) understanding and predicting ID scratching.

A successful pilger modeling effort would provide insight into the above phenomenon. In particular, the pilger modeling effort could make a significant contribution to this field if results would allow for semi-quantitative comment on the role of design parameters (reduction in area, Q-factor, mandrel shape, die side relief, etc.) or operational parameters (stroke rate, feed rate, mandrel-to-die position, etc.) on surface flaw generation. An additional idealized goal would be to have the capability to predict material properties such as microstructure effects, cold work distributions, or texture development.

Future research and development work may also consider combining the strains from a simulated cross-section of the tube during the pilger process with the results of the longitudinal forming simulation in an effort to establish a failure criterion. Comparison of these results with material specific strain fatigue data may quantify the impact of any proposed process changes on the introduction of defects to the finished product. Additional consideration can be made to the examination of grain boundary effects using an approach similar to the Yada formulations reported in (Reference [g]). Lastly the need for qualification testing or data availability may be identified to confirm the quantitative values reported by these modelling efforts.

REFERENCES

[a] "Zirconium Alloy Cold Pilgering Process Control by Modelling", Zirconium in the Nuclear Industry: Twelfth International Symposium, ASTM STP 1354 , J.Aubin et. al. GP Sabol and GD Moan eds, ASTM West Conshohocken, PA 2000 pp 460-481

[b] "A Fully 3D Finite Element Simulation of Cold Pilgering" , S. Mulot; A. Hacquin; P. Montmitonnet, JL Aubin; Journal of Material Processing Technology Vol 60, (1996) pg 505-512

[c] "Anisotropic Yield Locus Evolution During Cold Pilgering of Titanium Alloy Tubing"; R. W. Davies, MA Khaleel, WC Kinsel, HM Zbib; ASME Journal of Engineering Materials and Technology, April 2002, Vol 124 pg 125-134

[d] MSC Marc Software, Version MSC MARC 2001, MSC Software Corporation, SGI Version IRIX 6.2-6.5 2002

[e] "An Overview of the Use of Finite Element Methods for Simulation of Manufacturing and Fabrication Processes" by JA Martin ASME Pressure Vessel and Piping Conference Proceedings ASME PVP-400 Emerging Technologies pg 207-221

[f] Fundamental Finite Element Evaluations of a Three Dimensional Rolled Thread Form: Modelling and Experimental Results by JA Martin, ASME Pressure Vessel and Piping Conference Proceedings ASME PVP-373 Fatigue, Fracture, and Residual Stress 1998 pg 457-467

[g] "Prediction of Microstructural Changes and Mechanical Properties in Hot Strip Rolling" by H. Yada, as presented in the Proceedings of the International Symposium on Accelerated Cooling of Rolled Steel, Winnipeg Canada, August 24-25, 1987, Pergamon Press edited by GE Ruddle & AF Crawley

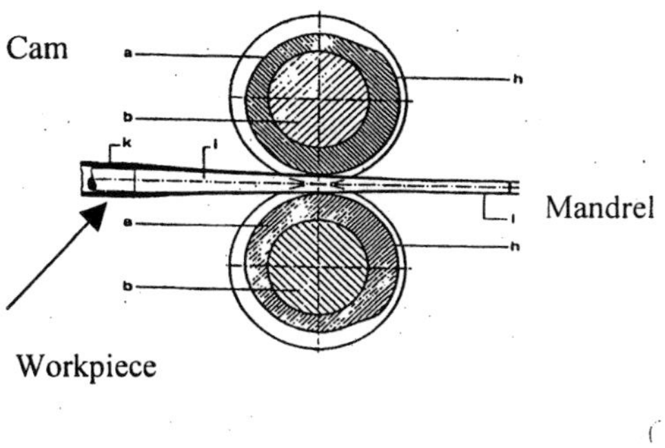

Figure 1 Sketch of a typical Pilger Process

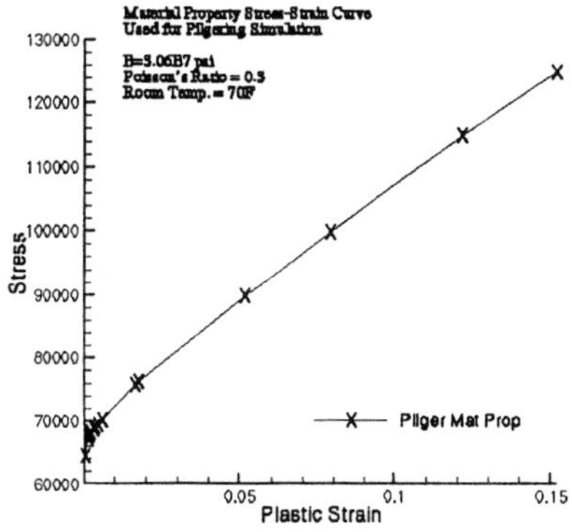

Figure 2 Material Property Stress (psi)-Strain (in/in) Curve Currently used in Finite Element Model Simulations

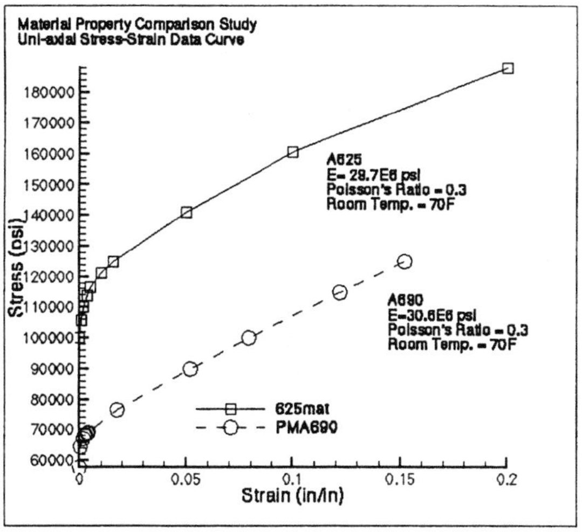

Figure 3 Properties used in Material Comparison

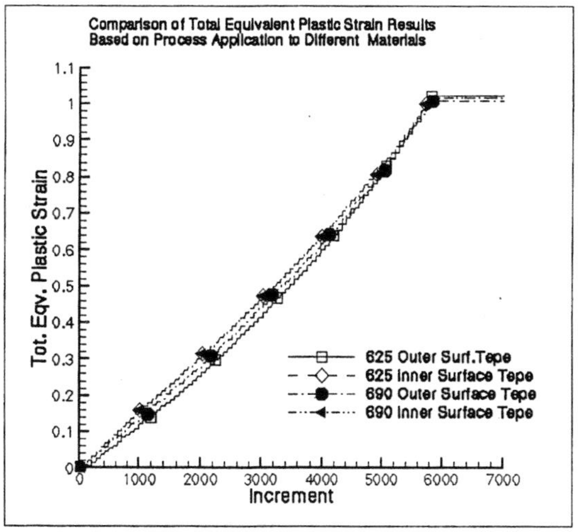

Figure 4 Materials Comparison (time vs in/in)

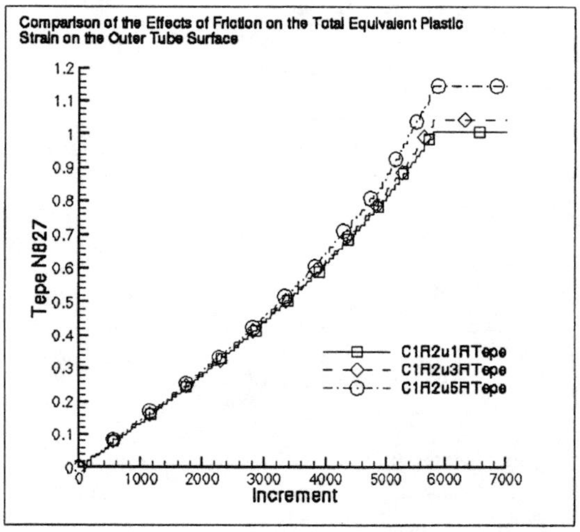

Figure 5 Effects of Varying Coefficient of Friction at Roller-Tube Interface

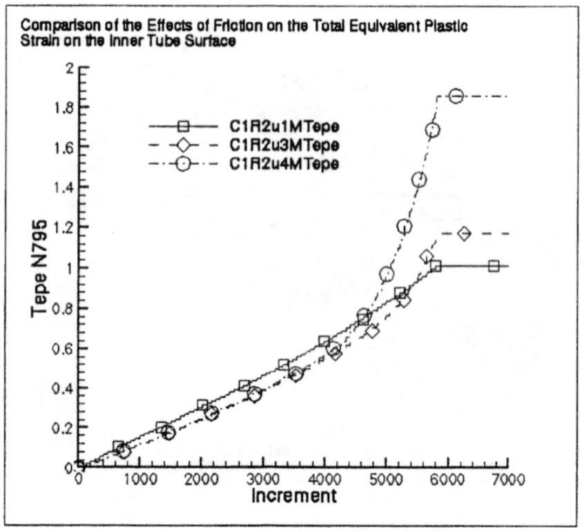

Figure 6 Effects of Varying Coefficient of Friction at the Inner Tube-Mandrel Interface

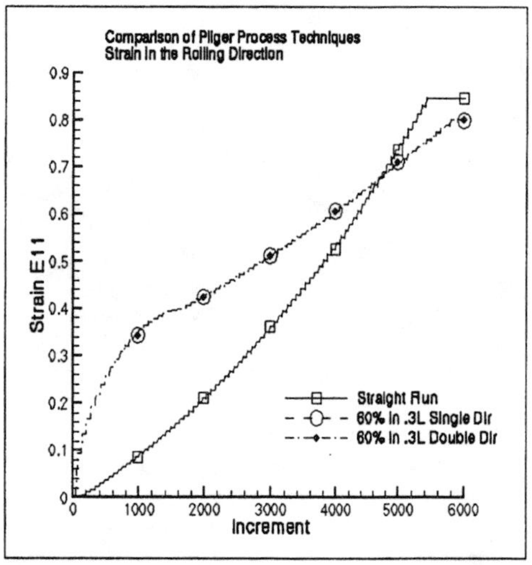

Figure 7 Mandrel Geometry Comparison

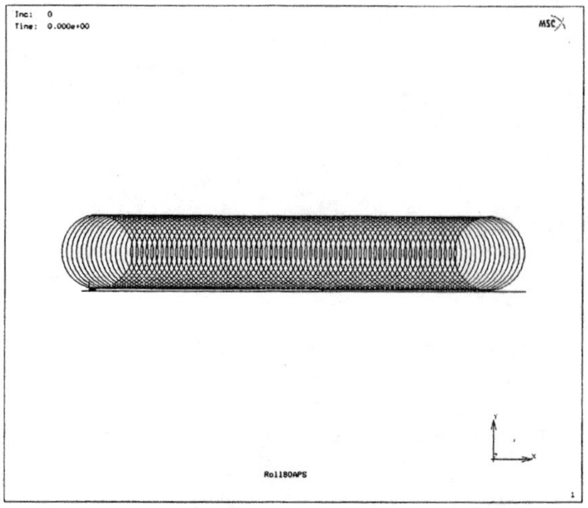

Figure 8 Overall view of the finite element model representative of the rolling portion of the pilger process

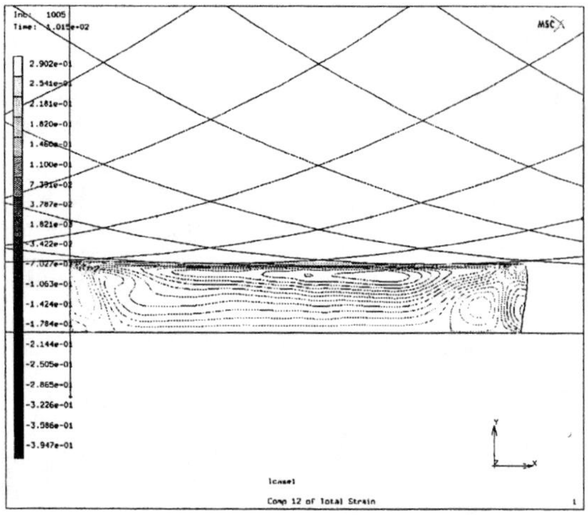

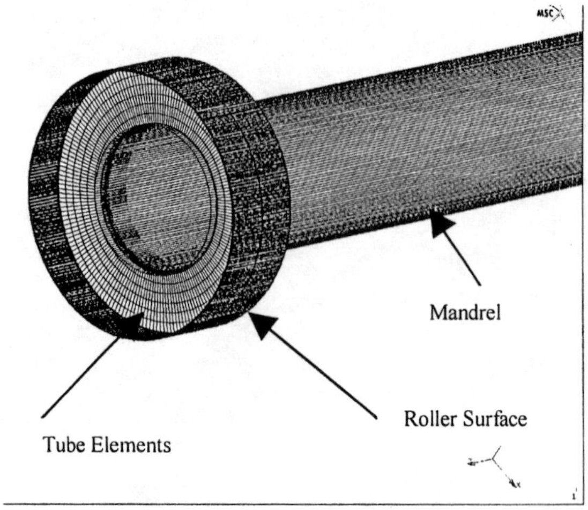

Figure 9 – Shear stress under a single roller at a random time during the process

Figure 10 Cross-section model representing relative roller cam, mandrel, and workpiece positional relationships during the pilger process

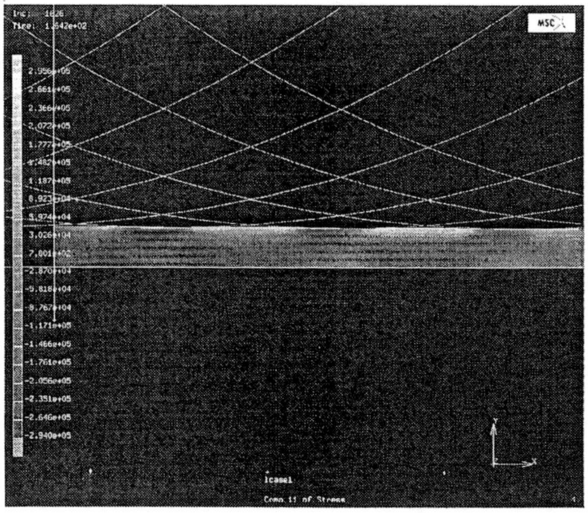

Figure 11 S_{11} stress component of a rolling portion simulation of the pilger process

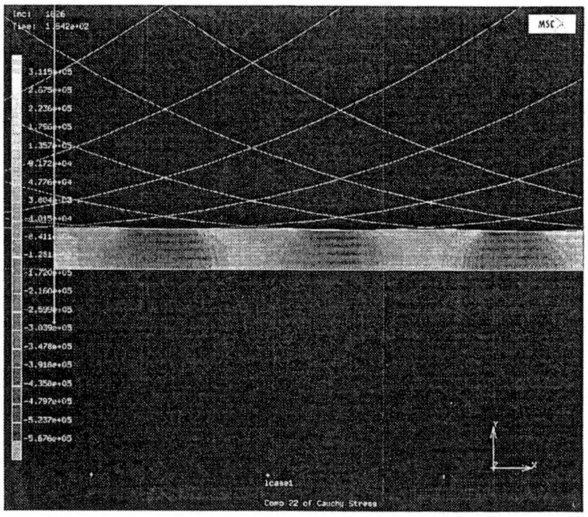

Figure 12 S_{22} stress component of the rolling portion simulation of the pilger process

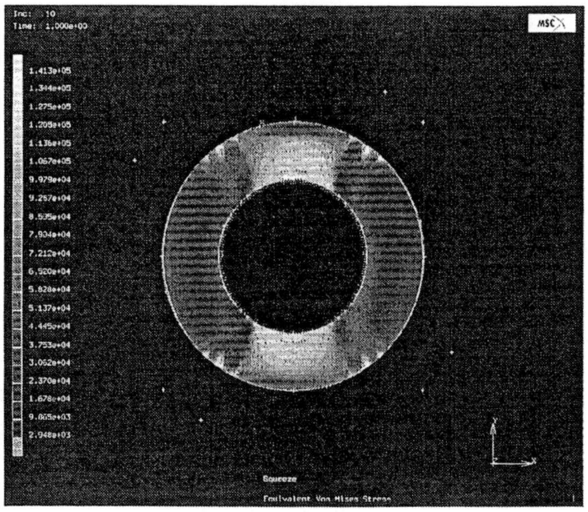

Figure 13 Cross-section model of the pilger process Von Mises stress during first loading step

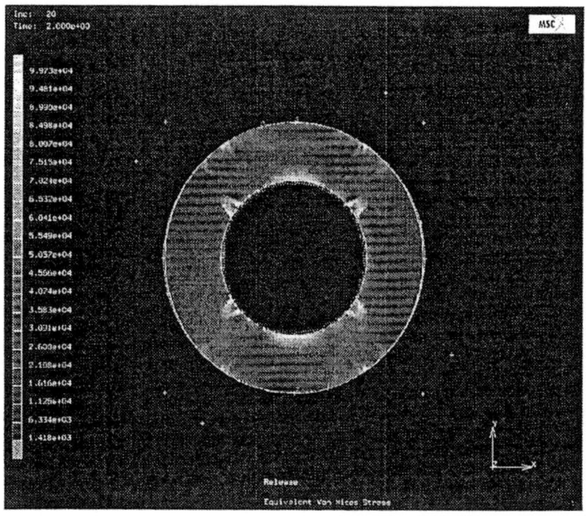

Figure 14 Cross-section model of the pilger process Von Mises stress during first unloading step

NEW AND EMERGING COMPUTATIONAL METHODS

Introduction

Jun Tang and B.N. Rao
The University of Iowa
Iowa City, Iowa

And

Y. H. Park
New Mexico State University
Las Cruces, New Mexico

The new and emerging computational methods are critically important. Therefore, the theory and computational methods have undergone significant development over the last decade, such as meshfree methods and so on. Furthermore, it has also expended the subjects of investigation from theory range to applicable algorithm. To elucidate and expedite the development and application, exchange the views and approaches of the results obtained from various studies are need. The *Computer Technology and Materials and Fabrication committees* have jointly developed the technical sessions on computational methods in pressure boundary integrity for several years, and this is the forth-consecutive year. In this year, three sessions consisting of 12 papers were developed. These papers focus on NEW and Emerging Computational Methods for fracture, fatigue, reliability, meshfree, nonlinear FEA, and elasto-plastic behavior of mesoscopic level. A brief summary of these papers is given below.

First paper presents a methodology for predicting the elasto-plastic behavior of polycrystalline steel at mesoscopic and macroscopic levels. Mesoscopic response of grains is modeled with anisotropic elasticity and crystal plasticity. Possible applications of the proposed approach include the estimation of (a) the minimum component/specimen size needed for the homogeneity assumption to become valid and (b) the correlation lengths in the resulting mesoscopical stress and strain fields, which may be used in well-established macroscopical material models.

The second paper proposes an efficient methodology for fatigue reliability assessment and its corresponding fatigue life prediction of mechanical components using the First-Order Reliability Method (FORM). In exploring the ability to predict spectral fatigue life and assessing the corresponding reliability under a specified dynamics environment, the theoretical background and the algorithm of a simple approach for reliability analysis will first be introduced based on fatigue failure modes of mechanical components. By using this proposed methodology, mechanical component fatigue reliability can be predicted according to different mission requirements.

In the third paper a temperature field had been investigated on a 350x200x400 mm block casting—the so-called "stone"—with a riser of 400 mm using a numerical model with graphical input and output. The paper provides results of the initial computation of the temperature field, which prove that the transfer of heat is solvable, and also that using the numerical model it is possible to optimize the technology of production of this ceramic material, which enhances its utilisation. The results are complemented with an approximated measurement of the chemical heterogeneity of a corundo-badelleyit material, EUCOR.

Fourth paper presents a simple and efficient computational procedure using a geometry based logic and surface mesh generation technique for automated modeling and crack growth simulation for residual strength estimation in pressurized thin-walled shells. The method of analysis presented is relevant to pressurized thin shell piping systems.

Fifth paper presents a new method for continuum shape sensitivity analysis of a crack in an isotropic, linear-elastic functionally graded material. The method involves the material derivative concept of continuum mechanics, domain integral representation of the *J*-integral and direct differentiation.

The sixth paper presents a practical fatigue reliability assessment method for stress-based life design. In this method, an efficient search algorithm for FORM has been used to evaluate reliability factors. The proposed method enables the fatigue reliability model to be used for stress-based fatigue life design applications.

In the seventh paper presents a coupled finite element/meshfree tool for manufacturing problem simulation and its performance compared with traditional finite element analysis.

In the eight paper the behavior and capacity of embedded steel plates in reinforced concrete structures, are studied using a finite-element model developed for non-linear analysis. Strength interaction diagrams and moment-rotation charts useful for analysis and design of such plates, are developed for eccentric compressive and eccentric tensile loading, at failure and collapse. Capacities for a common class of embedded plates with variation in its thickness are computed for the cases of combined compression and shear.

Ninth paper presents two new interaction integrals for calculating stress-intensity factors (SIFs) for a stationary crack in two-dimensional orthotropic functionally graded materials of arbitrary geometry. The method involves the finite element discretization, where the material properties are smooth functions of spatial co-ordinates and two newly developed interaction integrals for mixed-mode fracture analysis.

In the tenth paper, material processing simulation is carried out using a meshfree method. With the use of a meshfree method, the domain of the workpiece is discretized by a set of particles without using a structured mesh to avoid mesh distortion difficulties, which occurred during the course of large plastic deformation. The proposed meshfree method is formulated for rigid-plastic material. This approach uses the flow formulation based on the assumption that elastic effects are insignificant in the metal forming operation. In the rigid-plastic analysis, the main variable of the problem becomes flow velocity rather than displacement.

The eleventh paper presents cellular neural network (CNN) for real-time finite element method, useful as calculating temperature field and thermal stress field for rotor of turbine and so on. The comparability between template of CNN and the stiffness matrix of finite element method is analyzed, and the conception of finite element template (FMT) of CNN is discussed. The FMT can be suitable for finite element grid with arbitrary shape.

In the twelfth paper development of accurate post-processing technique (Modified Virtual Crack Closure Integral) to estimate strain energy release rates, and simple numerical method to simulate crack shape development in single and multiple interacting cracks (till they merge into single dominant crack) is presented.

PVP-Vol. 458, Computer Technology and Applications

PVP2003-1896

ELASTO-PLASTIC BEHAVIOR OF POLYCRYSTALLINE STEEL AT MESOSCOPIC AND MACROSCOPIC LEVELS

Marko Kovač, Igor Simonovski and Leon Cizelj
Jožef Stefan Institute
Reactor Engineering Division
Ljubljana, Slovenia

ABSTRACT

An important drawback of the classical continuum mechanics is idealization of inhomogenous microstructure of materials. Approaches, which model material behavior on mesosocopic level and can take inhomogenous microstructure of materials into the account, typically appeared over the last decade. Nevertheless, entirely anisotropic approach towards material behavior of a single grain is still not widely used.

The proposed approach divides the polycrystalline aggregate into a set of grains by utilizing Voronoi tessellation (random grain structure). Each grain is assumed to be a monocrystal with random orientation of crystal lattice. Mesoscopic response of grains is modeled with anisotropic elasticity and crystal plasticity. Strain and stress fields are calculated using finite element method. Material parameters for pressure vessel steel 22 NiMoCr 3 7 are used in analysis. The analysis is limited to 2D models.

Applications of the proposed approach include (a) the estimation of the minimum component/specimen size needed for the homogeneity assumption to become valid and (b) the estimation of the correlation lengths in the resulting mesoscopical stress fields, which may be used in well-established macroscopical material models. Both applications are supported with numerical examples and discussion of numerical results.

INTRODUCTION

Reliable estimation of the extreme deformations of material (steel) is needed to estimate whether the load will cause the component to fail or not. Therefore a lot of efforts were made during the past few years to determine mechanical properties of polycrystalline aggregates of different sizes.

Inhomogeneity of materials is stretching over whole length scale, from microscale (level of single atoms with typical size up to 1 nm) and mesoscale (crystal grains with typical size of 10–100 μm) to macroscale (machine parts with typical size over 10 mm). Important drawback of classical continuum mechanics, which was not entirely overcome by the existing continuum damage models, is idealization of inhomogenous microstructure of materials [1], [2]. Classical continuum mechanics therefore cannot predict accurately the differences between measured responses of specimens, which are different in size but geometrical similar.

As a result some approaches appeared, which tried to avoid the imperfections of classical continuum mechanics by consideration of material microstructure (e.g., [2]). Approaches, which use stochastic methods to represent microstructure of material and anisotropic material model, were introduced only recently [2], [3], [4].

The main goal of the paper is to propose an approach, which models elastic-plastic behavior on the mesoscopic level. This approach can be used to determine polycrystalline aggregates properties, which may be used in well-established macroscopical material models. Two applications of proposed approach are:

- The estimation of the minimum component/specimen size needed for the homogeneity assumption to become valid.
- The estimation of the correlation lengths in the resulting mesoscopical stress fields.

The analysis is limited to two-dimensional models. Material parameters for pressure vessel steel 22 NiMoCr 3 7 with bainitic microstructure with b.c.c. crystals were used in analysis.

THEORETICAL BACKGROUND

The main idea of proposed approach is to divide continuum (e.g., polycrystalline aggregate) into a set of sub-continua (grains). Response of monocrystal grains is modeled with anisotropic elasticity and crystal plasticity. Overall properties of the polycrystalline aggregate depend on the properties of randomly shaped and oriented grains. Analysis of polycrystalline aggregate is divided into modeling the random grain structure (using Voronoi tessellation and random orientation of crystal lattice) and calculation of strain/stress field. Finite element method, which proved as suitable [2], is used to obtain numerical solutions of strain and stress fields.

VORONOI TESSELLATION

The concept of Voronoi tessellation has recently been extensively used in materials science, especially for modeling random microstructures like aggregates of grains in polycrystals, patterns of intergranular cracks, and composites [4], [5], [6]. A Voronoi tessellation represents a cell structure constructed from a Poisson point process by introducing planar cell walls perpendicular to lines connecting neighboring points. This results in a set of convex polygons/polyhedra (Figure 1) embedding the points and their domains of attraction, which completely fill up the underlying space. All Voronoi tessellations used for the purpose of this paper were generated by the code VorTess [7].

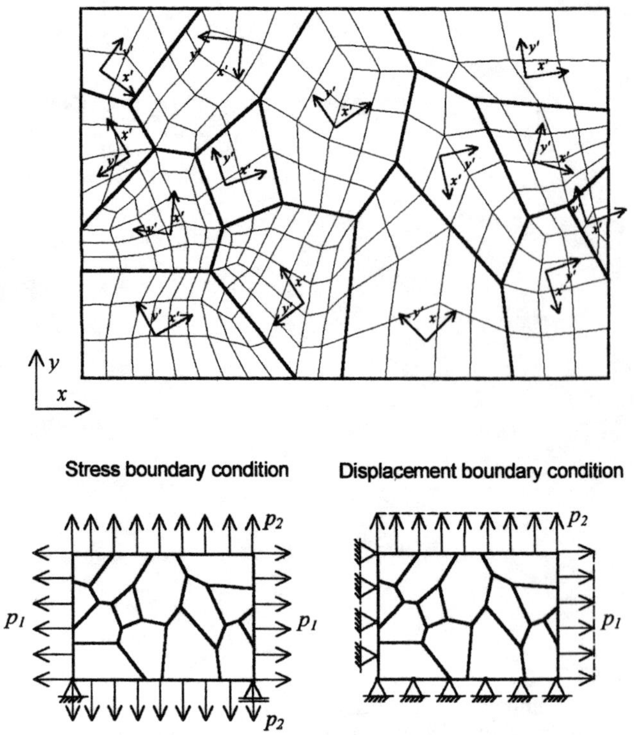

Figure 1: Voronoi tessellation of 14-grains aggregate with grain boundaries, orientations of crystal lattices, finite element mesh and boundary conditions

ANISOTROPIC ELASTICITY AND CRYSTAL PLASTICITY

Selected material 22 NiMoCr 3 7 has body-centered cubic crystal lattice with rather pronounced orthotropic elasticity. Each crystal grain is assumed to behave as a randomly oriented anisotropic continuum [8], [9]. The elastic properties (e.g., stiffness and compliance tensor) of the polycrystalline aggregate are completely defined by the properties of, and interaction between, the crystal grains.

Crystal plasticity assumes that plastic deformation is a result of crystalline slip only and therefore strongly depends on orientation of crystal lattice (for details see [5], [10] and references therein). Slip planes and direction are defined by orientations of crystal lattices and differ from grain to grain (random orientation). Rate-independent plasticity with Peirce et al. [11] and Asaro [12] hardening law was used in our research.

Material parameters for pressure vessel steel 22 NiMoCr 3 7 were obtained from literature (e.g., [8], [9], [13]) and from results of simple tensile test of pressure vessel steel 22 NiMoCr 3 7 [14]. Further details are available in [15], [16] and references therein.

ESTIMATION OF REPRESENTATIVE VOLUME ELEMENT SIZE

Large enough and geometrical similar components have equal macroscopic material response. On the macroscale these components appear homogeneous in statistical sense, regardless of mesoscopic material inhomogeneity [13]. However, this is not the case with small size components, where microstructure might play important role. Estimating minimal component size needed for the homogeneity assumption to become valid is therefore essential to determine effective range of classical continuum mechanics. This minimal component size is usually called representative volume element (RVE) [17]. The homogeneity assumption becomes valid (i.e., component/specimen is larger than RVE) when [17]:

$$C^*_{ijkl} \cong \left(D^*_{ijkl}\right)^{-1}, \qquad (1)$$

where C^*_{ijkl} and D^*_{ijkl} are macroscopic stiffness and compliance tensors of a polycrystalline aggregate) [17].

Equation (1) in general is not valid for components smaller than RVE. The different behavior of both tensors is governed by the size of the aggregate and macroscopic boundary conditions [10]: macroscopic stiffness tensor therefore assumes stress boundary condition, while macroscopic compliance tensor assumes displacement boundary condition. With general relations between stresses and strains in mind, eq. (1) can be simplified by use of macroscopic equivalent (Von Mises) stresses $<\sigma_{eq}>$:

$$\left\langle \sigma_{eq_s} \right\rangle \cong \left\langle \sigma_{eq_d} \right\rangle, \qquad (2)$$

where indexes $_s$ and $_d$ denote stress and displacement driven boundary conditions, respectively. Macroscopic equivalent stresses $<\sigma_{eq}>$ are calculated as:

$$\left\langle \sigma_{eq} \right\rangle = \frac{1}{V} \int_V \sigma_{eq} \, dV, \qquad (3)$$

where σ_{eq} stands for equivalent stress and V for volume of polycrystalline aggregate [15].

A relation between macroscopic equivalent stresses for both boundary conditions for a polycrystalline aggregate smaller than RVE can be written as [5]:

$$\frac{\left\langle \sigma_{eq_s} \right\rangle}{\left\langle \sigma_{eq_d} \right\rangle} = 1 + O\left(i/i_{RVE}\right), \qquad (4)$$

where i_{RVE} represents number of grains in RVE and i number of grains size in polycrystalline aggregate smaller than RVE. A RVE is achieved, when residuum O is smaller than 1% [18].

CORRELATION LENGTH

The numerical effort to simulate elastic-plastic behavior of polycrystalline aggregates approaching RVE may easily outgrow the existing computational capabilities. This fact severely limits the size of the specimen to be simulated.

The required numerical effort could in principle be reduced if the essential inhomogenities are identified and appropriately transferred to the macroscopic models. A frequently used and promising method is to calculate the domain of influence of crystal grains [19]. The correlation length [20] is one of the criteria for estimating statistical dependency of random variable (e.g., [21], [22]). The correlation length determines the decay of the mutual influence of two different random field locations and is a measure for the number of uncorrelated random variables, which are required to describe the random field with a given quality [23]. The correlation length in our research is calculated from the equivalent stress field, which is determined for every Gaussian integration point of the finite elements (Figure 2). Since stress is a 2D variable, the vector of data for the correlation length calculation has to be extracted. The correlation length is calculated for the selected direction at calculation point, with the length of the vector determined by the search radius. Predefined directions were used covering the range of 360° with the 30° intervals.

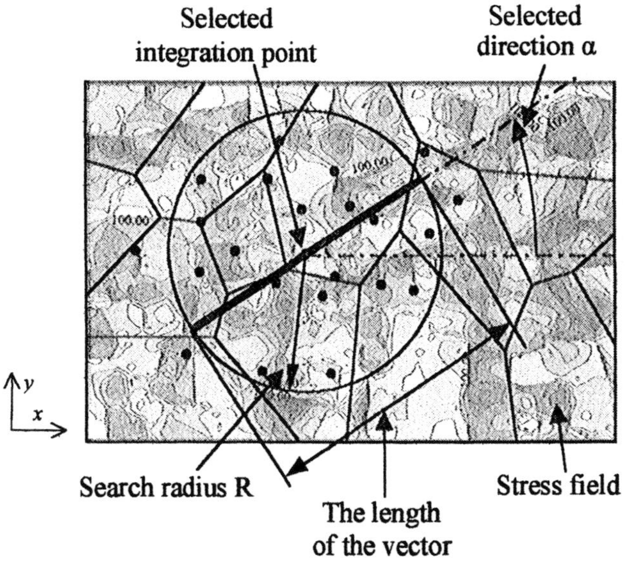

Figure 2: Procedure of correlation length calculation

The search radius R (Figure 2) determines the length of the vector of data and therefore the information contained in the vector. If the search radius is very small, the amount of information contained in the data vector could be to small for a meaningful estimation of the autocorrelation (covariance) function. On the other hand, if the search radius is large, the local nature of the calculated autocorrelation function is lost. The search radius is therefore a compromise between these two aspects. The search radius was therefore set to two times the average grain size (2×0.023 mm) to obtain a meaningful estimation of autocorrelation function and not to lose its local nature. Further details are available in [19], [16] and references therein.

RESULTS AND DISCUSSION

Proposed approach was used to calculate mesoscopic stress and strain fields in polycrystalline aggregates. Two macroscopic properties were then estimated, RVE size and correlation lengths in stress fields.

ESTIMATION OF RVE SIZE

RVE size was estimated for elasticity and plasticity. Analyses were carried out on polycrystalline aggregates with 14, 23, 53, 110 and 212 grains with sizes from 0.1 mm × 0.07 mm to 0.4 mm × 0.28 mm. 30 different random orientations of crystal lattices and 2 boundary conditions (stress and displacement boundary conditions) were analyzed for each polycrystalline aggregate. Analyses were carried out at biaxial loads p_1 = 200 MPa and p_2 = 100 MPa for elasticity and p_1 = 1000 MPa and p_2 = 500 MPa for plasticity (Figure 1).

Equivalent macroscopic strains and stresses are shown in Figure 3 (elasticity) and Figure 4 (plasticity) where *d* in the legend refers to displacement boundary condition, *s* refers to stress boundary condition and *ave* refers to average values (averaged over 30 different randomly orientated crystal lattices with displacement or stress boundary conditions). The number following abbreviation denotes number of grains of polycrystalline aggregates with separate values for each boundary condition. Analytical solution for elasticity for macroscopically homogenous material ($<\varepsilon_{eq}>$ = 0.0515% and $<\sigma_{eq}>$ = 96.2 MPa) is also shown in Figure 3 and Figure 4 obtained by continuum elasticity with material parameters: E = 210 GPa and ν = 0.29 [4].

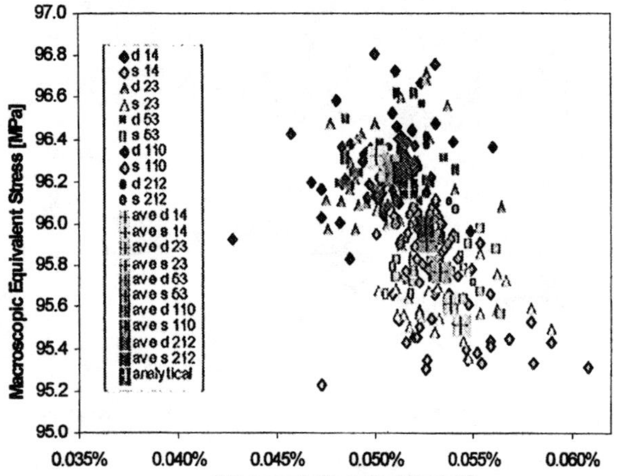

Figure 3: Scatter of macroscopic equivalent strain/stress in elasticity

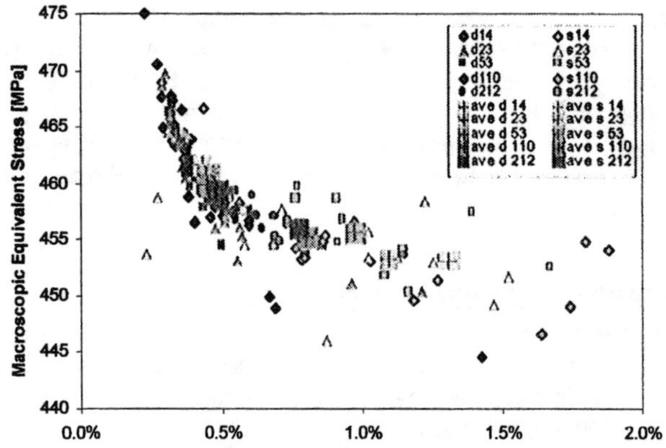

Figure 4: Scatter of macroscopic equivalent strain/stress in plasticity

A tendency towards decrease of scatter as number of grains in the aggregates increases can be observed. Average values of macroscopic strains and stresses (for both boundary conditions) show a clear trend towards analytical solution with increasing number of grains in the aggregate.

RVE size was estimated according to eq. (4). Macroscopic equivalent stresses were taken at macroscopic equivalent strain $<\varepsilon_{eq}>$ of 0.0515 % (elasticity) and 1% (plasticity). Figure 5 (elasticity) and Figure 6 (plasticity) show macroscopic equivalent stresses and scatter depending on number of grains in polycrystalline aggregate for displacement (denoted as *d*) and stress (denoted as *s*) driven boundary conditions. Extrapolation lines for average values are drawn in line with eq. (4).

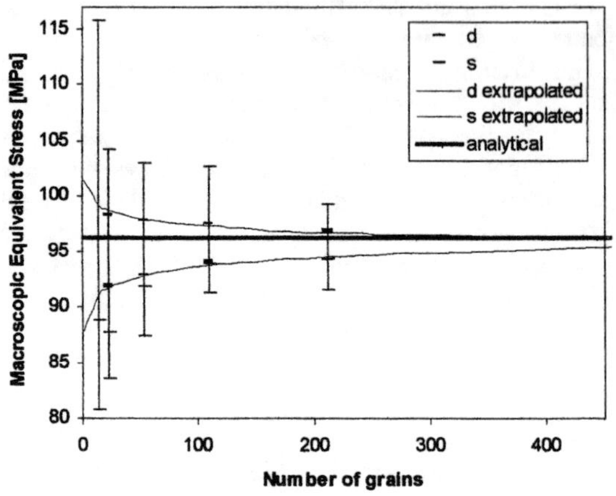

Figure 5: Convergence of macroscopic equivalent stresses in elasticity

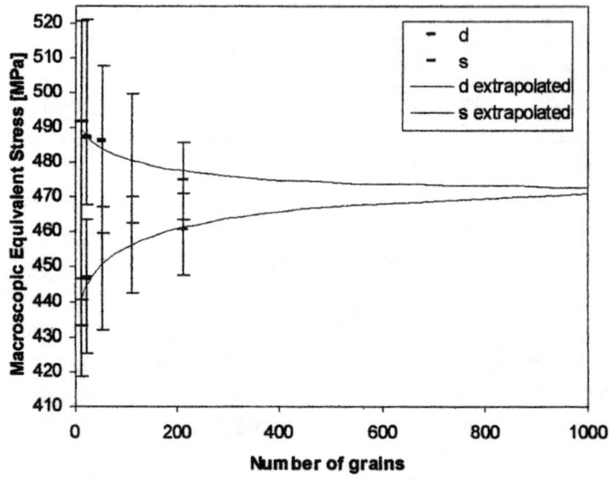

Figure 6: Convergence of macroscopic equivalent stresses in plasticity

With residuum O for 212-grains aggregate over 2% for elasticity and 5% for plasticity, one can conclude that RVE has not been achieved. However, trend toward analytical solution and decrease of scatter with increasing number of grains is clearly visible. The RVE size in elasticity is estimated to 380 grains, which corresponds to a polycrystalline aggregate of about 0.45 mm in size. This is comparable with results from literature for aluminum oxide [18]. The RVE size in plasticity is estimated to 800 grains, which corresponds to a polycrystalline aggregate of about 0.66 mm in size.

Minimal component size needed for the homogeneity assumption to become valid suggests that the assumption of homogeneity of material is valid when dealing with elastic-plastic response of component/specimen larger than 0.66 mm. However, consideration of grain structure of material might be important, when dealing with smaller component or when local processes, which develop at mesoscopic level, are of relatively large importance (e.g., initialization and growth of microcracks etc.).

CORRELATION LENGTHS

Correlation lengths for a 212-grains polycrystalline aggregate with displacement boundary condition were calculated at different biaxial loads. Figure 7 and Figure 8 show correlation lengths fields for elasticity ($p_1 = 210$ MPa, $p_2 = 105$ MPa) and plasticity ($p_1 = 1400$ MPa, $p_2 = 700$ MPa), respectively. Grain structure is also presented.

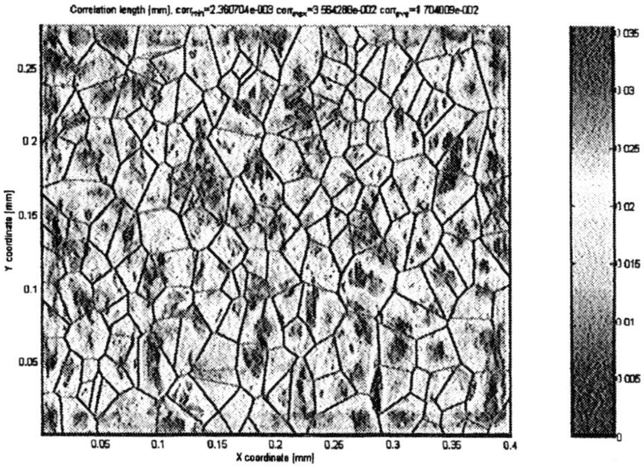

Figure 7: Correlation length for 212-grains polycrystalline aggregate in elasticity

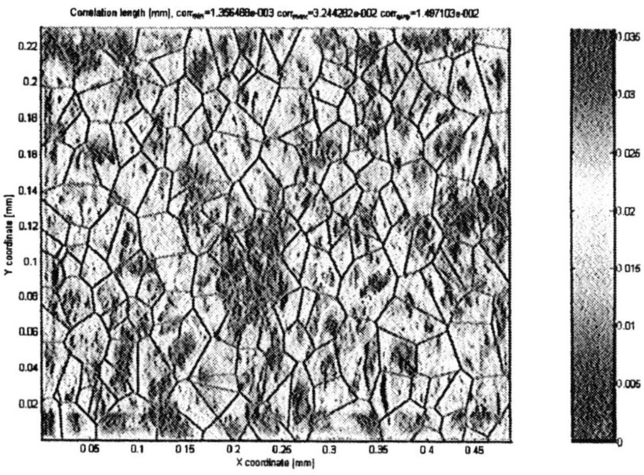

Figure 8: Correlation length for 212-grains polycrystalline aggregate in plasticity

It can be observed from Figure 7 and Figure 8 that correlation lengths in plasticity are smaller than correlation lengths in elasticity. Similar can be observed in Figure 9, which shows histogram of correlation length (regarding area of polycrystalline aggregate) for different biaxial loads.

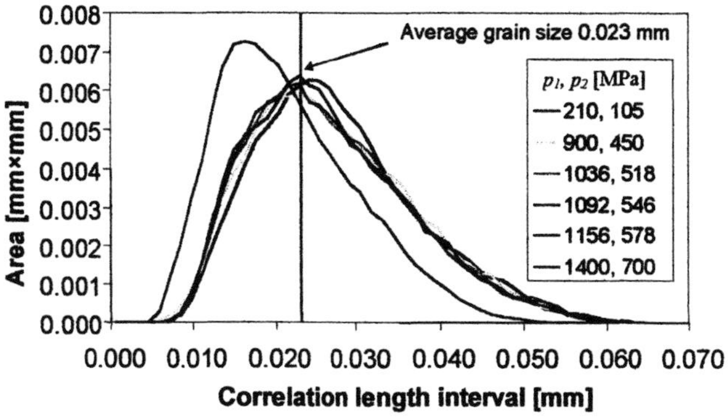

Figure 9: Histogram of correlation length for different biaxial loads

The maximal correlation length in elasticity ($p_1 = 210$ MPa, $p_2 = 105$ MPa) is 0.065 mm, corresponding to 2.83× the average size of a crystal grain. The average value of the correlation length is 0.027 mm, which is 18 % larger than the average crystal grain size. The average correlation length in plasticity ($p_1 = 1400$ MPa and $p_2 = 700$ MPa) is 0.021 mm, which is just below the average grain size. Correlation lengths in plasticity (with macroscopic equivalent stresses larger than yield strength $\sigma_Y = 440$ MPa) tend to decrease with increase of loads. This suggests that the equivalent stress fields in the search areas become increasingly more random as the load and the macroscopic equivalent stress increase. The area of influence of the finite elements and crystal grains therefore reduces as the loads increase.

Correlation lengths for used material calculated in elasticity and plasticity are in the order of crystal grain size. The correlation length enables estimation of the domain of influence of the individual crystal grain and could be used as a measure for essential inhomogeneities in the appropriate macroscopic models.

SUMMARY

A numerical approach, which models elastic-plastic behavior on mesoscopic level, was proposed. Presented approach combines the most important mesoscale features and compatibility with conventional continuum mechanics to model elastic-plastic behavior. Explicit modeling of the random grain structure is used. Grains are regarded as monocrystals and modeled with anisotropic elasticity and crystal plasticity.

Proposed approach was used to estimate two properties of the selected material, which can be used in well-established macroscopic material models. The minimum component/specimen size needed for the homogeneity assumption to become valid was estimated to be around 0.45 mm in elasticity and around 0.66 mm in plasticity.

The estimation of domain of influence of the individual crystal grain was based on calculation of correlation lengths. Equivalent stresses were used to obtain correlation lengths. Average correlation length is 0.027 mm in elasticity and 0.021 mm in plasticity. Correlation length is in the order of grain size, which is within the expectations. The area of influence of crystal grains reduces as the loads increases.

Future work will include broadening proposed approach with simulation of initialization and growth of microcracks and its effect on estimations of minimum component/specimen size for the homogeneity assumption validity and correlation lengths.

REFERENCES

[1] Needleman, A. Computational Mechanics at the Mesoscale. Acta Materialia. 2000; 48:105-124.

[2] Watanabe, Osamu; Zbib, Hussein M., and Takenouchi, Eiji. Crystal Plasticity, Micro-shear Banding in Polycrystals using Voronoi Tessellation. International Journal of Plasticity. 1998; 14(8):771-778.

[3] Nygards, M. and Gudmundson, P. Three-dimensional periodic Voronoi grain models and micromechanical FE-simulations of a two-phase steel. Computational Materials Science. 2002; 24(4):513-519.

[4] Kovač, Marko and Cizelj, Leon. Mesoscopic Approach to Modeling Elastic-Plastic Polycrystalline Material Behavior. Proc. of Int. Conf. Nuclear Energy in Central Europe 2001; Portorož, Slovenia. 2001.

[5] Kovač, M.; Simonovski, I., and Cizelj, L. Estimating Minimum Polycrystalline Aggregate Size for Macroscopic Material Homogeneity. Proc. of Int. Conf. Nuclear Energy for New Europe; Kranjska Gora, Slovenia. in print.

[6] Aurenhammer, F. Voronoi Diagrams-A Survey of a Fundamental Geometric Data Structure. ACM Computing Surveys. 1991; 23(3):345-405.

[7] Riesch-Oppermann, H. VorTess, Generation of 2-D random Poisson-Voronoi Mosaics as Framework for the Micromechanical Modelling of Polycristalline Materials. Karlsruhe, Germany: Forschungszentrum Karlsruhe; 1999; Report FZKA 6325.

[8] Nye, J. F. Physical Properties of Crystals. Oxford: Clarendon Press; 1985.

[9] Grimvall, G. Thermophysical Properties of Materials. Amsterdam: North-Holland; 1999.

[10] Huang, Y. A User-material Subroutine Incorporating Single Crystal Plasticity in the ABAQUS Finite Element Program. Cambridge, Massachusetts: Harvard University; 1991; MECH-178.

[11] Peirce, D.; Asaro, R. J., and Needleman, A. Material Rate Dependence and Localized Deformation in Crystalline Solids. Acta Metallurgica. 1982; 31:1951.

[12] Asaro, R. J. Micromechanics of Crystals and Polycrystals. Advances in Applied Mechanics. 1983; 23:1-115.

[13] Nemat-Nasser, S.; Okinaka, T., and Ni, L. A Physically-based Constitutive Model for BCC Crystals with Application to Polycrystalline Tantalum. Journal of the Mechanics and Physics of Solids. 1998; 46 (6):1009-1038.

[14] Tensile Tests, Specimens 5 mm for LISSAC. EMPA; 2001; Test Report, Nr. 201951/01.

[15] Kovač, M.. Influence of Microstructure on Development of Large Deformations in Reactor Pressure Vessel Steel. Ljubljana: University of Ljubljana; to be submitted.

[16] Cizelj, L.; Kovač, M.; Simonovski, I.; Petrič, Z.; Fabjan, L., and Mavko, B. Elastic-Plastic Behavior of Polycrystalline Aggregate with Stochastic Arrangement of Grains. Project LISSAC: Final Report. Rev 0. Ljubljana, Slovenia: Jožef Stefan Institute; 2002; IJS-DP-8667.

[17] Nemat-Nasser, S. and Hori, M. Micromechanics: Overall Properties of Heterogeneous Materials. Amsterdam: North-Holland; 1993.

[18] Weyer, S. Experimentelle Untersuchung und mikromechanische Modellierung des Schädigungsverhaltens von Aluminiumoxid unter Druckbeanspruchung. Karlsruhe, Germany: Universität Karlsruhe; 2001.

[19] Simonovski, I.; Kovač, M., and Cizelj, L. Correlation Length as Estimate of the Domain of Influence of Crystal Grain. Proc. of Int. Conf. Nuclear Energy for New Europe 2002; Kranjska Gora; in print.

[20] Grabec, I and Gradisek, J. Opis naključnih pojavov. Ljubljana: University of Ljubljana, Faculty of Mechanical Engineering; 2000.

[21] Carmeliet, J. and de Borst, R. Stochastic approaches for damage evolution in standard and non-standard continua. International Journal of Solids and Structures. 1995; 32(8):1149-1160.

[22] Borbély, A; Biermann, H, and Hartmann, O. FE investigation of the effect of particle distribution on the uniaxial stress-strain behaviour of particulate reinforced metal-matrix composites. Material Science and Engineering . 2001; A13:34-45.

[23] Schenk, C. A. and Schuëller, G. I. Buckling analysis of cylindrical shells with random geometric imperfections. International Journal of Non-Linear Mechanics. 2003; 38(7):1119-1132.

PVP-Vol. 458, Computer Technology and Applications

PVP2003-1897

STUDY OF EFFECTIVE ELASTIC MODULI OF CRACKED SOLID

Young H. Park
Assistant Professor
Mechanical Engineering Department
New Mexico State University
Las Cruces, NM 88003-8001
e-mail: ypark@nmsu.edu

Wesley Morgan
Graduate Student
Mechanical Engineering Department
New Mexico State University
Las Cruces, NM 88003-8001

ABSTRACT

In this paper, effective moduli of cracked solid material were investigated. An analytical approach is discussed for a cracked solid containing randomly oriented inclusions by using elastic potential and a standard tensorial basis. A numerical simulation of the testing of mechanical responses of samples of cracked solid material (porous material) is also carried out. The numerical scheme in this work will focus mainly on numerical modeling of the observed behavior, in particular, the dependence of the macroscopic material properties on the porocity.

1. INTRODUCTION

Many structural materials show microhetrogeneities which can significantly influence the macroscopic properties. These materials are found in nature as well as man-made situations. Typical examples are metal alloy systems, polymer blends, porous and cracked solid, and composite materials. Determination of the macroscopic material properties of such materials requires consideration of microheterogeneities. The development of analytical techniques for estimating the macro properties of structured media has taken varied paths over the past several decades, and various approximation methods have been developed (Hill, 1963; Budansky, 1965; Willis, 1977; Devries et al., 1989; Kachanov, 1992; Sevostianov and Kachanov, 2002).

In many engineering applications, microstructural fields within a macroscopic body often require the mathematically sound numerical scheme for their determination. Along with the development of computational methods so called unit cell methods become widely used. These methods are based on the concept of a representative volume element (RVE), originally introduced by Hill (1963). The unit cell methods allow to account easily for a complex microstructural morphology and enable the investigation of the influence of different geometrical features on the overall response (Christman et al., 1989; Tvergaard, 1990; McHugh et al., 1993). The latest work consists of describing the aspects of numerical or computational testing in addition to finding ways to reduce the complexities of such testing or modeling (Zohdi and Wriggers, 1999; Zohdi and Wriggers, 2001; Kouznetsova, Brekelmans, and Baaijens, 2001).

The problem of effective elastic properties of cracked solids is of interest in this paper. The effects of the microstructure on the mechanical properties of the porous material under the assumption of noninterating cracks will be reviewed and the numerical simulation of the testing of mechanical responses of samples of cracked solid material (porous material) will be carried out. The numerical scheme in this work will focus mainly on numerical modeling of the observed behavior, in particular, the dependence of the macroscopic material properties on the porocity.

2. STIFFNESS AND COMPLIANCE CONTRIBUTION TENSORS OF AN ICLUSION

2.1 Elastic Fields for Isotropic Ellipsoidal Inclusion

Consider a reference volume V_0 of an infinite three-dimensional medium containing an inclusion V_1. The compliance contribution tensor $\boldsymbol{H}$ of an inclusion can be defined by the following relation for the overall strain per volume V_0 (Sevostianov and Kachanov, 2002):

$$\varepsilon = \boldsymbol{M}_0\sigma + \boldsymbol{H}:\sigma \tag{2.1}$$

where the second term represents the strain change $\Delta\varepsilon$ due to the presence of the inclusion. The compliance contribution tensor $\boldsymbol{H}$ depends on the inclusion shape and its elastic properties. Similary, the stiffness contribution tensor $\boldsymbol{N}$ of an inclusion is defined by

$$\sigma = \boldsymbol{C}_0:\varepsilon + \boldsymbol{N}:\varepsilon \tag{2.2}$$

For the ellipsoidal inclusion, if a uniform stress (or strain) field at infinity is prescribed, then the resulting uniform strains and stresses inside the inclusion can be represented as (Kunin, 1983):

$$\left[\boldsymbol{J} + \boldsymbol{P}:(\boldsymbol{C}_1 - \boldsymbol{C}_0)\right]:\varepsilon_{in}=\varepsilon_0 \tag{2.3}$$

if strains ε_0 are prescribed at infinity, and

$$\left[\boldsymbol{J} + \boldsymbol{Q}:(\boldsymbol{M}_1 - \boldsymbol{M}_0)\right]:\sigma_{in}=\sigma_0 \tag{2.4}$$

if stresses σ_0 are prescribed at infinity. In Eqs. (2.3) and (2.4), $\mathcal{J}$ is unit forth rank tensor defined by

$$J_{ijkl} = (\delta_{ik}\delta_{jl}+\delta_{il}\delta_{jk})/2 \quad (2.5)$$

and $\boldsymbol{P}$ and $\boldsymbol{Q}$ can be expressed in terms of the famed Eshelby's tensor $\boldsymbol{S}$:

$$\boldsymbol{P} = \boldsymbol{S}:\boldsymbol{M}_0 \quad (2.6)$$

$$\boldsymbol{Q} = \boldsymbol{C}_0 : (\boldsymbol{J} - \boldsymbol{P} : \boldsymbol{C}_0) \quad (2.7)$$

Since stiffness tensor $\boldsymbol{C}$ and compliance tensor $\boldsymbol{M}$ for the elastic body have the following relation:

$$\boldsymbol{C}:\boldsymbol{M} = \boldsymbol{J} = \boldsymbol{M}:\boldsymbol{C} \quad (2.8)$$

the tensor $\boldsymbol{Q}$ can be rewritten, using Eshelby tensor $\boldsymbol{S}$, as

$$\boldsymbol{Q} = \boldsymbol{C} : (\boldsymbol{J} - \boldsymbol{P}:\boldsymbol{C}) = \boldsymbol{C} : (\boldsymbol{J} - \boldsymbol{S}: \boldsymbol{M}:\boldsymbol{C})$$

$$= \boldsymbol{C}: (\boldsymbol{J} - \boldsymbol{S}) \quad (2.9)$$

Using these results, the tensors of compliance and stiffness contribution of the inclusion can be obtained. Consider the volume average strain and stress relation:

$$\begin{aligned}\bar{\varepsilon} &= \boldsymbol{M}^*:\bar{\sigma} \\ &= \eta_0 \boldsymbol{M}_0 :< \sigma >_0 + \eta_1 \boldsymbol{M}_1 :< \sigma >_1\end{aligned} \quad (2.10)$$

where η_i is volume fraction of i-th phase and volume average stress $<\sigma>$ is defined as

$$< \sigma > = \frac{\int_V \sigma dV}{\int_V dV} \quad (2.11)$$

Equation (2.10) be rewritten as follows:

$$\boldsymbol{M}^*:\bar{\sigma} = \boldsymbol{M}_0 : (\bar{\sigma} - \eta_1 < \sigma >_1) + \eta_1 \boldsymbol{M}_1 :< \sigma >_1$$

$$(\boldsymbol{M}^* - \boldsymbol{M}_0):\bar{\sigma} = \eta_1 (\boldsymbol{M}_1 - \boldsymbol{M}_0) :< \sigma >_1 \quad (2.12)$$

Using Eq. (2.4) and the following relations

$$\boldsymbol{H} \equiv \boldsymbol{M}^* - \boldsymbol{M}_0$$

$$\eta_1 = \frac{V_*}{V} \quad (2.13)$$

Eq. (2.12) can be rearranged as

$$\frac{V_*}{V}(\boldsymbol{M}_1 - \boldsymbol{M}_0)^{-1}:\boldsymbol{H} = \left[\boldsymbol{J}\text{-}+\boldsymbol{Q} : (\boldsymbol{M}_1 - \boldsymbol{M}_0)\right]^{-1} \quad (2.14)$$

The compliance contribution tensor $\boldsymbol{H}$ is then given by

$$\boldsymbol{H} = \frac{V_*}{V}\left[(\boldsymbol{M}_1 - \boldsymbol{M}_0)^{-1} + \boldsymbol{Q}\right] \quad (2.15)$$

2.2 Elastic Potential

The approach used in this paper is based on the elastic potential in stresses (complementary energy density) of a representative volume element. The potential energy f for a solid with one cavity can be represented as a sum of two terms:

$$\begin{aligned}\mathrm{f} &= \frac{1}{2}\sigma : \varepsilon = \frac{1}{2}\sigma : \boldsymbol{M}_0 : \sigma + \frac{1}{2}\sigma : \boldsymbol{H} : \sigma \\ &= \mathrm{f}_0 + \Delta\mathrm{f}\end{aligned} \quad (2.16)$$

where f_0 is the potential in the absence of a cavity and Δf is the change in potential due to the cavity.

Derivation of the elastic potential in terms of proper parameters of cavity density was studied by Kachanov (Kachanov, 1992). Consider an ellipsoidal cavity with the surface boundary Γ and axes $2a_1$, $2a_2$, $2a_3$. Δf is defined as (Kachanov, 1992)

$$\begin{aligned}\Delta\mathrm{f} = \frac{V^*}{V}\{&W_1(\mathrm{tr}\,\sigma)^2 + W_2 \mathrm{tr}(\sigma \bullet \sigma) + [W_3(\mathrm{tr}\,\sigma)\sigma + W_4(\sigma \bullet \sigma)]:\mathbf{nn} \\ &+ W_5 \sigma:\mathbf{nnnn}:\sigma\}\end{aligned} \quad (2.17)$$

where

$$\begin{aligned}W_1 &= h_1 - h_2/2 \\ W_1 &= h_1 - h_2/2 \\ W_2 &= h_2 \\ W_3 &= 2h_3 - 2h_1 + h_2 \\ W_4 &= h_5 - 2h_2 \\ W_5 &= h_1 + h_2/2 - 2h_3 - h_5 + h_6\end{aligned} \quad (2.18)$$

and h_i are components of the compliance tensor H_{ijkl}.

2.3 Tensorial Basis

The operations of analytic inversion and multiplication of fourth rank tensors are conveniently done in terms of special tensorial basis $\boldsymbol{T}^{(1)}$ that are formed by combinations of unit tensor δ_{ij} and one or two unit vectors (Knin, 1983):

$$S_{ijkl} = \Sigma s_m T^{(m)}_{ijkl},\ Q_{ijkl} = \Sigma q_m T^{(m)}_{ijkl},\ C_{ijkl} = \Sigma c_m T^{(m)}_{ijkl} \quad (2.19)$$

so that finding these tensors reduces to calculation of factors s_m, q_m, and c_m. For example, general transversely isotropic fourth-rank tensor S_{ijkl} being represented in this basis has the following components:

$$s_1 = (S_{1111} + S_{1122})/2\,;\ s_2 = 2S_{1212}\,;\ s_3 = S_{1133}\,;\ s_4 = S_{3311} \quad (2.20)$$

$$s_5 = 4S_{1313}\,;\ s_6 = S_{3333}$$

Using the solution of Eshelby's problem (Eshelby, 1957), one can find components of Eshelby's tensor S_{ijkl} for penny shape inclusion ($a_1 = a_2 >> a_3$; see Fig. 1) as follows:

$$S_{1111} = S_{2222} = \frac{13 - 8\nu}{32(1-\nu)}\pi\frac{a_3}{a_1},\quad S_{3333} = 1 - \frac{1-2\nu}{1-\nu}\frac{\pi}{4}\frac{a_3}{a_1}$$

$$S_{1122} = S_{2211} = \frac{8\nu - 1}{32(1-\nu)}\pi\frac{a_3}{a_1},\quad S_{1133} = S_{2233} = \frac{2\nu - 1}{8(1-\nu)}\pi\frac{a_3}{a_1}$$

$$S_{3311} = S_{3322} = \frac{\nu}{1-\nu}\left(1 - \frac{4\nu + 1}{8\nu}\right)\pi\frac{a_3}{a_1}$$

$$S_{1212}=\frac{7-8\nu}{32(1-\nu)}\pi\frac{a_3}{a_1},\ S_{1313}=S_{2323}=\frac{1}{2}\left(1+\frac{\nu-2}{1-\nu}\frac{\pi}{4}\frac{a_3}{a_1}\right) \quad (2.21)$$

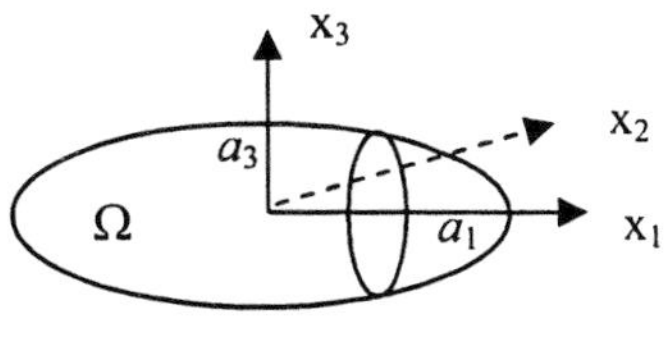

Figure 1. An ellipsoidal inclusion with principal half axes a_1, a_2, and a_3

Using Eq. (2.21) and the representations for elastic stiffness tensor and unit tensor in terms of the tensorial basis, the coefficiencts q_i for $\boldsymbol{Q}$ can be obtained as

$$q_1=\frac{2G}{1-2\nu}\left(\frac{1}{2}-\frac{3\pi}{16(1-\nu)}\frac{a_3}{a_1}\right)+\frac{2G}{1-2\nu}\left(\frac{-\nu}{1-\nu}\left(1-\frac{4\nu+1}{8\nu}\right)\pi\frac{a_3}{a_1}\right)$$

$$q_2=2G\left(1-\frac{7-8\nu}{16(1-\nu)}\pi\frac{a_3}{a_1}\right)$$

$$q_3=\frac{2G}{1-2\nu}\left(-\frac{2\nu-1}{8(1-\nu)}\pi\frac{a_3}{a_1}\right)+\frac{2G}{1-2\nu}\left(\frac{1-2\nu}{1-\nu}\frac{\pi}{4}\frac{a_3}{a_1}\right)$$

$$q_4=\frac{4G}{1-2\nu}\left(\frac{1}{2}-\frac{3\pi}{16(1-\nu)}\frac{a_3}{a_1}\right)+\frac{2G(1-\nu)}{1-2\nu}\left(\frac{-\nu}{1-\nu}\left(1-\frac{4\nu+1}{8\nu}\right)\pi\frac{a_3}{a_1}\right)$$

$$q_5=2G\left(\frac{2-\nu}{1-\nu}\frac{\pi}{4}\frac{a_3}{a_1}\right)$$

$$q_6=\frac{2G(1-\nu)}{1-2\nu}\left(\frac{1-2\nu}{1-\nu}\frac{\pi}{4}\frac{a_3}{a_1}\right)+8G\left(-\frac{2\nu-1}{8(1-\nu)}\pi\frac{a_3}{a_1}\right) \quad (2.22)$$

Note that compliance contribution tensor $\boldsymbol{H}$ for a crack – inclusion problem becomes

$$\boldsymbol{H}=\frac{V_*}{V}\boldsymbol{Q}^{-1} \quad (2.23)$$

since $(\boldsymbol{M}_1)^{-1}=0$. Inverse tensor $\boldsymbol{Q}^{-1}$ in Eq. (2.23) can be calculated as follows:

$$\boldsymbol{Q}^{-1}=\frac{q_6}{2\Delta}T^{(1)}+\frac{1}{q_2}T^{(2)}-\frac{q_3}{\Delta}T^{(3)}-\frac{q_4}{\Delta}T^{(4)}+\frac{4}{q_5}T^{(5)}+\frac{2q_1}{\Delta}T^{(6)} \quad (2.24)$$

where $\Delta=2(q_1q_6-q_3q_4)$. Compliance contribution tensor $\boldsymbol{H}$ is then

$$\boldsymbol{H}=\frac{V_*}{V}\left(\frac{q_6}{2\Delta}T^{(1)}+\frac{1}{q_2}T^{(2)}-\frac{q_3}{\Delta}T^{(3)}-\frac{q_4}{\Delta}T^{(4)}+\frac{4}{q_5}T^{(5)}+\frac{2q_1}{\Delta}T^{(6)}\right)$$
$$=\Sigma h_m T^{m}_{ijkl} \quad (2.25)$$

In the limiting case of a penny shape crack, this expression is simplified as follows:

$$h_1\to 0,\ h_2\to 0,\ h_3\to 0,\ h_4\to 0$$
$$h_5=\frac{16}{3}\frac{1-\nu}{G(2-\nu)}\frac{a^3}{V} \quad (2.26)$$
$$h_6=\frac{8}{3}\frac{1-\nu}{G}\frac{a^3}{V}$$

Thus, for a single penny shape rack, Δf becomes

$$\Delta f=\frac{a^3}{V}\frac{8}{3}\frac{1-\nu}{G}\left[\frac{1}{1-\nu/2}(\sigma\cdot\sigma):\mathbf{nn}+\left(1-\frac{1}{1-\nu/2}\right)\sigma:\mathbf{nnnn}:\sigma\right] \quad (2.27)$$

and the compliance tensor $\boldsymbol{H}$ can be determined, using $\Delta H_{ijkl}=\partial^2\Delta f/\partial\sigma_{ij}\partial\sigma_{kl}$, as

$$\boldsymbol{H}=\Sigma\Delta H=\frac{8}{3}\frac{1-\nu}{G(1-\nu/2)}\left[\frac{1}{2}(\boldsymbol{J}:\alpha+\alpha:\boldsymbol{J})-\frac{\nu}{2}\frac{1}{V}\Sigma(a^3\mathbf{nnnn})^{i}\right] \quad (2.28)$$

In Eq. (2.28), α is the crack density tensor defined by

$$\alpha=\frac{1}{V}\Sigma(a^3\mathbf{nn})^{i} \quad (2.29)$$

If the material is isotropic (i.e., inclusions are randomly oriented), both the crack density tensor α and the fourth rank tensor $(1/V)\Sigma(a^3\mathbf{nnnn})^{i}$ are isotropic. Since $tr\,\alpha=\rho$,

$$\alpha=(\rho/3)\,\boldsymbol{I} \quad (2.30)$$

where $\boldsymbol{I}=\delta_{ij}\delta_{kl}$. The fourth rank tensor $(1/V)\Sigma(a^3\mathbf{nnnn})^{i}$ can be determined using its symmetric property with respect to all rearrangements of indices in addition to the isotropic property. The resulting expression of the $\boldsymbol{H}$ tensor is

$$\boldsymbol{H}=\frac{8(1-\nu)\rho}{3G(1-\nu/2)}\left[\frac{5-\nu}{15}\boldsymbol{J}-\frac{\nu}{30}\boldsymbol{I}\right] \quad (2.31)$$

and the effective muduli readily follows:

$$E=E_0\left[1+\frac{16(1-\nu^2)(1-3\nu/10)}{9(1-\nu/2)}\rho\right]^{-1} \quad (2.32)$$

For the 2-D case, Δf has the following form:

$$\Delta f=\frac{\pi}{E_0'}\left[(\sigma\cdot\sigma):\mathbf{nn}\right] \quad (2.33)$$

where $E_0'=E_0$ for plane stress and $E_0'=E_0/(1-\nu_0^2)$ for plane strain. Using the similar steps used for the 3-D case, the effective moduli are readily obtained as

$$E=E_0'(1+\pi\rho)^{-1} \quad (2.34)$$

3. NUMERICAL SIMULATION

3.1 Representative Volume Element (RVE)

If one attempts to directly numerically simulate the response of an entire macroscopic structure by incorporating all of the heterogeneities, the problem becomes computationally complex and very time-consuming. The main interest of this paper is the effective stress-strain relationship over a representative volume element (RVE). To be exact, the goal is to determine an effective macroscopic linear elasticity tensor, E*, a relation between averages,

$$<\sigma>_\Omega = E^* : <\varepsilon>_\Omega \tag{3.1}$$

where

$$<\bullet>_\Omega \equiv \frac{1}{|\Omega|}\int_\Omega \bullet \, d\Omega \tag{3.2}$$

and σ and ε are the stress and strain tensor fields within a statistically represtative volume element (RVE) within volume $|\Omega|$ as shown in Fig. 2. The basic idea is to consider the RVE as small enough to be considered as a material point with respect to the size of the macrodomain under analysis, but large enough to be a statistically representative sample of the microstructure as shown in Fig. 2.

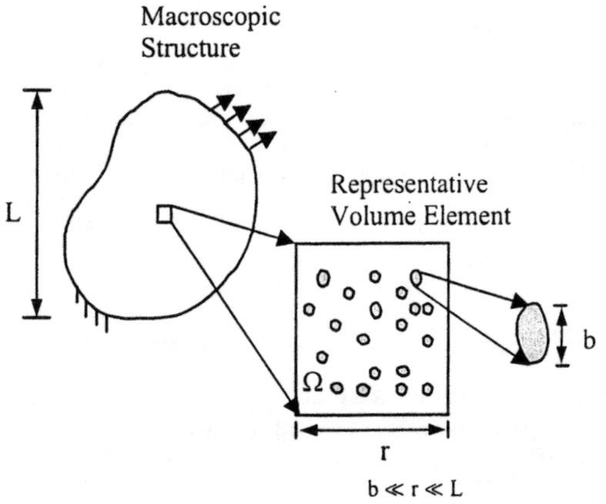

Figure 2. Illustration of representative volume element (RVE)

3.2 Testing Microheterogeneous Material Samples

In this paper, randomly oriented 2-D cracks are generated for numerical calculation of effective Young's moduli. The general purpose finite element analysis code ABAQUS was used to carry out 2-D samples with 3, 5, 8, and 12 randomly oriented cracks. Figures 3 shows a finite element model of unit cell containing 8cracks.

A compact micro/macro-energy equivalence statement, $<\sigma:\varepsilon>_\Omega = <\sigma>_\Omega : <\varepsilon>_\Omega$, known as Hill's condition (Hills, 1963) dictates the size requirements placed on the RVE. This micro/macro-energy condition must be realizable by the fields within the RVE for the sample to admit to anenergetically sensible homogenization (Zohdi and Wriggers, 2001). For any perfectly bonded heterogeneous body, in the absence of body force, two physically important loading states satisfy Hill's condition. The first loading state is pure linear displacements of the form as $\bar{\varepsilon}$

$$\mathbf{u}|_{\partial\Omega} = \bar{\varepsilon} \cdot \mathbf{x} \Rightarrow <\varepsilon>_\Omega = \bar{\varepsilon} \tag{3.3}$$

where $\bar{\varepsilon}$ is constant strain tensor. In Eq.1, the symbol $\bullet|_{\partial\Omega}$ represents boundary values. The second loading state is pure traction in the form as $\bar{\sigma}$

$$\mathbf{t}|_{\partial\Omega} = \bar{\sigma} \cdot \mathbf{x} \Rightarrow <\sigma>_\Omega = \bar{\sigma} \tag{3.4}$$

where $\bar{\sigma}$ is constant stress tensor. Since the effective response is assumed isotropic, only one test loading is necessary to determine the effective Young's modulus. The effective Young's moduli which are numerically calculated and theoretically predicted are liasted in Table 1. Both results are consistent in terms of accuracy and it can be concluded that for randomly oriented cracks, the approximation of noninteracting cracks provide reasonable results.

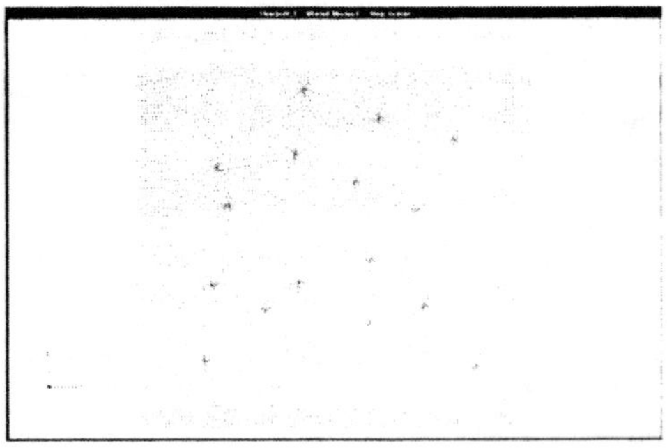

Figure 3. Finite element model containing 8-cracks

Number of cracks	Theoretical Young's modulus (Pa)	Computed Young's mudulus (Pa)
0	2.07×10^{11}	2.07×10^{11}
3	1.8917×10^{11}	1.9324×10^{11}
5	1.7890×10^{11}	1.8575×10^{11}
8	1.6542×10^{11}	1.5760×10^{11}
12	1.5033×10^{11}	1.4214×10^{11}

Table 1. Comparison of the effective Young's moduli

CONCLUSION

The effects of the microstructure on the mechanical properties of the porous material is reviewed and the numerical simulation of the testing of mechanical responses of samples of cracked solid material is carried out. Numerical results are compared with analytical results. For randomly oriented cracks, the approximation of noninteracting cracks provide reasonable agreement between two results.

4. ACKNOWLEDGEMENT

Special thanks to Professor I. Sevostianov for his comments and helpful suggestions.

REFERENCES

Bensousson, A., Lionis, J.L., PapanicolaouBudiansky, B., On the elastic moduli of some heterogeneous materials, *J. Mech. Phys. Solids*, 13 (4), 223-227, 1965.

Budiansky, B., On the elastic moduli of some heterogeneous materials, *J. Mech. Phys. Solids*, 13 (4), 223-227, 1965.

Christman, T., Needleman, A., and Suresh, S., An experimental and numerical study of deformation in metal-ceramic composites, *Acta Metall. Mater.*, 37(11), 3029-3050, 1989.

Devries, F., Dumontet, H., Duvaut, G., and Lene, F., Homogenization and damage for composite structures, *Int. J. Numer. Meth. Engrg.* 27, 285-298, 1989

Eshelby, J.D., The determination of the elastic field of an ellipsoidal inclusion, and related problems, *Proc. Roy. Soc.*, A241(1226), 376-396, 1957.

Hill, R., Elastic properties of reinforced solids: some theoretical principles, *J. Mech. Phys. Solids*, 11(5), .357-372, 1963.

Kachanov, M., Effective elastic properties of cracked solids: critical review of some basic concepts, Appl. Mech. Rev., 45(8), 304-335, 1992.

Kouznetsova, Brekelmans, and Baaijens, An approach to micro-macro modeling of heterogeneous materials, *Compu. Mech.* , 27, 37-48, 2001.

Kunin, I.A., *Elastic Media with Microstructure*, Springer, Berlin, 1983.

McHugh, P.E., Asaro, R.J., and Shin, C.F., Computational modeling of metal matrix composite materials-II. Isothermal stress-strain behavior, *Acta Metall. Mater.*, 41(5), 1477-1488, 1993.

Sevostianov, I. and Kachanov, M., Explicit cross-property correlations for anisotropic two-phase composite materials, *J. Mech. Phys. Solids*, 50, 253-282, 2002.

Tvergaard, V., Analysis of tensile properties for wisker-reinforced metal-matrix composites, *Acta Metall. Mater.*, 38(2), 185-194, 1990.

Willis, J.R., Bounds amd self-consistent estimates for the overall properties of anisotropic composites, *J. Mech. Phys. Solids*, 25(3), 185-202, 1977.

Zohdi, T. and Wriggers, P., A Domain Decomposition Method for Bodies with Heterogeneous Microstructure Based on Material Regularization, *Int. J. Sol. Struc.*, 36, 2507-2525, 1999.

Zohdi, T. and Wriggers, P., Aspects of the computational testing of the mechanical properties of microheterogeneous material samples, *Int. J. Numer. Meth. Engrg.* Vol. 50, pp. 2573-2599, 2001.

PVP-Vol. 458, Computer Technology and Applications

PVP2003-1898

CALCULATION OF THE TEMPERATURE FIELD OF THE SOLIDIFYING CERAMIC MATERIAL EUCOR

Frantisek Kavicka
Technical University of Brno
Technická 2, Brno 616 69,
Czech Republic

Josef Stetina
Technical University of Brno
Technická 2, Brno 616 69,
Czech Republic

Jaromir Heger
ALSTOM ˇPower technology Centre
Cambridge Road, Whetstone,
Leicester, UK

Bohumil Sekanina
Technical University of Brno
Technická 2, Brno 616 69,
Czech Republic

Pavel Ramik
Technical University of Brno
Technická 2, Brno 616 69,
Czech Republic

Jana Dobrovska
VŠB - Technical University of Ostrava
17. listopadu 15, 708 33 Ostrava,
Czech Republic

ABSTRACT

EUCOR, a corundo-badelleyit material, which is not only resistant to wear but also to extremely high temperatures, is seldom discussed in literature. The solidification and cooling of this ceramic material in a non-metallic mould is a very complicated problem of heat and mass transfer with a phase and structure change. Investigation of the temperature field, which can be described by the three-dimensional (3D) Fourier equation, is not possible without the employing of a numerical model of the temperature field of the entire system—comprising the casting, the mould and the surroundings.

The temperature field had been investigated on a 350x200x400 mm block casting—the so-called "stone"—with a riser of 400 mm, and using a numerical model with graphical input and output. The computation included the automatic generation of the mesh, and the successive display of the temperature field using iso-zones and iso-lines. The thermophysical properties of the cast, as well as the mould materials, were gathered, and the initial derivation of the boundary conditions was conducted on all boundaries of the system. The initial measurements were conducted using thermocouples in a limited number of points. The paper provides results of the initial computation of the temperature field, which prove that the transfer of heat is solvable, and also, using the numerical model, it is possible to optimise the technology of production of this ceramic material, which enhances its utilisation. The results are complemented with an approximated measurement of the chemical heterogeneity of EUCOR.

NOMENCLATURE

c	specific heat capacity	[J.kg^{-1}.K^{-1}]
c_v	specific volume heat capacity c_v=c.ρ	[J.m^{-3}.K^{-1}]
HTC	heat transfer coefficient	[W. m^{-2}.K^{-1}]
h	specific enthalpy	[J.kg^{-1}]
h_v	specific volume enthalpy $h_v = h\,\rho$	[J.m^{-3}]
k	heat conductivity	[W. m^{-1}.K^{-1}]
L	latent heat	[J.kg^{-1}]
τ	time	[s]
x,y,z	axes in given directions	
Q	heat flow in given direction	[W]
Q_{SOURCE}	latent heat of the phase/structural change	[W.m^{-3}]
QX,QY,QZ	heat flow in each given direction	[W]
V	volume	[m^3]
VX, VY, VZ	unitary heat conductivity	[W.K^{-1}]
t	temperature	[K]
ρ	density	[kg.m^{-3}]

1. INTRODUCTION

Corundo-badelleyit material (CBM) is a modern, electrically cast, heat- and wear-resistant material. It is resistant to corrosion and wear, even at very high temperatures. This material belongs to the not too well known area of the Al_2O_3-SiO_2-ZrO_2 system. It is produced in several plants throughout the world—under different trademarks—in three types, differing mainly in the ZrO_2 content. In the Czech Republic this material is produced with a 32-33% ZrO_2 content.

CBMs are used mainly in the construction of glass furnaces. In various steel-works aggregates they are used mostly in heating furnaces, etc. They have a high resistance to liquid glass as well as metal, and are also suitable for great temperature changes. Slabs from this material are therefore very suitable for the walls and floors of melting aggregates, linings, pouring filters, insulation plates, and for a number of other uses, which can be accessible after optimising the technology of their production and utility properties.

The requirements on the properties of CBM are usually determined:

a) For the internal walls of glass furnaces: resistance to liquid glass and the creation of bubbles when in contact with liquid glass.
b) For the production of wear-resistant products: resistance to wear, low porousness, crystalline structure, and resistance to temperature shocks.

From the foundry viewpoint, it is possible to compare the properties of EUCOR with those of commonly cast metals, especially steels and cast steel. In a sand mould, the solidification coefficient of steel is approximately 0.07 $m.h^{1/2}$ and that of EUCOR is 0.065 $m.h^{1/2}$; in a cast-iron mould it is 0.13 $m.h^{1/2}$ and that of EUCOR is 0.163 $m.h^{1/2}$, etc.

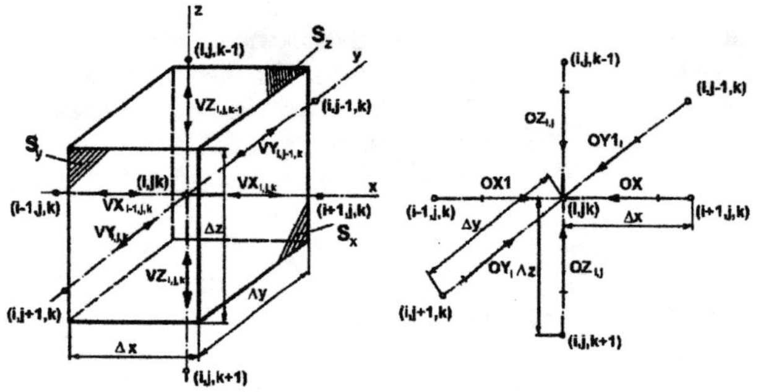

Figure 1 The thermal equilibrium diagram of a general nodal point of the network

2. A NUMERICAL MODEL OF SOLIDIFICATION, COOLING AND HEATING

Solidification—crystallization—and cooling belong to the most important technological processes. They are cases of one-to-three-dimensional transfer of heat and mass. In systems, which comprise the casting, the mould and surroundings, all three kinds of heat transfer take place. Since these problems cannot be solved analytically—even using the second-order partial differential Fourier Eq. (1) (where mass transfer is neglected and conduction is considered as the most important of the three kinds of heat transfer)—it is necessary to employ numerical methods.

$$\frac{dt}{d\tau} = \frac{k}{\rho \cdot c}\left(\frac{\delta^2 t}{\delta x^2} + \frac{\delta^2 t}{\delta y^2} + \frac{\delta^2 t}{\delta z^2}\right) + \frac{Q_{SOURCE}}{\rho \cdot c} \qquad (1)$$

A numerical model of solidification, cooling and heating had been applied for investigating one-to-three-dimensional stationary and transient temperature fields of systems comprising the casting, the mould and the surroundings. It is possible to investigate either the system as a whole, or any of its parts—during any industrial technological process whose individual sub-processes can be solidification, cooling, heating, refrigerating and others—in any sequence or individually. The model also enables the simulation of traditional and non-traditional technologies of casting in foundries, metallurgical plants, forging operations, heat-treatment processes, etc.

Equation (1), which describes the three-dimensional (3D) transient temperature field of a gravitationally-cast casting, was used for the application of the numerical model. Upon reaching the stationary state, the derivative of the temperature by time becomes zero. Eq. (1) includes the temperature field of a casting in all its three phases: in the melt, in the mushy zone and in the solid phase. Here it is necessary to introduce the specific volume enthalpy $h_v = c.\rho.t$, which is dependent on temperature. Eq. (1) then takes on the form

$$\frac{\partial h_v}{\partial t} = \frac{\partial}{\partial x}\left(k\frac{\partial t}{\partial x}\right) + \frac{\partial}{\partial y}\left(k\frac{\partial t}{\partial x}\right) + \frac{\partial}{\partial z}\left(k\frac{\partial t}{\partial z}\right) \qquad (2)$$

The specific heat capacity c, density ρ and heat conductivity k are also functions of temperature.

The 3D transient temperature field of the mould is described by Eq. (1) without the member $Q_{SOURCE}/\rho.c$. Figure 1 illustrates the thermal equilibrium of an elementary volume—a general nodal point i,j,k. The unitary heat conductivities and heat flows, in the directions of all the main axes, are also indicated here. In the z-direction it is

$$VZ_{i,j,k} = k(t)\frac{S_z}{\Delta z} \quad \text{or} \quad VZ_{i,j,k-1} = k(t)\frac{S_z}{\Delta z} \qquad (3)$$

The heat flows through the point i,j,k in the z-direction are

$$\dot{Q}Z1_{i,j} = VZ_{i,j,k}(t^{(\tau)}_{i,j,k-1} - t^{(\tau)}_{i,j,k}) \qquad (4)$$

$$\dot{Q}Z_{i,j} = VZ_{i,j,k+1}(t^{(\tau)}_{i,j,k+1} - t^{(\tau)}_{i,j,k}) \qquad (5)$$

The unknown temperature at an arbitrary nodal point of the mould, in the course of a time step of $\Delta\tau$, is

$$t^{(\tau+\Delta\tau)}_{i,j,k} = t^{(\tau)}_{i,j,k} + (\dot{Q}Z1_{i,j} + \dot{Q}Z_{i,j} + \dot{Q}Y1_i + \dot{Q}Y_i + \dot{Q}X1 + \dot{Q}X)\frac{\Delta\tau}{c.\Delta x.\Delta y.\Delta z} \qquad (6)$$

where $t^{\tau}_{i,j,k}$ is the temperature in the previous time.

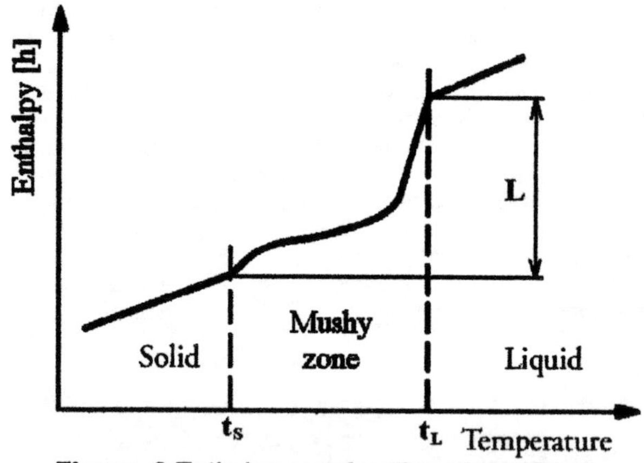

Figure 2 Enthalpy as a function of temperature

The unknown enthalpy at an arbitrary nodal point of the casting in the course of $\Delta\tau$ is

$$h^{(\tau+\Delta\tau)}_{v,i,j,k} = h^{(\tau)}_{v,i,j,k} + (\dot{Q}Z1_{i,j} + \dot{Q}Z_{i,j} + \dot{Q}Y1_i + \dot{Q}Y_i + \dot{Q}X1 + \dot{Q}X)\frac{\Delta\tau}{c.\Delta x.\Delta y.\Delta z} \qquad (7)$$

where $h^{\tau}_{v,i,j,k}$ is the enthalpy in the previous time.

The basic principle of the method is based on the method of control volumes, solving the equilibrium on an element of the network (Figure 1). The enthalpy-temperature function must be known for each cast material to be investigated. The calculated enthalpy at an arbitrary nodal point (Eq. 7) is then transformed to temperature via the function in Figure 2.

The explicit difference method had been chosen for this investigation because it enables the application of the most convenient method of numerical simulation of the release of latent heat of phase and structural changes using the thermodynamic enthalpy function.

The susceptibility of the explicit method to oscillations is minimized by a series of longer and shorter time steps. Another variant is one that evaluates the attained stationary state of the modeled process via three shorter time steps.

The software also considers the non-linearity of the task, i.e.:

- The dependence of the thermophysical properties of all materials entering the system, and
- The dependence of the heat-transfer coefficients (*HTC*s)—on all boundaries of the system—on the temperature of the surface of the casting and mould.

The software also performs all the necessary tasks—from the automatic generation of the mesh, through the computation of the thermophysical properties and the definition of boundary conditions, to the actual numerical simulation of the temperature field.

3. THE ASSIGNMENT

The assignment was aimed at investigating a transient 3D temperature field of a system comprising a casting-and-riser, the mould and the surroundings, using a numerical model. Figure 3 illustrates the main dimensions. The dimensions of the actual casting—the "stone" of the special ceramic material EUCOR—were 400 x 350 x 200 mm, the riser comprised a 300-mm-high truncated four-sided pyramid. The initial temperature of the mould was 20°C, which equalled the room temperature. The pouring temperature was 1800°C, the pouring time was 10 s, the liquidus temperature was 1775°C and the solidus 1765°C. The mould material was made from a mixture of water-glass and sand, and the bottom of the mould was made from a layer of magnesite of a thickness of 50 mm. The level of the riser remained untreated. It was necessary to compare the results of the calculation with the temperatures measured within the casting and the mould using thermocouples. It was also necessary to prepare the numerical model and the mesh in order to successively monitor the effects of the pouring temperature, the shape of the riser, the insulation of its level, etc., in order to optimise the technology of the pouring of the stone.

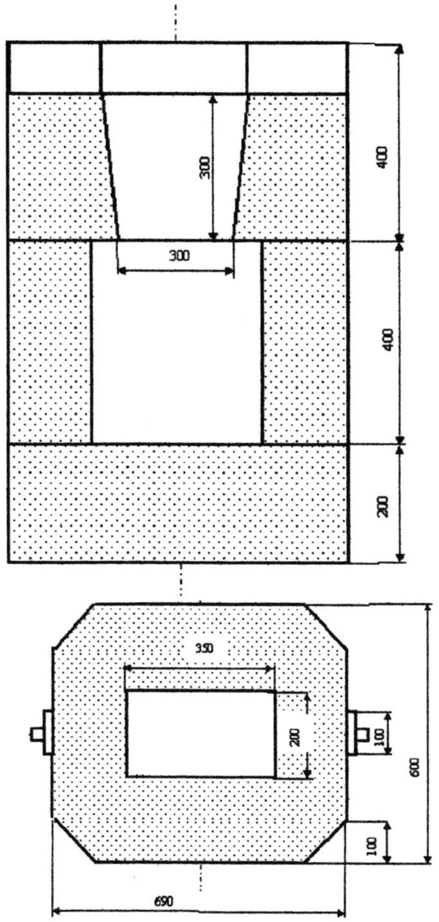

Figure 3 The casting-riser-mould system

4. THE PREPARATION FOR SIMULATION

The mesh is generated automatically—with nodes 20 mm apart. The temperature field is symmetrical along the axes. It is therefore sufficient to investigate the temperature field of one quadrant only. Figure 4 shows the mesh for the casting-riser-mould system. The resultant heat flow through both longitudinal sections is equal to zero. The longer time step was 10 s.

The thermophysical properties of the ceramic material were considered dependent on temperature. The dependence of the heat capacity on temperature is shown in Figure 5, heat conductivity in Figure 6, and relative decrease of density in Figure 7. The density of the mould material was assumed to be 1600 kg/m^3. The heat conductivity was 3.3 W/mK within the temperature range 0-100°C and 0.88 W/mK above 100°C; the heat capacity was 800 J/kgK within the temperature range 0-100°C and 1000 J/kgK above 100°C.

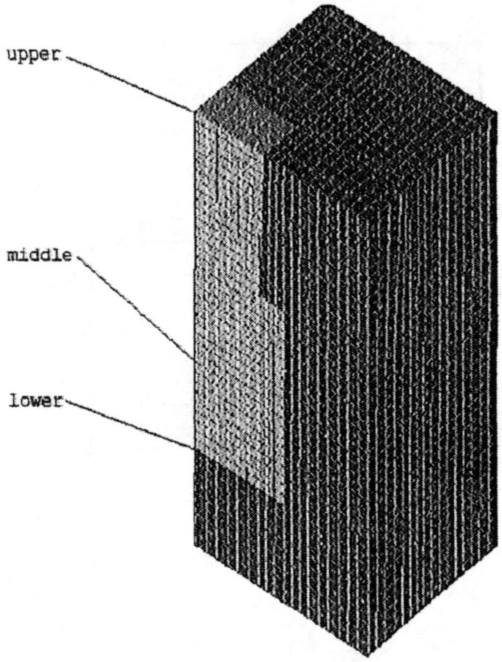

Figure 4 The 3D computational network

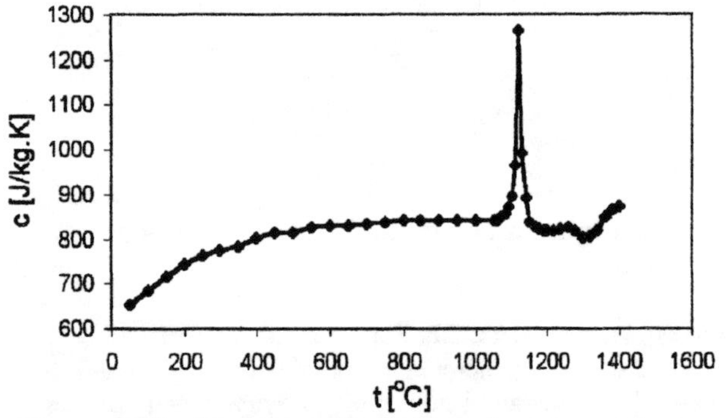

Figure 5 The heat capacity-temperature dependence

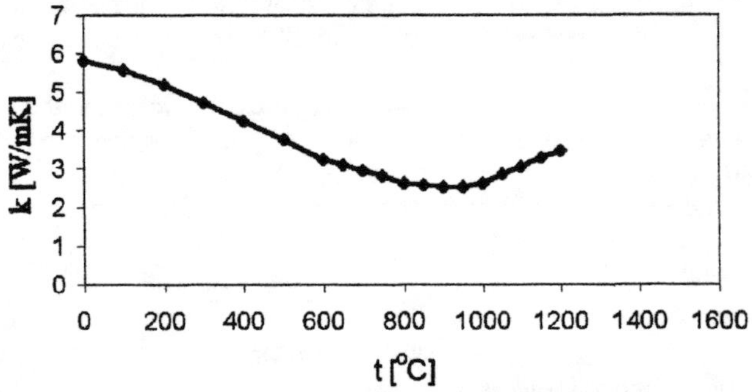

Figure 6 The heat conductivity-temperature dependence

The boundary conditions were defined on all planes bordering the system. Heat transfer by radiation and convection into the surroundings was considered from the top of the mould, and also from the level of the riser after pouring, from the base and the frame. The resultant *HTC*s were defined using theory. This means natural convection around the planar plates. Ideal physical contact was presumed between the mould and the casting, and between the mould and the riser.

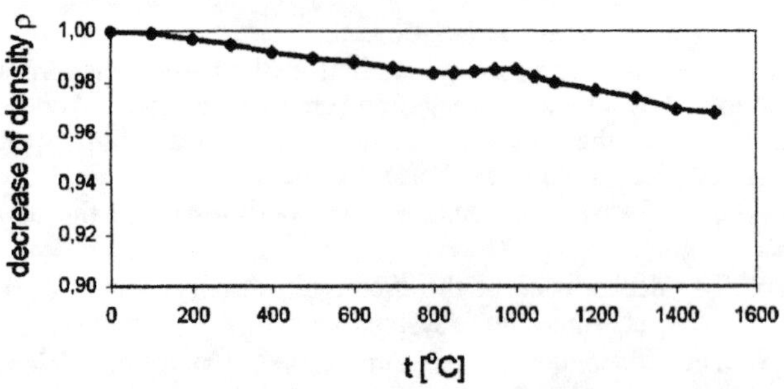

Figure 7 The relative decrease of density-temperature dependence

5. THE ACTUAL EXPERIMENT

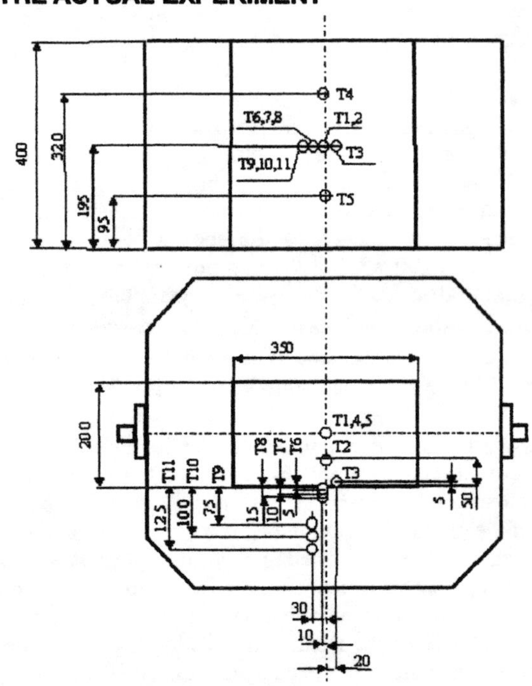

Figure 8 The measurement points

The numerical model of the temperature field of the casting was confronted with experimental measurements and corrected. The temperatures were measured in the actual casting and also in the mould. Special tungsten-rhenium thermocouples had to be used—for the measurement within the actual casting—in order to withstand the high pouring temperature, which is approximately 300°C higher than the pouring temperature of steel. The measurement lasted about 19 hours. The measurement points are shown in Figure 8. Three GRANT measurement stations were used for the recording of data throughout the process. Figures 9 and 10 show the measured and computed temperature history in point T5 in the casting and in point T6 in the mould. The differences between the measured course and the calculated course of temperatures

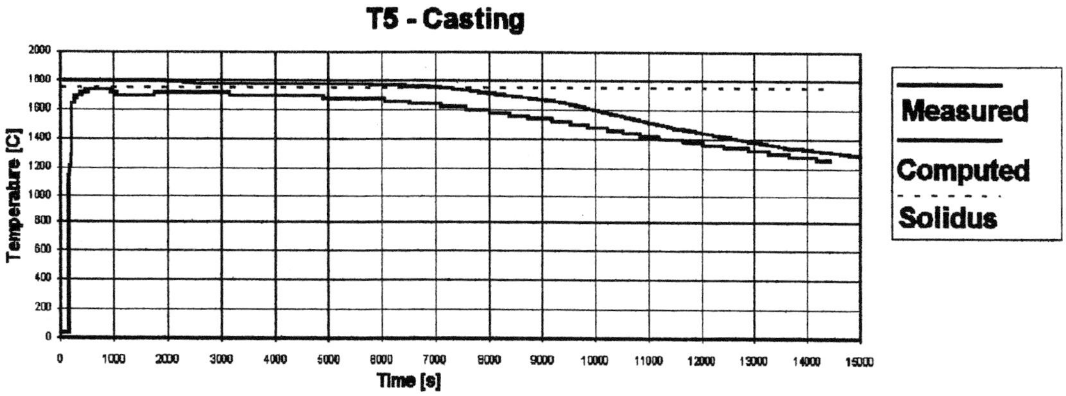

Figure 9 The measured and computed temperature history in the point T5 in the casting

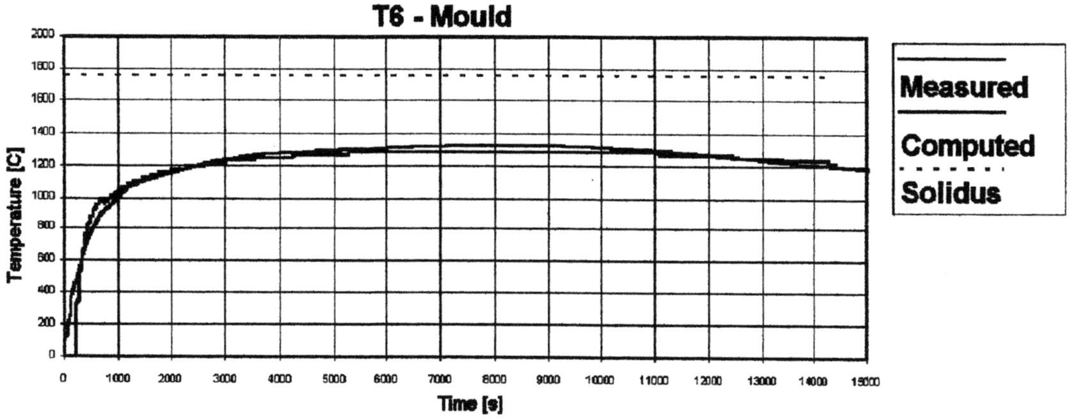

Figure 10 The measured and computed temperature history in the point T6 in the mould

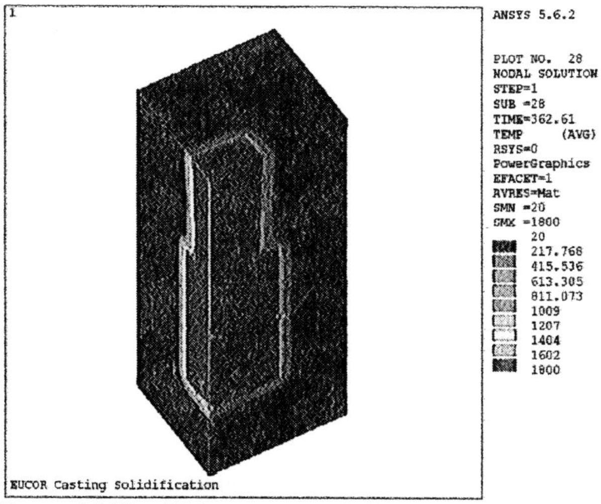

Figure 11 The 3D temperature field of the system after 6 min

within the casting are caused by the difficulty to maintain the positions of the thermocouples within the casting during the filling of the mould (Figure 9). It is possible to achieve a more acceptable congruence by exact positioning of the thermocouple in the mould (Figure 10). The differences at the lower temperatures are caused by the difficulty to exactly define the amount of moisture in the sand mould.

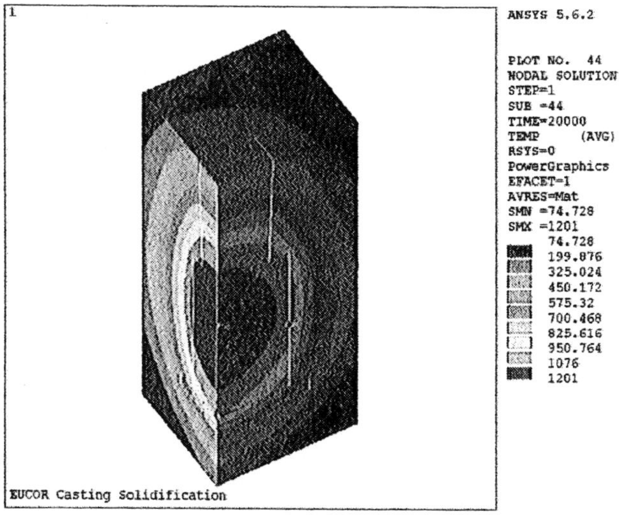

Figure 12 The 3D temperature field of the system after 5.6 hours

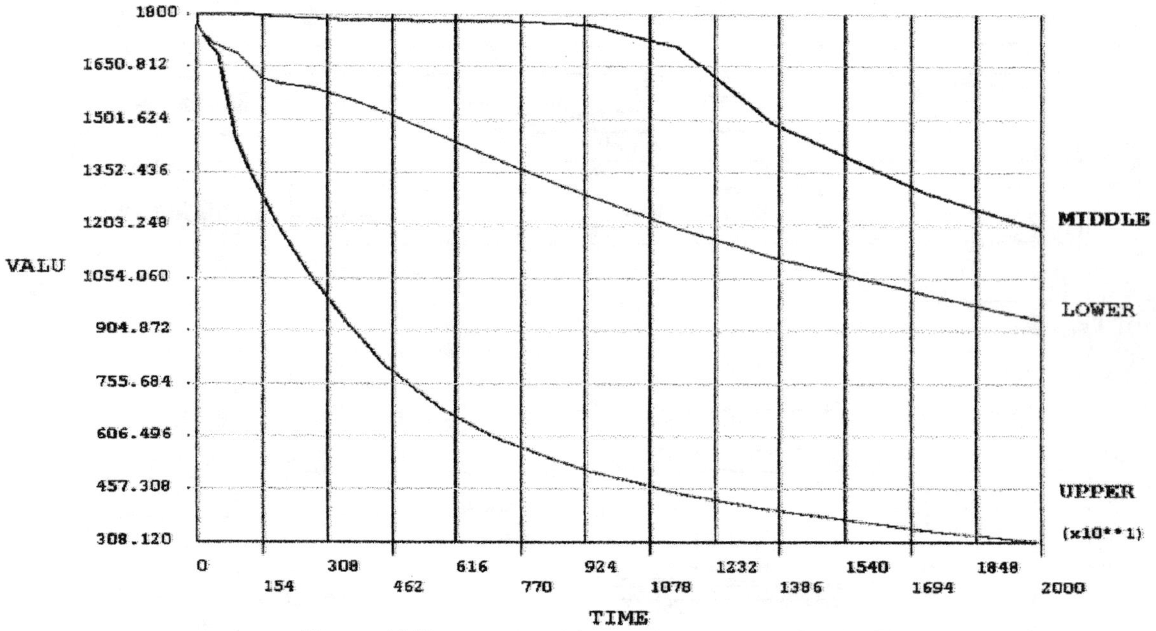

Figure 13 Temperature history of the three points in Figure 4

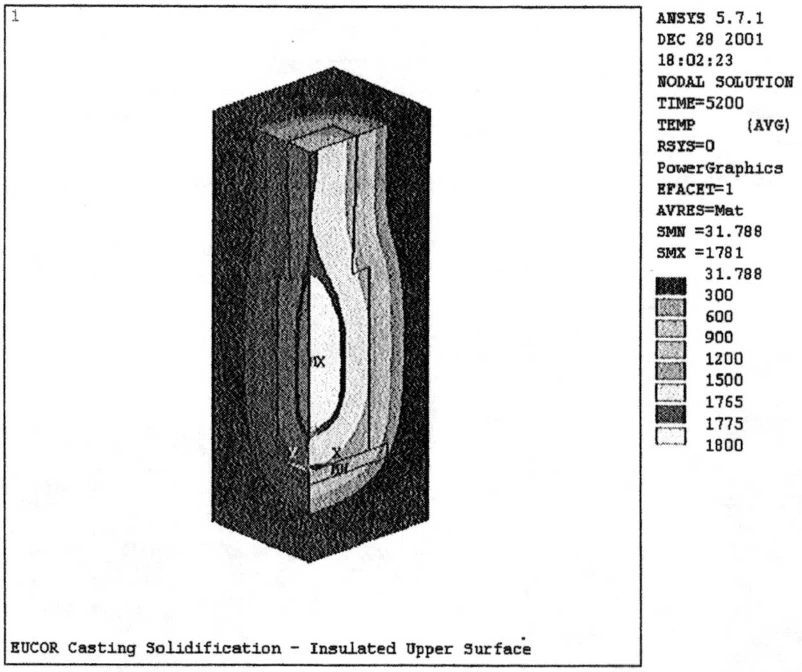

Figure 14 The Freezing of the upper part of the riser at 5200 s

6. RESULTS

The results attained from the analysis of the temperature field of a solidifying casting and the heating of the mould represent only one quadrant of the system in question. The thermokinetics of the phenomenon were monitored over a five-day period when the casting was kept inside the mould in order to cool completely.

Figures 11 and 12 show the temperature field after 6 minutes and 5.6 hours respectively. These figures are entered merely as an example—after calculation it is possible to obtain these graphs for arbitrary times. Furthermore, it is possible to plot the temperature history of any point of the mesh (Figure 13).

7. A NEW RISER

The primary condition for a healthy casting (i.e. one without internal faults) is directed solidification. Figure 14 proves that the current casting of EUCOR is not optimal, because the 'refilling' of the casting from the riser is cut off—such a case represents so-called undirected solidification. The riser is the first to solidify—at 5,200 s.

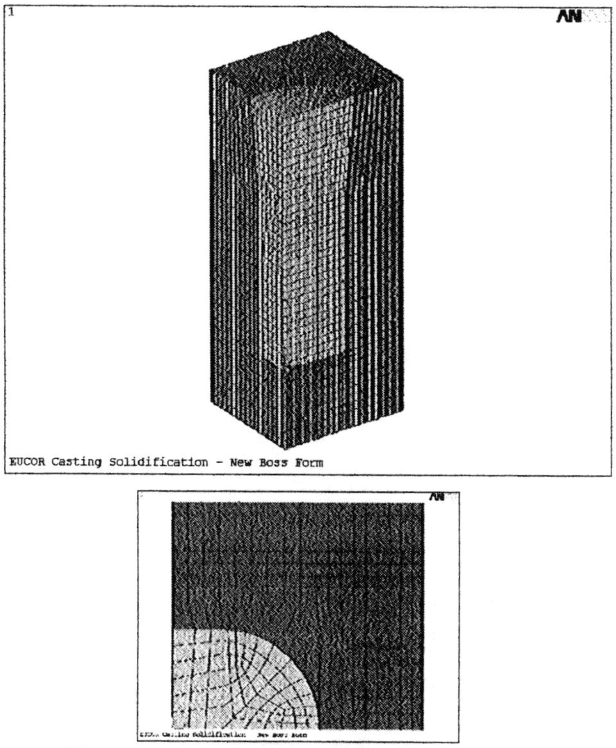

Figure 15 The Network of the optimised riser

The new shape of the riser was optimised according to the latest findings. This new riser had been designed using empirical equations with the temperatures calculated within the casting with the former riser. Figure 15 shows the computed mesh of the casting and the mould with its optimised riser, together with a detail of the shape of the riser.

The new—optimised—riser (Figure 16) achieves directed solidification, where the casting is being refilled from the riser. The next step is to use numerical optimisation methods. For advanced optimisation techniques and related literature, see [6].

8. CONCLUSIONS

The investigation of the temperature field had two objectives:

1. Directed solidification, as the primary condition for a healthy casting.
2. Optimisation of the technology of casting, together with the preservation of optimum utility properties of the product.

The achievement of these objectives depends on the ability main factors which characterise the solidification process or accompany it.

The results of the investigation of the quantities should reveal the causes of heterogeneities within the casting with respect to phase and structural changes. It should also focus on the thermokinetics of the creation of shrinkage porosities and cavities and at the prediction of their creation and, therefore, to control the optimisation of the shape and sizes of the risers, the method of insulation, the treatment of the level, etc. The main economic criteria to be observed are the saving of liquid material, mould and insulation materials, the saving of energy, the optimisation of the casting process, and the properties of the cast product.

The paper provides results of the initial calculation of the temperature field, which prove that the transfer of heat is solvable, and also that—using the numerical model—it is possible to optimise the technology of production of this ceramic material. This will enhance the production in a number of industries, namely the glass, energy, foundry and metallurgical.

ACKNOWLEDGMENTS

This analysis was conducted using a program devised within the framework of grant projects GACR No. 106/01/1464 and 106/01/1164, No. 106/03/0271 , No. 106/01/0379, COST-APOMAT-OC526.10, EUREKA-COOP No.2716, KONTAKT No. 2001/015 and 23-2003-04 and CZ 323001/2201.

REFERENCES

[1] Kavicka F. and Stetina J., "A numerical model of heat transfer in a system "plate-mould-surroundings, Proceedings of the 6th International conference on advanced computational methods in heat transfer," Madrid, Wessex Institute of technology, 2000

[2] Kavicka F. and Stransky K. and Buzek Z. and Dobrovska J. and Dobrovska V. and Salva O. and Ticha J. , Corundum-badelleyite ceramics - the perspective cast refractory material, Proceedings of the International conference Industrial furnaces and refractory materials, Kosice, Slovakia, p. 82-86, 2000.

[3] Ticha J. and Spousta V. , Measurement of the course of primary cooling of the corrundum-baddeleyite material EUCOR, Proceedings and CD ROM of the 9th International symposium METAL 2000, Czech Republic, Ostrava, paper No. 137, 2000.

[4] Salva O. and Buzek Z. and Husar I. , About CBC application in metallurgy, Proceedings of abstracts and CD ROM of the 10th International symposium METAL 2001, Czech Republic, Ostrava, p. 36, paper No. 182, 2000.

[5] Ptackova M. and Janova D. and Buchal A. and Kavicka F. and Stransky K. , Structure and phase characteristics of a corundum-baddeleyite cast ceramics, Acta Metallurgica Slovaca 7, p. 484-486, 2001.

[6] Popela P., Application of stochastic programming in foundry, Folia Fac. Sci. Nat. Univ. Masarykianae Brunensis, Vol. 7, pages 117-139, (1998).

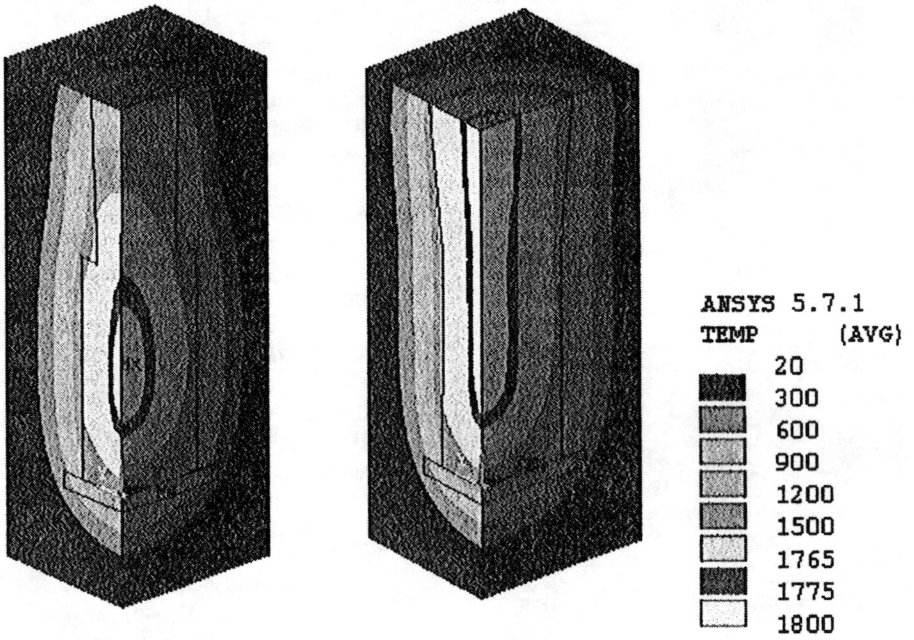

Figure 16 The 3D Temperature fields before and after optimisation

PVP-Vol. 458, Computer Technology and Applications

PVP2003-1899

Automated Modeling and Crack Growth Simulation in Pressurized Thin Shell Panels with Multi-site Damage

K. V. N. Gopal and B. Dattaguru

Department of Aerospace Engineering
Indian Institute of Science,
Bangalore, India - 560012

Abstract

The residual strength estimation of thin-walled pressurized shell structures is of relevance to pressurized thin piping systems and aged airframes for life extension programs. This requires powerful and efficient computational techniques using finite element method and numerical fracture mechanics for elastic or inelastic stress analysis and crack growth simulation. For this purpose, detailed modeling and finite element mesh generation of built-up structures like stiffened cracked thin shells is necessary and it is a computationally intensive task. Automating the entire process from geometric modeling to stress analysis and crack growth simulation (requires remeshing) vastly improves the efficiency of the computational analysis and reduces the chances of modeling and simulation errors. A geometric primitive based technique has been developed for automated modeling and meshing. The work is carried out primarily on aged fuselage shell panels but the method is applicable to other pressurized thin piping systems.

This paper presents a simple and efficient computational procedure using a geometry based logic and surface mesh generation technique for automated modeling and crack growth simulation in pressurized thin-walled shells. The approach has been used to develop a structural integrity evaluator software. Finite element analysis is carried out using a commercial software, and these results are fed to the structural integrity evaluator. Some results of the nonlinear finite element analysis and elastic-plastic stable crack growth simulation in pressurized stiffened fuselage panels using this approach are presented.

1. INTRODUCTION

Efficient and reliable structural integrity assessment procedures are necessary for maintenance of safety and extension of operational life of pressure vessels such as airframes and thin piping systems. The problem of widespread fatigue damage in pressurized airframes, causing rapid degradation in the residual strength and damage tolerance capability of the structure, has become an issue of great concern during the last decade. Efficient computational techniques using finite element methods and numerical fracture mechanics together with non-destructive inspections offer a viable and economical methodology for residual strength estimation and structural integrity assessment. The developments in the field of fracture mechanics and vast improvements in computational power have aided these efforts and facilitated the wide applicability of damage tolerance based maintenance and life extension programs. This helps in planning and minimizing expensive experimental testing. The present work is carried out on aged fuselage shell panels, but the method of approach is applicable to pressurized thin piping systems.

The modeling of built-up structures such as damaged stiffened airframes is a computationally intensive task requiring large man-hours. Efficient techniques are needed to quickly generate the geometric model and finite element mesh and perform crack growth simulation with adaptive remeshing for the prescribed loading. GUI-based modeling approaches are attractive but require frequent user interaction and can be extremely cumbersome for problems like crack growth modeling in stiffened fuselage panels. Automating the entire process from modeling to stress analysis and crack growth simulation can substantially reduce user intervention and minimize modeling and simulation errors, thus improving the efficiency and accuracy of the analysis. Graphical display tools can be used to display the results at various stages of the analysis. Based on these procedures a software has been developed for automated modeling and crack growth simulation in pressurized stiffened airframes. In the following sections the software framework is outlined and some results of the nonlinear finite element analyses and elastic-plastic stable crack growth simulation are presented.

2. STRUCTURAL INTEGRITY EVALUATOR

The Structural Integrity Evaluator (STIEV) software framework is essentially modular and is designed to automate the entire process from development of the geometric and finite element model to structural integrity evaluation. The software framework consists essentially of a STIEV shell with a FE core sitting on a fracture mechanics foundation (Fig. 1). It consists of the following modules (Fig. 2)

A *'solid modeler'* operating from a minimum geometric definition of the various components with damage (MSD/MED, corrosion) generates the geometric representation of the structure. The *'mesh generator'* module uses automatic surface mesh generation techniques to develop the finite element mesh from the *'geometric model'* together with the boundary conditions and loading information and includes the cracks and other damage. The *'analysis'* module interfaces the finite element model with an external finite element analysis package for stress analysis. The *'fracture'* module (damage assessment and damage growth) computes the required fracture parameters from the stress analysis results and simulates the crack growth. This is then processed for residual strength estimation. This involves local remeshing around the crack regions for each stage of crack growth. The user input is limited to a set of input files with minimum required data on the geometry, material, loading and boundary conditions and a control file for controlling the analysis and output. The output at various stages of analysis can be stored for further processing and display of results.

3. MODELLING & MESHING ALGORITHM

Commercially available finite element analysis software packages are very efficient for one time analysis of new structures or a simple fracture analysis of cracked structures. Modeling damage growth and particularly elastic-plastic stable crack growth is beyond the scope of these programs. Current geometric modeling procedures are generally based on approaches like the bezier, b-spline and NURBS techniques. These approaches generate curves/surfaces from a set of user-specified control points lying outside the curve/surface & algebraic polynomials, defining their shape & geometry. They are not suited for the task of regenerating a surface from an existing numerical definition in terms of points lying on the surface.

In this paper a simple alternative method using interpolation approach is presented. The approach combines (i) an efficient automated geometric representation of the structural details and damage information (ii) a logical procedure to generate boundary nodes based on user desired total FE model size and (iii) a surface meshing algorithm, to develop the finite element model of the damaged structure.

Every structural component including attachments, damages etc. can be represented by a simple set of 0D (points), 1D (curve), 2D (surface) & 3D(solid) primitives. A damaged/repaired built-up structure can thus be modeled as a synthesis of these primitives. In contrast to bezier and b-spline type of geometric definition, the entities run through the user defined points and need no additional polynomial parameter information. This definition of the structure enables quick, accurate and virtually complete representation of the structural details to begin with and convenient automatic updating of the geometry during damage growth simulation. For stiffened panel type of structures, a library of 0D, 1D & 2D primitives can be defined as shown in Fig. 3. Thickness and material can be associated with each of the 2D entities.

Meshing algorithms based on techniques such as modified quadtree and octree approaches [1], advancing front technique [2], delaunay triangulation [3, 4] and mapping methods, cannot be directly used for problems involving crack growth in stiffened shell structures. For crack growth problems, a meshing algorithm with efficient remeshing and adaptive properties is necessary. For built-up shell structures, 2D meshing algorithms cannot be applied directly. Instead, each surface can be mapped from 3D space to a parametric 2D domain and meshed using a planar meshing algorithm. The mesh in the parametric domain is mapped back to the original 3D surface to get the surface mesh. Meshes of individual components are assembled after ensuring consistency to generate the mesh for the entire built-up structure.

In the present software, a hybrid algorithm [5] to generate all-quad or all-triangular meshes in arbitrary two-dimensional domains, based on a combination of quadtree and advancing front techniques has been extended for surface meshing. The algorithm follows a recursive spatial decomposition procedure [5] to determine internal element density from boundary nodal density; a boundary contraction procedure [5] to generate an initial mixed-element mesh and a polygon-splitting procedure to produce an all-quadrilateral mesh. The meshing algorithm was interfaced with the modeler by defining a characteristic length parameter based on user-defined finite element size to compute the inter-nodal spacing on the boundary and automatically generate the boundary nodes from the geometric model. The boundary nodes and edges along with any mandatory internal nodes are the input to the module. The algorithm allows smooth transition in meshing cracked regions and facilitates proper application of numerical fracture mechanics approaches to evaluate crack tip parameters. The flow chart of the algorithm is shown in Fig. 4.

4. CRACK GROWTH SIMULATION AND RESIDUAL STRENGTH ANALYSIS

Residual strength analysis is necessary for residual life estimation and structural integrity evaluation of a damaged structural component. The stress analysis for the developed finite element model is performed through an interface to an external FEA package. The fracture module computes the fracture parameters using the FEA output and performs crack growth simulation as required.

Structural components such as pressurized fuselage panels are generally made of thin ductile aluminum alloys that experience inelastic deformation as the crack size increases. For more accurate estimates of residual strength and life, it is necessary to model the elastic-plastic stable crack growth in such structures. The generally used fracture criteria like stress intensity factor K, J-integral etc are not valid for large scale yielding. A suitable fracture criterion valid for elastic-plastic multiple crack growth is need for crack growth simulation. Based on the results from a number of experimental and numerical studies conducted in recent years, the *Critical Crack Tip Opening Angle Criterion* (CTOA) [6, 7] has been used as the fracture criterion for modeling elastic-plastic crack growth in thin metallic structures in the present approach. A nodal release procedure for self-similar crack growth and also a remeshing procedure (after local deletion of mesh) was used for numerical crack growth modeling. For crack growth using remeshing, the state variables need to be accurately transferred from the old finite element model to the new finite element model to retain the history information. In addition, the nodal displacements must be transferred from the old node locations to the new node locations. In this work, the technique of inverse iso-parametric mapping method [8, 9], based on the inversion of the element shape functions in the distorted mesh, has been used for state variable mapping. The resulting nonlinear equations are solved using a direct Newton-Raphson iteration scheme [8].

Since analyses based on CTOA fracture criterion are direct simulations of realistic crack growth, multiple crack growth interaction and link-up can be automatically captured as the crack propagates. During crack growth analysis for MSD cracks, all the crack tips are checked for the critical fracture criterion. Residual strength of a damaged structure is obtained directly from the crack growth data.

5. CASE STUDIES

The examples below illustrate the application of the proposed methodology for modeling crack growth in flat and built-up shell panels.

Example 1: Stiffened fuselage skin panel

To illustrate the working of the software a cylindrical shell panel of size 40" x 40" x 0.036"; radius =100" (Fig. 5a) has been modeled. A hat shaped longitudinal stiffener of 1" x 1" x 1" x 1" x1" x 0.036" is riveted at midsection with a single row of rivets (dia = 0.15") at a pitch of 1.0". The stiffener has a large aspect ratio (>10) and is automatically treated like a pseudo 1D entity. The automatic meshing procedure is used to generate the mesh (Fig. 5b). Examples with a crack, cutout and bonded attachments are shown in Figs. 5((e) - (f)). The panels and the stiffeners are modeled as swept surfaces while the rivet rows and cracks are modeled as line entities. The interface between the stiffener and skin attachments has to be checked to ensure consistency and eliminate superfluous nodes. In the region around the cracks and the rivets, the algorithm generates a fine mesh and transitions to a relatively coarse mesh away from the cracks.

Example 2: Numerical Studies of Flat Panel Specimens

As a preliminary application of the methodology, crack growth in unstiffened center-cracked tension (CCT) specimens made of 2024-T3 Al alloy (Fig. 6) was simulated for varying width and height. Numerical simulation of 1.5m wide panels with same 2a/W ratio was also performed to study panel size effects. Two-dimensional elastic-plastic finite element analysis based on incremental flow theory with Von Mises yield criterion and small strain assumption was performed. A piecewise linear representation was used for the uniaxial stress-strain curve for 2024-T3 Al alloy. A $CTOA_c = 5.25^o$ measured at 0.04 in behind the crack tip was used in this study based on experimental results [7]. Comparisons between numerical results and experimental measurements [6] for the applied stress versus half crack extension (Fig. 7) show that as width of the panel increases, the relative difference between experimental results and numerical simulations marginally increases due to the three-dimensional nature of the stresses around the crack tip, a result of the constraint effects due to the finite thickness of the panels. The analysis demonstrated the ability of the methodology to effectively model CTOA controlled elastic-plastic stable crack growth in thin-sheet Al alloys.

Example 3: Residual Strength Analysis of Stiffened Fuselage Skin Panels

The STIEV methodology was applied to model self-similar crack growth and estimate the residual strength in a stiffened fuselage panel with single and MSD cracks. A 3-stringer wide, 3-frame long fuselage panel configuration having a radius of curvature of 72" with a lap joint at the central stringer without tear straps was chosen for the analysis [10]. The lap joint was a three-row configuration with 3/16 dia countersunk-head rivets. The other two stringers were spot-welded. The upper and lower skins were made of 0.04-inch thick, 2024-T3 Al alloy. The stringers and frames were made of 7075-T6 Al alloy. Frames were simply connected to stringers by rivets. The lead crack of 10.0-in was symmetrically located around the central frame line. The MSD cracks of 0.046-in length were symmetric to the lead crack at all the three rivets in front of the lead crack. Rivet holes are not explicitly modeled in the finite element model and hence the MSD crack was modeled as a small crack with a length equal to the rivet diameter plus the MSD length.

The entire skin panel, stiffeners and frames were modeled as swept surfaces. The cracks and rivet rows were modeled as line entities. A mesh pattern with 0.04" size crack tip elements was used. The $CTOA_c$ used in this problem was 5.7 ^{0} measured 0.04" behind the crack tip. The finite element software NISA was used for stress analysis. The skin panel, stiffeners and frames were modeled with quadrilateral shell elements and rivets were modeled as spring elements connecting the finite element nodes in the upper and lower skins. Pressure loading was applied on the skin. A piecewise linear representation was used for the uniaxial stress-strain curves for 2024-T3 and 7075-T6 Al alloys. The panel configuration and the FE mesh with MSD cracks are shown in Fig. 8 and Fig. 9. Both geometric and material non-linearities were included in the analysis to account for the out-of-plane bulging deformation and plasticity. Self-similar crack growth was simulated using STIEV with remeshing at each stage of crack growth. The MSD cracks significantly reduce the residual strength of the panel. For the case with only the 10.0-in lead crack a crack growth of 4.0-in and a corresponding residual strength of 15.33 psi was obtained. In the presence of MSD cracks the crack growth was only 3.0-in with the residual strength dropping to 11.31 psi.

6. CONCLUSIONS

Efficient computation techniques and a software framework for automated modeling and crack growth simulation for in built-up shell structures have been presented. These techniques enable quick and efficient residual strength estimation of pressurized airframes and thin piping systems for life extension and maintenance programs. The tools being modular can easily be adapted to various graphic display packages and FEA programs. Case studies presented demonstrate the application of the methodology and computational techniques for elastic-plastic stable crack growth simulation and residual strength estimation required for life extension of thin walled pressurized shell structures.

REFERENCES

1. P.L. Baehmann, S.L. Witchen, M.S. Shephard, K.R. Grice and M.A. Yerry, '*Robust, geometrically based, automatic two-dimensional mesh generation*', Int. j. numer. methods eng., 24, 1043-1078 (1987).

2. T. D. Blacker and M.B. Stephenson, 'Paving: '*A new approach to automated quadrilateral mesh generation*', Int. j. numer. methods eng., 32, 811-847 (1991).

3. J.Z. Zhu and O.C.Zienkiewicz, '*Adaptive finite element analysis with quadrilaterals*', Comput. Struct., 40, 1097-1104 (1991).

4. D.O. Potyondy, Paul A. Wawrzynek and Anthony R. Ingraffea, '*An algorithm to generate quadrilateral or triangular surface meshes in arbitrary domains with applications to crack propagation*', Int. j. numer. methods eng., 38, 2677-2701 (1995).

5. J.C. Cavendish, '*Automatic triangulation of arbitrary planar domains for the finite element method*', Int. j. numer. methods. Eng., 8, 679-697 (1974).

6. J. C. Newman, Jr., D. C. Dawicke, and C. A. Bigelow '*Finite element analysis and fracture simulation in thin-sheet aluminum alloy*' Proc. of Intl. Workshop on Structural Integrity of Aging Airplane, 1992.

7. D. S. Dawicke. '*Residual Strength Predictions Using a Crack Tip Opening Angle Criterion*', Proceedings of the NASA Symposium on the Continued Air-worthiness of Aircraft Structures, pp. 555-566, Atlanta, Georgia, 1996

8. K. Kato, N. S. Lee and K. J. Bathe, '*Adaptive Finite lement Analysis of Large Strain Elastic Response*', Comput. Struct., 47, No. 4/5, pp829-855 (1993).

9. I. L. Lim, I. W. Johnston, S. K. Choi and V. Murti, '*An Improved Numerical Inverse Isoparametric Mapping echnique For 2D Mesh Rezoning*', Eng. Frac. Mech., 1 (3), 417-435 (1992).

10. D. Cope '*Corrosion Damage Assessment FrameWork:Corrosion/Fatigue Effects on Structural Integrity*',Technical Report, D500-13008-1, The Boeing Defenceand Space Group (1998).

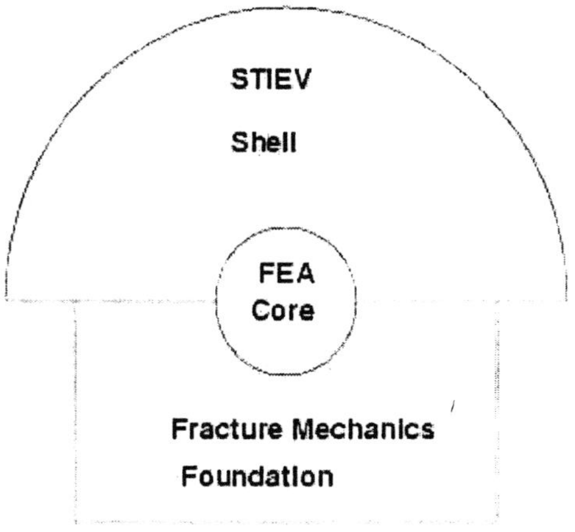

Fig. 1. STIEV Concept

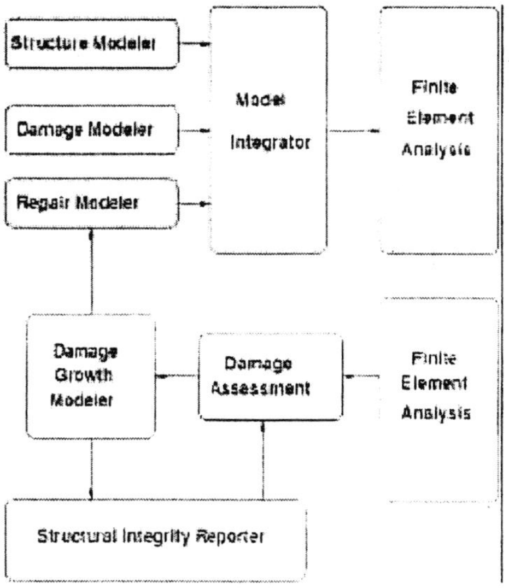

Fig. 2. STIEV architecture

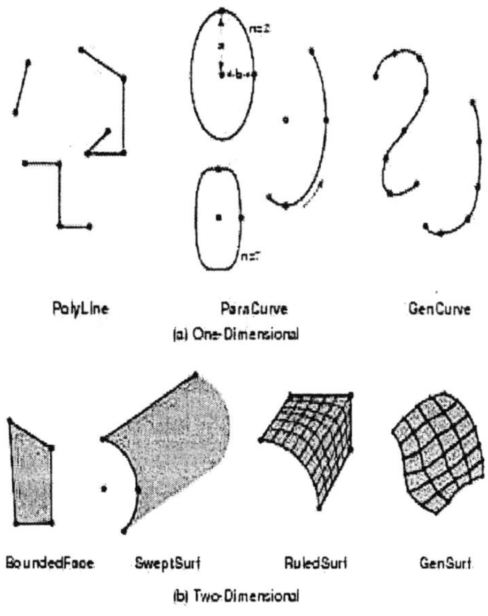

Fig. 3. 1-D and 2-D geometric primitives

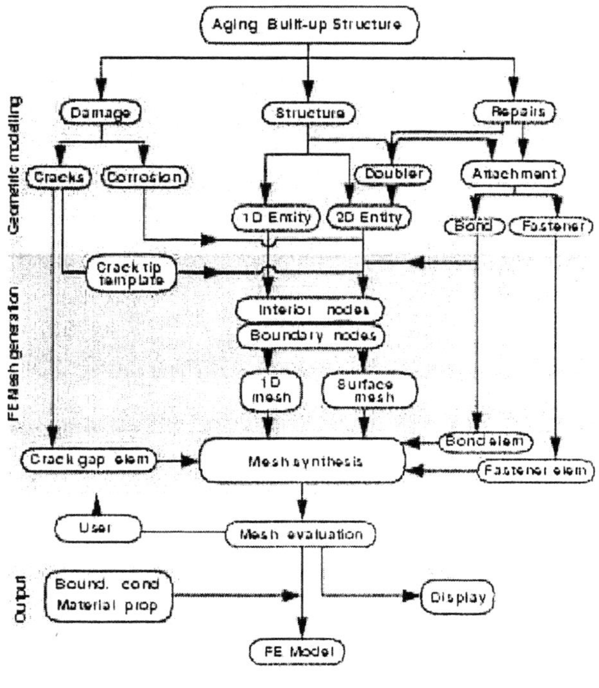

.Fig. 4. Flow chart of STIEV architecture

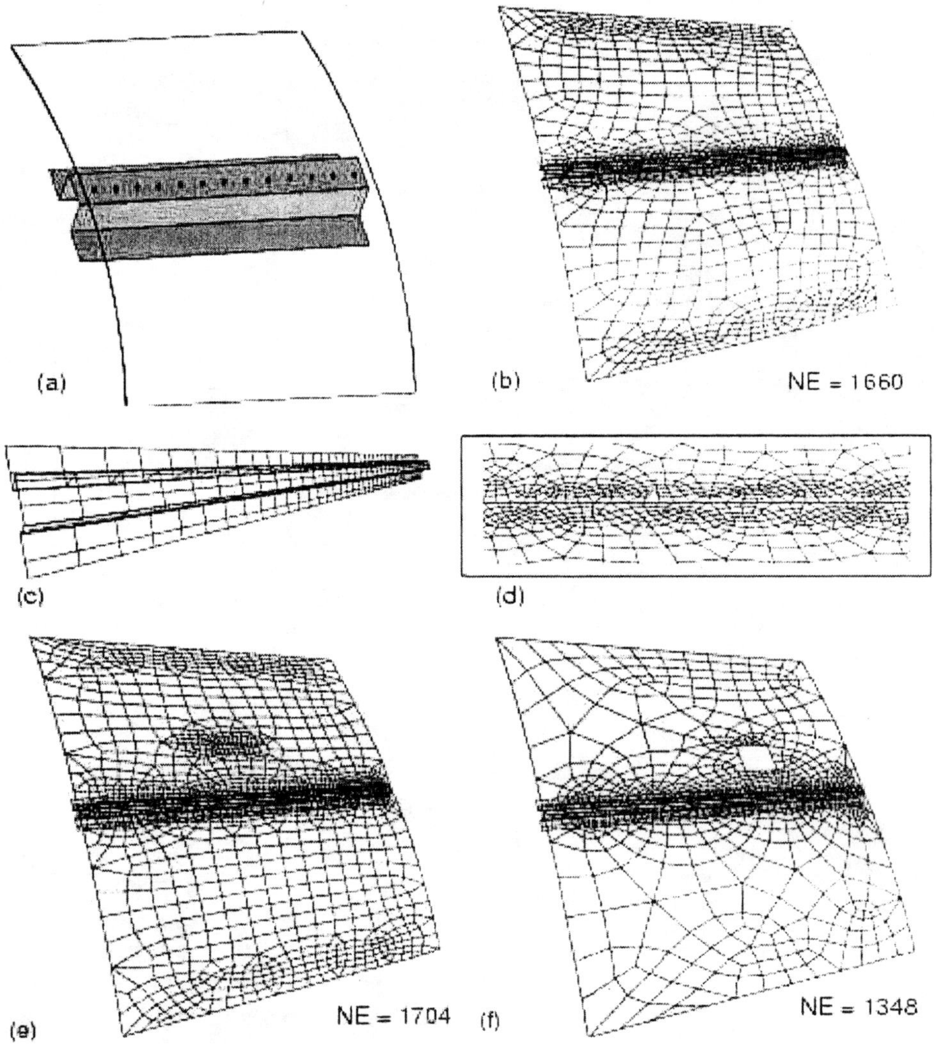

Fig. 5. Stiffened fuselage skin panel: (a) Configuration (b) Fine Mesh © Stiffener Mesh and (d) Sheet mesh at stiffener attachment (e) a crack (f) a cutout

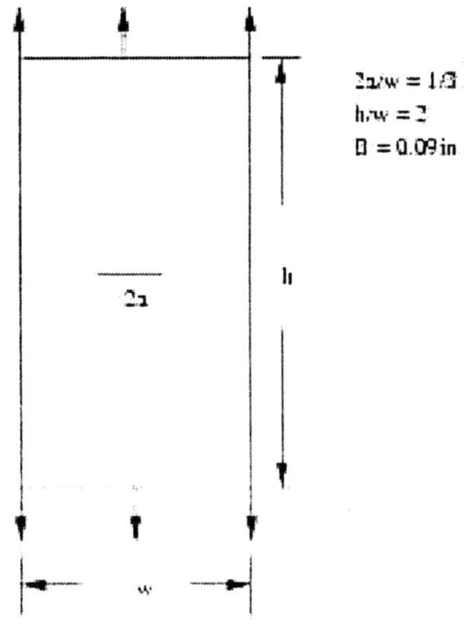

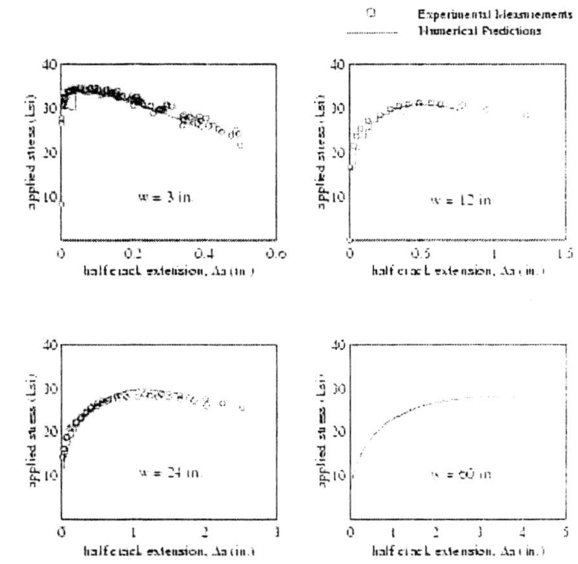

Fig. 6. CCT Specimen

Fig.7. Applied stress vs half crack extension for various specimen sizes

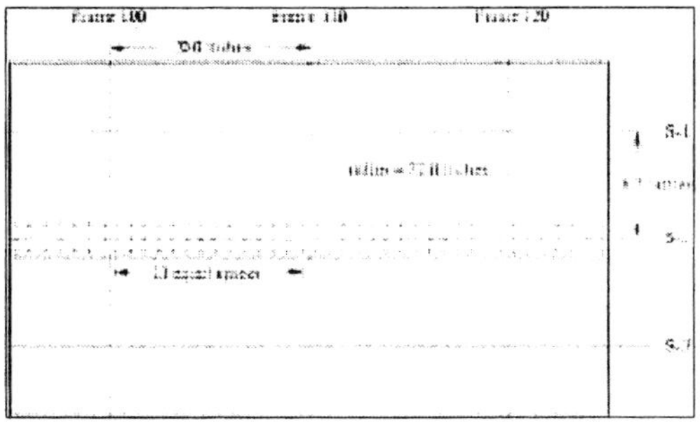

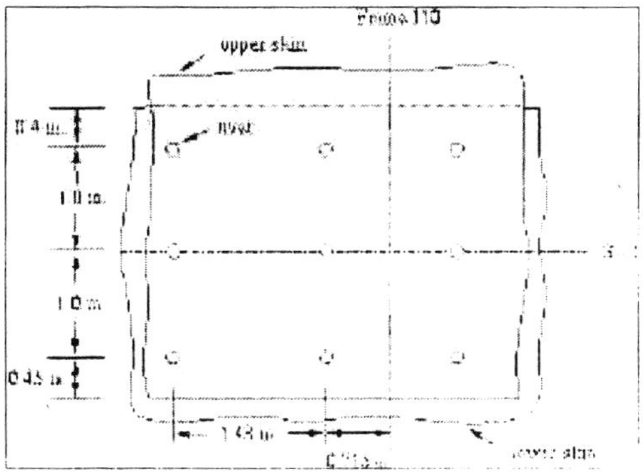

Fig. 8. (a)Dimensions of a generic narrow body fuselage panel (from [10]) and (b) detailed rivet spacing

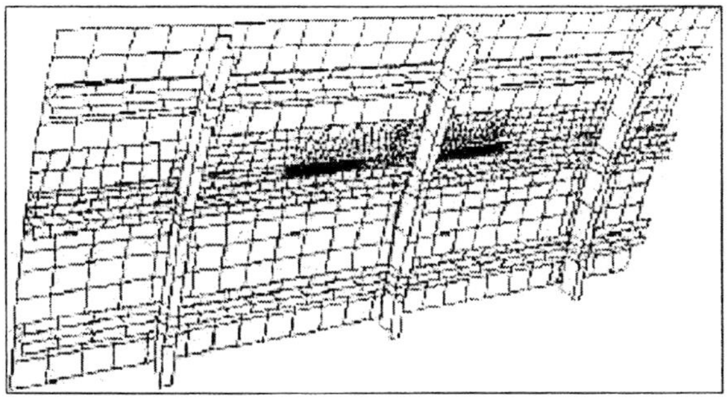

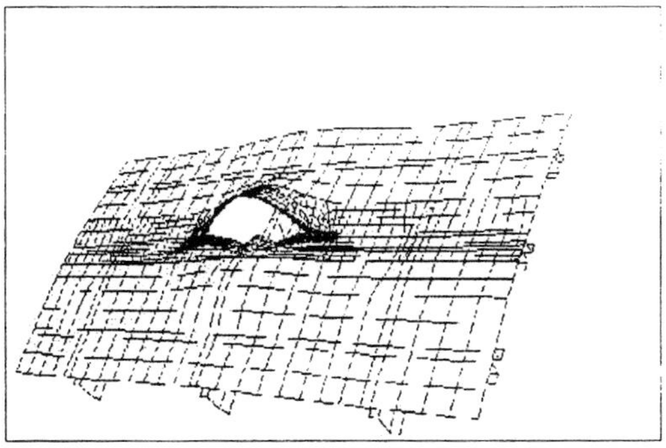

Fig. 9. (a) Finite element mesh with lead and MSD cracks (b) Deformed mesh

PVP-Vol. 458, Computer Technology and Applications

PVP2003-1900

TRANSIENT ANALYSIS OF HEAT CONDUCTION IN ORTHOTROPIC MEDIUM BY THE DQEM AND EDQ BASED TIME INTEGRATION SCHEMES

Chang-New Chen
Department of Naval Architecture and Marine Engineering
National Cheng Kung University, Tainan, Taiwan
E-mail: cchen@mail.ncku.edu.tw

ABSTRACT

The transient heat conduction in orthotropic medium is solved by using the DQEM to the spacial discretization and EDQ to the temporal discretization. In the DQEM discretization, DQ is used to define the discrete element model. Discrete transient equations defined at interior nodes in all elements, transition conditions defined on the inter-element boundary of two adjacent elements and boundary conditions at the structural boundary form a transient equation system at a specified time stage. The transient equation system is solved by the direct time integration schemes of time-element by time-element method and stages by stages method which are developed by using EDQ and DQ. Numerical results obtained by the developed numerical algorithms are presented. They demonstrate the developed numerical solution procedure.

INTRODUCTION

The analysis of transient heat conduction in orthotropic medium is frequently necessary for modern engineering design. Certain numerical methods can be used in the analysis of this type of engineering structures. A rather efficient method that can be used to develop solution algorithms for the analysis of transient heat conduction in orthotropic medium is to use the DQEM [1] for spacial discretization and EDQ for temporal discretization.

The DQEM adopts the DQ related discretization techniques. The method of DQ approximates a partial derivative of a variable function with respect to a coordinate at a node as a weighted linear sum of the function values at all nodes along that coordinate direction [2]. The original DQ can only be used to solve simple problems [3,4].

The DQ can be generalized which results in obtaining the generic differential quadrature (GDQ) [5,6]. The weighting coefficients for a grid model defined by a coordinate system having arbitrary dimensions can also be generated. The configuration of a grid model can be arbitrary. In the GDQ, a certain order derivative or partial derivative of the variable function with respect to the coordinate variables at a node is expressed as the weighted linear sum of the values of function and/or its possible derivatives or partial derivatives at all nodes.

DQ and GDQ have also been extended which results in the extended differential quadrature (EDQ) [7,8]. In the EDQ discretization, the number of total degrees of freedom attached to the nodes are the same as the number of total discrete fundamental relations required for solving the problem. A discrete fundamental relation can be defined at a point which is not a node. The points for defining fundamental relations are discrete points. A node can also be a discrete point. For the DQ and GDQ, a node is also a discrete point. Then a certain order derivative or partial derivative, of the variable function existing in a fundamental relation, with respect to the coordinate variables at an arbitrary discrete point can be expressed as the weighted linear sum of the values of function and/or its possible derivatives or partial derivatives at all nodes. Time can also be a coordinate variable [6]. Thus in solving a problem, a discrete fundamental relation can be defined at a discrete point which is not a node. If a point used for defining discrete fundamental relations is also a node, it is not necessary that the number of discrete fundamental relations at that node equals the number of degrees of freedom attached to it. This concept has been used to construct the discrete inter-element transition conditions and boundary conditions in the differential quadrature element analyses of beam bending problem and warping torsion bar problem [9,10]. Thus, DQ and GDQ can be extended to get these two discretization techniques more flexible in treating the boundary conditions or transition conditions defined on the inter-element boundaries of two adjacent elements when they are applied to the DQEM and generalized differential quadrature element method (GDQEM) [5,11].

The author has proposed a discretization method for solving a generic engineering or scientific problem having an arbitrary domain configuration [1]. Like the finite element method (FEM), in this method the analysis domain of a problem is first separated into a certain number of subdomains or elements. Then the DQ or GDQ discretization is carried out on an element-basis. The governing differential or partial differential equations defined on the elements, the transition conditions on inter-element boundaries and the boundary conditions on the analysis domain boundary are in computable algebraic forms after the DQ or GDQ discretization. By assembling all discrete fundamental equations the overall algebraic system can be obtained which is used to solve the problem. The interior elements can be regular. However, in order to solve the problem having an arbitrary analysis domain configuration elements connected to or near the analysis domain boundary might need to be irregular. The mapping technique can be used to develop irregular elements. It results in the DQEM. The GDQ can also be used to develop the irregular elements. It results in the GDQEM. The theoretical basis of DQEM and GDQEM is rigorous since all fundamental relations are locally satisfied. Consequently, the convergence properties of these two discrete element analysis methods is excellent.

The refinement procedure can be used to the DQEM and GDQEM analyses [1,5]. There are two refinement methods. One is to increase the elements which is the h refinement while the other one is to increase the order of the assumed variable function which is the p refinement. The convergence can be assured by successively carrying out the refinement analysis which adopts a certainly defined refinement indicator and a convergence criterion. The refinement indicator can be the absolute or relative local error of the variable function or a certain physical quantity defined by the derivatives or partial derivatives of the variable function, or an absolute or relative error norm defined by the error of the variable function or a physical quantity defined by the derivatives or partial derivatives of the variable function. The h refinement can be achieved by either the enrichment of mesh or the design of a new mesh. The p refinement can be achieved by either raising the number of element nodes or possibly adding a certain correction function to the assumed variable function. The adaptive concept can also be introduced into the refinement procedure. There are three methods for solving the overall algebraic system. The first one is to use the direct method to all refinement stages. The second one is to use the iterative method to all refinement stages. The last one is to use the direct method to the initial refinement stage following the use of the iterative method to the other refinement stages. From the computation cost point of view, the two iterative refinement techniques are effective for solving generic problems. They are especially effective for solving nonlinear problems. The adaptive h refinement procedure, and the adaptive p refinement procedure by raising the number of element nodes was also introduced at the time when the DQEM was proposed [1]. In addition, the repositioning techniques can also be used to improve the solutions.

In generating the mesh, if the external cause is composed of various locally distributed causing functions in order to better approximate the true distribution of the external cause the mesh must be designed in such a way that the external cause in one element will not have two or more locally different distributed functions. The adoption of this adaptive discretization technique will lead to a better solution since that locally different causing functions will lead to locally different response functions. Without adopting this technique it will result in a poor approximation of the element-basis external cause if significantly different causing functions, in quantity or order of distribution, coexist in an element. Moreover, the analysis will be more efficient by adopting this adaptive discretization technique since in subdomains having locally higher order distributed causing functions higher order elements can be used while in subdomains having locally lower order distributed causing functions lower order elements can be used [1]. If the refinement procedure is adopted, it will result in an adaptive refinement analysis.

In treating a concentrated external cause existing in the problem domain, the mesh can be designed in such a way that the concentrated external cause is located on some inter-element boundaries and included in the natural transition conditions or kinematic transition conditions. If the external cause is a force related quantity, it can also be located in some element domains and approximated by the composition of certain continuous functions based on the rule of force equivalence. However, the solution will not be able to reflect the locally transition response behavior.

It has been proved that DQEM and GDQEM are efficient [1,5-13]. The convergence rate of DQEM is excellent. The DQEM and GDQEM also have the same advantage as the finite element method of general geometry and systematic boundary treatment. For solving vibration problems, the mass matrix is diagonal which requires a little storage space. Since zeros appear on-diagonal, it is positive semidefinite. The lines with zero mass can be eliminated by some mathematical manipulations. The mass matrix is simpler to form and cheaper to use as compared to the consistent mass matrix used in the FEM analysis. It also gives greater accuracy and fewer spurious oscillations than a consistent mass matrix.

The transient response of heat conduction in orthotropic medium was solved by using the DQEM to the spacial discretization and EDQ based direct time integration method [15] to the transient response analysis. Numerical procedures are summarized and presented. Sample results are also presented. They demonstrate the developed dynamic response analysis model.

EXTENDED DIFFERENTIAL QUADRATURE

In using the EDQ to solve a problem, the number of total degrees of freedom attached to the nodes is the same as the number of total discrete fundamental relations required for solving the problem. A discrete fundamental relation can be defined at a point which is not a node. Then a certain order of derivative or partial derivative, of the variable function existing in a fundamental relation, at an arbitrary point with respect to the coordinate variables can be expressed as the weighted linear sum of the values of variable function and/or its possible derivatives at all nodes [7-8]. Thus in solving a problem, a discrete fundamental relation can be defined at a point which is not a node. If a point used for defining discrete fundamental relations is also a node, it is not necessary that the number of discrete fundamental relations at that node equals the number of degrees of freedom attached to it. This concept has been used to construct the discrete inter-element transition conditions and boundary conditions in the differential quadrature element analyses of beam bending problem, warping torsion bar problem.

Let $\pi(\xi_i)$ denote the variable function associated with a problem with ξ_i the space coordinates or time variable. The EDQ discretization for a derivative of order m at discrete point α can be expressed by

$$\frac{d^m \pi_\alpha}{d\xi_i^m} = D_{\alpha i}^{\xi_i^m} \bar{\pi}_i, \quad i = 1, 2, ..., \bar{N} \tag{1}$$

where $\bar{N}$ is the number of degrees of freedom and $\bar{\pi}_{\bar{\alpha}}$ the values of variable function and/or its possible derivatives at the N nodes. The variable function can be a set of appropriate analytical functions denoted by $\Upsilon_p(\xi_i)$. The substitution of $\Upsilon_p(\xi_i)$ in Eq. (1) leads to a linear algebraic system for determining the weighting coefficients $D_{\alpha i}^{\xi_i^m}$. The variable function can also be approximated by

$$\pi(\xi_i) = \psi_p(\xi_i)\bar{\pi}_p, \quad p = 1, 2, ..., \bar{N} \tag{2}$$

where $\psi_p(\xi_i)$ are the corresponding interpolation functions of $\bar{\pi}_p$. Adopting $\psi_p(\xi_i)$ as the variable function $\pi(\xi_i)$ and substituting it into Eq. (1), a linear algebraic system for determining $D_{\alpha i}^{\xi_i^m}$ can be obtained. And the mth order differentiation of (2) at discrete point α also leads to the extended GDQ discretization equation (1) in which $D_{\alpha i}^{\xi_i^m}$ is expressed by

$$D_{\alpha i}^{\xi_i^m} = \frac{d^m \psi_i}{d\xi_i^m} |_\alpha \tag{3}$$

Using this equation, the weighting coefficients can be easily obtained by simple algebraic calculations.

The variable function can also be approximated by

$$\pi(\xi_i) = \Upsilon_p(\xi_i) c_p, \quad p = 1, 2, ..., \bar{N} \tag{4}$$

where $\Upsilon_p(\xi_i)$ are appropriate analytical functions and c_p are unknown coefficients. The constraint conditions at all nodes can be expressed as

$$\bar{\pi}_p = \chi_{p\bar{p}} c_{\bar{p}} \tag{5}$$

where $\chi_{p\bar{p}}$ are composed of the values of $\Upsilon_p(\xi_i)$ and/or their possible derivatives at all nodes. Solving Eq. (5) for $c_{\bar{p}}$ then substituting it in Eq. (4), the variable function can be rewritten as

$$\phi(\xi_i) = \Upsilon_p(\xi_i) \chi_{\bar{p}p}^{-1} \bar{\pi}_{\bar{p}} \tag{6}$$

Using the above equation, the weighting coefficients can also be obtained

$$D_{\alpha i}^{\zeta^m} = \frac{\partial^m \Upsilon_{\bar{p}}}{\partial \zeta^m} |_\alpha \chi_{i\bar{p}}^{-1} \tag{7}$$

Various analytical functions such as sinc functions, Lagrange polynomials, Chebyshev polynomials, Bernoulli polynomials, Euler polynomials, rational functions, ..., etc. can be used to define the weighting coefficients. To solve problems having singularity properties, certain singular functions can be used for the EDQ discretization. The problems having infinite domains can also be treated.

Consider the one-dimensional discretization using only one DOF representing the variable function at the node to define the DQ. For this DQ model, N equals $\bar{N}$ and Lagrange interpolation functions can be used to explicitly express the weighting coefficients. Let $L_\beta(\xi)$ denote the Lagrange interpolation functions represented by

$$L_\beta(\xi) = \frac{M(\xi)}{M^{(1)}(\xi_\beta)} \tag{8}$$

where

$$M(\xi) = \prod_{\gamma=1}^{n+1}(\xi - \xi_\gamma), \quad M^{(1)}(\xi_\beta) = \frac{dM(\xi_\beta)}{d\xi}$$

with n the order of approximation. Letting $M^{(2)}(\xi_\beta) = d^2M(\xi_\beta)/d\xi^2$. The weighting coefficients $D^\xi_{\alpha\beta}$ for the first order derivative can be derived

$$\begin{aligned} D^\xi_{\alpha\beta} &= \frac{dL_\beta}{d\xi}|_\alpha \\ &= \frac{(\xi_\alpha-\xi_\beta)M^{(1)}(\xi_\beta)M^{(1)}(\xi_\alpha)-M(\xi_\alpha)[M^{(1)}(\xi_\beta)+(\xi_\alpha-\xi_\beta)M^{(2)}(\xi_\beta)]}{[(\xi_\alpha-\xi_\beta)M^{(1)}(\xi_\beta)]^2} \\ &= \begin{cases} \frac{M^{(1)}(\xi_\alpha)}{(\xi-\xi_\beta)M^{(1)}(\xi_\beta)}, & \text{for } \alpha \neq \beta \\ \frac{M^{(2)}(\xi_\alpha)}{2M^{(1)}(\xi_\beta)} = -\sum_{\beta=1,\beta\neq\alpha}^N D_{\alpha\beta}, & \text{for } \alpha = \beta \end{cases} \end{aligned} \tag{9}$$

When the uniform grid is used, Eq. (9) is reduced to

$$D^\xi_{\alpha\beta} = (-1)^{\alpha+\beta}\frac{(\alpha-1)!(N-\alpha)!}{\Delta\xi(\alpha-\beta)(\beta-1)!(N-\beta)!} \quad \text{for } \alpha \neq \beta \tag{10}$$

where $\Delta\xi = \xi_\alpha - \xi_\beta$. The mth order weighting coefficients $D^{\xi^m}_{\alpha\beta}$ can be similarly calculated. Assume that the nodes are symmetric with respect to the middle point and let $r = N - \alpha + 1$ and $s = N - \beta + 1$. Then $D^{\xi^m}_{\alpha\beta} = D^{\xi^m}_{rs}$ if m is even, while $D^{\xi^m}_{\alpha\beta} = -D^{\xi^m}_{rs}$ if m is odd. The number of arithmetic operations can thus be reduced in calculating the weighting coefficients.

Because only the variable functions at nodes are used to define the DQ, the higher order weighting coefficients can also be calculated by the following recurrent procedure by using the first order weighting coefficients $D^\xi_{\alpha\beta}$

$$D^{\xi^2}_{\alpha\beta} = \sum_{\gamma=1}^N D^\xi_{\alpha\gamma}D^\xi_{\gamma\beta},$$

$$D^{\xi^3}_{\alpha\beta} = \sum_{\gamma=1}^N D^{\xi^2}_{\alpha\gamma}D^\xi_{\gamma\beta},$$

$$\vdots$$

$$D^{\xi^m}_{\alpha\beta} = \sum_{\gamma=1}^N D^{\xi^{m-1}}_{\alpha\gamma}D^\xi_{\gamma\beta} \tag{11}$$

The above recurrent computation procedure is equivalent to the procedure developed by Shu and Richard [14]. However, the above procedure involves matrix multiplication while Shu and Richard's procedure involves component expansion. The related computer algorithms are different.

SOLUTION OF DISCRETE TRANSIENT EQUATION SYSTEM

The direct integration method adopting the EDQ is used to solve the discrete transient equation system. There are two different approaches for developing the integration algorithms. The numerical procedures of these two methods are illustrated by the schemes for the transient response of problems having a second order temporal derivative.

Time-Element by Time-Element Integration Algorithm

Consider that the transient equation system, at a stage of the tth incremental step, of the transient problems is expressed by

$$M_{rs}\ddot{U}^t_s + C_{rs}\dot{U}^t_s + K_{rs}U^t_s = F^t_r \tag{12}$$

where M_{rs}, C_{rs} and K_{rs} are coefficient matrices, U^t_s the vector of response variables and F^t_r the vector of external causes. Let $\Delta t$and τ denote the time increment or the size of time-element and the natural coordinate with respect to the time t. Then, by using the EDQ to discretize $\dot{U}^t_s$ and $\ddot{U}^t_s$ in the above equation, the discrete equation at the time stage p of the tth incremental step can be expressed by the following equation

$$\left(\frac{1}{\Delta t^2}M_{rs}D^{\tau^2}_{pq} + \frac{1}{\Delta t}C_{rs}D^\tau_{pq} + K_{rs}\Psi_{pq}\right)\tilde{U}^{t,q}_s = F^{t,p}_r \tag{13}$$

where $\tilde{U}^{t,q}_s$ are response variables and/or their derivatives with respect to t, and Ψ_{pq} are the corresponding interpolation functions of the EDQ discretization. In order to solve the above equation, two initial conditions are required. The transient responses can be updated by increasing the time, step by step. Each step represents a time-element. Two initial conditions are required for solving each incremental step. Let $\bar{U}_s^t$ denote the initial response variables of the tth incremental step. The initial condition of response gradients is expressed as

$$U_s^{t,1} = \bar{U}_s^t \tag{14}$$

Let $\bar{\dot{U}}_s^t$ denote the initial response gradients of the tth incremental step. The initial condition of response gradients is expressed as

$$\frac{1}{\Delta t} D_{1q}^{\tau} \bar{U}_s^{t,q} = \bar{\dot{U}}_s^t \tag{15}$$

The values of $\bar{U}_s^t$ and $\bar{\dot{U}}_s^t$ can be obtained from the solutions of the $(t-1)$th incremental step. The response histories can be updated by a time-element by time-element procedure.

Various time-elements can be used to develop the direct integration schemes. Consider the Lagrange time-element having L stage nodes. Since no time derivative of response variable is adopted for the time-element, discrete time stages for defining the discrete transient equations coincide the node stages of the element, and Ψ_{pq} represents the Kronecker delta δ_{pq}. Use the DOF assigned to the first node stage to define the condition of initial response variables, and use the DOF assigned to the second node stages to define the initial condition of response gradients. Then, by using Eqs. (12), (14) and (15), the following matrix equation can be obtained

$$\bar{U}_s^t \{\bar{K}_{RI}\} + [\bar{K}_{RR}]\{U^{t,R}\} = \{\bar{F}^{t,R}\} \tag{16}$$

where

$$\{U^{t,R}\} = \lfloor U_s^{t,2}\ U_s^{t,3} \cdots U_s^{t,L} \rfloor^T,$$

$$\{\bar{F}^{t,R}\} = \lfloor \bar{\dot{U}}_s^t\ F_s^{t,3} \cdots F_s^{t,L} \rfloor^T,$$

$$\{\bar{K}_{RI}\} = \lfloor \tfrac{1}{\Delta t}\delta_{rs}D_{11}^{\tau}\ \tfrac{1}{\Delta t^2}M_{rs}D_{31}^{\tau^2} + \tfrac{1}{\Delta t}C_{rs}D_{31}^{\tau} \cdots \tfrac{1}{\Delta t^2}M_{rs}D_{L1}^{\tau^2} + \tfrac{1}{\Delta t}C_{rs}D_{L1}^{\tau} \rfloor^T,$$

and

$$[\bar{K}_{RR}] = \begin{bmatrix} \frac{1}{\Delta t}\delta_{rs}D_{12}^{\tau} & \frac{1}{\Delta t}\delta_{rs}D_{13}^{\tau} & \cdots & \frac{1}{\Delta t}\delta_{rs}D_{1L}^{\tau} \\ \frac{1}{\Delta t^2}M_{rs}D_{32}^{\tau^2} + \frac{1}{\Delta t}C_{rs}D_{32}^{\tau} & \frac{1}{\Delta t^2}M_{rs}D_{33}^{\tau^2} + \frac{1}{\Delta t}C_{rs}D_{33}^{\tau} + K_{rs} & \cdots & \frac{1}{\Delta t^2}M_{rs}D_{3L}^{\tau^2} + \frac{1}{\Delta t}C_{rs}D_{3L}^{\tau} \\ \vdots & \vdots & \cdots & \vdots \\ \frac{1}{\Delta t^2}M_{rs}D_{L2}^{\tau^2} + \frac{1}{\Delta t}C_{rs}D_{L2}^{\tau} & \frac{1}{\Delta t^2}M_{rs}D_{L3}^{\tau^2} + \frac{1}{\Delta t}C_{rs}D_{L3}^{\tau} & \cdots & \frac{1}{\Delta t^2}M_{rs}D_{LL}^{\tau^2} + \frac{1}{\Delta t}C_{rs}D_{LL}^{\tau} + K_{rs} \end{bmatrix} \tag{17}$$

Using Eq. (16), displacements $\{U^{t,R}\}$ of the remaining $L-1$ node stages can be found

$$\{U^{t,R}\} = [\bar{K}_{RR}]^{-1}(\{\bar{F}^{t,R}\} - \bar{U}_s^t\{\bar{K}_{RI}\}) \tag{18}$$

It should be noted that if the problem is linear and that the size of the time-element is constant, then $[\bar{K}_{RR}]$ is a constant matrix. Consequently, only one decomposition is necessary for updating the response histories if a direct solution scheme is used.

Consider a $C^1 - C^0$ EDQ time-element having E stage nodes and L DOF. Assume that the first stage node has two DOF representing the displacement and velocity of the stage node, and that each of the remaining $E-1$ stage nodes has one DOF representing the displacement of the stage node. The two DOF assigned to the first stage node are used to define the two initial conditions of an element step, while the DOF assigned to each other stage node is used to define a discrete equation of motion. By using Eqs. (12), (14) and (15), the following matrix equation can be obtained

$$[\tilde{K}_{RI}]\{\tilde{\bar{U}}^{t,I}\} + [\tilde{K}_{RR}]\{\tilde{U}^{t,R}\} = \{\tilde{F}^{t,R}\} \tag{19}$$

where

$$\{\tilde{\bar{U}}^{t,I}\} = \lfloor \bar{U}_s^t \; \dot{\bar{U}}_s^t \rfloor^T,$$

$$\{\tilde{U}^{t,R}\} = \lfloor U_s^{t,2} \; U_s^{t,3} \cdots U_s^{t,L-1} \rfloor^T,$$

$$\{\tilde{F}^{t,R}\} = \lfloor F_s^{t,2} \; F_s^{t,3} \cdots F_s^{t,L-1} \rfloor^T,$$

$$[\tilde{K}_{RI}] = \begin{bmatrix} \frac{1}{\Delta t^2} M_{rs} D_{21}^{\tau^2} + \frac{1}{\Delta t} C_{rs} D_{21}^{\tau} & \frac{1}{\Delta t} M_{rs} D_{22}^{\tau^2} + C_{rs} D_{22}^{\tau} \\ \cdot & \cdot \\ \cdot & \cdot \\ \cdot & \cdot \\ \frac{1}{\Delta t^2} M_{rs} D_{(L-1)1}^{\tau^2} + \frac{1}{\Delta t} C_{rs} D_{(L-1)1}^{\tau} & \frac{1}{\Delta t} M_{rs} D_{(L-1)2}^{\tau^2} + C_{rs} D_{(L-1)2}^{\tau} \end{bmatrix}$$

and

$$[\tilde{K}_{RR}] = \begin{bmatrix} \frac{1}{\Delta t^2} M_{rs} D_{23}^{\tau^2} + \frac{1}{\Delta t} C_{rs} D_{23}^{\tau} + K_{rs} & \cdots & \frac{1}{\Delta t^2} M_{rs} D_{2L}^{\tau^2} + \frac{1}{\Delta t} C_{rs} D_{2L}^{\tau} \\ \cdot & \cdots & \cdot \\ \cdot & \cdots & \cdot \\ \cdot & \cdots & \cdot \\ \frac{1}{\Delta t^2} M_{rs} D_{(L-1)3}^{\tau^2} + \frac{1}{\Delta t} C_{rs} D_{(L-1)3}^{\tau} & \cdots & \frac{1}{\Delta t^2} M_{rs} D_{(L-1)L}^{\tau^2} + \frac{1}{\Delta t} C_{rs} D_{(L-1)L}^{\tau} + K_{rs} \end{bmatrix} \quad (20)$$

Using Eq. (19), displacements $\{U^{t,R}\}$ of the remaining $L-2$ node stages can be found

$$\{\tilde{U}^{t,R}\} = [\tilde{K}_{RR}]^{-1}(\{\tilde{F}^{t,R}\} - [\tilde{K}_{RI}]\{\tilde{\bar{U}}^{t,I}\}) \quad (21)$$

It should be noted that if $L-1$ in the superscripts implies that the $C^1 - C^0$ time-element has $L-1$ stage nodes, while L in the subscripts implies that the time-element has L DOF.where C_{rs} is the damping matrix, U_s^t the displacement vector and F_r^t the load vector.

Stages by Stages Integration Algorithm

A different algorithm was developed to solve the discrete transient equation system. In this algorithm, the equally spaced grid Lagrange DQ is used to discretize $\dot{U}_s^t$ and $\ddot{U}_s^t$. One incremental step solution similar to the time-element by time-element solution procedure of the previous algorithm is first carried out. Since no rotational DOF is assigned to the Lagrange DQ model, the DOF assigned to the second node is used to define the initial condition of response gradient. After this solution step, a time increment is increased. Assume that the Lagrange DQ model has L nodes and that the value of the time increment equals the distance spanning $N+1$ consecutive stages (nodes) of the initial solution step. N further time stages are thus defined. Then, by using this newly increased N stages and the previous L-N stages, the same DQ model can be used to discretize $\dot{U}_s^t$ and $\ddot{U}_s^t$ at the newly increased N stages. The discretized equation is expressed by

$$\left(\frac{1}{\Delta t^2} M_{rs} D_{\alpha m}^{\tau^2} + \frac{1}{\Delta t} C_{rs} D_{\alpha m}^{\tau} + K_{rs}\delta_{\alpha m}\right) \hat{U}_s^m = \hat{F}_r^{\alpha}, \quad \alpha = L-N+1, L-N+2, ..., L \quad (22)$$

where $\hat{U}_s^m$ is the response vector of the DQ temporal discretization and $\hat{F}_r^{\alpha}$ the vector of external cause of the newly increased N stages . Since only response variables at the newly increased N stages are unknowns, matrix partition technique can be used to obtain an equation system with unknowns the response variables at the newly increased N stages. This solution system is smaller than the solution system of the time-element by time-element solution algorithm and the first incremental step solution of the current algorithm which has the response variables of L-2 stages as unknowns. In this direct integration algorithm, the response histories can be updated by a stages by stages solution procedure. For solving transient problems, the maximum value of N will be L-2. If the value of N is larger than 1, the numerical stability is rather poor. The approach of adopting $N = 1$ is a stage by stage method. For the solution of linear problem, both the two solution systems need only one decomposition if a direct solution scheme is used. However, the previous algorithm has better numerical stability.

PROBLEM DESCRIPTION

Let ρ and σ denote the density and specific heat of the medium, respectively. For the two-dimensional nonuniform problem with orthotropic medium, the governing equation is expressed by

$$(\tilde{k}_x T_{,x})_{,x} + (\tilde{k}_y T_{,y})_{,y} + \tilde{Q} - \rho\sigma\frac{\partial T}{\partial t} = 0 \tag{23}$$

where T is the temperature, $\tilde{k}_x = k_x h$, $\tilde{k}_y = k_y h$, $\tilde{Q} = Qh$, h the thickness of the medium, k_x and k_y thermal conductivities and Q the energy generation rate. And the Neumann boundary condition is

$$\bar{q} = k_x l T_{,x} + k_y m T_{,y} \tag{24}$$

where $\bar{q}$ is the conduction heat flux into the domain, and l and m are the direction cosines of the outward unit normal, while on the Dirichlet boundary the temperature is prescribed. Solution of the boundary value problem of the governing equation provides components of the heat flux through the following definition

$$q_x = -k_x T_{,x}, \quad q_y = -k_y T_{,y} \tag{25}$$

NUMERICAL DISCRETIZATION

For the spacial discretization, element configuration would change from element to element in the mesh. The mapping technique is used to develop irregular elements. By introducing an invertible transformation between a master element $\tilde{\Omega}$ of regular shape and an arbitrary physical element Ω^e it should be possible to transform the partial differential operations on Ω^e so that they hold on $\tilde{\Omega}$. The mapping of $\tilde{\Omega}$ onto Ω^e is defined by the following coordinate transformations

$$x_{\bar{i}} = x_{\bar{i}}(\xi_r) \tag{26}$$

where $x_{\bar{i}}$ are physical coordinates in Ω^e and ξ_r are natural coordinates in $\tilde{\Omega}$. Then the transformations of the first order partial derivatives of the temperature are

$$T_{,\bar{i}} = \xi_{r,\bar{i}} T_{,r} \tag{27}$$

where $\xi_{r,\bar{i}}$ is the inverse matrix of the Jacobian matrix $x_{\bar{i},r}$. The first order partial derivatives of $\tilde{k}_x$ and $\tilde{k}_y$ can similarly be derived. And the transformations of the second order partial derivatives of the temperature are

$$T_{,\bar{i}\bar{j}} = \xi_{r,\bar{i}}\xi_{s,\bar{j}} T_{,rs} + \xi_{t,\bar{i}\bar{j}} T_{,t} \tag{28}$$

The outlined mapping transformations are generic which hold good for adopting any kinds of appropriate analytical functions. Thus various domain configurations and mapping techniques can be adopted. The simulation for transformation adopting the polynomial is carried out. The transformation relations are expressed by

$$x_{\bar{i}} = N_\gamma(\xi_r)\tilde{x}_{\bar{i}\gamma}, \quad \gamma = 1, 2, ..., N_c \tag{29}$$

where $\tilde{x}_{\bar{i}\gamma}$ are $x_{\bar{i}}$ and their possible partial derivatives with respective to ξ_r at nodes used to define the transformations, $N_\gamma(\xi_r)$ are the corresponding shape functions and N_c is the number of $N_\gamma(\xi_r)$. Using Eq. (18), the first order partial derivatives of the physical coordinates with respect to the natural coordinates can be obtained.

$$x_{\bar{i},\xi_{\bar{j}}} = N_{\gamma,\xi_{\bar{j}}}(\xi_r)\tilde{x}_{\bar{i}\gamma} \tag{30}$$

And the second order partial derivatives of the physical coordinates with respect to the natural coordinates are

$$x_{\bar{i},\xi_{\bar{j}}\xi_{\bar{k}}} = N_{\gamma,\xi_{\bar{j}}\xi_{\bar{k}}}(\xi_r)\tilde{x}_{\bar{i}\gamma} \tag{31}$$

Using the natural coordinates ξ and η as independent variables. Then, the substitution of the transformation equations for $T_{,x}$, $T_{,y}$, $T_{,xx}$, $T_{,yy}$, $\tilde{k}_{x,x}$ and $\tilde{k}_{y,y}$ into Eq. (23) leads to the following governing equation:

$$F_1(\xi,\eta)T_{,\xi\xi} + F_2(\xi,\eta)T_{,\xi\eta} + F_3(\xi,\eta)T_{,\eta\eta} + F_4(\xi,\eta)T_{,\xi} + F_5(\xi,\eta)T_{,\eta} + \tilde{Q} - \rho c\frac{\partial T}{\partial t} = 0 \tag{32}$$

where

$$F_1(\xi,\eta) = \tilde{k}_x\xi_{,x}^2 + \tilde{k}_y\xi_{,y}^2, \quad F_2(\xi,\eta) = 2(\tilde{k}_x\xi_{,x}\eta_{,x} + \tilde{k}_y\xi_{,y}\eta_{,y}), \quad F_3(\xi,\eta) = \tilde{k}_x\eta_{,x}^2 + \tilde{k}_y\eta_{,y}^2, \quad F_4(\xi,\eta) = \tilde{k}_x\xi_{,xx} + \tilde{k}_y\xi_{,yy} + \tilde{k}_{x,\xi}\xi_{,x}^2 + \tilde{k}_{x,\eta}\xi_{,x}\eta_{,x} + \tilde{k}_{y,\xi}\xi_{,y}^2 + \tilde{k}_{y,\eta}\xi_{,y}\eta_{,y}, \quad F_5(\xi,\eta) = \tilde{k}_x\eta_{,xx} + \tilde{k}_y\eta_{,yy} + \tilde{k}_{x,\xi}\xi_{,x}\eta_{,x} + \tilde{k}_{x,\eta}\eta_{,x}^2 + \tilde{k}_{y,\xi}\xi_{,y}\eta_{,y} + \tilde{k}_{y,\eta}\eta_{,y}^2 \tag{33}$$

The use of Eq. (27) in Eq. (24) leads to the following transformed equation

$$(k_x l\xi_{,x} + k_y m\xi_{,y})T_{,\xi} + (k_x l\eta_{,x} + k_y m\eta_{,y})T_{,\eta} = \bar{q} \tag{34}$$

Equation (25) can also be transformed by using Eq. (27)

$$q_x = -k_x(\xi_{,x}T_{,\xi} + \eta_{,x}T_{,\eta}),\, q_y = -k_y(\xi_{,y}T_{,\xi} + \eta_{,y}T_{,\eta}) \tag{35}$$

The kinematic transition condition on the inter-element boundary $\partial\Omega^{r,s}$ of two adjacent elements r and s is the continuity of temperature which is expressed as

$$T^r = T^s, \quad \text{on} \quad \partial\Omega^{r,s} \tag{36}$$

or the assumption of temperature which is expressed as

$$T^r = T^s = \bar{T}^{r,s} \quad \text{on} \quad \partial\Omega^{r,s} \tag{37}$$

where $\bar{T}^{r,s}$ is the prescribed temperature. By using Eq. (34), the natural transition condition on the inter-element boundary can also be written as

$$\begin{aligned}&\left(k_x l^r \xi^r_{,x} + k_y m^r \xi^r_{,y}\right) T^r_{,\xi} + \left(k_x l^r \eta^r_{,x} + k_y m^r \eta^r_{,y}\right) T^r_{,\eta} + \\ &\left(k_x l^s \xi^s_{,x} + k_y m^s \xi^s_{,y}\right) T^s_{,\xi} + \left(k_x l^s \eta^s_{,x} + k_y m^s \eta^s_{,y}\right) T^s_{,\eta} \\ &= \tilde{q}^{r,s}\end{aligned} \tag{38}$$

where $\tilde{q}$ is the conduction heat flux into the domain.

In the temporal discretization, the EDQ is used. Let τ and Δt denote the corresponding natural coordinate of time t and the period of EDQ temporal discretization. The period is the element length of the time-element or the time increment for the time-element by time-element method. It is a step by step direct integration scheme. Assume that a discrete time point at which the time discretization needs to be defined coincides a node of the time element. Also assume that the temporal discretization uses a $\bar{N}^t$-DOF EDQ model. In the spacial discretization, quadrilateral element adopting DQ is used. Then, the discrete transient equation at the time stage p of the tth incremental step at an interior node (α, β) of a quadrilateral element e can be obtained from Eq. (32)

$$\begin{aligned}&F_1(\xi_\alpha, \eta_\beta) D^{e\xi^2}_{\alpha m} \tilde{T}^{et,p}_{m\beta} + F_2(\xi_\alpha, \eta_\beta) D^{e\eta}_{\beta n} D^{\xi}_{\alpha m} \tilde{T}^{et,p}_{mn} \\ &+ F_3(\xi_\alpha, \eta_\beta) D^{e\eta^2}_{\beta n} \tilde{T}^{et,p}_{\alpha n} + F_4(\xi_\alpha, \eta_\beta) D^{e\xi}_{\alpha m} \tilde{T}^{et,p}_{m\beta} \\ &+ F_5(\xi_\alpha, \eta_\beta) D^{e\eta}_{\beta n} \tilde{T}^{et,p}_{\alpha n} + \tilde{Q}^e_{\alpha\beta} - \frac{(\rho\sigma)^e_{(\alpha)(\beta)}}{\Delta t} D^{\tau}_{pq} \tilde{T}^{et,q}_{\alpha\beta} \\ &= 0\end{aligned} \tag{39}$$

where $D^{\xi^2}_{e\alpha m}$ is the weighting coefficient for the temporal discretization, $D^{\xi^2}_{e\alpha m}$ and $D^{e\eta^2}_{\beta n}$ are the weighting coefficients for the second order spacial discretizations in ξ and η directions, respectively, and $D^{e\xi}_{\alpha m}$ and $D^{e\eta}_{\beta n}$ are the corresponding weighting coefficients for the first order spacial discretizations. Consider the inter-element boundary which is the $\xi = 1$ side of element r and the $\xi = 0$ side of element s. Using Eq. (36), the discrete continuity conditions of temperature are expressed by

$$T^{tt,p}_{N_\xi \beta} = T^{st,p}_{1\beta}, \quad \beta = 1, 2, ..., N_\eta \tag{40}$$

and the condition of prescribed temperature at a node on the inter-element boundary is

$$T^{rt,p}_{N_\xi \beta} = T^{st,p}_{1\beta} = \bar{T}^{(r,s)t,p}_{\beta} \tag{41}$$

And using Eq. (38), the natural transition condition at a node on the inter-element boundary can also be obtained

$$\begin{aligned}&\left(k_{xN_\xi(\beta)} l^r_{N_\xi(\beta)} \xi^r_{N_\xi(\beta),x} + k_{yN_\xi(\beta)} m^r_{N_\xi(\beta)} \xi^r_{N_\xi(\beta),y}\right) D^{r\xi}_{N_\xi m} T^{rt,p}_{m\beta} \\ &+ \left(k_{xN_\xi(\beta)} l^r_{N_\xi(\beta)} \eta^r_{N_\xi(\beta),x} + k_{yN_\xi(\beta)} m^r_{N_\xi(\beta)} \eta^r_{N_\xi(\beta),y}\right) D^{r\eta}_{\beta n} T^{rt,p}_{N_\xi n} \\ &+ \left(k_{x1(\beta)} l^s_{1(\beta)} \xi^s_{1(\beta),x} + k_{y1(\beta)} m^s_{1(\beta)} \xi^s_{1(\beta),y}\right) D^{s\xi}_{1m} T^{st,p}_{m\beta} \\ &+ \left(k_{x1(\beta)} l^s_{1(\beta)} \eta^s_{1(\beta),x} + k_{y1(\beta)} m^s_{1(\beta)} \eta^s_{1(\beta),y}\right) D^{s\eta}_{\beta n} T^{st,p}_{1n} = \tilde{q}^{(r,s)t,p}_{\beta}\end{aligned} \tag{42}$$

Letting element n be an element with $\xi = 1$ side being the Neumann boundary, the discrete Neumann boundary condition at a node can be obtained by using Eq. (42)

$$\begin{aligned}&\left(k_{xN_\xi(\beta)} l^n_{N_\xi(\beta)} \xi^n_{N_\xi(\beta),x} + k_{yN_\xi(\beta)} m^n_{N_\xi(\beta)} \xi^n_{N_\xi(\beta),y}\right) \\ &\times D^{n\xi}_{N_\xi m} T^{nt,p}_{m\beta} + \left(k_{xN_\xi(\beta)} l^n_{N_\xi(\beta)} \eta^n_{N_\xi(\beta),x}\right. \\ &\left. + k_{yN_\xi(\beta)} m^n_{N_\xi(\beta)} \eta^n_{N_\xi(\beta),y}\right) D^{n\eta}_{\beta n} T^{nt,p}_{N_\xi n} = \bar{q}^{nt,p}_{\beta}\end{aligned} \tag{43}$$

More details regarding the DQEM irregular element formulation can be seen in Reference 12.

With the kinematic transition conditions (40) in mind, then assemble the discrete transient equations (39) for all elements, the discrete natural transition conditions (42), and the discrete natural boundary conditions (43), a discrete transient equation system with the coefficients $M_{rs} = 0$ can be obtained.

Like FEM, the assemblage is based on an element by element procedure. However, in assembling the discrete equations of element e, the discrete governing equations

(39), the discrete natural transition conditions, expressed by temperatures, at nodes on the inter-element boundary of two adjacent elements at the two element boundary nodes are directly assembled to the overall discrete equation system [1]. An element basis explicit matrix equation, containing the discrete governing equations and the discrete element boundary fluxes, is not necessary to be formed in the assemblage process. Consequently, the DQEM is different from FEM which needs to form the element stiffness equation, and which neglects the exact natural transition and Neumann condition.

NUMERICAL EXAMPLES

The problem solved involves the two-dimensional transient heat conduction of a medium having two different orthotropic materials which is shown in Fig. 1. There are two subdomains having the two different materials and two different uniformly distributed heat generation rates. The values of $\sigma\rho$ of the left medium and right medium are 1 and 2, respectively. The shape of each subdomain can be described by the quadratic serendipity shape functions. The problem has both the Neumann and Dirichlet boundaries. There is also a constant heat flux into the medium on the inter-subdomain boundary. The DQEM with the element grid, 11 × 11, is used to the spacial discretization. The mapping technique is used to transform all fundamental equations defined on the irregular physical elements into fundamental equations defined on the regular natural elements. It is also used to generate grid points and calculate the direction cosines of the outward unit normal at discrete points on the element boundary [10]. The DQ discretizations are carried on the regular elements. In this analysis, the Chebyshev polynomials are used to define the DQ discretization. Chebyshev polynomials of the natural coordinates ξ_i can be generated from the following recursion formula: $T_{n+1}(\xi_i) = 2\xi_i T_n(\xi_i) - T_{n-1}(\xi_i)$, with the two initial members $T_0(\xi_i) = 1$ and $T_1(\xi_i) = \xi_i$. Certain procedures for adopting an analytical function to define the DQ discretization can be used to calculate the weighting coefficients [2]. In an element, the DQ nodes are defined by the roots of Chebyshev polynomials. With the range being $-1 \leq \xi_i \leq 1$, the roots of Chebyshev polynomials are: $\xi_{i1} = -1$, $\xi_{i\alpha_i} = -cos\frac{\alpha_i \pi}{N_i^e+1}$ for $\alpha_i = 2, 3, ..., N_i^e - 1$ and $\xi_{iN_i^e} = 1$, where N_i^e is the number of discrete points in the ξ_i direction. At a corner node, one discrete condition equation is considered. At the intersection of the Neumann boundary and the inter-subdomain boundary, the discrete Neumann condition defined on the left subdomain is considered. At D and F, the considered conditions are Dirichlet conditions. At A and C, the considered Dirichlet condition is $\bar{T} = 0$. Various techniques concerning the setup of discrete transition conditions on an inter-subdomain boundary and a corner node have been proposed in developing the DQEM analysis model [10]. The transient analysis only involves the first order temporal derivative. The Lagrange time-element by time-element procedure is used to direct integrate the discrete transient equation system. Consequently, only initial temperatures at each incremental time step are necessary for updating the response history. Numerical results of the temperature $T_{,E}$ at E of four different time stages for the solutions using various orders of time integration and sizes of time increment are summarized and listed in Table 1. It shows fast convergence.

CONCLUSIONS

A numerical analysis model for solving the transient response of continuum mechanics problems was used to solve the transient heat conduction in a medium with two orthotropic materials. The DQEM was used to the spacial discretization of the medium, while the EDQ-basis time-element by time-element method was used to carry out the direct time integration solution of the discrete transient equation system resulting from the numerical discretization. The numerical model was summarized and presented. Numerical results proved that the developed numerical algorithm has excellent convergence properties.

REFERENCES

1. Chen, C. N., 1995, "A Differential Quadrature Element Method," Proceedings of the 1st International Conference on Engineering Computation and Computer Simulation, Changsha, CHINA, Vol. 1, pp. 25-34.

2. Bellman, R. E. and Casti, J., 1971, "Differential Quadrature and Long-term Integration", J. Math. Anal. Appls., Vol. 34, pp. 235-238.

3. Bert, C. W. and Malik, M., 1996, "Differential Quadrature Method in Computational Mechanics: a Review," Appl. Mechs. Rev., Vol. 49, pp. 1-28.

4. Du, H., Lim, M. K. and Lin, R. M., 1994, "Application of Generalized Differential Quadrature Method to Structural Problems," Intl. J. Numer. Methods Engr., Vol. 37, pp. 1881-1896.

5. Chen, C. N., 1998, "A Generalized Differential Quadrature Element Method," Proceedings of the International Conference on Advanced Computational Methods in Engineering, Gent, BELGIUM, pp. 721-728.

6. Chen, C. N., 1999, "Generalization of Differential Quadrature Discretization," Numer. Algor., Vol. 22, pp. 167-182.

7. Chen, C. N., 1998, "Extended Differential Quadrature," Applied Mechanics in the Americas (eds D. Pam-

plona *et al.*), American Academy of Mechanics, Vol. 6, pp. 389-392.

8. Chen, C. N., 1999, "Differential Quadrature Element Analysis Using Extended Differential Quadrature," Comput. Math. Appls., Vol. 39, pp. 65-79.

9. Chen, C. N., 1998, "The Warping Torsion Bar Model of the Differential Quadrature Element Method," Comput. Struct., Vol. 66, pp. 249-257.

10. Chen, C. N., 1998, "Solution of Beam on Elastic Foundation by DQEM," J. Engr. Mechs., ASCE, Vol. 124, pp. 1381-1384.

11. Chen, C. N., 2000, "A Generalized Differential Quadrature Element Method," Comput. Methods Appl. Mechs. Engr., Vol. 188, pp. 553-566.

12. Chen, C. N., 1998, "Potential Flow Analysis by Using the Irregular Elements of the Differential Quadrature Element Method," Proceedings of the 17th International Conference on Offshore Mechanics and Arctic Engineering, Lisbon, PORTUGAL, ASME, Paper No. OMAE98-0361, in CD-ROM.

13. Chen, C. N., 1999,"The Differential Quadrature Element Method Irregular Element Torsion Analysis Model," Appl. Math. Model., Vol. 23, pp. 309-328.

14. Shu, C. and Richards, B. E., 1992, "Application of Generalized Differential Quadrature to Solve Two-dimensional Incompressible Navier-Stokes Equations," Intl. J. Numer. Methods Fluids, Vol. 15, 791-798.

15. Chen, C. N., 2002, "Extended GDQ and Related Discrete Element Analysis Methods for Transient Analyses of Continuum Mechanics Problems," Proc. Pressure Vessels & Piping Conference, ASME, PVP-Vol. 441, pp. 37-44, Vancouver, CANADA.

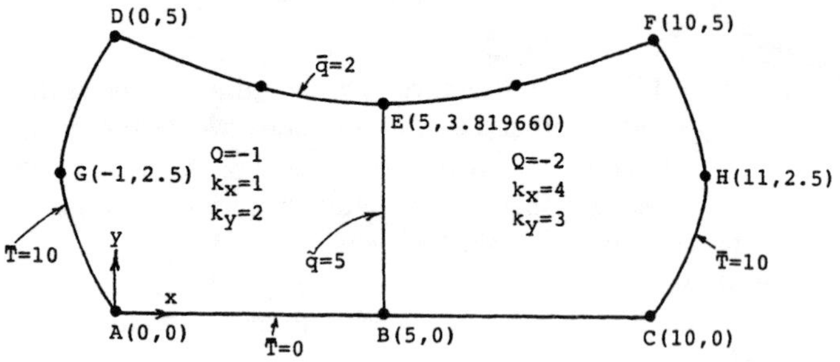

Fig. 1. Transient heat conduction in an orthotropic medium

Table 1. Temperatures of the point E at four different time stages Convergence of the DQEM analysis of the heat conduction in an orthotropic medium by using the curved quadrilateral element

DOF per step	Time increment	$t = .02$	$t = .04$	$t = .06$	$t = .08$
2	.02	$.296029\times10^0$	$.418709\times10^0$	$.501445\times10^2$	$.577310\times10^0$
	.01	$.329958\times10^0$	$.420525\times10^0$	$.506143\times10^2$	$.587442\times10^0$
	.005	$.344314\times10^0$	$.416052\times10^0$	$.510106\times10^2$	$.594914\times10^0$
3	.02	$.369079\times10^0$	$.411212\times10^0$	$.516184\times10^2$	$.602999\times10^0$
	.01	$.348732\times10^0$	$.409533\times10^0$	$.517565\times10^2$	$.604012\times10^0$

PVP-Vol. 458, Computer Technology and Applications

PVP2003-1901

CONTINUUM SHAPE SENSITIVITY ANALYSIS OF MODE-I FRACTURE IN FUNCTIONALLY GRADED MATERIALS

B. N. Rao and S. Rahman
Center for Computer-Aided Design
The University of Iowa
Iowa City, IA 52242

ABSTRACT

This paper presents a new method for continuum shape sensitivity analysis of a crack in an isotropic, linear-elastic functionally graded material. The method involves the material derivative concept of continuum mechanics, domain integral representation of the *J*-integral and direct differentiation. Unlike virtual crack extension techniques, no mesh perturbation is needed in the proposed method to calculate the sensitivity of stress-intensity factors. Since the governing variational equation is differentiated prior to the process of discretization, the resulting sensitivity equations are independent of approximate numerical techniques, such as the meshless method, finite element method, boundary element method, or others. In addition, since the *J*-integral is represented by domain integration, only the first-order sensitivity of the displacement field is needed. Numerical results show that first-order sensitivities of *J*-integral obtained by using the proposed method are in excellent agreement with the reference solutions obtained from finite-difference methods for the structural and crack geometries considered in this study.

INTRODUCTION

In recent years, functionally graded materials (FGMs) have been introduced and applied in the development of structural components subject to non-uniform service requirements. FGMs, which possess continuously varying microstructure and mechanical and/or thermal properties, are essentially two-phase particulate composites, such as ceramic and metal alloy phases, synthesized such that the composition of each constituent changes continuously in one direction, to yield a predetermined composition profile [1]. The absence of sharp interfaces in FGM largely reduces material property mismatch, which has been found to improve resistance to interfacial delamination and fatigue crack propagation [2]. However, the microstructure of FGM is generally heterogeneous, and the dominant type of failure in FGM is crack initiation and growth from inclusions. The extent to which constituent material properties and microstructure can be tailored to guard against potential fracture and failure patterns is relatively unknown. Such issues have motivated much of the current research into the numerical computation of stress intensity factors (SIFs) and its impact on fracture of FGMs [3,4]. However, in many applications of fracture mechanics, derivatives of SIF with respect to crack size are also needed for predicting stability and arrest of crack propagation in FGM. Another major use of the derivatives of SIF is in the reliability analysis of cracked structures. For example, the first- and second-order reliability methods [5], frequently used in probabilistic fracture mechanics [6-12], require the gradient and Hessian of the performance function with respect to the crack length. In linear-elastic fracture, the performance function builds on SIF. Hence, both first and second-order derivatives of SIF are needed for probabilistic analysis.

For predicting sensitivities of SIF under mode-I condition, some methods have already appeared for homogenous materials. In 1988, Lin and Abel [13] employed a virtual crack extension technique [14-17] and the variational formulation in conjunction with the Finite element method (FEM) to calculate the first-order derivative of SIF for a structure containing a single crack. Subsequently, Hwang *et al.* [18] generalized this method in 1998 to calculate both first and second-order derivatives for structures involving multiple crack systems, axisymmetric stress state, and crack-face and thermal loading. However, these methods require a mesh perturbation, a fundamental requirement of all virtual crack extension techniques. For second-order derivatives, the number of elements surrounding the crack tip that are effected by mesh perturbation has a significant effect on the solution accuracy. To overcome this problem, Chen *et al*, [19-21] recently applied concepts of shape sensitivity analysis to calculate the first-order derivative of SIFs. In this new method, the domain integral representation of the *J*-integral (mode-I) or the interaction integral (mixed-mode) is invoked and then the material derivative concept of continuum mechanics is used to obtain the first-order sensitivity of SIFs. Since the governing variational equation is differentiated before the process of discretization, the resulting sensitivity equations are independent of any approximate numerical techniques, such as the FEM, boundary element method, and others. However, the presently available sensitivity equations are valid only for analyzing cracks in homogenous materials. Hence, there is a clear need to develop new sensitivity equations for cracks in FGMs.

This paper presents a new method for continuum shape sensitivity analyses of a crack in an isotropic, linear-elastic FGM. The method involves the material derivative concept of continuum mechanics, domain integral representation of a J-integral, and direct differentiation. Unlike virtual crack extension techniques, no mesh perturbation is needed in the proposed method to calculate the sensitivity of stress-intensity factors. Since the governing variational equation is differentiated prior to the process of discretization, the resulting sensitivity equations are independent of approximate numerical techniques, such as the meshless method, finite element method, boundary element method, or others. In addition, since the J-integral is represented by domain integration, only the first-order sensitivity of the displacement field is needed. Numerical results show that first-order sensitivities of J-integral or stress intensity factors obtained by using the proposed method are in excellent agreement with the reference solutions obtained from finite-difference methods for the structural and crack geometries considered in this study.

SHAPE SENSITIVITY ANALYSIS

Velocity Field

Consider a general three-dimensional body with a specific configuration, referred to as the reference configuration, with domain Ω, boundary Γ, and a body material point identified by position vector $x \in \Omega$. Consider the motion of the body from the configuration with domain Ω and boundary Γ into another configuration with domain Ω_τ and boundary Γ_τ, as shown in Figure 1. This process can be expressed as

$$T: x \rightarrow x_\tau, \quad x \in \Omega, \tag{1}$$

where x and x_τ are the position vectors of a material point in the reference and perturbed configurations, respectively, T is a transformation mapping, and τ is a time-like parameter with

$$\begin{aligned} x_\tau &= T(x,\tau) \\ \Omega_\tau &= T(\Omega,\tau). \\ \Gamma_\tau &= T(\Gamma,\tau) \end{aligned} \tag{2}$$

A velocity field V can then be defined as

$$V(x_\tau,\tau) = \frac{dx_\tau}{d\tau} = \frac{dT(x,\tau)}{d\tau} = \frac{\partial T(x,\tau)}{\partial \tau}. \tag{3}$$

In the neighborhood of an initial time $\tau = 0$, assuming a regularity hypothesis and ignoring high-order terms, T can be approximated by

$$T(x,\tau) = T(x,0) + \tau\frac{\partial T(x,0)}{\partial \tau} + O(\tau^2) \equiv x + \tau V(x,0), \tag{4}$$

where $x = T(x,0)$ and $V(x) = V(x,0)$.

Shape Sensitivity Analysis

The variational governing equation for a structural component with domain Ω can be formulated as [22,23]

$$a_\Omega(z,\bar{z}) = \ell_\Omega(\bar{z}), \quad \text{for all } \bar{z} \in Z \tag{5}$$

where z and $\bar{z}$ are the actual displacement and virtual displacement fields of the structure, respectively, Z is the space of kinematically admissible virtual displacements, and $a_\Omega(z,\bar{z})$ and $\ell_\Omega(\bar{z})$ are energy bilinear and load linear forms, respectively. The subscript Ω in Equation 5 is used to indicate the dependency of the governing equation on the shape of the structural domain.

The pointwise material derivative at $x \in \Omega$ is defined as [22,23]

$$\dot{z} = \lim_{\tau \to 0}\left[\frac{z_\tau(x + \tau V(x)) - z(x)}{\tau}\right]. \tag{6}$$

If z_τ has a regular extension to a neighborhood of Ω_τ, then

$$\dot{z}(x) = z'(x) + \nabla z^T V(x), \tag{7}$$

where

$$z' = \lim_{\tau \to 0}\left[\frac{z_\tau(x) - z(x)}{\tau}\right] \tag{8}$$

is the partial derivative of z and $\nabla = \{\partial/\partial x_1, \partial/\partial x_2, \partial/\partial x_3\}^T$ is the vector of gradient operators. One attractive feature of the partial derivative is that, given the smoothness assumption, it commutes with the derivatives with respect to x_i, $i = 1$, 2, and 3, since they are derivatives with respect to independent variables, *i.e.*,

$$\left(\frac{\partial z}{\partial x_i}\right)' = \frac{\partial}{\partial x_i}(z'), \quad i = 1,2, \text{ and } 3. \tag{9}$$

Let ψ_1 be a domain functional, defined as an integral over Ω_τ, i.e.,

$$\psi_1 = \int_{\Omega_\tau} f_\tau(x_\tau)\, d\Omega_\tau, \tag{10}$$

where f_τ is a regular function defined on Ω_τ. If Ω is C^k regular, then the material derivative of ψ_1 at Ω is [22,23]

$$\dot{\psi}_1 = \int_\Omega \left[f'(x) + \text{div}\,(f(x)V(x))\right] d\Omega. \tag{11}$$

For a functional form of

$$\psi_2 = \int_{\Omega_\tau} g(z_\tau, \nabla z_\tau)\, d\Omega_\tau, \tag{12}$$

the material derivative of ψ_2 at Ω using Equations 9 and 11 is

$$\dot{\psi}_2 = \int_\Omega \begin{bmatrix} g_{,z_i}\dot{z}_i - g_{,z_i}\left(z_{i,j}V_j\right) + g_{,z_{i,j}}\dot{z}_{i,j} - g_{,z_{i,j}}\left(z_{i,j}V_j\right)_{,j} \\ +\mathrm{div}\left(gV\right) \end{bmatrix} d\Omega \,, (13)$$

for which, a comma is used to denote partial differentiation, *e.g.*, $z_{i,j} = \partial z_i / \partial x_j$, $\dot{z}_{i,j} = \partial \dot{z}_i / \partial x_j$, $g_{,z_i} = \partial g / \partial z_i$, $g_{,z_{i,j}} = \partial g / \partial z_{i,j}$ and V_j is the jth component of $\boldsymbol{V}$. In Equation 13, the material derivative $\dot{z}$ is the solution of the sensitivity equation obtained by taking the material derivative of Equation 5.

If no body force is involved, the variational equation (Equation 5) can be written as

$$a_\Omega(z,\bar{z}) \equiv \int_\Omega \sigma_{ij}(z)\varepsilon_{ij}(\bar{z})d\Omega \; = \ell_\Omega(\bar{z}) \equiv \int_\Gamma T_i \bar{z}_i d\Gamma \qquad (14)$$

where $\sigma_{ij}(z)$ and $\varepsilon_{ij}(\bar{z})$ are the stress and strain tensors of the displacement z and virtual displacement $\bar{z}$, respectively, T_i is the ith component of the surface traction, and $\bar{z}_i$ is the ith component of $\bar{z}$. Taking the material derivative of both sides of Equation 14 and using Equation 9,

$$a_\Omega(\dot{z},\bar{z}) = \ell'_V(\bar{z}) - a'_V(z,\bar{z}), \;\; \forall\, \bar{z} \in Z \qquad (15)$$

where the subscript V indicates the dependency of the terms on the velocity field. The terms $\ell'_V(\bar{z})$ and $a'_V(z,\bar{z})$ can be further derived as [22,23]

$$\ell'_V(\bar{z}) = \int_\Gamma \left\{ -T_i\left(z_{i,j}V_j\right) + \left[(T_i\bar{z}_i)_{,j} n_j + \kappa_\Gamma (T_i\bar{z}_i) \right](V_i n_i) \right\} d\Gamma \quad (16)$$

and

$$a'_V(z,\bar{z}) \;\; = \;\; -\int_\Omega \begin{bmatrix} \sigma_{ij}(z)\left(\bar{z}_{i,k}V_{k,j}\right) + \sigma_{ij}(\bar{z})\left(z_{i,k}V_{k,j}\right) - \\ \varepsilon^T(\bar{z})\nabla C\varepsilon(z)V - \sigma_{ij}(z)\varepsilon_{ij}(\bar{z})\mathrm{div}V \end{bmatrix} d\Omega \;\; (17)$$

where n_i is the ith component of unit normal vector $\boldsymbol{n}$, κ_Γ is the curvature of the boundary, $\bar{z}_{i,j} = \partial \bar{z}_i / \partial x_j$, and $V_{i,j} = \partial V_i / \partial x_j$.

To evaluate the sensitivity expression of Equation 13, a numerical method is needed to solve Equation 14; for which, a standard finite element method (FEM) was used in this study. If the solution z of Equation 14 is obtained using a FEM code, the same code can be used to solve Equation 15 for $\dot{z}$. This solution of $\dot{z}$ can be obtained efficiently since it requires only the evaluation of the same set of FEM matrix equations with a different fictitious load, which is the right hand side of Equation 15.

THE *J*-INTEGRAL AND ITS SENSITIVITY

The *J*-integral

Consider a body with a crack of length a, subject to mode-I loading. Using an arbitrary counter-clockwise path Γ around the crack tip, as shown in Figure 2(a), the path independent J-integral under mode-I condition for a cracked body is given by [24]

$$J = \int_\Gamma \left(W\delta_{1j} - \sigma_{ij}\frac{\partial z_i}{\partial x_1} \right) n_j \, d\Gamma \qquad (18)$$

where $W = \int \sigma_{ij} d\varepsilon_{ij}$ is the strain energy density with σ_{ij} and ε_{ij} representing components of stress and strain tensors, respectively, z_i is the ith component of the displacement vector, n_j is the jth component of the outward unit vector normal to an arbitrary contour Γ enclosing the crack tip. For linear elastic material models it can shown that $W = \sigma_{ij}\varepsilon_{ij}/2 = \varepsilon_{ij}D_{ijkl}\varepsilon_{kl}/2$, where D_{ijkl} is a component of constitutive tensor. Applying the divergence theorem, the contour integral in Equation 18 can be converted into an equivalent domain form, given by [29]

$$J = \int_A \left(\sigma_{ij}\frac{\partial z_i}{\partial x_1} - W\delta_{1j} \right) \frac{\partial q}{\partial x_j} dA + \int_A \frac{\partial}{\partial x_j}\left(\sigma_{ij}\frac{\partial z_i}{\partial x_1} - W\delta_{1j} \right) q dA \; (19)$$

where A is the area inside the contour and q is a weight function chosen such that it has a value of *unity* at the crack tip, *zero* along the boundary of the domain, and arbitrary elsewhere. By expanding the second integrand, Equation 19 reduces to

$$\begin{aligned} J = & \int_A \left(\sigma_{ij}\frac{\partial z_i}{\partial x_1} - W\delta_{1j} \right) \frac{\partial q}{\partial x_j} dA \\ & + \int_A \left(\frac{\partial \sigma_{ij}}{\partial x_j}\frac{\partial z_i}{\partial x_1} + \sigma_{ij}\frac{\partial^2 z_i}{\partial x_j \partial x_1} - \sigma_{ij}\frac{\partial \varepsilon_{ij}}{\partial x_1} - \frac{1}{2}\varepsilon_{ij}\frac{\partial D_{ijkl}}{\partial x_1}\varepsilon_{kl} \right) q dA \end{aligned} \; .(20)$$

Using equilibrium ($\partial\sigma_{ij}/\partial x_j = 0$) and strain-displacement ($\varepsilon_{ij} = \partial u_i/\partial x_j$) conditions and noting that $\partial D_{ijkl}/\partial x_1 = 0$ in homogenous materials, the second integrand of Equation 20 vanishes, yielding

$$J = \int_A \left(\sigma_{ij}\frac{\partial z_i}{\partial x_1} - W\delta_{1j} \right) \frac{\partial q}{\partial x_j} dA \,, \qquad (21)$$

which is the classical domain form of the J-integral in homogenous materials.

For non-homogeneous materials, even though the equilibrium and strain-displacement conditions are satisfied, the material gradient term of the second integrand of Equation 12 does not vanish. So Equation 20 reduces to a more general integral, henceforth referred to as the $\tilde{J}$-integral [25], which is

$$\tilde{J} = \int_A \left(\sigma_{ij}\frac{\partial z_i}{\partial x_1} - W\delta_{1j} \right) \frac{\partial q}{\partial x_j} dA - \int_A \frac{1}{2}\varepsilon_{ij}\frac{\partial D_{ijkl}}{\partial x_1}\varepsilon_{kl} q dA \,. \quad (22)$$

By comparing Equation 22 to the classical J-integral (see Equation 21), the presence of material non-homogeneity results in the addition of the second domain integral. Although this integral is negligible for a path very close to the crack tip, it must be accounted for with relatively large integral domains, so that the $\tilde{J}$-integral can be accurately calculated.

Sensitivity of the $\tilde{J}$-integral

For two-dimensional plane stress or plane strain problems, once the stress-strain, and strain-displacement relationships are applied, Equation 22 can be expressed as

$$J = \int_A h dA \tag{23}$$

where,

$$h = h_1 + h_2 + h_3 + h_4 - h_5 - h_6 - h_7 - h_8 - h_9 - h_{10}\,. \tag{24}$$

The explicit expressions of h, i = 1,...,10 are given in Appendix A for both plane stress and plane strain conditions. In accordance with Equation 11, the material derivative of J-integral can be expressed as

$$\dot{J} = \int_A [h' + \text{div}(hV)] dA \ , \tag{25}$$

where,

$$h' = h_1' + h_2' + h_3' + h_4' - h_5' - h_6' - h_7' - h_8' - h_9' - h_{10}' \tag{26}$$

and $V = \{V_1, V_2\}^T$. Assuming the crack length a to be the variable of interest, a change in crack length in the x_1 direction (mode-I) only, *i.e.*, $V = \{V_1, 0\}^T$, results in the expression of Equation 25 as

$$\dot{\tilde{J}} = \int_A (H_1 + H_2 + H_3 + H_4 - H_5 - H_6 - H_7 - H_8 - H_9 - H_{10}) dA \ , \tag{27}$$

where,

$$H_i = h_i' + \frac{\partial (h_i V_1)}{\partial x_1}, \ i = 1,...,10 \tag{28}$$

For illustration purpose, consider the term H_1 under plane stress conditions,

$$H_1 = h_1' + \frac{\partial (h_1 V_1)}{\partial x_1}, \tag{29}$$

which can be expanded as

$$\begin{aligned} H_1 &= \frac{1}{2(1-\nu^2)}\left[\left(E\varepsilon_{11}^2 \frac{\partial q}{\partial x_1}\right)' + \frac{\partial}{\partial x_1}\left(E\varepsilon_{11}^2 \frac{\partial q}{\partial x_1} V_1\right)\right] \\ &= \frac{E\varepsilon_{11}\varepsilon_{11}'}{(1-\nu^2)}\frac{\partial q}{\partial x_1} + \frac{E\varepsilon_{11}^2}{2(1-\nu^2)}\frac{\partial q'}{\partial x_1} \\ &+ \frac{\varepsilon_{11}^2 V_1}{2(1-\nu^2)}\frac{\partial E}{\partial x_1}\frac{\partial q}{\partial x_1} + \frac{E\varepsilon_{11}V_1}{(1-\nu^2)}\frac{\partial \varepsilon_{11}}{\partial x_1}\frac{\partial q}{\partial x_1} \\ &+ \frac{E\varepsilon_{11}^2 V_1}{2(1-\nu^2)}\frac{\partial^2 q}{\partial x_1^2} + \frac{E\varepsilon_{11}^2}{2(1-\nu^2)}\frac{\partial q}{\partial x_1}\frac{\partial V}{\partial x_1} \end{aligned} \tag{30}$$

In this study the velocity field V, is chosen in such a way that the FEM mesh in the domain over which the $\tilde{J}$-integral in the Equation 22 is evaluated has a virtual rigid body translation along with the crack tip. The velocity field V, used in the current study will be explained in more detail in the next section. Since q is defined around the crack tip in the domain over which the $\tilde{J}$-integral is evaluated, if the crack tip moves, the value of q around the new crack tip will be same as that of the old crack tip, hence $\dot{q} = 0$. Therefore,

$$q' = \dot{q} - \nabla^T q V = -\nabla^T q V = -\frac{\partial q}{\partial x_1}V_1 - \frac{\partial q}{\partial x_2}V_2\,. \tag{31}$$

For the case, $V_2 = 0$,

$$q' = -\frac{\partial q}{\partial x_1}V_1\,. \tag{32}$$

From Equation 32, it follows that

$$\frac{\partial q'}{\partial x_1} = -\frac{\partial^2 q}{\partial x_1^2}V_1 - \frac{\partial q}{\partial x_1}\frac{\partial V_1}{\partial x_1} \tag{33}$$

$$\frac{\partial q'}{\partial x_2} = -\frac{\partial^2 q}{\partial x_1 \partial x_2}V_1 - \frac{\partial q}{\partial x_1}\frac{\partial V_1}{\partial x_2}\,. \tag{34}$$

As in this study the elastic modulus $E(\boldsymbol{x})$, is independent of the crack length change,

$$E' = 0\,. \tag{35}$$

Hence for the case, $V_2 = 0$,

$$\dot{E} = \nabla^T E V = \frac{\partial E}{\partial x_1}V_1\,. \tag{36}$$

Using Equation 7, and strain-displacement relationship, for the case, $V_2 = 0$, it can be shown that,

$$\varepsilon_{11}' = \frac{\partial \dot{z}_1}{\partial x_1} - \frac{\partial^2 z_1}{\partial x_1^2}V_1 - \frac{\partial z_1}{\partial x_1}\frac{\partial V_1}{\partial x_1} \tag{37}$$

$$\varepsilon_{22}' = \frac{\partial \dot{z}_2}{\partial x_2} - \frac{\partial^2 z_2}{\partial x_2 \partial x_1}V_1 - \frac{\partial z_2}{\partial x_1}\frac{\partial V_1}{\partial x_2} \tag{38}$$

$$\varepsilon_{12}' = \frac{1}{2}\left(\frac{\partial \dot{z}_1}{\partial x_2} + \frac{\partial \dot{z}_2}{\partial x_1} - \frac{\partial z_1}{\partial x_1}\frac{\partial V_1}{\partial x_2} - \frac{\partial z_2}{\partial x_1}\frac{\partial V_1}{\partial x_1} - \frac{\partial \varepsilon_{12}}{\partial x_1}V_1\right). \tag{39}$$

Substituting Equations 31-39, in the Equation 30 and subsequent simplification leads to,

$$H_1 = \frac{E\varepsilon_{11}\dot{\varepsilon}_{11}}{(1-\nu^2)}\frac{\partial q}{\partial x_1} - \frac{E\varepsilon_{11}\varepsilon_{11}}{(1-\nu^2)}\frac{\partial q}{\partial x_1}\frac{\partial V_1}{\partial x_1} + \frac{\varepsilon_{11}\varepsilon_{11}V_1}{2(1-\nu^2)}\frac{\partial E}{\partial x_1}\frac{\partial q}{\partial x_1} \tag{40}$$

Similar procedure can be carried out to obtain the expressions of H_i, i = 1,10, for plane stress and plane strain conditions, respectively. Equations B1-B10 and B11-B20 in Appendix B provide explicit expressions of H_i, i = 1,10, for plane stress and plane strain conditions, respectively. These expressions of H_i, i = 1,10, when inserted in Equation 27, yield the first-order sensitivity of $\tilde{J}$ with respect to crack size. Note, when the velocity field is unity at the crack tip, $\dot{\tilde{J}}$ is equivalent to $\partial\tilde{J}/\partial a$.

The integral in Equation 27 is independent of the domain size A and can be calculated numerically using standard Gaussian quadrature. A 2 × 2 or higher integration rule is recommended to calculate $\dot{\tilde{J}}$. A flow diagram for calculating the sensitivity of J is shown in Figure 3.

Velocity Field

Velocity field definition is an important step in continuum shape sensitivity analysis. Applying an inappropriate velocity field for shape sensitivity analysis will yield inaccurate sensitivity results. The design velocity field must meet numerous, stringent theoretical and practical criteria [26]. A number of methods have been proposed in the literature to compute the design velocity field [26]. The following of this section illustrates the velocity field that was adopted in this study for the continuum shape sensitivity analysis of mode-I fracture in FGMs by using an example.

Consider an edge-cracked plate with length $2L$ units, width W (= $w_1 + w_2$) units and a crack length a. Due to the symmetry of geometry consider only half of the plate, as shown in Figure 4. In the Figure 4, $ABCD$ is the domain of size $2b_1 \times b_2$ units over which $\tilde{J}$-integral and its sensitivity is evaluated, L' is equal to half the length of plate (L) minus the length of the quadrilateral finite element along the natural boundary in the x_2–direction. The velocity field V used in this study is defined as follows in Equations 41-43:

$$V(\boldsymbol{x}) = \begin{Bmatrix} C_1(x_1)C_2(x_2) \\ 0 \end{Bmatrix} \tag{41}$$

where,

$$C_1(x_1) = \begin{cases} V_1 & \text{if } |x_1| \le b_1 \\ \dfrac{V_1(x_1 - w_2)}{(b_1 - w_2)} & \text{if } x_1 > b_1 \\ \dfrac{V_1(x_1 + w_1)}{(w_1 - b_1)} & \text{if } x_1 < -b_1 \end{cases} \tag{42}$$

and

$$C_2(x_2) = \begin{cases} 1 & \text{if } x_2 \le b_2 \\ \dfrac{L' - (x_2 - b_2)}{L'} & \text{if } x_2 > b_2 \text{ and } x_2 \le L' \\ 0 & \text{if } x_2 > L' \end{cases} \tag{43}$$

In all the mode-I fracture numerical examples presented in the next section the velocity field in Equations 41-43 was employed.

NUMERICAL EXAMPLES

Example 1: Sensitivity Analysis of Edge-Cracked Plate

Consider an edge-cracked plate with length L = 8 units, width W = 1 unit, and crack length a, as shown in Figure 5(a). Three loading conditions including the uniform fixed grip loading (constant strain), the membrane loading (constant tensile stress), and pure bending (linear stress) were considered. Figures 5(a), 5(b), and 5(c) show the schematics of the three loading conditions. The elastic modulus was assumed to follow an exponential function, given by

$$E(x_1) = E_1 \exp(\eta x_1), \quad 0 \le x_1 \le W \tag{44}$$

where $E_1 = E(0)$, $E_2 = E(W)$, and $\eta = \ln(E_2/E_1)$. In Equation 44, E_1 and η are two independent material parameters that characterize the elastic modulus variation. The following numerical values were used: $E_1 = 1$ unit, $E_2/E_1 = \exp(\eta) = 0.1, 0.2, 5$, and 10, and $a/W = 0.1, 0.2, 0.3, 0.4, 0.5$ and 0.6. The Poisson's ratio was held constant with ν = 0.3. A plane strain condition was assumed. Erdogan and Wu [27], who originally studied this example, provided a theoretical solution for normalized mode-I stress intensity factors.

Due to the symmetry of geometry and load, only half of the plate, as shown in Figure 5(d), was analyzed. The FEM models used in the analysis for $a/W = 0.1, 0.2, 0.3, 0.4, 0.5$ and 0.6 are shown in the Figures 6(a)-6(f), respectively. A 2 × 2 Gaussian integration was employed. The size of the domain ($2b_1 \times b_2$ units) around the crack tip used for evaluation of $\tilde{J}$-integral and its sensitivity for $a/W = 0.1, 0.2, 0.3, 0.4, 0.5$ and 0.6 are respectively as follows: 0.1 × 0.05 units, 0.2 × 0.1 units, 0.4 × 0.2 units, 0.6 × 0.3 units, 0.8 × 0.4 units, and 0.6 × 0.3 units.

Tables 1-3 show normalized mode-I stress intensity factors $K_I/\sigma_0\sqrt{\pi a}$, $K_I/\sigma_t\sqrt{\pi a}$, and $K_I/\sigma_b\sqrt{\pi a}$, and $\partial\tilde{J}/\partial a$ under fixed grip, membrane loading, and bending, respectively for various a/W ratios and for $E_2/E_1 = 0.1$, where $\sigma_0 = E_1\varepsilon_0/(1-\nu^2)$, $\varepsilon_0 = 1$, $\sigma_t = \sigma_b = 1$ unit. The results show that the predicted normalized SIF obtained in the present study agree very well with the analytical results of Erdogan and Wu [27], for all three types of loading and for various a/W ratios. Tables 1-3 also presents the numerical results of $\partial\tilde{J}/\partial a$. Two sets of results are shown for $\partial\tilde{J}/\partial a$, the first computed using the proposed method and second calculated using the finite-difference method. A perturbation of 10^{-5} times the crack length was used in the finite-difference calculations. The results in Tables 1-3 demonstrate that continuum shape sensitivity analysis provides accurate estimates of $\partial J/\partial a$ as compared with corresponding results from the finite-difference method. Unlike the virtual crack extension technique, no mesh perturbation is required using the proposed method. Similar results were presented in Tables 4-6 for $E_2/E_1 = 0.2$, in Tables 7-9 for $E_2/E_1 = 5.0$, and in Tables 10-12 for $E_2/E_1 = 10.0$. The results in Tables 1-12 demonstrate that continuum shape sensitivity analysis provides accurate estimates of $\partial\tilde{J}/\partial a$ for various combinations of loading conditions, a/W, and E_2/E_1 ratios.

Example 2: Sensitivity Analysis of Three-Point Bend Specimen under Mode-I

Consider a three-point bend specimen with length L = 54 units, depth $2H$ = 10 units, and thickness t = 1 unit, as shown in Figure 7(a). A concentrated load P = 1 unit was applied at the middle of the beam of span L_S = 50 units and two supports were symmetrically placed with respect to an edge crack of length a. In the depth direction, the beam consists of $2h$ units deep FGM sandwiched between two distinct homogeneous materials, each of which has depth $H-h$. E_1 and E_2 represent the elastic moduli of the bottom and top layers. Two types of elastic modulus variations: (1) linear variation and, (2) exponential variation of the FGM layer were considered, with the end values matching the properties of the bottom and top layers. Mathematically, linear variation is defined as

$$E(x_1) = \begin{cases} E_2, & x_1 \geq h \\ \dfrac{E_1+E_2}{2} + \dfrac{E_2-E_1}{2h}x_1, & -h \leq x_1 \leq h \\ E_1, & x_1 \leq -h \end{cases}, \tag{45}$$

and, exponential variation is defined as

$$E(x_1) = \begin{cases} E_2, & x_1 \geq h \\ \sqrt{E_1E_2}\exp\left(\dfrac{x_1}{2h}\ln\dfrac{E_2}{E_1}\right), & -h \leq x_1 \leq h \\ E_1, & x_1 \leq -h \end{cases}, \tag{46}$$

where E_1, E_2, and $2h$ are material parameters. The following numerical values were chosen: $2h = 1$ unit, $E_1 = 1$ unit, and $E_2/E_1 = 0.05, 0.1, 0.2, 0.5, 1, 2, 5, 10,$ and 20. For each E_2/E_1 ratio, three different crack lengths with $a/2H = 0.45, 0.5$ and 0.55 were selected such that the crack tips were either at the middle of the FGM layer ($a/2H = 0.5$) or at the material interfaces ($a/2H = 0.45$ or 0.55). The Poisson's ratio was held constant with $\nu = 0.3$. A plane stress condition was assumed.

Due to symmetric geometry and loading with respect to the crack, only a half model of the beam, as shown in Figure 7(b), was analyzed. The FEM discretization for the half beam model used in the analysis for $a/2H = 0.45, 0.5$ and 0.55 are shown in the Figures 8(a)-8(c), respectively. A 2 × 2 Gaussian integration was employed. A domain of size ($2b_1 \times b_2$ units) 8 × 5 units around the crack tip used for evaluation of $\tilde{J}$-integral and its sensitivity for $a/2H = 0.45, 0.5$ and 0.55.

Tables 13 and 14 shows the predicted normalized mode-I SIF $K_I\sqrt{H}/P$, obtained in the present study for various combinations of E_2/E_1 and $a/2H = 0.45$ for linear and exponential elastic modulus variation, respectively. Tables 13 and 14 also present the numerical results of $\partial\tilde{J}/\partial a$ for linear and exponential elastic modulus variation, respectively. Two sets of results are shown for $\partial\tilde{J}/\partial a$, the first computed using the proposed method and second calculated using the finite-difference method. A perturbation of 10^{-6} times the crack length was used in the finite-difference calculations. The results in Tables 13 and 14 demonstrate that continuum shape sensitivity analysis provides accurate estimates of $\partial\tilde{J}/\partial a$ as compared with corresponding results from the finite-difference method for various combinations of E_2/E_1 under both linear and exponential elastic modulus variations. Similar results were presented in Tables 15 and 16 for $a/2H = 0.50$, and in Tables 17 and 18 for $a/2H = 0.55$. The results in Tables 13-18 demonstrate that continuum shape sensitivity analysis provides accurate estimates of $\partial\tilde{J}/\partial a$ for various combinations of elastic modulus variations, E_2/E_1, and $a/2H$ ratios.

Example 3: Sensitivity Analysis of Composite Strip

Consider the square composite strip configuration studied by Eischen [28] with size L = 1 unit, $2h_1$ = 0.6 units and $2h_2$ = 0.4 units, as shown in Figure 9(a). A crack of length a = 0.4 units is located on the line $x_2 = 0$. The Poisson's ratio was held constant at ν = 0.3. The elastic modulus was assumed to vary smoothly according to a hyperbolic tangent function, given by

$$E(x_1) = \frac{E_1+E_2}{2} + \frac{E_1-E_2}{2}\tanh\left[\eta(x_1+0.1)\right], \quad -0.5 \leq x_1 \leq 0.5 \tag{47}$$

where E_1 and E_2 are the bounds of $E(x_1)$, and η is a non-homogeneity parameter that controls the variation of $E(x_1)$ from E_1 to E_2, as shown in Figure 9(a). When $\eta \to \infty$, a sharp discontinuity occurs in the slope of $E(x_1)$ across the interface at $x_1 = -0.1$. A tensile load corresponding to $\sigma_{22}(x_1,1) = \bar{\varepsilon}E(x_1)/(1-\nu^2)$ was applied at the top edge, which results in a uniform strain $\varepsilon_{22}(x_1,x_2) = \bar{\varepsilon}$ in the corresponding uncracked structure. The following numerical values were used: $E_1 = 1$ unit, $E_2 = 3$ units, $\eta a = 0, 2, 4, 6,$ and 20 units, and $\bar{\varepsilon} = 1$. A plane strain condition was assumed.

The FEM discretization used in the analysis is shown in Figure 4(b). A 2 × 2 Gaussian integration was employed. A domain of size $2b_1 \times b_2$ with $b_1 = b_2 = 0.3$ units, around the crack tip is used for evaluation of $\tilde{J}$-integral and its sensitivity.

Table 19 compares the predicted normalized mode-I SIF $K_I/\left[\bar{\varepsilon}E(-0.5)\sqrt{\pi a}\right]$ obtained in the present study with Eischen's results [28] for several values of ηa. The normalized SIF results obtained in the present study agree very well with the reference solution. Table 19 also presents the numerical results of $\partial\tilde{J}/\partial a$ for several values of ηa. Two sets of results are shown for $\partial\tilde{J}/\partial a$, the first computed using the proposed method and second calculated using the finite-difference method. problem. A perturbation of 10^{-5} times the crack length was used in the finite-difference calculations. The results in Table 19 demonstrate that continuum shape sensitivity analysis provides accurate estimates of $\partial\tilde{J}/\partial a$ as compared with corresponding results from the finite-difference method for small values of ηa. However, for large values of ηa

difference between the continuum shape sensitivity analysis and the finite difference increases. The reason for this discrepancy at large ηa values is due to sharp discontinuity occurs in the slope of $E(x_1)$.

SUMMARY AND CONCLUSIONS

A new method for continuum shape sensitivity analyses of a crack in an isotropic, linear-elastic functionally graded materials. The method involves the material derivative concept of continuum mechanics, domain integral representation of a *J*-integral or an interaction integral and direct differentiation. Unlike virtual crack extension techniques, no mesh perturbation is needed in the proposed method to calculate the sensitivity of stress-intensity factors. Since the governing variational equation is differentiated prior to the process of discretization, the resulting sensitivity equations are independent of approximate numerical techniques, such as the meshless method, finite element method, boundary element method, or others. In addition, since the *J*-integral is represented by domain integration, only the first-order sensitivity of the displacement field is needed. Numerical results show that first-order sensitivities of *J*-integral obtained by using the proposed method are in excellent agreement with the reference solutions obtained from finite-difference methods for the structural and crack geometries considered in this study.

ACKNOWLEDGMENTS

The authors would like to acknowledge the financial support of the U.S. National Science Foundation (NSF) under Award No. CMS-9900196. The NSF program director was Dr. Ken Chong.

Appendix A. The h-functions

For plane stress,

$$h_1 = \frac{E\varepsilon_{11}^2}{2(1-\nu^2)}\frac{\partial q}{\partial x_1}, \quad (A1)$$

$$h_2 = \frac{E\varepsilon_{12}}{(1+\nu)}\frac{\partial z_2}{\partial x_1}\frac{\partial q}{\partial x_1}, \quad (A2)$$

$$h_3 = \frac{E\varepsilon_{11}\varepsilon_{12}}{(1+\nu)}\frac{\partial q}{\partial x_2}, \quad (A3)$$

$$h_4 = \frac{E(\varepsilon_{22}+\nu\varepsilon_{11})}{1-\nu^2}\frac{\partial z_2}{\partial x_1}\frac{\partial q}{\partial x_2}, \quad (A4)$$

$$h_5 = \frac{E\varepsilon_{22}^2}{2(1-\nu^2)}\frac{\partial q}{\partial x_1}, \quad (A5)$$

$$h_6 = \frac{E\varepsilon_{12}^2}{(1+\nu)}\frac{\partial q}{\partial x_1}, \quad (A6)$$

$$h_7 = \frac{\varepsilon_{11}^2 q}{2(1-\nu^2)}\frac{\partial E}{\partial x_1}, \quad (A7)$$

$$h_8 = \frac{\nu\varepsilon_{11}\varepsilon_{22} q}{1-\nu^2}\frac{\partial E}{\partial x_1}, \quad (A8)$$

$$h_9 = \frac{\varepsilon_{22}^2 q}{2(1-\nu^2)}\frac{\partial E}{\partial x_1}, \quad (A9)$$

and

$$h_{10} = \frac{\varepsilon_{12}^2 q}{(1+\nu)}\frac{\partial E}{\partial x_1}. \quad (A10)$$

For plane strain,

$$h_1 = \frac{(1-\nu)E\varepsilon_{11}^2}{2(1+\nu)(1-2\nu)}\frac{\partial q}{\partial x_1}, \quad (A11)$$

$$h_2 = \frac{E\varepsilon_{12}}{(1+\nu)}\frac{\partial z_2}{\partial x_1}\frac{\partial q}{\partial x_1}, \quad (A12)$$

$$h_3 = \frac{E\varepsilon_{11}\varepsilon_{12}}{(1+\nu)}\frac{\partial q}{\partial x_2}, \quad (A13)$$

$$h_4 = \frac{E((1-\nu)\varepsilon_{22}+\nu\varepsilon_{11})}{(1+\nu)(1-2\nu)}\frac{\partial z_2}{\partial x_1}\frac{\partial q}{\partial x_2}, \quad (A14)$$

$$h_5 = \frac{(1-\nu)E\varepsilon_{22}^2}{2(1+\nu)(1-2\nu)}\frac{\partial q}{\partial x_1}, \quad (A15)$$

$$h_6 = \frac{E\varepsilon_{12}^2}{(1+\nu)}\frac{\partial q}{\partial x_1}, \quad (A16)$$

$$h_7 = \frac{(1-\nu)\varepsilon_{11}^2 q}{2(1+\nu)(1-2\nu)}\frac{\partial E}{\partial x_1}, \quad (A17)$$

$$h_8 = \frac{\nu\varepsilon_{11}\varepsilon_{22} q}{(1+\nu)(1-2\nu)}\frac{\partial E}{\partial x_1}, \quad (A18)$$

$$h_9 = \frac{(1-\nu)\varepsilon_{22}^2 q}{2(1+\nu)(1-2\nu)}\frac{\partial E}{\partial x_1}, \quad (A19)$$

and

$$h_{10} = \frac{\varepsilon_{12}^2 q}{(1+\nu)}\frac{\partial E}{\partial x_1}. \quad (A20)$$

Appendix B. The H-functions

For plane stress,

$$H_1 = \frac{E\varepsilon_{11}}{(1-\nu^2)}\frac{\partial \dot{z}_1}{\partial x_1}\frac{\partial q}{\partial x_1} - \frac{E\varepsilon_{11}\varepsilon_{11}}{(1-\nu^2)}\frac{\partial q}{\partial x_1}\frac{\partial V_1}{\partial x_1} + \frac{\varepsilon_{11}\varepsilon_{11}V_1}{2(1-\nu^2)}\frac{\partial E}{\partial x_1}\frac{\partial q}{\partial x_1}, \quad \text{(B1)}$$

$$\begin{aligned} H_2 = {} & \frac{E}{2(1+\nu)}\left(\frac{\partial \dot{z}_1}{\partial x_2}+\frac{\partial \dot{z}_2}{\partial x_1}\right)\frac{\partial z_2}{\partial x_1}\frac{\partial q}{\partial x_1} + \frac{E\varepsilon_{12}}{(1+\nu)}\frac{\partial \dot{z}_2}{\partial x_1}\frac{\partial q}{\partial x_1} \\ & - \frac{E\varepsilon_{11}}{2(1+\nu)}\frac{\partial z_2}{\partial x_1}\frac{\partial q}{\partial x_1}\frac{\partial V_1}{\partial x_2} - \frac{E}{(1+\nu)}\frac{\partial z_2}{\partial x_1}\frac{\partial z_2}{\partial x_1}\frac{\partial q}{\partial x_1}\frac{\partial V_1}{\partial x_1}, \\ & - \frac{E}{2(1+\nu)}\frac{\partial z_1}{\partial x_2}\frac{\partial z_2}{\partial x_1}\frac{\partial q}{\partial x_1}\frac{\partial V_1}{\partial x_1} + \frac{\varepsilon_{12}V_1}{(1+\nu)}\frac{\partial E}{\partial x_1}\frac{\partial z_2}{\partial x_1}\frac{\partial q}{\partial x_1} \end{aligned} \quad \text{(B2)}$$

$$\begin{aligned} H_3 = {} & \frac{E\varepsilon_{11}}{2(1+\nu)}\left(\frac{\partial \dot{z}_1}{\partial x_2}+\frac{\partial \dot{z}_2}{\partial x_1}\right)\frac{\partial q}{\partial x_2} + \frac{E\varepsilon_{12}}{(1+\nu)}\frac{\partial \dot{z}_1}{\partial x_1}\frac{\partial q}{\partial x_2} \\ & - \frac{E\varepsilon_{11}\varepsilon_{11}}{2(1+\nu)}\frac{\partial q}{\partial x_2}\frac{\partial V_1}{\partial x_2} - \frac{E\varepsilon_{11}}{2(1+\nu)}\frac{\partial z_2}{\partial x_1}\frac{\partial q}{\partial x_2}\frac{\partial V_1}{\partial x_1}, \\ & - \frac{E\varepsilon_{11}\varepsilon_{12}}{(1+\nu)}\frac{\partial q}{\partial x_1}\frac{\partial V_1}{\partial x_2} + \frac{\varepsilon_{11}\varepsilon_{12}V_1}{(1+\nu)}\frac{\partial E}{\partial x_1}\frac{\partial q}{\partial x_2} \end{aligned} \quad \text{(B3)}$$

$$\begin{aligned} H_4 = {} & \frac{E}{1-\nu^2}\frac{\partial \dot{z}_2}{\partial x_2}\frac{\partial z_2}{\partial x_1}\frac{\partial q}{\partial x_2} + \frac{\nu E}{1-\nu^2}\frac{\partial \dot{z}_1}{\partial x_1}\frac{\partial z_2}{\partial x_1}\frac{\partial q}{\partial x_2} \\ & - \frac{E}{1-\nu^2}\frac{\partial z_2}{\partial x_1}\frac{\partial z_2}{\partial x_1}\frac{\partial q}{\partial x_2}\frac{\partial V_1}{\partial x_2} - \frac{\nu E\varepsilon_{11}}{1-\nu^2}\frac{\partial z_2}{\partial x_1}\frac{\partial q}{\partial x_2}\frac{\partial V_1}{\partial x_1} \\ & + \frac{E\varepsilon_{22}}{1-\nu^2}\frac{\partial \dot{z}_2}{\partial x_1}\frac{\partial q}{\partial x_2} - \frac{E\varepsilon_{22}}{1-\nu^2}\frac{\partial z_2}{\partial x_1}\frac{\partial q}{\partial x_1}\frac{\partial V_1}{\partial x_2}, \\ & + \frac{\nu E\varepsilon_{11}}{1-\nu^2}\frac{\partial \dot{z}_2}{\partial x_1}\frac{\partial q}{\partial x_2} - \frac{\nu E\varepsilon_{11}}{1-\nu^2}\frac{\partial z_2}{\partial x_1}\frac{\partial q}{\partial x_1}\frac{\partial V_1}{\partial x_2} \\ & + \frac{\varepsilon_{22}V_1}{1-\nu^2}\frac{\partial E}{\partial x_1}\frac{\partial z_2}{\partial x_1}\frac{\partial q}{\partial x_2} + \frac{\nu\varepsilon_{11}V_1}{1-\nu^2}\frac{\partial E}{\partial x_1}\frac{\partial z_2}{\partial x_1}\frac{\partial q}{\partial x_2} \end{aligned} \quad \text{(B4)}$$

$$\begin{aligned} H_5 = {} & \frac{E\varepsilon_{12}}{(1+\nu)}\left(\frac{\partial \dot{z}_1}{\partial x_2}+\frac{\partial \dot{z}_2}{\partial x_1}\right)\frac{\partial q}{\partial x_1} - \frac{E\varepsilon_{12}\varepsilon_{11}}{(1+\nu)}\frac{\partial V_1}{\partial x_2}\frac{\partial q}{\partial x_1}, \\ & - \frac{E\varepsilon_{12}}{(1+\nu)}\frac{\partial z_2}{\partial x_1}\frac{\partial V_1}{\partial x_1}\frac{\partial q}{\partial x_1} + \frac{\varepsilon_{12}^2 V_1}{(1+\nu)}\frac{\partial E}{\partial x_1}\frac{\partial q}{\partial x_1} \end{aligned} \quad \text{(B5)}$$

$$H_6 = \frac{E\varepsilon_{22}}{1-\nu^2}\frac{\partial \dot{z}_2}{\partial x_2}\frac{\partial q}{\partial x_1} - \frac{E\varepsilon_{22}}{1-\nu^2}\frac{\partial z_2}{\partial x_1}\frac{\partial q}{\partial x_1}\frac{\partial V_1}{\partial x_2} + \frac{\varepsilon_{22}\varepsilon_{22}V_1}{2(1-\nu^2)}\frac{\partial E}{\partial x_1}\frac{\partial q}{\partial x_1}, \quad \text{(B6)}$$

$$\begin{aligned} H_7 = {} & \frac{\varepsilon_{11}q}{1-\nu^2}\frac{\partial \dot{z}_1}{\partial x_1}\frac{\partial E}{\partial x_1} - \frac{\varepsilon_{11}\varepsilon_{11}q}{1-\nu^2}\frac{\partial E}{\partial x_1}\frac{\partial V_1}{\partial x_1} \\ & + \frac{\varepsilon_{11}\varepsilon_{11}q}{2(1-\nu^2)}\frac{\partial E}{\partial x_1}\frac{\partial V_1}{\partial x_1} + \frac{\varepsilon_{11}\varepsilon_{11}qV_1}{2(1-\nu^2)}\frac{\partial^2 E}{\partial x_1^2}, \end{aligned} \quad \text{(B7)}$$

$$\begin{aligned} H_8 = {} & \frac{\nu\varepsilon_{22}q}{1-\nu^2}\frac{\partial \dot{z}_1}{\partial x_1}\frac{\partial E}{\partial x_1} + \frac{\nu\varepsilon_{11}q}{1-\nu^2}\frac{\partial \dot{z}_2}{\partial x_2}\frac{\partial E}{\partial x_1} \\ & - \frac{\nu\varepsilon_{11}q}{1-\nu^2}\frac{\partial E}{\partial x_1}\frac{\partial z_2}{\partial x_1}\frac{\partial V_1}{\partial x_2} + \frac{\nu\varepsilon_{11}\varepsilon_{22}qV_1}{1-\nu^2}\frac{\partial^2 E}{\partial x_1^2}, \end{aligned} \quad \text{(B8)}$$

$$\begin{aligned} H_9 = {} & \frac{\varepsilon_{22}q}{1-\nu^2}\frac{\partial \dot{z}_2}{\partial x_2}\frac{\partial E}{\partial x_1} - \frac{\varepsilon_{22}q}{1-\nu^2}\frac{\partial E}{\partial x_1}\frac{\partial z_2}{\partial x_1}\frac{\partial V_1}{\partial x_2} \\ & + \frac{\varepsilon_{22}\varepsilon_{22}qV_1}{2(1-\nu^2)}\frac{\partial^2 E}{\partial x_1^2} + \frac{\varepsilon_{22}\varepsilon_{22}q}{2(1-\nu^2)}\frac{\partial E}{\partial x_1}\frac{\partial V_1}{\partial x_1}, \end{aligned} \quad \text{(B9)}$$

and

$$\begin{aligned} H_{10} = {} & \frac{\varepsilon_{12}q}{(1+\nu)}\left(\frac{\partial \dot{z}_1}{\partial x_2}+\frac{\partial \dot{z}_2}{\partial x_1}\right)\frac{\partial E}{\partial x_1} - \frac{\varepsilon_{12}\varepsilon_{11}q}{(1+\nu)}\frac{\partial E}{\partial x_1}\frac{\partial V_1}{\partial x_2} \\ & - \frac{\varepsilon_{12}q}{(1+\nu)}\frac{\partial E}{\partial x_1}\frac{\partial z_2}{\partial x_1}\frac{\partial V_1}{\partial x_1} + \frac{\varepsilon_{12}\varepsilon_{12}q}{(1+\nu)}\frac{\partial E}{\partial x_1}\frac{\partial V_1}{\partial x_1}. \\ & + \frac{\varepsilon_{12}\varepsilon_{12}qV_1}{(1+\nu)}\frac{\partial^2 E}{\partial x_1^2} \end{aligned} \quad \text{(B10)}$$

For plane strain,

$$\begin{aligned} H_1 = {} & \frac{(1-\nu)E\varepsilon_{11}}{(1+\nu)(1-2\nu)}\frac{\partial \dot{z}_1}{\partial x_1}\frac{\partial q}{\partial x_1} - \frac{(1-\nu)E\varepsilon_{11}\varepsilon_{11}}{(1+\nu)(1-2\nu)}\frac{\partial q}{\partial x_1}\frac{\partial V_1}{\partial x_1} \\ & + \frac{(1-\nu)\varepsilon_{11}\varepsilon_{11}V_1}{2(1+\nu)(1-2\nu)}\frac{\partial E}{\partial x_1}\frac{\partial q}{\partial x_1} \end{aligned}, \quad \text{(B11)}$$

$$\begin{aligned} H_2 = {} & \frac{E}{2(1+\nu)}\left(\frac{\partial \dot{z}_1}{\partial x_2}+\frac{\partial \dot{z}_2}{\partial x_1}\right)\frac{\partial z_2}{\partial x_1}\frac{\partial q}{\partial x_1} + \frac{E\varepsilon_{12}}{(1+\nu)}\frac{\partial \dot{z}_2}{\partial x_1}\frac{\partial q}{\partial x_1} \\ & - \frac{E\varepsilon_{11}}{2(1+\nu)}\frac{\partial z_2}{\partial x_1}\frac{\partial q}{\partial x_1}\frac{\partial V_1}{\partial x_2} - \frac{E}{(1+\nu)}\frac{\partial z_2}{\partial x_1}\frac{\partial z_2}{\partial x_1}\frac{\partial q}{\partial x_1}\frac{\partial V_1}{\partial x_1}, \\ & - \frac{E}{2(1+\nu)}\frac{\partial z_1}{\partial x_2}\frac{\partial z_2}{\partial x_1}\frac{\partial q}{\partial x_1}\frac{\partial V_1}{\partial x_1} + \frac{\varepsilon_{12}V_1}{(1+\nu)}\frac{\partial E}{\partial x_1}\frac{\partial z_2}{\partial x_1}\frac{\partial q}{\partial x_1} \end{aligned} \quad \text{(B12)}$$

$$\begin{aligned} H_3 = {} & \frac{E\varepsilon_{11}}{2(1+\nu)}\left(\frac{\partial \dot{z}_1}{\partial x_2}+\frac{\partial \dot{z}_2}{\partial x_1}\right)\frac{\partial q}{\partial x_2} + \frac{E\varepsilon_{12}}{(1+\nu)}\frac{\partial \dot{z}_1}{\partial x_1}\frac{\partial q}{\partial x_2} \\ & - \frac{E\varepsilon_{11}\varepsilon_{11}}{2(1+\nu)}\frac{\partial q}{\partial x_2}\frac{\partial V_1}{\partial x_2} - \frac{E\varepsilon_{11}}{2(1+\nu)}\frac{\partial z_2}{\partial x_1}\frac{\partial q}{\partial x_2}\frac{\partial V_1}{\partial x_1}, \\ & - \frac{E\varepsilon_{11}\varepsilon_{12}}{(1+\nu)}\frac{\partial q}{\partial x_1}\frac{\partial V_1}{\partial x_2} + \frac{\varepsilon_{11}\varepsilon_{12}V_1}{(1+\nu)}\frac{\partial E}{\partial x_1}\frac{\partial q}{\partial x_2} \end{aligned} \quad \text{(B13)}$$

$$\begin{aligned} H_4 = {} & \frac{\nu E}{(1+\nu)(1-2\nu)}\frac{\partial \dot{z}_1}{\partial x_1}\frac{\partial z_2}{\partial x_1}\frac{\partial q}{\partial x_2} - \frac{\nu E\varepsilon_{11}}{(1+\nu)(1-2\nu)}\frac{\partial z_2}{\partial x_1}\frac{\partial q}{\partial x_2}\frac{\partial V_1}{\partial x_1} \\ & + \frac{(1-\nu)E}{(1+\nu)(1-2\nu)}\frac{\partial \dot{z}_2}{\partial x_2}\frac{\partial z_2}{\partial x_1}\frac{\partial q}{\partial x_2} - \frac{(1-\nu)E}{(1+\nu)(1-2\nu)}\frac{\partial z_2}{\partial x_1}\frac{\partial z_2}{\partial x_1}\frac{\partial q}{\partial x_2}\frac{\partial V_1}{\partial x_2} \\ & + \frac{(1-\nu)E\varepsilon_{22}}{(1+\nu)(1-2\nu)}\frac{\partial q}{\partial x_2}\frac{\partial \dot{z}_2}{\partial x_1} - \frac{(1-\nu)E\varepsilon_{22}}{(1+\nu)(1-2\nu)}\frac{\partial z_2}{\partial x_1}\frac{\partial q}{\partial x_1}\frac{\partial V_1}{\partial x_2} \\ & + \frac{\nu E\varepsilon_{11}}{(1+\nu)(1-2\nu)}\frac{\partial q}{\partial x_2}\frac{\partial \dot{z}_2}{\partial x_1} - \frac{\nu E\varepsilon_{11}}{(1+\nu)(1-2\nu)}\frac{\partial z_2}{\partial x_1}\frac{\partial q}{\partial x_1}\frac{\partial V_1}{\partial x_2} \\ & + \frac{(1-\nu)\varepsilon_{22}V_1}{(1+\nu)(1-2\nu)}\frac{\partial E}{\partial x_1}\frac{\partial z_2}{\partial x_1}\frac{\partial q}{\partial x_2} + \frac{\nu\varepsilon_{11}V_1}{(1+\nu)(1-2\nu)}\frac{\partial E}{\partial x_1}\frac{\partial z_2}{\partial x_1}\frac{\partial q}{\partial x_2} \end{aligned}, \quad \text{(B14)}$$

$$\begin{aligned} H_5 = {} & \frac{E\varepsilon_{12}}{(1+\nu)}\left(\frac{\partial \dot{z}_1}{\partial x_2}+\frac{\partial \dot{z}_2}{\partial x_1}\right)\frac{\partial q}{\partial x_1} - \frac{E\varepsilon_{12}\varepsilon_{11}}{(1+\nu)}\frac{\partial V_1}{\partial x_2}\frac{\partial q}{\partial x_1}, \\ & - \frac{E\varepsilon_{12}}{(1+\nu)}\frac{\partial z_2}{\partial x_1}\frac{\partial V_1}{\partial x_1}\frac{\partial q}{\partial x_1} + \frac{\varepsilon_{12}^2 V_1}{(1+\nu)}\frac{\partial E}{\partial x_1}\frac{\partial q}{\partial x_1} \end{aligned} \quad \text{(B15)}$$

$$H_6 = \frac{(1-\nu)E\varepsilon_{22}}{(1+\nu)(1-2\nu)}\frac{\partial \dot{z}_2}{\partial x_2}\frac{\partial q}{\partial x_1} - \frac{(1-\nu)E\varepsilon_{22}}{(1+\nu)(1-2\nu)}\frac{\partial z_2}{\partial x_1}\frac{\partial q}{\partial x_1}\frac{\partial V_1}{\partial x_2} + \frac{(1-\nu)\varepsilon_{22}\varepsilon_{22}V_1}{2(1+\nu)(1-2\nu)}\frac{\partial E}{\partial x_1}\frac{\partial q}{\partial x_1}, \quad \text{(B16)}$$

$$H_7 = \frac{(1-\nu)\varepsilon_{11}q}{(1+\nu)(1-2\nu)}\frac{\partial \dot{z}_1}{\partial x_1}\frac{\partial E}{\partial x_1} - \frac{(1-\nu)\varepsilon_{11}\varepsilon_{11}q}{(1+\nu)(1-2\nu)}\frac{\partial E}{\partial x_1}\frac{\partial V_1}{\partial x_1} + \frac{(1-\nu)\varepsilon_{11}\varepsilon_{11}q}{2(1+\nu)(1-2\nu)}\frac{\partial E}{\partial x_1}\frac{\partial V_1}{\partial x_1} + \frac{(1-\nu)\varepsilon_{11}\varepsilon_{11}qV_1}{2(1+\nu)(1-2\nu)}\frac{\partial^2 E}{\partial x_1^2}, \quad \text{(B17)}$$

$$H_8 = \frac{\nu\varepsilon_{22}q}{(1+\nu)(1-2\nu)}\frac{\partial \dot{z}_1}{\partial x_1}\frac{\partial E}{\partial x_1} + \frac{\nu\varepsilon_{11}q}{(1+\nu)(1-2\nu)}\frac{\partial \dot{z}_2}{\partial x_2}\frac{\partial E}{\partial x_1} - \frac{\nu\varepsilon_{11}q}{(1+\nu)(1-2\nu)}\frac{\partial E}{\partial x_1}\frac{\partial z_2}{\partial x_1}\frac{\partial V_1}{\partial x_2} + \frac{\nu\varepsilon_{11}qV_1}{(1+\nu)(1-2\nu)}\frac{\partial^2 E}{\partial x_1^2}\frac{\partial z_2}{\partial x_2}, \quad \text{(B18)}$$

$$H_9 = \frac{(1-\nu)\varepsilon_{22}q}{(1+\nu)(1-2\nu)}\frac{\partial \dot{z}_2}{\partial x_2}\frac{\partial E}{\partial x_1} - \frac{(1-\nu)\varepsilon_{22}q}{(1+\nu)(1-2\nu)}\frac{\partial E}{\partial x_1}\frac{\partial z_2}{\partial x_1}\frac{\partial V_1}{\partial x_2} + \frac{(1-\nu)\varepsilon_{22}\varepsilon_{22}q}{2(1+\nu)(1-2\nu)}\frac{\partial E}{\partial x_1}\frac{\partial V_1}{\partial x_1} + \frac{(1-\nu)\varepsilon_{22}\varepsilon_{22}qV_1}{2(1+\nu)(1-2\nu)}\frac{\partial^2 E}{\partial x_1^2}, \quad \text{(B19)}$$

and

$$H_{10} = \frac{\varepsilon_{12}q}{(1+\nu)}\left(\frac{\partial \dot{z}_1}{\partial x_2} + \frac{\partial \dot{z}_2}{\partial x_1}\right)\frac{\partial E}{\partial x_1} - \frac{\varepsilon_{12}\varepsilon_{11}q}{(1+\nu)}\frac{\partial E}{\partial x_1}\frac{\partial V_1}{\partial x_2} - \frac{\varepsilon_{12}q}{(1+\nu)}\frac{\partial E}{\partial x_1}\frac{\partial z_2}{\partial x_1}\frac{\partial V_1}{\partial x_1} + \frac{\varepsilon_{12}\varepsilon_{12}q}{(1+\nu)}\frac{\partial E}{\partial x_1}\frac{\partial V_1}{\partial x_1} + \frac{\varepsilon_{12}\varepsilon_{12}qV_1}{(1+\nu)}\frac{\partial^2 E}{\partial x_1^2}. \quad \text{(B20)}$$

REFERENCES

1. Suresh, S. and Mortensen, A., "Fundamentals of Functionally Graded Materials," *IOM Communications Ltd.*, London, 1998.

2. Erdogan, F., "Fracture Mechanics of Functionally Graded Materials," *Composites Engineering*, Vol. 5, No. 7, pp. 753-770, 1995.

3. Gu, P., Dao, M., and Asaro, R, J., "A Simplified Method for Calculating the Crack Tip Field of functionally Graded Materials Using the Domain Integral," *Journal of Applied Mechanics*, Vol. 66, pp. 101-108, 1999.

4. Rao, B. N., and Rahman, S., "Meshfree Analysis of Cracks in Isotropic Functionally Graded Materials," *Engineering Fracture Mechanics*, Vol. 70, pp. 1-27, 2003.

5. Madsen, H. O., Krenk, S., and Lind, N. C., Methods of Structural Safety, Prentice-Hall, Inc., Englewood Cliffs, New Jersey, 1986.

6. Grigoriu, M., Saif, M. T. A., El-Borgi, S., and Ingraffea, A., "Mixed-Mode Fracture Initiation and Trajectory Prediction under Random Stresses," *International Journal of Fracture*, Vol. 45, pp. 19-34, 1990.

7. Provan, James, W., Probabilistic Fracture Mechanics and Reliability, Martinus Nijhoff Publishers, Dordrecht, The Netherlands, 1987.

8. Besterfield, G.H. , Liu, W.K., Lawrence, M. A. and Belytschko, T., "Fatigue Crack Growth Reliability by Probabilistic Finite Elements," *Computer Methods in Applied Mechanics and Engineering*, Vol. 86, pp. 297-320, 1991.

9. Besterfield, G. H., Lawrence, M. A., and Belytschko, T., "Brittle Fracture Reliability by Probabilistic Finite Elements," *ASCE Journal of Engineering Mechanics*, Vol. 116, No. 3, pp. 642-659, 1990.

10. Rahman, S., "A Stochastic Model for Elastic-Plastic Fracture Analysis of Circumferential Through-Wall-Cracked Pipes Subject to Bending," *Engineering Fracture Mechanics*, Vol. 52, No. 2, pp. 265-288, 1995.

11. Rahman, S. and Kim, J-S., "Probabilistic Fracture Mechanics for Nonlinear Structures," *International Journal of Pressure Vessels and Piping*, Vol. 78, No. 4, pp. 9-17, 2000.

12. Rahman, S., "Probabilistic Fracture Mechanics by *J*-estimation and Finite Element Methods," *Engineering Fracture Mechanics*, Vol. 68, pp. 107-125, 2001.

13. Lin, S. C. and Abel, J., "Variational Approach for a New Direct-Integration Form of the Virtual Crack Extension Method," *International Journal of Fracture*, Vol. 38, pp. 217-235, 1988.

14. deLorenzi, H. G., "On the Energy Release Rate and the J-integral for 3-D Crack Configurations," *International Journal of Fracture*, Vol. 19, pp. 183-193, 1982.

15. deLorenzi, H. G., "Energy Release Rate Calculations by the Finite Element Method," *Engineering Fracture Mechanics*, Vol. 21, pp. 129-143, 1985.

16. Haber, R. B. and Koh, H. M., "Explicit Expressions for Energy Release Rates using Virtual Crack Extensions," *International Journal of Numerical Methods in Engineering*, Vol. 21, pp. 301-315, 1985.

17. Barbero, E. J. and Reddy, J. N., "The Jacobian Derivative Method for Three-Dimensional Fracture Mechanics," *Communications in Applied Numerical Methods*, Vol. 6, pp. 507-518, 1990.

18. Hwang, C. G., Wawrzynek, P. A., Tayebi, A. K., and Ingraffea, A. R., "On the Virtual Crack Extension Method for Calculation of the Rates of Energy Release Rate," *Engineering Fracture Mechanics*, Vol. 59, pp. 521-542, 1998.

19. Chen, G., Rahman, S., and Y. H. Park, "Shape Sensitivity Analysis of Linear-Elastic Cracked Structures," *ASME Journal of Pressure Vessel Technology*, Vol. 124, No. 4, pp. 476-482, 2002.

20. Chen, G., Rahman, S., and Y. H. Park, "Shape Sensitivity and Reliability Analyses of Linear-Elastic Cracked Structures," *International Journal of Fracture*, Vol. 112, No. 3, 2001, pp. 223-246.

21. Chen, G., Rahman, S. and Park, Y. H., "Shape Sensitivity Analysis in Mixed-Mode Fracture Mechanics," *Computational Mechanics*, Vol. 27, No. 4, 2001, pp. 282-291.

22. Choi, K.K. and Haug, E.J., "Shape Design Sensitivity Analysis of Elastic Structures," *Journal of Structural Mechanics*, Vol. 11, No. 2, pp. 231-269, 1983.

23. Haug, E. J., Choi, K. K., and Komkov, V., Design Sensitivity Analysis of Structural Systems, Academic Press, New York, NY, 1986.

24. Rice, J. R., "A Path Independent Integral and the Approximate Analysis of Strain Concentration by Notches and Cracks," *Journal of Applied Mechanics*, 35, pp. 379-386, 1968.

25. Moran, B. and Shih, F., "Crack Tip and Associated Domain Integrals from Momentum and Energy Balance," *Engineering Fracture Mechanics*, Vol. 27, pp. 615-642, 1987.

26. Choi, K.K. and Chang, K.H., "A Study of Design Velocity Field Computation for Shape Optimal Design," *Finite Elements in Analysis and Design*, Vol. 15, pp. 317-341, 1994.

27. Erdogan, F., and Wu, B, H., "The Surface Crack Problem for a Plate with Functionally Graded Properties," *Journal of Applied Mechanics*, Vol. 64, pp. 449-456, 1997.

28. Eischen, J, W., "Fracture of Nonhomogeneous Materials," *International Journal of Fracture*, Vol. 34, pp. 3-22, 1987.

Table 1. Sensitivity of J for an edge-cracked plate under fixed grip loading by the proposed and finite difference methods ($E_2/E_1 = 0.1$)

a/W	$K_I/\sigma_0\sqrt{\pi a}$		Sensitivity of J-integral ($\partial\tilde{J}/\partial a$)		
	Present Results	Erodogan & Wu [27]	Proposed Method	Finite Difference	Difference[(a)] (percent)
0.1	1.1577	1.1648	8.0745	8.0745	0.0002
0.2	1.2892	1.2963	18.0526	18.0527	0.0006
0.3	1.4965	1.5083	41.5662	41.5667	0.0012
0.4	1.8009	1.8246	99.9351	99.9367	0.0016
0.5	2.2583	2.3140	259.7135	259.7208	0.0028
0.6	2.9997	3.1544	772.5495	772.5706	0.0027

(a) Difference = ($\partial\tilde{J}/\partial a$ by finite difference method - $\partial\tilde{J}/\partial a$ by proposed method)×100/$\partial\tilde{J}/\partial a$ by finite difference method

Table 2. Sensitivity of J for an edge-cracked plate under Membrane Loading by the proposed and finite difference methods ($E_2/E_1 = 0.1$)

a/W	$K_I/\sigma_t\sqrt{\pi a}$		Sensitivity of J-integral ($\partial\tilde{J}/\partial a$)		
	Present Results	Erodogan & Wu [27]	Proposed Method	Finite Difference	Difference[(a)] (percent)
0.1	0.8096	0.8129	5.4641	5.4648	0.0120
0.2	1.2925	1.2965	23.0400	23.0408	0.0038
0.3	1.8488	1.8581	71.7504	71.7516	0.0017
0.4	2.5454	2.5699	206.2253	206.2301	0.0023
0.5	3.5001	3.5701	601.9136	601.9302	0.0028
0.6	4.9645	5.1880	1939.6908	1939.7545	0.0033

(a) Difference = ($\partial\tilde{J}/\partial a$ by finite difference method - $\partial\tilde{J}/\partial a$ by proposed method)×100/$\partial\tilde{J}/\partial a$ by finite difference method

Table 3. Sensitivity of J for an edge-cracked plate under Bending by the proposed and finite difference methods ($E_2/E_1 = 0.1$)

a/W	$K_I/\sigma_b\sqrt{\pi a}$		Sensitivity of J-integral ($\partial\tilde{J}/\partial a$)		
	Present Results	Erodogan & Wu [27]	Proposed Method	Finite Difference	Difference[(a)] (percent)
0.1	2.0289	2.0427	15.3509	15.3494	−0.0096
0.2	1.8907	1.9040	21.1042	21.1033	−0.0039
0.3	1.8662	1.8859	35.2488	35.2489	0.0005
0.4	1.9437	1.9778	66.9626	66.9627	0.0001
0.5	2.1468	2.2151	145.0595	145.0640	0.0031
0.6	2.5544	2.7170	373.0417	373.0470	0.0014

(a) Difference = ($\partial\tilde{J}/\partial a$ by finite difference method - $\partial\tilde{J}/\partial a$ by proposed method)×100/$\partial\tilde{J}/\partial a$ by finite difference method

Table 4. Sensitivity of J for an edge-cracked plate under fixed grip loading by the proposed and finite difference methods (E_2/E_1 = 0.2)

a/W	$K_I/\sigma_0\sqrt{\pi a}$		Sensitivity of J-integral ($\partial\tilde{J}/\partial a$)		
	Present Results	Erodogan & Wu [27]	Proposed Method	Finite Difference	Difference[(a)] (percent)
0.1	1.1627	1.1670	7.2425	7.2425	0.0003
0.2	1.3020	1.3058	15.1677	15.1678	0.0008
0.3	1.5269	1.5330	33.3988	33.3993	0.0014
0.4	1.8636	1.8751	77.6551	77.6564	0.0017
0.5	2.3775	2.4031	196.5079	196.5134	0.0028
0.6	3.2208	3.2981	571.9119	571.9329	0.0037

(a) Difference = ($\partial\tilde{J}/\partial a$ by finite difference method - $\partial\tilde{J}/\partial a$ by proposed method)×100/ $\partial\tilde{J}/\partial a$ by finite difference method

Table 5. Sensitivity of J for an edge-cracked plate under Membrane Loading by the proposed and finite difference methods (E_2/E_1 = 0.2)

a/W	$K_I/\sigma_t\sqrt{\pi a}$		Sensitivity of J-integral ($\partial\tilde{J}/\partial a$)		
	Present Results	Erodogan & Wu [27]	Proposed Method	Finite Difference	Difference[(a)] (percent)
0.1	1.0519	1.0553	6.2970	6.2969	−0.0007
0.2	1.3925	1.3956	18.5807	18.5809	0.0010
0.3	1.8338	1.8395	48.9682	48.9688	0.0012
0.4	2.4314	2.4436	127.5675	127.5700	0.0020
0.5	3.2959	3.3266	348.8805	348.8897	0.0026
0.6	4.6717	4.7614	1073.4914	1073.5312	0.0037

(a) Difference = ($\partial\tilde{J}/\partial a$ by finite difference method - $\partial\tilde{J}/\partial a$ by proposed method)×100/ $\partial\tilde{J}/\partial a$ by finite difference method

Table 6. Sensitivity of J for an edge-cracked plate under Bending by the proposed and finite difference methods (E_2/E_1 = 0.2)

a/W	$K_I/\sigma_b\sqrt{\pi a}$		Sensitivity of J-integral ($\partial\tilde{J}/\partial a$)		
	Present Results	Erodogan & Wu [27]	Proposed Method	Finite Difference	Difference[(a)] (percent)
0.1	1.6675	1.6743	9.4842	9.4843	0.0017
0.2	1.5893	1.5952	12.4843	12.4843	0.0003
0.3	1.6037	1.6122	20.2284	20.2288	0.0020
0.4	1.7070	1.7210	37.4875	37.4879	0.0011
0.5	1.9264	1.9534	79.4636	79.4662	0.0033
0.6	2.3419	2.4037	200.3289	200.3363	0.0037

(a) Difference = ($\partial\tilde{J}/\partial a$ by finite difference method - $\partial\tilde{J}/\partial a$ by proposed method)×100/ $\partial\tilde{J}/\partial a$ by finite difference method

Table 7. Sensitivity of J for an edge-cracked plate under fixed grip loading by the proposed and finite difference methods ($E_2/E_1 = 5.0$)

a/W	$K_I/\sigma_0\sqrt{\pi a}$		Sensitivity of J-integral ($\partial\tilde{J}/\partial a$)		
	Present Results	Erodogan & Wu [27]	Proposed Method	Finite Difference	Difference[(a)] (percent)
0.1	1.2407	1.2372	4.9757	4.9757	0.0001
0.2	1.4965	1.4946	8.7112	8.7112	0.0005
0.3	1.9133	1.9118	16.7389	16.7391	0.0011
0.4	2.5748	2.5730	35.3423	35.3429	0.0016
0.5	3.6601	3.6573	83.2082	83.2101	0.0023
0.6	5.5760	5.5704	229.0280	229.0357	0.0034

(a) Difference = ($\partial\tilde{J}/\partial a$ by finite difference method - $\partial\tilde{J}/\partial a$ by proposed method)×100/ $\partial\tilde{J}/\partial a$ by finite difference method

Table 8. Sensitivity of J for an edge-cracked plate under Membrane Loading by the proposed and finite difference methods ($E_2/E_1 = 5.0$)

a/W	$K_I/\sigma_t\sqrt{\pi a}$		Sensitivity of J-integral ($\partial\tilde{J}/\partial a$)		
	Present Results	Erodogan & Wu [27]	Proposed Method	Finite Difference	Difference[(a)] (percent)
0.1	0.9937	0.9908	2.4673	2.4673	0.0002
0.2	1.1334	1.1318	3.5414	3.5414	−0.0004
0.3	1.3709	1.3697	6.0340	6.0341	0.0013
0.4	1.7498	1.7483	11.6301	11.6303	0.0016
0.5	2.3678	2.3656	25.4610	25.4615	0.0021
0.6	3.4494	3.4454	66.0007	66.0028	0.0032

(a) Difference = ($\partial\tilde{J}/\partial a$ by finite difference method - $\partial\tilde{J}/\partial a$ by proposed method)×100/ $\partial\tilde{J}/\partial a$ by finite difference method

Table 9. Sensitivity of J for an edge-cracked plate under Bending by the proposed and finite difference methods ($E_2/E_1 = 5.0$)

a/W	$K_I/\sigma_b\sqrt{\pi a}$		Sensitivity of J-integral ($\partial\tilde{J}/\partial a$)		
	Present Results	Erodogan & Wu [27]	Proposed Method	Finite Difference	Difference[(a)] (percent)
0.1	0.6405	0.6385	0.9514	0.9514	0.0002
0.2	0.6882	0.6871	1.0564	1.0564	−0.0019
0.3	0.7787	0.7778	1.4992	1.4992	0.0017
0.4	0.9246	0.9236	2.4829	2.4830	0.0016
0.5	1.1595	1.1518	4.7736	4.7737	0.0018
0.6	1.5620	1.5597	11.0239	11.0242	0.0029

(a) Difference = ($\partial\tilde{J}/\partial a$ by finite difference method - $\partial\tilde{J}/\partial a$ by proposed method)×100/ $\partial\tilde{J}/\partial a$ by finite difference method

Table 10. Sensitivity of J for an edge-cracked plate under fixed grip loading by the proposed and finite difference methods (E_2/E_1 = 10.0)

a/W	$K_I/\sigma_0\sqrt{\pi a}$		Sensitivity of J-integral ($\partial\tilde{J}/\partial a$)		
	Present Results	Erodogan & Wu [27]	Proposed Method	Finite Difference	Difference[(a)] (percent)
0.1	1.2711	1.2664	4.7261	4.7261	0.0001
0.2	1.5762	1.5740	8.2725	8.2725	0.0006
0.3	2.0739	2.0723	15.7277	15.7279	0.0011
0.4	2.8752	2.8736	33.0064	33.0069	0.0015
0.5	4.2159	4.2140	77.4798	77.4817	0.0024
0.6	6.6340	6.6319	213.0891	213.0965	0.0035

(a) Difference = ($\partial\tilde{J}/\partial a$ by finite difference method - $\partial\tilde{J}/\partial a$ by proposed method)×100/ $\partial\tilde{J}/\partial a$ by finite difference method

Table 11. Sensitivity of J for an edge-cracked plate under Membrane Loading by the proposed and finite difference methods (E_2/E_1 = 10.0)

a/W	$K_I/\sigma_t\sqrt{\pi a}$		Sensitivity of J-integral ($\partial\tilde{J}/\partial a$)		
	Present Results	Erodogan & Wu [27]	Proposed Method	Finite Difference	Difference[(a)] (percent)
0.1	0.8664	0.8631	1.6960	1.6960	−0.0015
0.2	1.0035	1.0019	2.2681	2.2681	0.0001
0.3	1.2302	1.2291	3.6603	3.6604	0.0015
0.4	1.5896	1.5884	6.7545	6.7546	0.0007
0.5	2.1775	2.1762	14.2718	14.2722	0.0027
0.6	3.2138	3.2124	35.9205	35.9217	0.0035

(a) Difference = ($\partial\tilde{J}/\partial a$ by finite difference method - $\partial\tilde{J}/\partial a$ by proposed method)×100/ $\partial\tilde{J}/\partial a$ by finite difference method

Table 12. Sensitivity of J for an edge-cracked plate under Bending by the proposed and finite difference methods (E_2/E_1 = 10.0)

a/W	$K_I/\sigma_b\sqrt{\pi a}$		Sensitivity of J-integral ($\partial\tilde{J}/\partial a$)		
	Present Results	Erodogan & Wu [27]	Proposed Method	Finite Difference	Difference[(a)] (percent)
0.1	0.5102	0.5082	0.5625	0.5625	−0.0026
0.2	0.5658	0.5648	0.6124	0.6124	−0.0003
0.3	0.6595	0.6588	0.8508	0.8508	0.0021
0.4	0.8050	0.8043	1.3787	1.3787	−0.0004
0.5	1.0358	1.0350	2.5977	2.5978	0.0031
0.6	1.4294	1.4286	5.8906	5.8908	0.0037

(a) Difference = ($\partial\tilde{J}/\partial a$ by finite difference method - $\partial\tilde{J}/\partial a$ by proposed method)×100/ $\partial\tilde{J}/\partial a$ by finite difference method

Table 13. Sensitivity of J for three-point bend specimen under mode-I loading conditions by the proposed and finite difference methods (Linear Variation and $a/2H$ = 0.45)

E_2/E_1	Present Results ($K_I\sqrt{H}/P$)	Sensitivity of J-integral ($\partial\tilde{J}/\partial a$)		
		Proposed Method	Finite Difference	Difference[(a)] (percent)
0.05	32.4069	467.7523	467.7689	0.0035
0.1	23.3146	198.2592	198.2623	0.0016
0.2	17.3461	80.2695	80.2709	0.0017
0.5	11.6568	22.4226	22.4227	0.0004
1	8.1420	7.7004	7.7004	0.0009
2	5.2322	4.3617	4.3617	0.0004
5	2.5268	5.1632	5.1631	–0.0011
10	1.3240	4.9711	4.9709	–0.0027
20	0.6735	3.5769	3.5767	–0.0054

(a) Difference = ($\partial\tilde{J}/\partial a$ by finite difference method - $\partial\tilde{J}/\partial a$ by proposed method)×100/ $\partial\tilde{J}/\partial a$ by finite difference method

Table 14. Sensitivity of J for three-point bend specimen under mode-I loading conditions by the proposed and finite difference methods (Exponential Variation and $a/2H$ = 0.45)

E_2/E_1	Present Results ($K_I\sqrt{H}/P$)	Sensitivity of J-integral ($\partial\tilde{J}/\partial a$)		
		Proposed Method	Finite Difference	Difference[(a)] (percent)
0.05	31.6868	144.2164	144.2212	0.0033
0.1	23.0812	77.7745	77.7763	0.0023
0.2	16.9573	40.1238	40.1240	0.0005
0.5	11.2333	15.8866	15.8869	0.0017
1	8.1420	7.7004	7.7004	0.0009
2	5.8630	3.6311	3.6311	0.0002
5	3.8133	1.1306	1.1306	–0.0008
10	2.7871	0.2980	0.2980	–0.0003
20	2.0593	–0.0417	–0.0417	0.0179

(a) Difference = ($\partial\tilde{J}/\partial a$ by finite difference method - $\partial\tilde{J}/\partial a$ by proposed method)×100/ $\partial\tilde{J}/\partial a$ by finite difference method

Table 15. Sensitivity of J for three-point bend specimen under mode-I loading conditions by the proposed and finite difference methods (Linear Variation and $a/2H$ = 0.50)

E_2/E_1	Present Results ($K_I\sqrt{H}/P$)	Sensitivity of J-integral ($\partial\tilde{J}/\partial a$)		
		Proposed Method	Finite Difference	Difference[(a)] (percent)
0.05	31.1898	1411.8779	1411.9621	0.0060
0.1	23.9149	555.1371	555.1524	0.0027
0.2	18.3227	204.2048	204.2077	0.0014
0.5	12.5795	40.8503	40.8507	0.0009
1	9.4678	11.2628	11.2628	0.0006
2	7.3161	5.2616	5.2616	–0.0001
5	5.4968	4.4094	4.4093	–0.0008
10	4.6112	3.8706	3.8705	–0.0026
20	4.0328	2.6829	2.6828	–0.0055

(a) Difference = ($\partial\tilde{J}/\partial a$ by finite difference method - $\partial\tilde{J}/\partial a$ by proposed method)×100/ $\partial\tilde{J}/\partial a$ by finite difference method

Table 16. Sensitivity of J for three-point bend specimen under mode-I loading conditions by the proposed and finite difference methods (Exponential Variation and $a/2H$ = 0.50)

E_2/E_1	Present Results ($K_I\sqrt{H}/P$)	Sensitivity of J-integral ($\partial\tilde{J}/\partial a$)		
		Proposed Method	Finite Difference	Difference[(a)] (percent)
0.05	20.2541	369.3652	369.3809	0.0042
0.1	17.2310	169.3468	169.3481	0.0008
0.2	14.5336	75.7558	75.7565	0.0009
0.5	11.4374	25.7051	25.7053	0.0008
1	9.4678	11.2628	11.2628	0.0006
2	7.8416	4.8227	4.8227	0.0007
5	6.2115	1.3712	1.3712	0.0013
10	5.3110	0.3780	0.3780	0.0017
20	4.6126	–0.0024	–0.0024	–0.0987

(a) Difference = ($\partial\tilde{J}/\partial a$ by finite difference method - $\partial\tilde{J}/\partial a$ by proposed method)×100/ $\partial\tilde{J}/\partial a$ by finite difference method

Table 17. Sensitivity of J for three-point bend specimen under mode-I loading conditions by the proposed and finite difference methods (Linear Variation and $a/2H$ = 0.55)

E_2/E_1	Present Results ($K_I\sqrt{H}/P$)	Sensitivity of J-integral ($\partial\tilde{J}/\partial a$)		
		Proposed Method	Finite Difference	Difference[(a)] (percent)
0.05	15.2836	6695.6345	6696.1466	0.0076
0.1	13.8754	1366.1375	1366.1836	0.0034
0.2	12.8026	295.3975	295.4016	0.0014
0.5	11.7664	49.5108	49.5113	0.0008
1	11.1611	17.2040	17.2042	0.0011
2	10.6297	8.3144	8.3144	0.0009
5	9.9848	4.9074	4.9074	−0.0004
10	9.5873	3.3001	3.3000	−0.0014
20	9.3017	1.7388	1.7388	−0.0020

(a) Difference = ($\partial\tilde{J}/\partial a$ by finite difference method - $\partial\tilde{J}/\partial a$ by proposed method)×100/ $\partial\tilde{J}/\partial a$ by finite difference method

Table 18. Sensitivity of J for three-point bend specimen under mode-I loading conditions by the proposed and finite difference methods (Exponential Variation and $a/2H$ = 0.55)

E_2/E_1	Present Results ($K_I\sqrt{H}/P$)	Sensitivity of J-integral ($\partial\tilde{J}/\partial a$)		
		Proposed Method	Finite Difference	Difference[(a)] (percent)
0.05	11.5808	437.5786	437.5634	−0.0035
0.1	11.4819	208.7195	208.7213	0.0008
0.2	11.3826	99.0466	99.0452	−0.0014
0.5	11.2539	36.6429	36.6432	0.0008
1	11.1611	17.2040	17.2042	0.0011
2	11.0802	8.0782	8.0782	0.0004
5	11.0212	2.9177	2.9177	0.0004
10	11.0423	1.2681	1.2681	0.0002
20	11.1399	0.4919	0.4919	0.0001

(a) Difference = ($\partial\tilde{J}/\partial a$ by finite difference method - $\partial\tilde{J}/\partial a$ by proposed method)×100/ $\partial\tilde{J}/\partial a$ by finite difference method

Table 19. Sensitivity of J for for a composite strip configuration under mode-I by the proposed and finite difference methods ($a/W = 0.4$)

ηa	$K_I/\bar{\varepsilon}\bar{E}(-0.5)\sqrt{\pi a}$		Sensitivity of J-integral ($\partial\tilde{J}/\partial a$)		
	Eischen [28]	Present Results	Proposed Method	Finite Difference	Difference[(a)] (percent)
0.0	2.112	2.1127	79.4463	79.4594	0.0165
2.0	2.295	2.2967	236.9506	236.4278	−0.2211
4.0	2.571	2.5732	336.4751	330.9685	−1.6638
6.0	2.733	2.7355	412.9277	398.1843	−3.7026
20.0	3.228	3.2192	802.6819	675.8779	−18.7614

(a) Difference = ($\partial\tilde{J}/\partial a$ by finite difference method - $\partial\tilde{J}/\partial a$ by proposed method)×100/ $\partial\tilde{J}/\partial a$ by finite difference method

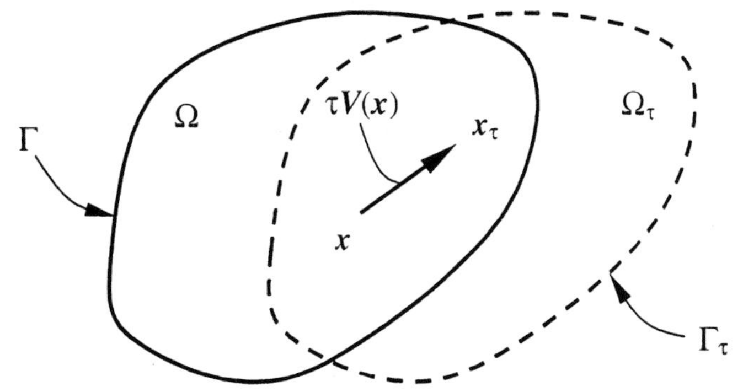

Figure 1. Variation of domain

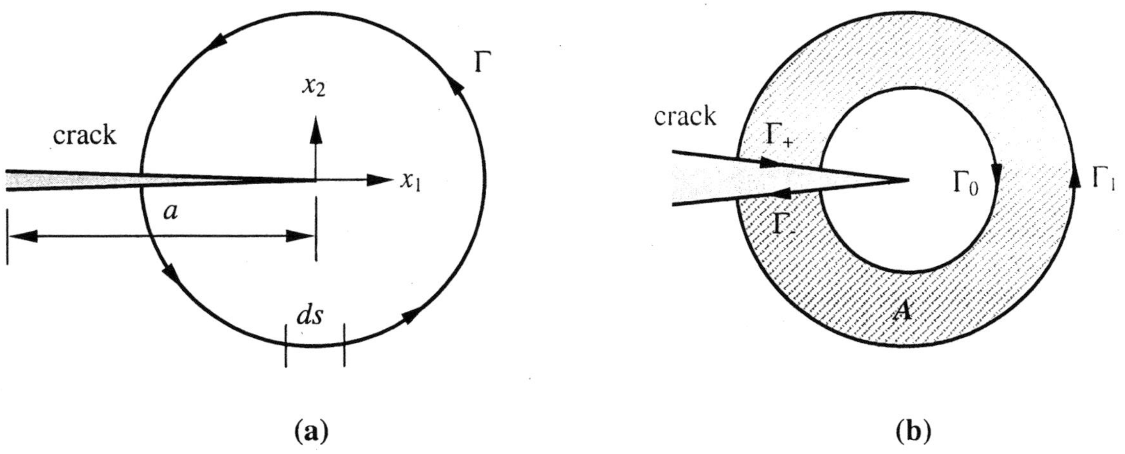

Figure 2. $\tilde{J}$ -integral fracture parameter; (a) Arbitrary contour around a crack; and (b) Inner and outer contours enclosing *A*

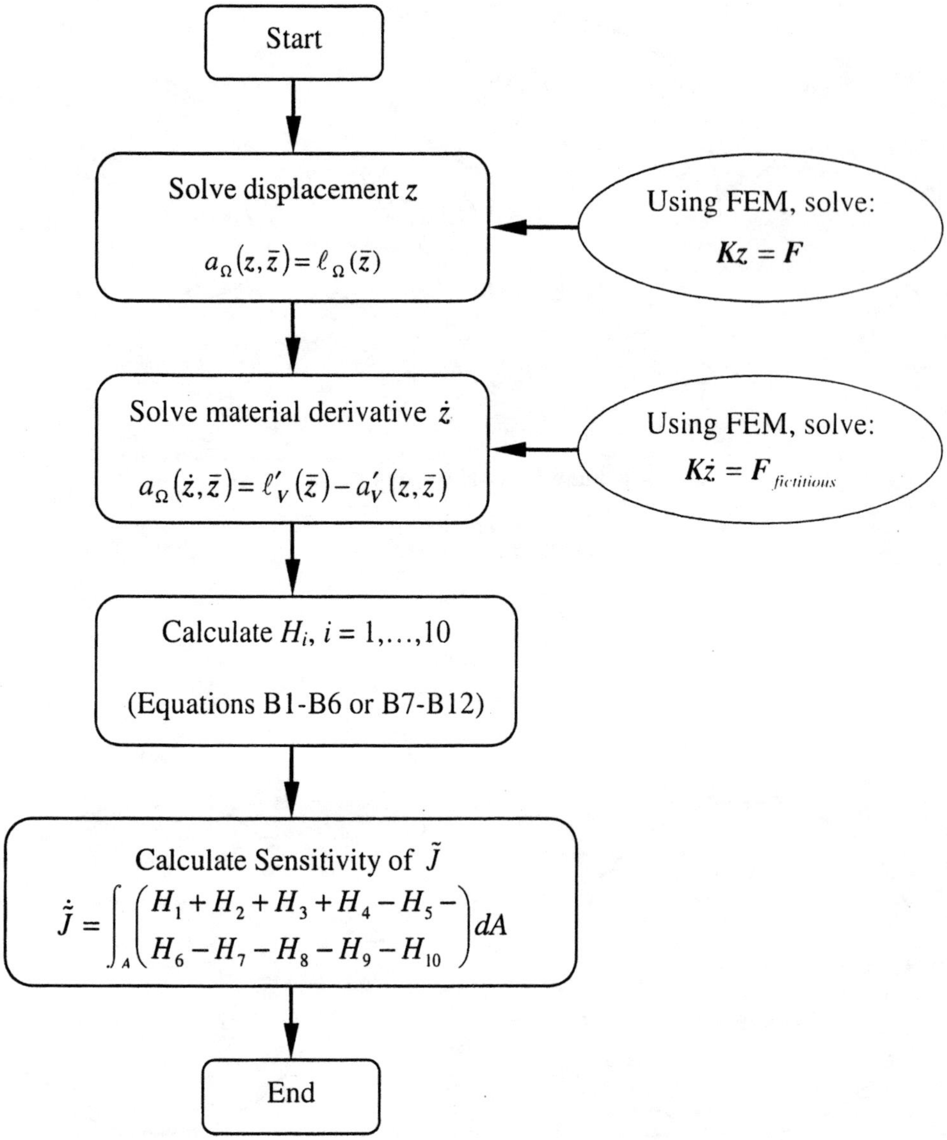

Figure 3. A flowchart for continuum sensitivity analysis of crack size

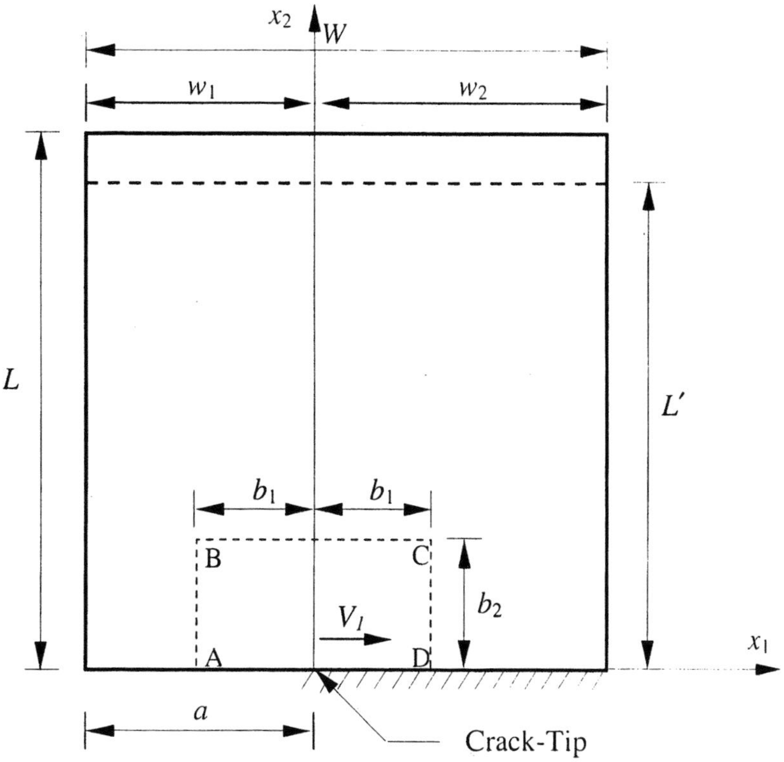

Figure 4. Schematic for velocity field definition

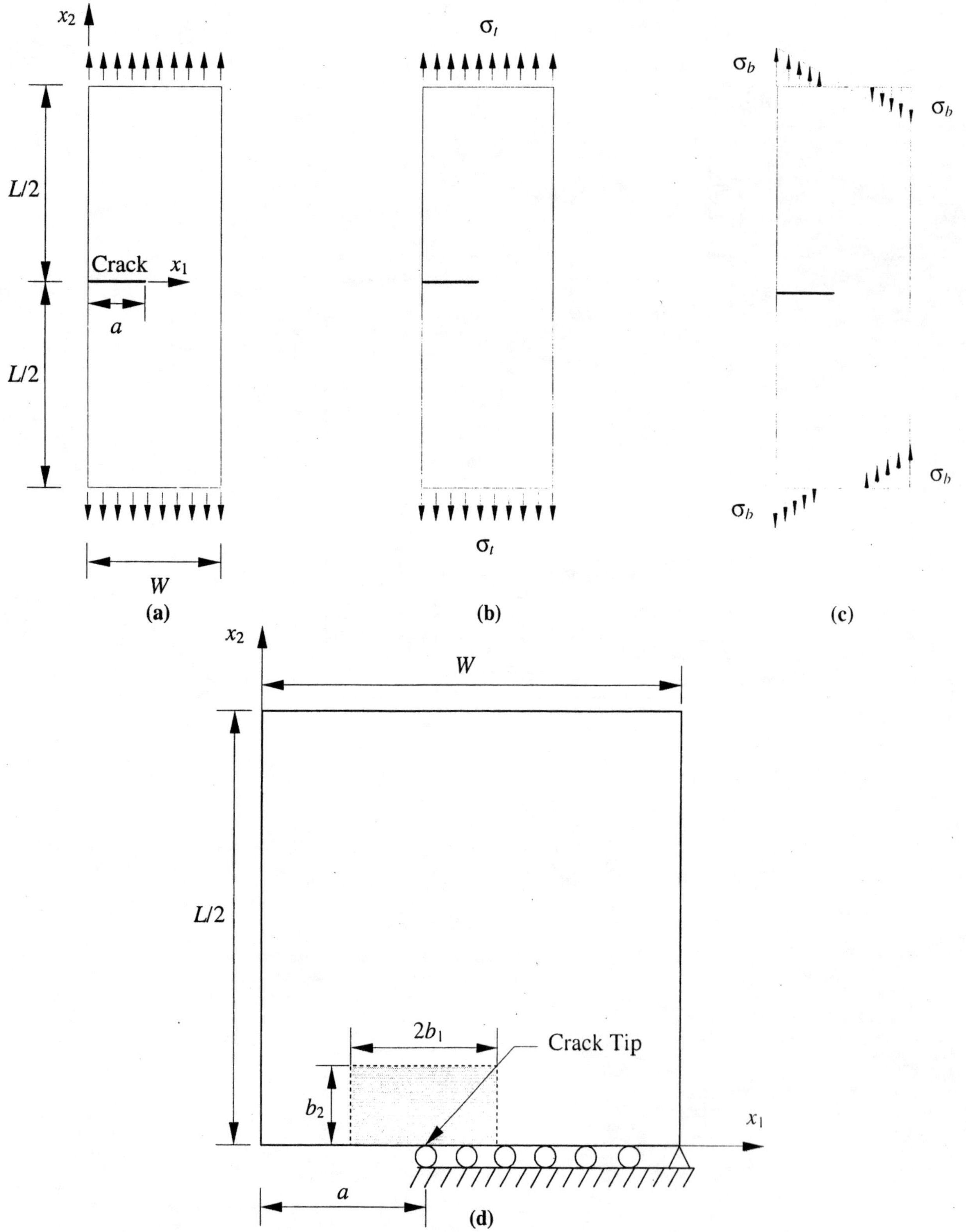

Figure 5. Edge-cracked plate under mode I loading; (a) Geometry and loads for fixed grip loading; (b) Membrane loading; (c) Bending; and (d) Half model

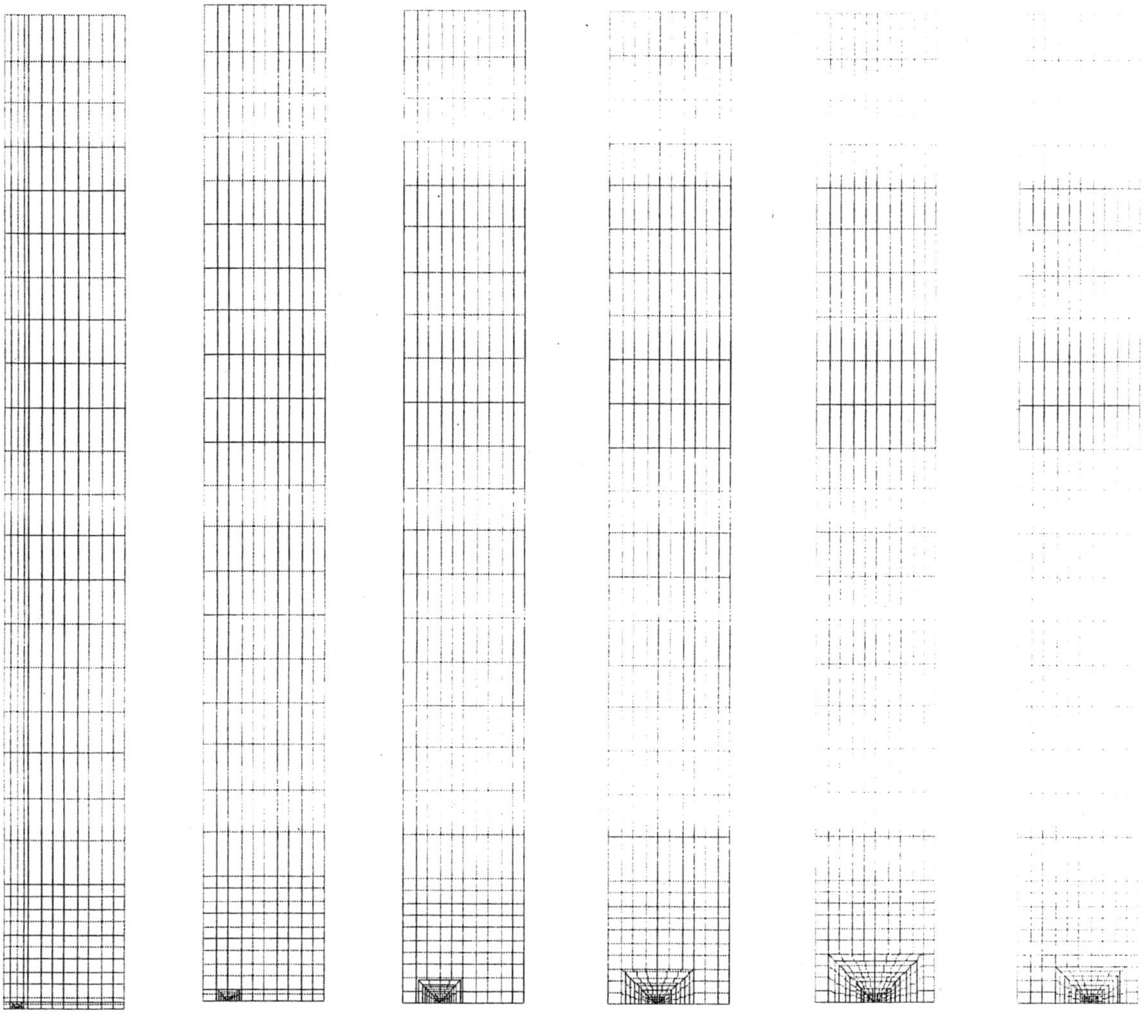

(a) (b) (c) (d) (e) (f)

Figure 6. FEM discretization of edge-cracked plate under mode I loading; (a) *a/W* = 0.1 (1421 nodes, 434 8-noded quadrilateral elements, 6 focused quarter-point 6-noded triangular elements); (b) *a/W* = 0.2 (1193 nodes, 360 8-noded quadrilateral elements, 6 focused quarter-point 6-noded triangular elements); (c) *a/W* = 0.3 (1205 nodes, 364 8-noded quadrilateral elements, 8 focused quarter-point 6-noded triangular elements); (d) *a/W* = 0.4 (1289 nodes, 390 8-noded quadrilateral elements, 12 focused quarter-point 6-noded triangular elements); (e) *a/W* = 0.5 (1361 nodes, 412 8-noded quadrilateral elements, 16 focused quarter-point 6-noded triangular elements); and (f) *a/W* = 0.6 (1289 nodes, 390 8-noded quadrilateral elements, 12 focused quarter-point 6-noded triangular elements)

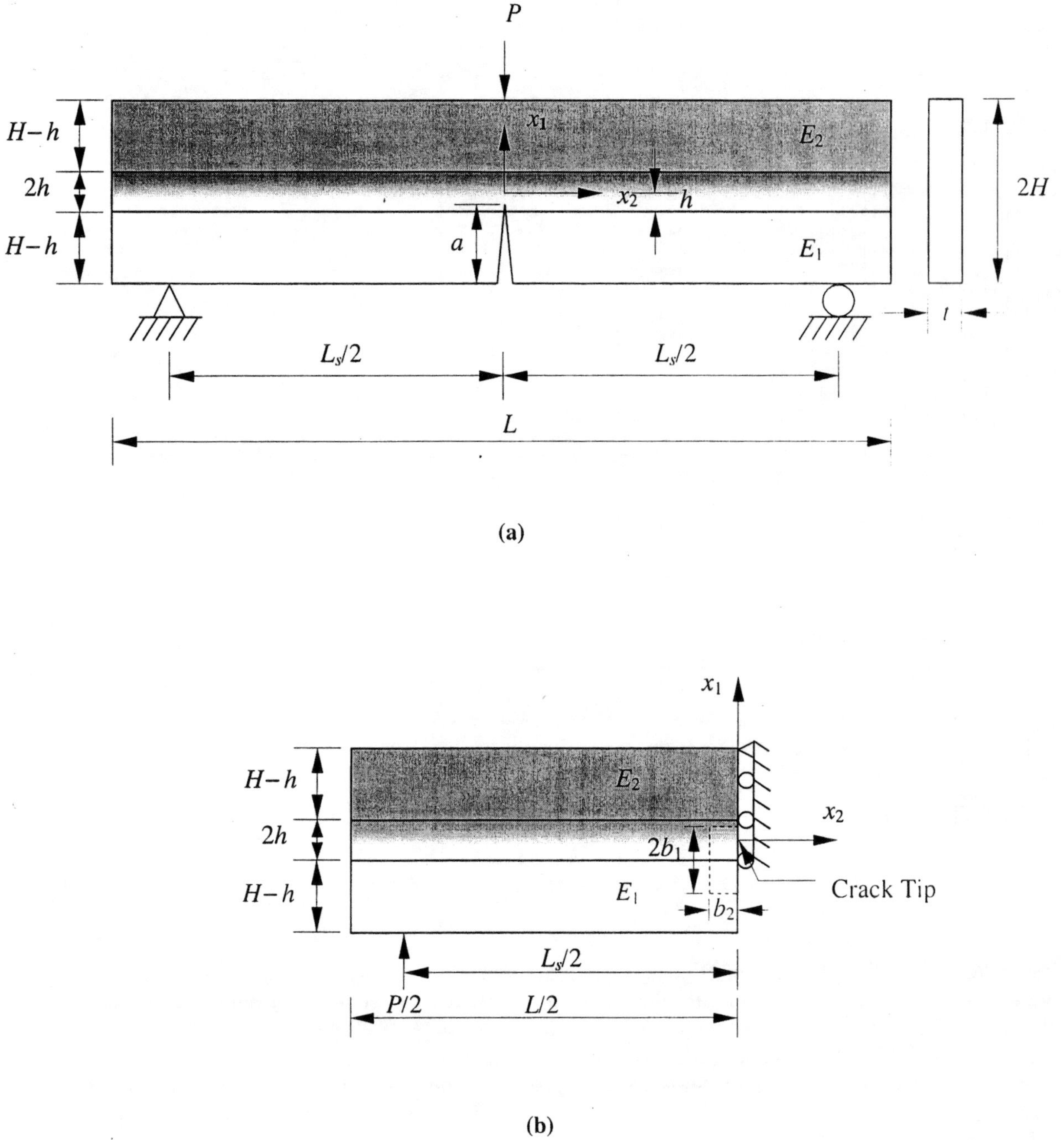

Figure 7. Three-point bend specimen under mode I loading; (a) Geometry and loads; and (b) Half model

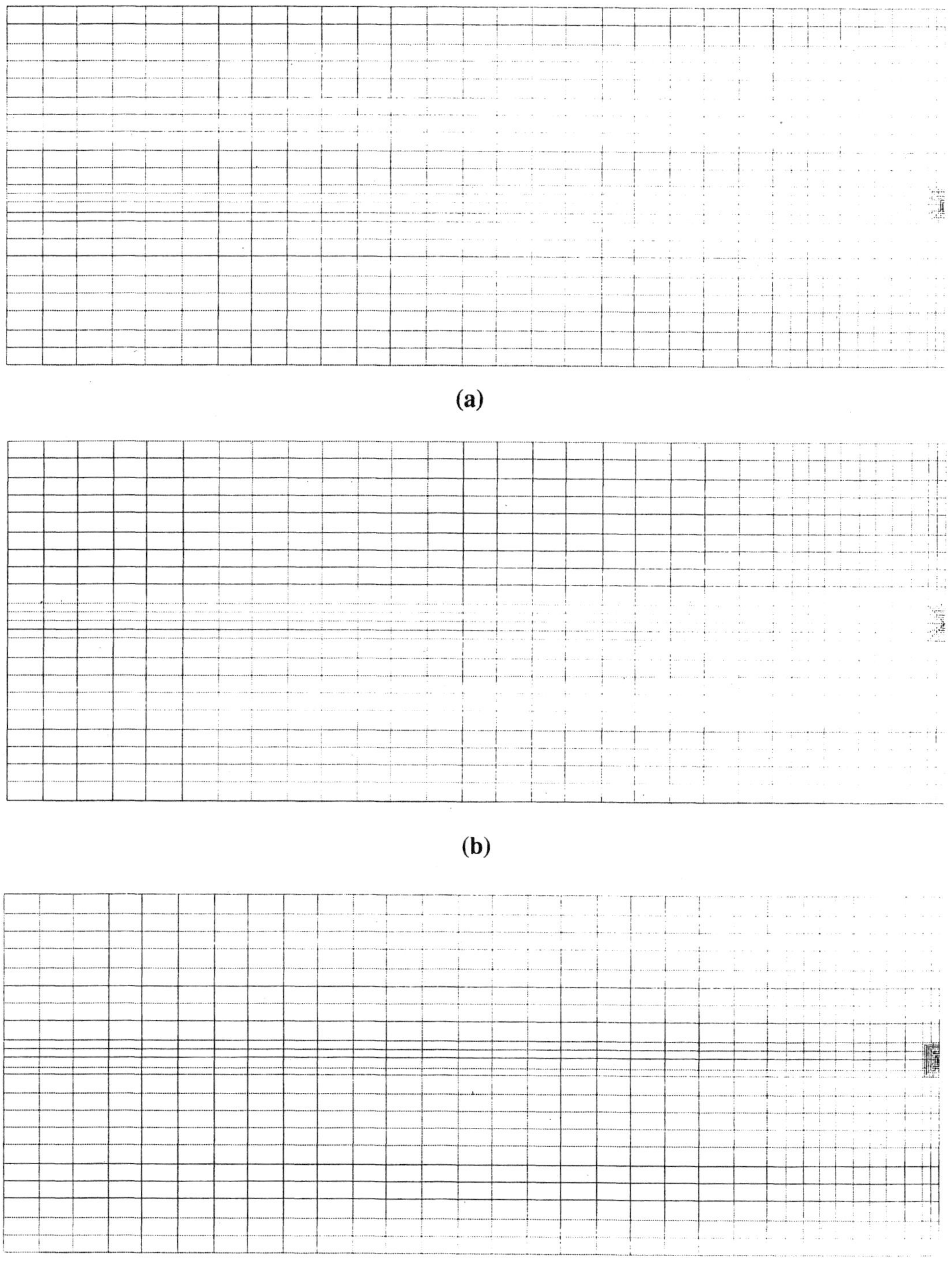

(a)

(b)

(c)

Figure 8. FEM discretization of three-point bend specimen under mode I loading; (a) *a*/2*H* = 0.45 **(2669 nodes, 838 8-noded quadrilateral elements, 8 focused quarter-point 6-noded triangular elements);** (b) *a*/2*H* = **0.50 (2669 nodes, 838 8-noded quadrilateral elements, 8 focused quarter-point 6-noded triangular elements); and** (c) *a*/2*H* = **0.55 (2669 nodes, 838 8-noded quadrilateral elements, 8 focused quarter-point 6-noded triangular elements)**

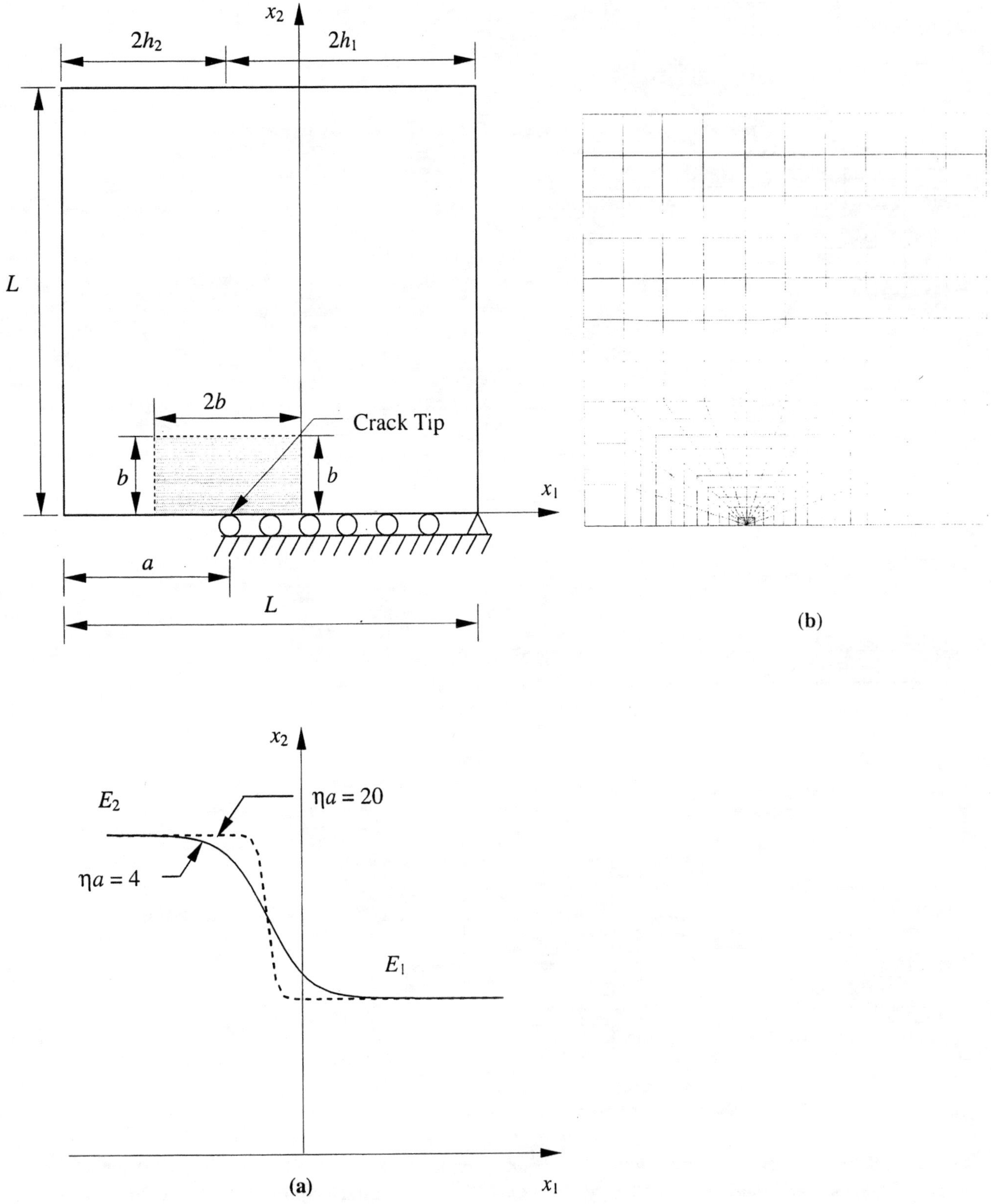

Figure 9. Composite strip under mode-I loading; (a) Geometry and Elastic modulus variation; (b) FEM discretization (649 nodes, 190 8-noded quadrilateral elements, 12 focused quarter-point 6-noded triangular elements)

PVP-Vol. 458, Computer Technology and Applications

PVP2003-1902

AN EFFICIENT METHODOLOGY FOR FATIGUE RELIABILITY ANALYSIS OF MECHANICAL COMPONENTS BASED ON THE STRESS-LIFE PREDICTION APPROACH

Jun Tang
CENTER FOR COMPUTER-AIDED DESIGN
College Of Engineering
UNIVERSITY OF IOWA
IOWA CITY, IA 52242, U.S.A.

Young Ho Park
DEPARTMENT OF MECHANICAL ENGINEERING
College Of Engineering
NEW MEXICO STATE UNIVERSITY
Las Cruces, NM 88003, U.S.A.

ABSTRACT

An efficient methodology for fatigue reliability assessment and its corresponding fatigue life prediction of mechanical components using the First-Order Reliability Method (FORM) is developed in this paper. Using the proposed method, a family of reliability defined S-N curves, called R-S-N curves, can be constructed. In exploring the ability to predict spectral fatigue life and assessing the corresponding reliability under a specified dynamics environment, the theoretical background and the algorithm of a simple approach for reliability analysis will first be introduced based on fatigue failure modes of mechanical components. It will then be explained how this integrated method will carry out the spectral fatigue damage and failure reliability analysis. By using this proposed methodology, mechanical component fatigue reliability can be predicted according to different mission requirements.

1. INTRODUCTION

In fatigue reliability analysis, the mathematical model belongs to the random process model, which is defined by a safety margin

$$\min G(X, Y(t)) \qquad (1.1)$$

where G is a failure function, X is a set of random variables, Y(t) is a set of random processes, and t is time. This is different from the random variable model. Generally, the failure probability is a first outcrossing probability (Bjerager, P., 1990). When compared to random variable reliability problems, the difficulty percieved is that the analytical computation methods for random process models are rather restricted. Even for some of the simplest cases there are few exact results for the first crossing probability existing. In a random variable reliability problem, which is defined by the safety margin G(X), the failure probability is an integral of the multivariate density function $f_X(x)$. The reliability analysis is to evaluate the failure probability. This reliability analysis can be approximately accomplished using the First-Order Reliability Method (FORM) with a rataionally invariant measure as the reliability (Bjerager, P., 1990). The theory and applications of FORM have been developed and improved during the last two decades. Meanwhile, in the fatigue analysis field, the stress-based fatigue life analysis has been developed and used for years. The stress-life approach is the most important method to be used, especially for High Cycle Fatigue (HCF) life. The fatigue failure is defined when the crack reaches certain length. The stress-life curves (S-N curves) are used to characterize material's response to cyclic loading. In this method, a single stress-life relation, S-N relation, which is applicable in HCF regimes, is constructed. Generally speaking, the Palmgren-Miner cumulative damage criterion is a definition of fatigue damage. With this definition through cycle counting methods and a mathematical characterization of the stress-life (S-N) fatigue curve, the fatigue life can be predicted. In this fatigue life prediction process, the cycle counting method plays a very important role. It is generally regarded as the method leading to the best estimators of fatigue life. It is designed to catch both slow and rapid variation in the load process by forming cycle amplitudes and by pairing high maxim with low minima even if they are separated by intermediate extremes. Due to the great importance of the cycle-count, a stress process can be decomposed into several cycles in which the stress time-histories are sinusoid.

This is enlightenment for the fatigue reliability analysis. To be precise, the cycle counting method may be used to deal with the random process model by using the random variable model. This means that through cycle counting the random process problem then becomes several random variable problems. Therefore, the purpose of this study is to develop an algorithm which combines the accumulate damage analysis with FORM to evaluate fatigue reliability. In this approach a reliability factor is evaluated using FORM for a specified reliability, then a reliability related S-N curve will be generated by using the reliability factor. A reliability related S-N curve family can then be constructed for evaluating the relationship between mission life and desired reliability. In fact, one of the popular engineering fatigue design methods, which can handle reliability assessment, is to use a reliability factor to modify the fatigue limit of a given material (Shigley and Mischke, 1989). This reliability factor is determined from the fatigue (endurance) limit and applied stress interference, where the fatigue limit is considered to be distributed and the applied stress is deterministic. This method makes the reliability assessment of infinite fatigue life design possible. Unfortunately, for the finite life design that corresponds to the applied stress level higher than the fatigue limit, this method does not provide enough information. To make the reliability assessment of the finite fatigue life design possible, different approaches have been taken, notably by Lambert (1976), Avakov (1991), Jia et al. (1993), Murty et al. (1995), and Baldwin & Thacker (1995). Each approach considers a different aspect of uncertainty, but all these approaches must adhere to various constraints for practical applications. Tang and Zhao et al. (1995, 1998, 1999, and 2000) developed an analytical method to evaluate the reliability factor and established a reliability defined S-N relationship (the R-S-N curve), which is independent of stress history. This method is limited to handling two random variables. Tang and Park (2001, 2002) recently developed a new algorithm, which combined the accumulate damage approach with FORM for evaluating fatigue reliability recently. This method, however, is based on strain-life prediction approaches. To aid in giving integrity to this methodology, this paper will focus on development of a new algorithm based on the stress-life approach.

This paper presents a probabilistic method for deciding the relationship between the allowable stress or endurance limit and fatigue reliability. The reliability of a component that considers the stress (loading), strength, endurance limit or other material properties as random variables can be assessed. To validate the proposed method, a numerical example is presented.

2. STRESS-BASED FATIGUE LIFE PREDICTION AND RELIABILITY RELATED STRESS-LIFE CURVES

In a typical stress-life curve of metal materials, the strength corresponding to the point of the S-N curve represented by the horizontal asymptote is called the endurance limit S_e. Suppose fatigue life $N = N_0$ is at the transition point of the S-N curve, for engineering purposes N_0 is usually considered to be 10^6, beyond which the curve shows the endurance limit.

The approximate power relationship can be used to estimate the S-N curve for metal as follows (Bannantine et al., 1990):

$$S = 10^C n^b \qquad (2.1)$$

where exponents C and b are determined by two designated points (an approximate expression) of the S-N curve. For most cases, C and b are given as:

(2.2)

where S_{1000} = stress at which 1000 fatigue cycle life is achieved, and can be estimated as 0.9 times of the ultimate stress, i.e., $S_{1000} = 0.9\,S_u$. The life is then a function of two stress parameters, ultimate stress S_u and endurance limit S_e as

$$N_p = f(S, S_u, S_e) \qquad (2.3)$$

where N_p is the predicted cycle life of the component.

For most metal, there is a fatigue ratio, S_e/S_u, to identify S_e using S_u and vise versa. The relationship between S_e and S_u is as shown in Eq. (2.4).

(2.4)

In this case, the fatigue life is the function of either endurance limit S_e or ultimate stress S_u, and Eq. (2.3) may be written as

$$N_p = f(S, S_u) \quad \text{or} \quad N_p = f(S, S_e) \qquad (2.5)$$

It can later be seen that this relationship is very helpful for keeping the evaluation of the reliability factor simple.

In the fatigue life prediction, however, the environmental loading on a component is varied in time in a random fashion; the stress variation in a component will also be varied in time and be a stochastic process. This means that a given varied amplitude history of operating stress is expressed as a function of time within certain time interval (called a "block"). In order to predict life using the S-N relationship in Eq. (2.1), it is often advantageous to consider a quasi-constant loading level life. Thus, the stress time history is always broken down into a series of defined constant amplitude cycles through the use of a cycle-counting approach, which in most cases, is the rainflow counting method. After a rainflow counting, an irregular stress history has been transformed through hysteresis curves (stress-strain loops) into an equivalent sequence of closed loop constant amplitude stress histories. Since the fatigue effect is likely to gradually weaken the capacity of the component, it does not consider those affected in a separated problem. To obtain the life of whole loading history, a widely accepted damage summation model, such as Palmgren-Miner rule, will then be used to combine the individual life found for each defined cycle into the predicted life of the whole history.

Using the Palmgren-Miner definition and a mathematical characterization of the stress-life (S-N) fatigue curve in Eq. (2.1), the fatigue life can be predicted. In the process, the cycle counting method, such as rainflow counting is used to decompose an irregular stress process into several constant amplitude cycles, and each can be predicted by using Eq. (2.1).

An additional problem is that a S-N curve is determined by fatigue tests or by empirical relations among certain basic material property parameters (Bannantine et al., 1990). However, the experimental data of the S-N diagram are actually scattered (American Society for Testing and Material, 1981), and the S-N curve is an "average" drawn through a scatter band of points plotting the failure of individual fatigue specimens. From a probabilistic point of view, the S-N curve should not be a single curve, but instead, a family of curves which correspond to a different probability of occurrence. To be exact, a reliability-related S-N curve family needs to be generated. Although a family of S-N curves with different probabilities of occurrence can be defined experimentally, such experiments have not been conducted for most engineering materials. As a result, experimentally defined S-N curves with different reliabilities are not available. Therefore, it is necessary to develop a statistical method to obtain more reliable designs, where the reliability can be quantified and easily measured.

In most cases, fatigue life is defined as the number of stress/strain cycles N, and is calculated by a given relationship (S-N/ε-N), such as in Eq. (2.1) for stress-based life. In actuality, the fatigue failure probability is related to the stress/strain level. Thus, the fatigue failure probability P_f should be a function of these random process variables. The failure probability corresponding to random stress S can be estimated as

$$P_f = P(G<0) = P(N(S) < N_0) \qquad (2.6)$$

where N_0 is the desired life (i.e. the number of cycles corresponding to a limiting value) and N(S) is the life corresponding to damage accumulation.

Equation (2.6) shows how failure reliability relates to an S-N curve. This implies that a reliability parameter with a specified reliability and associated stress level S exists. Therefore, Eq. (2.6) indicates that not only does the S-N curve exist, but also that the reliability-defined S-N curve, referred to as R-S-N curve and having an associated reliability or failure probability parameter, also exists.

The proposed method provides an algorithm to evaluate a reliability factor that is based on a specified reliability. Once the reliability factor is evaluated, the ultimate strength S_u and the endurance limit S_e can be corrected with reliability by following two equations:

$$\begin{aligned} S^*_u &= K_r S_u \\ S^*_e &= K_r S_e \end{aligned} \qquad (2.7)$$

where K_r is a reliability factor,

Since in stress-life prediction the life is a function of either the ultimate strength S_u or the endurance limit S_e, the reliability factor can be unified. The problem is then how to evaluate the reliability factor for either the the ultimate strength S_u or the endurance limit S_e level. This provides a simple way to use FORM for evaluating fatigue reliability.

It has been previously proven (Zhao and Tang, 1995 and Tang, 1999) that the reliability factor is independent of the life, meaning the reliability factor for a specified reliability is a constant. In fact, Eqs. (2.4) and (2.5) also indicate that if either the ultimate stress S_u or the endurance limit S_e for a specified reliability is given, the reliability related S-N curve is generated. Therefore, to evaluate the reliability factor for a specified reliability, the endurance limit S_e can be selected as one of the random variables to perform the reliability analysis using FORM. Once the reliability factors for different specified reliability have been evaluated, the R-S-N curve family has been generated and the reliability versus life curve is constructed. The fatigue life under the specified reliability or reliability based on mission life can then be predicted using these curves.

In order to use FORM, an empirical stress-life equation is used as the performance function, in which any design variable or combination of variables, such as material property parameters (including fatigue strength coefficient, fatigue ductility coefficient, etc.), applied stress, and geometric parameters, can be expressed as random variables. The proposed method can be formulated as the inverse of conventional reliability analysis, which determines reliability indices. The reliability factor is then calculated using the proposed method for a specified reliability index. Once the reliability factor for a specified reliability has been evaluated, a reliability-defined S-N curve (R-S-N curve) can be generated from an nominal S-N relationship by modifying the nominal S-N curve using a "unique" reliability factors for an assigned reliability. A family of R-S-N curves, which includes the conventional S-N curve, can then be obtained.

3. FIRST-ORDER RELIABILITY METHOD

Reliability analysis evaluates the failure (or safety) probability of a component or system under specific operating conditions during the mission period. For the random variable vector $\boldsymbol{X} = [X_i]^T$ (i = 1, 2,..., n), the performance criteria are described by the performance functions G($\boldsymbol{X}$) such that the system fails if G($\boldsymbol{X}$) < 0. The statistic description of G($\boldsymbol{X}$) is characterized by its cumulative distribution function (CDF) $F_G(g)$ as

$$F_G(g) = P(G(x) < g) = \int_{G(x)<g} \cdots \int f_X(x) dx_1 \ldots dx_n \qquad x^L \le x \le x^U \qquad (3.1)$$

where $f_X(x)$ is the joint probability density function of all random variables and g is named the probabilistic performance measure.

A generalized probability index β_G, which is a non-increasing function of g, is introduced (Madsen et al., 1986) as

$$F_G(g) = \Phi(-\beta_G) \qquad (3.2)$$

which can be expressed in two ways using the following inverse transformations (Rubinstein, 1981; Tu et al., 1999), respectively, as

$$\begin{aligned} \beta_G(g) &= -\Phi^{-1}(F_G(g)) \\ g(\beta_G) &= F_G^{-1}(\Phi(-\beta_G)) \end{aligned} \qquad (3.3)$$

At a given design $\mathbf{d}^k = [d_i^k]^T \equiv [\mu_i^k]^T$, the evaluation of reliability index $\beta_s(\mathbf{d}^k)$ is performed using the well-developed conventional reliability analysis (Madsen et al., 1986) as

$$\beta_s(\mathbf{d}^k) = -\Phi^{-1}\left(\int_{G(x)<0} \cdots \int f_x(x) dx_1 \cdots dx_n\right) \qquad x^L \le x \le x^U \qquad (3.4)$$

Also, the evaluation of $g^*(\mathbf{d}^k)$ can be performed in an inverse reliability analysis (Tu et al, 1999a) as

$$g^*(\mathbf{d}^k) = F_G^{-1}\left(\int_{G(x)<g^*} \cdots \int f_X(x) dx_1 \cdots dx_n\right. \qquad x^L \le x \le x^U \qquad (3.5)$$

In this research, Eq. (3.5) is used to determine the fatigue life performance function for a given reliability target. The evaluation of Eqs. (3.5) requires reliability analysis where the multiple integration is involved as shown in Eq. (3.5), and the exact probability integration is in general extremely complicated to compute. The FORM (first-order reliability method) often provides adequate accuracy and is widely accepted for reliability analysis (Wu and Wirsching, 1984; Wu et al, 1990; Wirsching et al., 1991).

In FORM, the transformation (Hohenbichler and Rackwitz, 1981; Madsen et al., 1986) from the nonnormal random system parameter **X** (*x*-space) to the independent and standard normal variable **U** (*u*-space) is required. If all system parameters are mutually independent, the transformations can be simplified as

$$\begin{aligned} u_i &= \Phi^{-1}(F_{x_i}(x_i)), \quad i = 1, 2, \cdots, n \\ x_i &= F_{x_i}^{-1}(\Phi(u_i)), \quad i = 1, 2, \cdots, n \end{aligned} \tag{3.6}$$

The performance function G(*x*) can then be represented as $G_U(\boldsymbol{u})$ in the *u*-space. The point on the hypersurface $G_U(\boldsymbol{u}) = 0$ with the maximum joint probability density is the point with the minimum distance from the origin and is named the most probable point (MPP) $\boldsymbol{u}^*_{g=0}$. The minimum distance, named the first-order reliability index $\beta_{s,FORM}$, is an approximation of the generalized probability index corresponding to g=0 as

$$\beta_{s,FORM} \approx \beta_s = \beta_G(0) \tag{3.7}$$

Inversely, the performance function value at the MPP $\boldsymbol{u}^*_{\beta=\beta_t}$ with the distance β_a from the origin is an approximation of the probabilistic performance measure g^* as

$$g^*_{FORM} = G_U\left(u^*_{\beta=\beta_t}\right) = g^* = g^*(\beta_t) \tag{3.8}$$

Thus, the first-order reliability analysis is to find the MPP on the hypersurface $G_U(\boldsymbol{u}) = g$ in the *u*-space, and the first-order inverse reliability analysis is to find the MPP that renders the minimum distance β_t from the origin.

For fatigue life reliability assessment of structural systems, the random variables may be time dependent. The performance function is also time-related. Since FORM cannot address time related reliability problems, the performance function must first be treated as time independent. However, fatigue life is predicted using the S-N curve. The time dependent stress can be represented as several time-independent strain levels using the cycle counting approach. The performance function is then:

$$G(\mathbf{S}) = N(\mathbf{S}) - N_0 \tag{3.9}$$

Similar to the fatigue damage summation routine, the problem now becomes how to summarize individual reliabilities into a single value that represents the reliability of the system or component for the applied time history. The method proposed herein develops the practical approach and algorithm to solve this problem.

The first-order probabilistic performance measure g^*_{FORM} is obtained from a sphere-constrained nonlinear optimization problem (Tu et al., 1999) in the u-space defined as

$$\begin{aligned} &\text{minimize} && G_U(\mathbf{u}) \\ &\text{subject to} && \|\mathbf{u}\| = \beta_t \end{aligned} \tag{3.10}$$

where the optimum point on a target reliability surface is identified as the MPP $\mathbf{u}^*_{\beta=\beta_t}$ with a prescribed reliability $\beta_t = \left\|\mathbf{u}^*_{\beta=\beta_t}\right\|$. Unlike the conventional reliability analysis for the reliability index calculation, only the direction vector $\mathbf{u}^*_{\beta=\beta_t} / \left\|\mathbf{u}^*_{\beta=\beta_t}\right\|$ needs to be determined by exploring an explicitly simple sphere constraint $\|\mathbf{u}\| = \beta_t$.

The Advanced Mean Value (AMV) method (Wu, 1994) is a efficient search method to obtain the first-order probabilistic performance function. For the first-order AMV, the Mean Value (MV) method is defined first as follows:

$$\mathbf{u}^*_{MV} = \beta_t \mathbf{n}(0) \tag{3.11}$$

where

$$\mathbf{n}(0) = -\frac{\nabla_X G(\mu)}{\|\nabla_X G(\mu)\|} = -\frac{\nabla G_U(0)}{\nabla G_U(0)} \tag{3.12}$$

and n is mean values of random parameters.

To minimize the performance function G(U), the normalized steepest descent direction **n**(0) is defined at the mean value, as in the MV method. The AMN method iteratively updates the unit vector of the steepest descent direction at a probable point

$$\mathbf{u}^{(1)}_{AMV} = \mathbf{u}^*_{MV} \tag{3.13}$$

$$\mathbf{u}^{(k+1)}_{AMV} = \beta_t \mathbf{n}(\mathbf{u}^{(k)}_{AMV}) \tag{3.14}$$

where

$$\mathbf{n}(\mathbf{u}^{(k+1)}_{AMV}) = -\frac{\nabla G_U(\mathbf{u}^{(k)}_{AMV})}{\left\|\nabla G_U(\mathbf{u}^{(k)}_{AMV})\right\|} \tag{3.15}$$

4. ALGORITHM OF PROPOSED METHOD AND RELATIONSHIP BETWEEN RELIABILITY AND FATIGUE LIFE

The fatigue reliability analysis is the same as the fatigue life prediction, without considerable past experience it is very difficult to evaluate the reliability. From the life prediction approach, a good deal of enlightenment was obtained for fatigue reliability analysis. With the similar procedure and a reliability-related S-N (R-S-N) curve, the life of each cycle based on specified reliability can then be evaluated. If a R-S-N curve family exists, the life versus reliability curve is generated. However, for each applied stress cycle counted, the stress amplitudes are constant, and the corresponding fatigue reliability can be assessed by FORM.

The fatigue failure mechanism where the state of component changes irreversibly with time to failure, is treated as rate process. This implies that the time dependence of reliability degradation can be modeled as a rate reaction. The Plamgren-Miner's linear damage rule is a heuristic idea that the use of cumulative fatigue damage can be substituted for the rate process. Therefore, the method of fatigue life prediction process can be used to reduce the fatigue random process problem to one of a time invariant.

By using the inversed FORM, a reliability index can be obtained with a specified reliability. Based on this index, a life corresponding to a given reliability will be evaluated. The life compares with the life based on the nominal S-N curve, and a reliability factor K_r for the given reliability has been defined. Therefore, by giving different reliabilities, the different reliability factors can be evaluated. Eventually, the reliability related S-N curves will be generated. As previously mentioned, in stress-based life prediction, the S-N curve is only related to the strength S_u or the endurance limit S_e and the reliability factor is unique. The endurance limit S_e (stress level) can then be used as one of the random variables to calculate the reliability factor with different materials so that the R-S-N curve family for these materials can be created easily whatever the applied stress histories are. Hence, the reliability versus life curve for different material can also be generated directly.

For desired reliability, the S-N curve must be modified using corresponding ultimate strength S_u or the endurance limit S_e from Eq. (2.7). The fatigue life at the desired reliability can then be obtained.

As mentioned, some methods have been developed to generate R-S-N curves, however, the reliability models on which these methods are based, are simplistic, inefficient, or divergent in nature. In the previous approach that developed by the authors (Tang and Park, 2001), a R-ε-N curve family was generated using FORM. One of the primary goals of this research is to constitute a complete set of fatigue reliability analysis method by developing an algorithm for more accurate and efficient generation of R-S-N curves.

The method uses the life prediction procedure in conjunction with the well-known FORM approach to solve fatigue reliability problems for an assigned constant stress amplitude history by obtaining the reliability indices and corresponding factors. Since the FORM method can address reliability factors of more complex problems, the number of random variables and corresponding distributions can be arbitrary. Once the different reliability factors have been evaluated, the reliability defined S-N family of curves can be generated. The fatigue life versus reliability curve can subsequently be created. Given this improvement, the reliability of the mechanical components becomes more accurate and realistic.

To assess reliability, the proposed method depends on establishing factors for the different reliability levels. These factors are initially unknown and must be obtained. By assigning a reliability value, a corresponding β_{FORM} can then be determined. The FORM can now be applied to the life corresponding to this β_{FORM} value. Since a unique reliability factor exists for each reliability value, the reliability factor K_r for an assigned reliability R can be evaluated by:

$$K_R = {(N_f)_R}/{(N_f)_{50}} \qquad (4.1)$$

where $(N_f)_{50}$ indicates the fatigue life with a 50% reliability, obtained from the nominal S-N curve, and $(N_f)_R$ is the fatigue life related to the assigned reliability. Once the reliability factor is evaluated, the reliability defined S-N curve is generated by multiplying this reliability factor by the nominal S-N curve factor, which is a constant based on the unique condition. The procedure is repeated until there are sufficient S-N curves to create the R-S-N curve family, see Fig. 4.1.

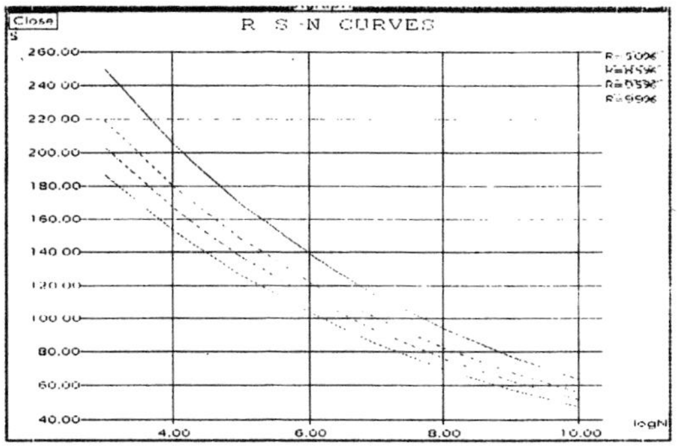

Fig 4.1 R-S- N curve family

Once the family of R-S-N curves is obtained, the fatigue life versus reliability curve is generated and employed to evaluate mission life or fatigue life reliability based on the desired reliability of the component.

The scenarios of the proposed method, therefore are as follows: For selected material, the nominal S-N curve is given, this is a 50% reliability related S-N curve. With the corresponding endurance limit S_e and a given reliability, the inversed FORM is used to evaluate the coincident life. The reliability factor which corresponds to a specified reliability can then be evaluated using Eq. (4.1). With this factor a new reliability related S-N curve can be generated. Iterated this procedure for different reliabilities, and a R-S-N curve family can be generated. From the curve family, a reliability versus life curve can be created.

5. EXAMPLE

In this example, oscillating stress of varied amplitude causes fatigue failure. A schematical example of cyclic stress time history and the corresponding cycle local strain time history is shown in Figs. 5.1a and b, respectively. Since the cyclic loading history has some uncertainty, the stress contains uncertainty as well. Hence, the stress history forms a random process. Also, in this example, it is assumed that the material properties, such as Young's Modulus E, fatigue strength exponent b, fatigue ductility exponent c, and ultimate stress S_u are random variables. For simplicity, it is also assumed that all these random variables are normally distributed. The mean values are E=30000ksi, b=–0.071, and c=–0.66, and the ultimate stress S_u = 146 kpsi. Their standard deviations are defined as σ_{S_u} =0.08 for ultimate stress, and σ_x =0.06 for other random variables, respectively. According to Eq. (2.4), the endurance limit will be 73 kpsi, and the reliability factors for this material can be evaluated based on this endurance limit value.

The rainflow counting method is utilized to transform an irregular stress history into an equivalent sequence of the closed loop constant amplitude stress.

For each counted applied stress cycle, a non-linear equation (2.1) can be solved to obtain the life N. After life for each cycle is computed, Palmgren-Miner's linear summation rule is used to predict the life for the entire stress history.

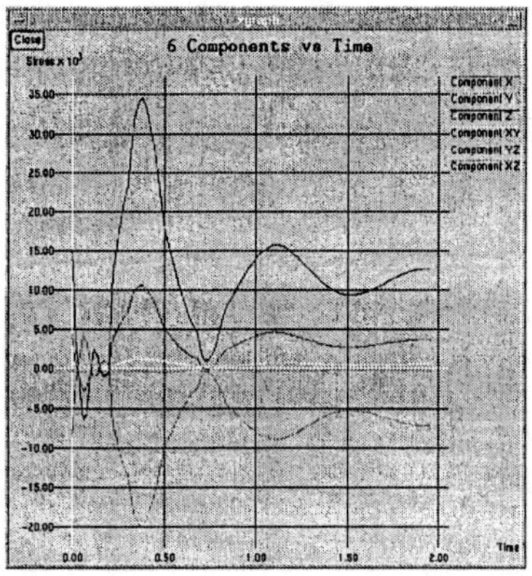

Figure 5.1 Cyclic stress Time history

When certain reliability is specified, the reliability index is known. A fatigue life corresponding to a given reliability index is calculated using the FORM given in section 3. Table 1 lists the given reliability, their corresponding reliability indices, and fatigue lives corresponding to different given reliability levels. Using Eq. (4.1), the reliability factors are then evaluated and the reliability related S-N curves can be generated. From Table 1, the life versus reliability curve can be generated directly as shown in Fig. 5.2.

Table 1 Results for Reliability and Life

Reliability	Life (N_f)	Reliability Index
50	4.064698e+08	---
65.17	2.624278e+08	0.39
80.23	1.532482e+08	0.85
85.08	1.218127e+08	1.04
90	8.965054e+07	1.288
95	5.683090e+07	1.645
99	2.260623e+07	2.326

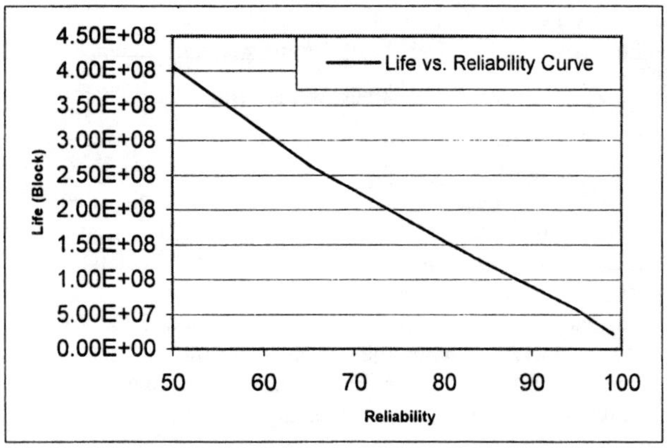

Figure 5.2 Life versus reliability curve

6. CONCLUSION

A practical reliability prediction method for stress-based fatigue life design is presented. A major feature of the proposed method is the use of unique reliability factors, defined using the FORM method. The reliability analysis enables the reliability model to be used for the general stress-based fatigue life design applications on which a complete set of reliability fatigue test data is unavailable.

The use of unique reliability factors enables engineers to directly employ this method in solving practical life design problems. As such, the developed method presents a practical approach to engineering design. Since all parameters can be determined using the standard fatigue theory and standard statistical test data methods, the application of this approach is straightforward. Therefore, this approach exhibits significant potential for engineering design applications in industry.

Compared with other methods, the computational effort required for this proposed method presents a very reasonable alternative and is more straightforward with respect to existing design requirements. The method developed herein also lends itself to inverse operations, i.e., determination of the life under a specified reliability for each counted cycle.

REFERENCES

Avakov, V. A., 1991, "Fatigue Reliability Functions in Semilogarthmic Coordinates," *Reliability, Stress Analysis, and Failure Prevention*, ASME, pp. 55-60.

Baldwin, J. D. and Thacker, J. G., 1995, "A Strain-Based Fatigue Reliability Analysis Method," Journal of Mechanical Design, Vol. 117, pp. 229-234.

Bannantine, J. A., Comer, J. J. and Handroch, 1990, *Fundamentals of Metal Fatigue Analysis*, Prentice Hall, NJ.

Bjerager, P., 1990, "Probability Computation Methods in Structural and Mechanical Reliability", *Computational Mechanics of Probabilistic and Reliability Analysis - Chap. 2*, Edited by Liu, W.K. and Belytschko, T., Elmepress Int'l., Lausanne, Switzerland.

Hohenbichler, M. and Rackwitz, R., 1981, "Nonnormal Dependent Vectors in Structural Reliability," *Journal of the Engineering Mechanics Division ASCE*, 107(6), pp. 1227-1238.

Jia, Y., Wang, Y., Shen, G. and Jia Z., 1993, "Equivalent fatigue load in machine-tool probabilistic reliability," Part I: Theoretical basis," Int J Fatigue 15, No. 6, pp. 474-477.

Lambert, R. G., 1976, "Analysis of Fatigue Under Random Vibration," *The Shock and Vibration Bulletin*, pp. 55-71.

Madsen, H.O., Krenk, S., and Lind, N.C., 1986, *Method of Structural Safety*, Prentice-Hall, Inc., New Jersey.

Murty, A. S. R., Gupta, U. C. and Krishna, A. R., 1995, "A New Approach to Fatigue Strength Distribution for Fatigue Reliability Evaluation," *Int'l. Journal of Fatigue*, 17, No. 2, pp. 85-89.

Rubinstein, R.Y., *Simulation and the Monte Carlo Method*, 1981, John Wiley and Sons, New York, NY.

Shigley, J.E., Mischke, C.R., 1989, *Mechanical Engineering Design*, fifth edition, McGraw-Hill Book Co., New York, NY.

Tang, J., Zhao, J., 1995, "A Practical Approach for Predicting Fatigue Reliability under Random Cyclic Loading ", *Reliability Engineering and System Safety*, Vol. 50, No. 1, pp. 7-15.

Tang, J., 1999, "An Efficient Asymptotic Approach For Assessing Fatigue Reliability," *ASME Trans. Journal of PVP*, Vol. 121, No. 2, pp. 220-224.

Tang, J., Park, Y. H., 2001, "A Methodology for Fatigue Reliability Analysis of Mechanical Components Using First Order Reliability Method", *ASME PVP Conference Proceedings*, July 2001, Vol. 417, pp. 169-174.

Tang, J., Park, Y. H., 2002, "An Efficient Algorithm Based On First Order Reliability Method For Fatigue Reliability Analysis Of Mechanical Components", *ASME PVP Conference Proceedings*, August 2002, Vol. 438, pp. 125-132.

Tu, J., Choi, K.K., and Park, Y.H., 1999, "A New Study on Reliability-Based Design Optimization," *Journal of Mechanical Design*, ASME, Vol. 121, No. 4, pp. 557-564.

Wirsching, P. H., Torng, T. Y. and Martin, W. S., 1991, "Advanced Fatigue Reliability Analysis," *Int'l Journal of Fatigue*, 13 No 5, pp. 389-394.

Wu Y.T., 1994, "Computational Methods for Efficient Structural Reliability and Reliability Sensitivity Analysis," *AIAA Journal*, Vol. 28, No. 9, pp. 1663-1669.

Wu, Y.T., Millwater, H.R., and Cruse, T.A., 1990, "Advanced Probabilistic Structural Analysis Method for Implicit Performance Functions," *AIAA Journal*, Vol. 28, No. 9, pp. 1663-1669.

Wu, Y. T. and Wirsching, P. H., 1984 "Advanced Reliability Method for Fatigue Analysis", *Journal of Engineering Mech, ASCE* 110, 4, pp. 536-553.

Youn, B. D., Choi, K. K., and Park, Y. H., 2002, "Hybrid Analysis For Reliability Based Design Optimization," *ASME Jounal of Mechanical Design*, (in press)

Zhao, J., Tang, J. and Wu, H. C., 1998, "A Reliability Assessment Method In Strain-Based Fatigue Life Analysis", *ASME Trans. Journal of PVP*, Vol. 120, No. 1, pp. 99-104.

Zhao, J., Tang, J. and Wu, H. C., 2000, " A Practical Reliability Prediction Method For Fatigue Life Design," *ASME Trans. Journal of PVP*, Vol. 122, No. 3, pp. 270-275.

PVP-Vol. 458, Computer Technology and Applications

PVP2003-1903

COUPLED FINITE ELEMENT/MESHFREE SIMULATION OF MANUFACTURING PROBLEMS

Hui-Ping Wang and Mark E. Botkin
GM R&D Center
Mail Code 480-106-256
30500 Mound Road
Warren, MI 48090-9055

Cheng-Tang Wu
LSTC
7374 Las Positas Road
Livermore, CA 94550

Sijun He
Analytical Design Service
Mail Code 480-400-111
30500 Mound Road
Warren, MI 48090-9055

ABSTRACT

This paper studies the performance of a newly developed coupled finite element/meshfree tool in manufacturing problem simulation. Two manufacturing problems, a battery tray forming and an axisymmetric forging, are used for demonstration. In both problems, the coupled finite element/meshfree analysis is compared with the traditional finite element analysis. The numerical studies show that the coupled finite element/meshfree solution is as accurate as the finite element solution for the problems without mesh distortion difficulties. For the ones involving excessive deformation, the coupled method provides more reliable solutions than the traditional finite element method, but higher CPU is required.

INTRODUCTION

The finite element (FE) method has been broadly used to solve partial differential equations (PDEs) in solid, thermal and fluid flow problems over the past four decades. It is a method well developed and refined for industrial applications. Successful FE-based commercial tools such as NASTRAN, ABAQUS and LS-DYNA are used extensively in the assessment of vehicle designs and the simulation of manufacturing processes in automotive industries. However, the accuracy of the FE method strongly depends on the quality of the FE meshes. When dealing with problems involving severe shape changes, this method can have deteriorated accuracy due to excessive mesh distortion. Research on remeshing and mesh adaptation [1-3] has therefore been performed to relieve the mesh distortion difficulties.

As alternatives, over the past 10 years various meshfree methods [4-9] have also been developed for solving PDEs in engineering and scientific problems. The common feature of the meshfree methods is that the approximation of the field variables is accomplished entirely based on nodal information without usage of element connectivity; therefore, these methods are naturally free of the mesh-related difficulties, and particularly effective in handling problems with large deformation and moving discontinuities. Extensive evidence shows that some meshfree methods, such as Element Free Galerkin (EFG) method and Reproducing Kernel Particle method (RKPM), can provide better solution accuracy than the conventional FE method for the same number of degrees of freedom.

Although promising, meshfree methods are computationally much more expensive than the FEM, which greatly limits their applications in industrial problems. To make the meshfree method affordable to complex engineering applications, a coupled FE/Meshfree method has been developed in LS-DYNA for the general dynamic explicit analysis in solids and shells [10, 11]. The method partitions the problem domain into FE and meshfree zones so that the meshfree method is only applied to the area where the flexibility of the meshfree method is needed while the FEM is used for the remainder. Thus, the computational time for the meshfree zones can be kept low for problems whose severe deformations are restricted to small areas.

In this paper, the general methodology of the coupling will be briefly introduced. The applications of this method in the simulation of two manufacturing problems will be demonstrated. Comparisons will be made with the traditional FEM to show the effectiveness of the coupled method.

COUPLED FINITE ELEMENT/MESHFREE METHOD

The purpose of coupling FEM and the meshfree method is to keep the computational cost due to the meshfree computation low so that the problems can be solved at a cost as low as possible. Figure 1 illustrates the main concept of this coupling. The problem domain Ω is divided into FE computation zones and meshfree computation zones, that is, $\Omega = \Omega_{FEM} \cup \Omega_{meshfree}$. The FE zones consist of non-overlapping and conforming elements. The meshfree zones are discretized into sets of particles with compact supports.

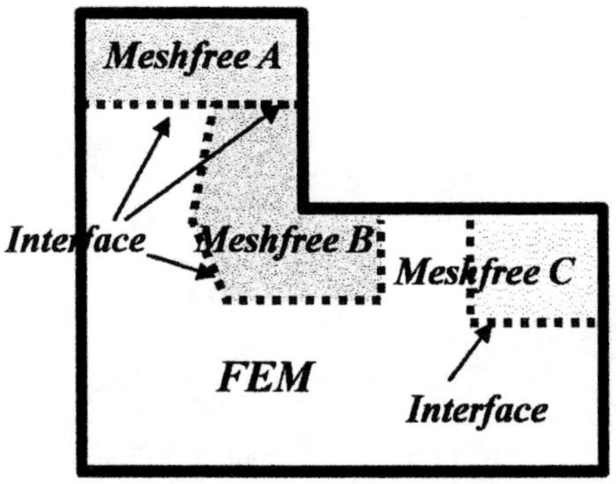

Figure 1 The Coupled FE/Meshfree Computation Zones

The FE approximation and the meshfree approximation are employed for the field variables in Ω_{FEM} and $\Omega_{meshfree}$, respectively. The discrete approximation of solution $\boldsymbol{u}(\boldsymbol{x})$ is

$$\boldsymbol{u}^h(\boldsymbol{x}) = \begin{cases} \sum\limits_{\substack{L\\ x_L\in\Omega_{FEM}}}^{KP} \Phi_L^{[m]}(\boldsymbol{x})\boldsymbol{d}_L;\ \forall \boldsymbol{x}\in\Omega_{FEM} \\ \sum\limits_{\substack{I\\ x_I\in\Omega_{meshfree}}}^{NP} \bar{w}_a^{[n]}(\boldsymbol{x};\boldsymbol{x}-\boldsymbol{x}_I)\boldsymbol{d}_I + \sum\limits_{\substack{L\\ x_L\in\Gamma_{Interface}\\ or\ x_L\in\Gamma_{Boundary}}}^{MP} \Phi_L^{[m]}(\boldsymbol{x})\boldsymbol{d}_L;\ \forall \boldsymbol{x}\in\Omega_{meshfree} \end{cases} \tag{1}$$

where $\Gamma_{Interface} = \Omega_{FEM} \cap \Omega_{meshfree}$, KP is the total number of nodes per element, NP is the total number of meshfree particles whose supports cover point $\boldsymbol{x}$, MP is the total number of the finite element nodes on the interface or the domain boundary that influence the approximation. $\bar{w}_a^{[n]}(\boldsymbol{x};\boldsymbol{x}-\boldsymbol{x}_I)$, the weight function, is defined as

$$\bar{w}_a^{[n]}(\boldsymbol{x};\boldsymbol{x}-\boldsymbol{x}_I) = \boldsymbol{H}^{[n]^T}(\boldsymbol{x}-\boldsymbol{x}_I)\boldsymbol{b}^{[n]}(\boldsymbol{x})w_a(\boldsymbol{x}-\boldsymbol{x}_I) \tag{2}$$

where $\boldsymbol{H}^{[n]}$ is a vector of the nth order monomial basis functions, and $\boldsymbol{b}^{[n]}$, the coefficient vector, is so constructed that the nth order reproducing conditions are satisfied:

$$\sum_{\substack{I\\ x_I\in\Omega_{meshfree}}}^{NP} \bar{w}_a^{[n]}(\boldsymbol{x};\boldsymbol{x}-\boldsymbol{x}_I)x_I^i y_I^j z_I^k + \sum_{\substack{L\\ x_L\in\Gamma_{Interface}\\ or\ x_L\in\Gamma_{Boundary}}}^{MP} \Phi_L^{[m]}(\boldsymbol{x})x_L^i y_L^j z_L^k = x^i y^j z^k,\ i+j+k=0,\cdots,n \tag{3}$$

To maintain the continuity of the approximation across the interface, the reproducing order n is chosen to be same as the finite element interpolation order m.

Now consider problems governed by the equations of motion $\rho\ddot{\boldsymbol{u}} = \nabla\cdot\sigma - f_b$ in $\Omega = \Omega_{FEM} \cup \Omega_{meshfree}$ with given boundary and initial conditions. To solve this set of PDEs, the corresponding weak form is constructed with the Galerkin method. The coupled FE/meshfree approximation expressed in Equation (1) is introduced to the field variables. The discrete system of equations can then be obtained. To ensure the linear exactness in the meshfree Galerkin approximation of Dirichlet boundary value problems, a Stabilized Conforming Nodal Integration Method is employed for the domain integration in the meshfree computation zones. For explicit dynamic computations, a new lumping method is developed for the mass and body force calculations. Detailed derivations and formulations can be found in [10, 11].

NUMERICAL EXAMPLES

Two manufacturing problems are demonstrated, a battery tray forming and an axisymmetric forging. In both problems, the coupled FE/meshfree method is compared with the conventional FE method.

A Battery Tray Forming

A battery tray forming is illustrated in Figure 2. This model is used to investigate the accuracy of the coupled FE/meshfree analysis as to the FE analysis. The model consists of a punch, a blank, a blank holder and a die. The punch pushes the blank down into the die to form the desired shape. The blank is a 422.5mmx331.0mm rectangular sheet with thickness equal to 0.8mm. The material properties for the blank are: mass density $RO = 0.783e\text{-}5\ kg/mm^3$, Young's modulus E=2.07e+2 GPa, Poisson's Ratio PR=0.3. The LS-DYNA anisotropic material Type 36, *MAT_3-PARAMETER_BARLAT, is employed to model the blank.

Two analyses are performed. The first is a coupled FE/meshfree analysis where the blank is modeled by the meshfree shell formulation with 101x81 meshfree particles. The second is the FE analysis where the blank is modeled by the Belytschko-Tsay shell formulation with 100x80 quadrilateral elements. Both analyses are successfully completed. Figure 3 gives the progressive deformations from the coupled FE/meshfree analysis. The final shapes of the blank from both analyses are shown in Figure 4. The resulted punch forces are plotted in Figure 5. These results show that the coupled FE/meshfree solution is as accurate as the FE solution for this forming problem. It is noticed that the CPU time used by the coupled FE/meshfree analysis is 5.2 times of the time used by the FE analysis.

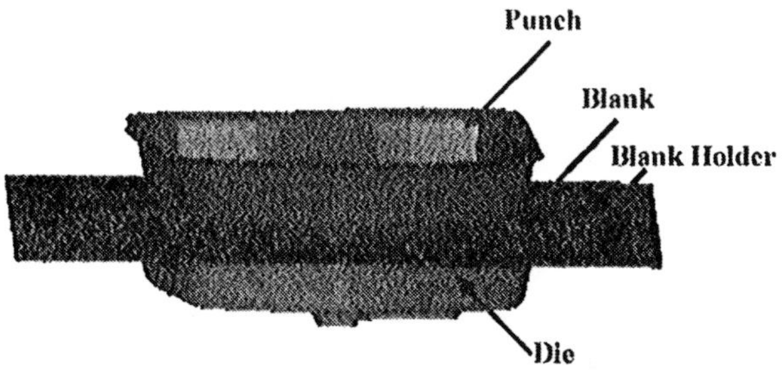

Figure 2 Problem Description Of A Battery Tray Forming

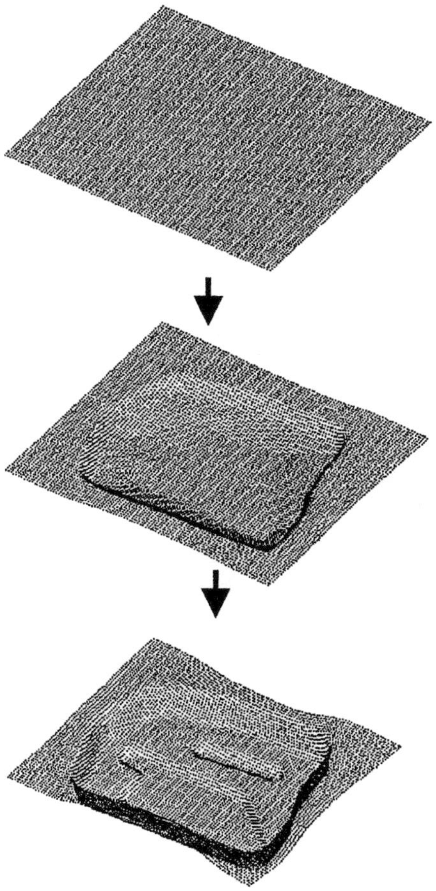

Figure 3 Progressive Deformations From The Coupled FE/Meshfree Analysis Of A Battery Tray Forming

(a) FE Analysis

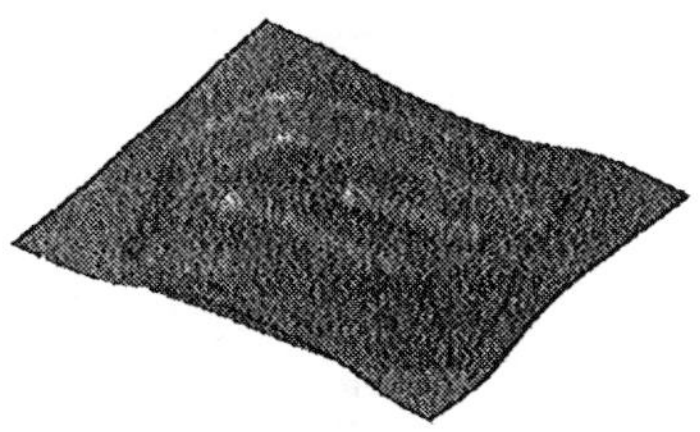

(b) Coupled FE/Meshfree Analysis

Figure 4 Final Shape Of The Blank From The FE And Coupled FE/Meshfree Analysis

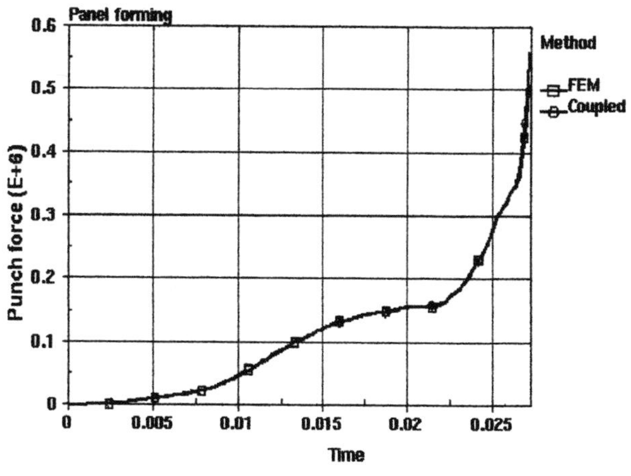

Figure 5 Comparisons Of Resulted Punch Forces In A Battery Tray Forming Simulation

An Axisymmetric Forging

Forging is a typical manufacturing process involving large material deformation. As illustrated in Figure 6, a cylindrical rigid die is pressed into a cylindrical deformable workpiece so that the workpiece deforms into a desired cup shape. The workpiece has a radius of 100mm and a height of 100mm, and is made of elastic-plastic material that follows a given stress-strain curve and strain rate dependency. The material properties are: mass density $RO = 0.783e\text{-}5\,kg/mm^3$, Young's modulus E=2.0e+2 GPa, Poisson's Ratio PR=0.3, and Yield stress SIGY =2.07e-1 GPa. The stress-strain curve and the strain rate effect are listed in Table 1 and Table 2, respectively. The die has a radius of 70mm and is moving down with a constant velocity of 1mm/ms. The friction coefficient μ between the die and the workpiece is 0.2.

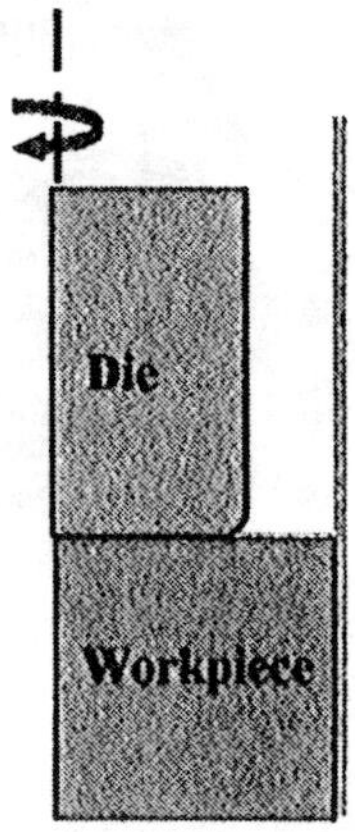

Figure 6 Problem Description Of An Axisymmetric Forging

Table 1. Plastic stress/strain curve

EPS1	*EPS2*	*EPS3*	*EPS4*	*EPS5*
0.00000	*0.00800*	*0.01600*	*0.04000*	*99.000*
ES1	*ES2*	*ES3*	*ES4*	*ES5*
0.20700	*0.02500*	*0.02750*	*0.02899*	*0.0300*

Table 2. Cowper/Symonds strain rate parameters

C	*P*
40.0	*5.0*

The problem is first analyzed by the conventional FE method. Figure 7 shows the discrete FE model, where 1300 nodes and 1198 axisymmetric quadrilateral elements are used. Both the selective reduced integration (SRI) and the reduced integration (RI) with Belytschko-Bindeman hourglass control are considered. The progressive deformations obtained from the FE SRI analysis and the FE RI analysis are illustrated in Figure 8 and Figure 9, respectively. As shown in the deformation plots, severe mesh distortion appears near the die corner as the die pushes down. The FE SRI analysis diverges at time t=18ms. The FE RI analysis completes, but the final deformation shows a big gap between the die and the workpiece, which is a non-physical deformation behavior resulted from the over-stiff material response.

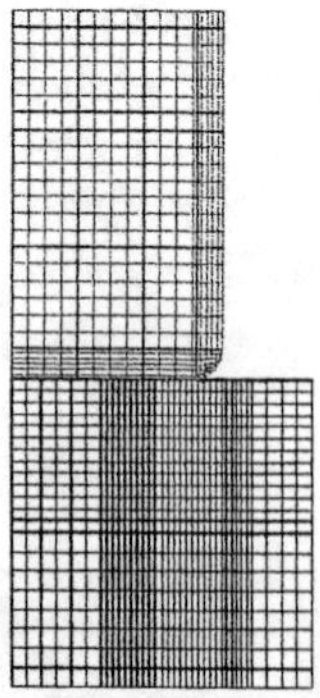

Figure 7 The Finite Element Model Of An Axisymmetric Forging

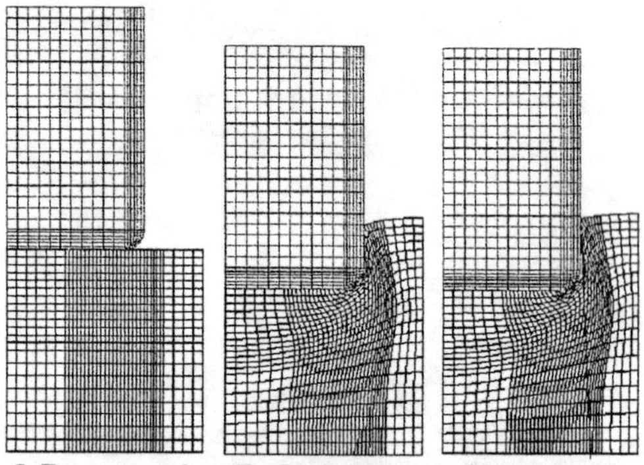

Figure 8 Progressive Defomations Obtained From The FE SRI Analysis Of The Forging Process

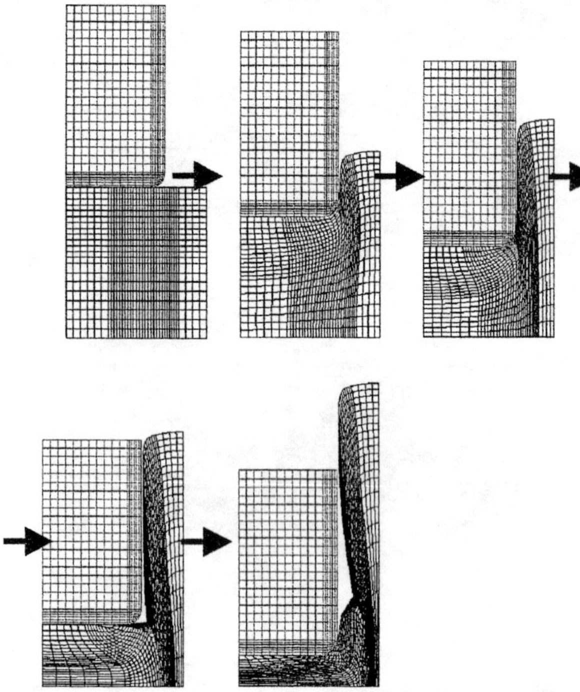

Figure 9 Progressive Defomations Obtained From The FE RI Analysis Of The Forging Process

The problem is then analyzed by the coupled FE/meshfree method. Noticing that most mesh distortion occurs near the die corner, we designate the area that is in contact with the die corner as the meshfree computation zone as shown in Figure 10(a). Figure 10(b) plots the discrete coupled FE/meshfree model, where 378 meshfree particles among 1300 nodes and 860 axisymmetric quadrilateral elements are used. Reasonable progressive deformations are obtained from this coupled FE/meshfree analysis as plotted in Figure 11. Table 3 gives the CPU comparison among two FE analyses and the coupled FE/meshfree analysis. In this particular model, about 8 times of CPU was consumed by the coupled FE/meshfree analysis compared with the CPU used by the FE RI analysis. The FE SRI analysis also consumes large CPU but still cannot converge. Figure 12 gives contact force comparisons among the three methods. It shows that the initial contact force obtained from the coupled FE/meshfree method agrees well with the FE SRI solution, and the FE RI method with the hourglass control generates a much stiffer response.

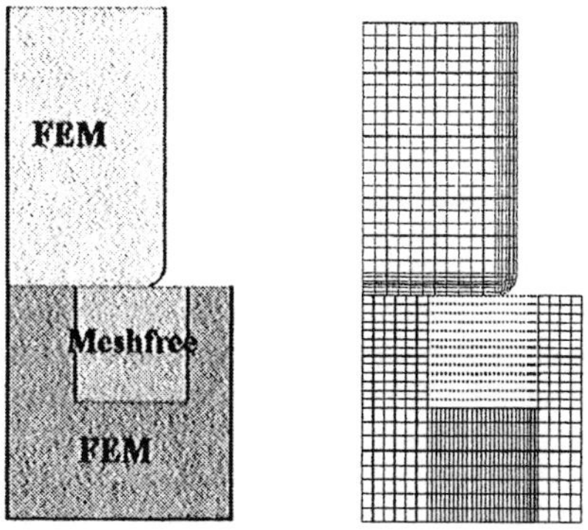

(a) Computation Zones (b) The discrete model
Figure 10 The Coupled FE/Meshfree Model Of The Forging Problem

Table 3. CPU Costs Of The Forging Analysis

SRI FEM	*RI FEM*	*Coupled Method*
*35703 seconds**	*5094 seconds*	*42182 seconds*

*diverges at 37% of the process

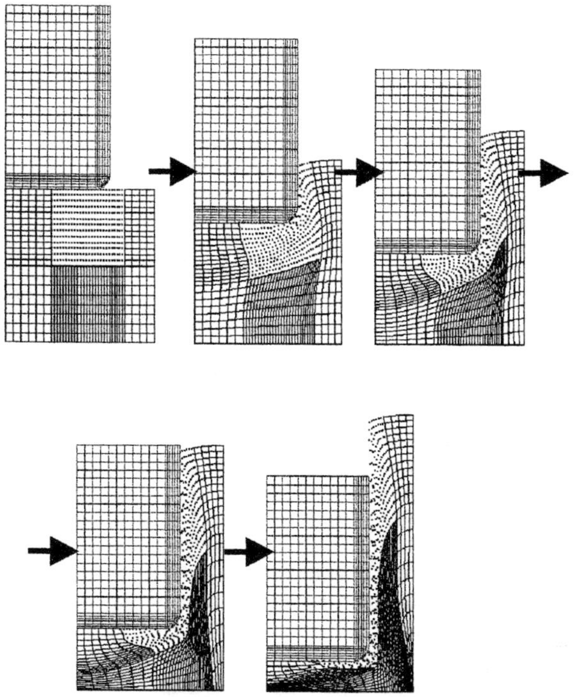

Figure 11 Progressive Deformations Obtained From The Coupled FE/Meshfree Analysis Of The Forging Process

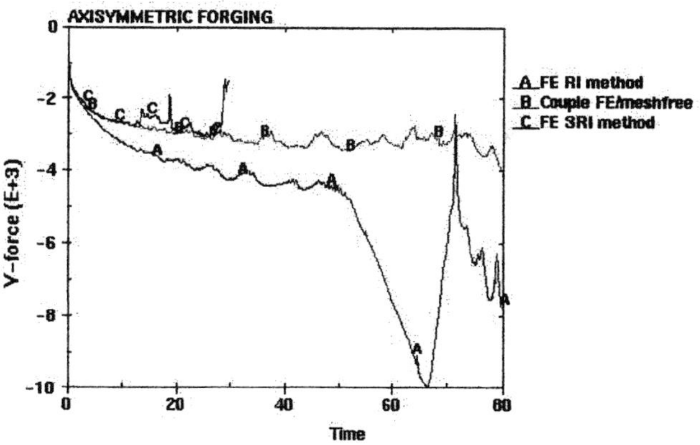

Figure 12 Comparisons Of Contact Forces In The Forging Simulation

SUMMARY AND CONCLUSIONS

This paper investigated the performance of a newly developed coupled finite element/meshfree analysis tool in LS-DYNA. This tool partitions the original solution domain into FE and meshfree computation zones so that the meshfree method can be only applied to the area where it performs more effectively than the FEM, such as the large deformation areas. Two manufacturing examples, a battery tray forming and an axisymmetric forging, were employed for the study. The first problem does not involve excessive deformation, and was used to verify the accuracy of the coupled FE/meshfree analysis. The second example involves local large deformation, therefore is suitable to apply the coupled FE/meshfree analysis tool. The solution comparison between the FE simulation and the coupled FE/meshfree analysis in the battery tray forming problem showed that the coupled FE/meshfree solution was as accurate as the FE solution. For the problem with excessive deformation, the coupled method was more robust. The FE analysis either diverged or led to wrong solutions due to the severe element distortions. The coupled FE/meshfree analysis successfully completed the simulation and provided reasonable solutions, but higher CPU was consumed.

ACKNOWLEDGMENTS

The coupled Finite Element/Meshfree analysis tool in LS-DYNA was developed and delivered by Livermore Software Technology Corporation as part of the *Meshfree Method for Design and Manufacturing* project sponsored by General Motors Corporation.

REFERENCES

1. Botkin M. E., Wentorf Rolf, Karamete B. Kaan and Raghupathy R. "Adaptive refinement of quadrilateral finite element shell meshes," AIAA-2001-1400 in Proceedings of AIAA SDM conference, Seattle, April 16-19, 2001.

2. Botkin M. E., Hui-Ping Wang and Raghupathy R. "Adaptive Refinement of Quadrilateral Finite Element Meshes Based Upon A Posteriori Error Estimation of Quantities of Interest," AIAA-2002-1656 in Proceedings of AIAA SDM conference, Colorado, April 22-25, 2002.

3. Ainsworth M. and Senior B. "An adaptive refinement strategy for hp-finite element computations," APPL NUMER MATH 26 (1-2): 165-178,1998.

4. Randles, P. W., and Libersky, L. D., "Smoothed particle hydrodynamics: some recent improvements and applications," Computational Methods in Applied Mechanical Engineering, 139, 375–408, 1996.

5. Belytschko, T., Lu, Y. Y., and Gu, L., "Element free Galerkin method," Int. J. Num. Methods Eng., 37, 229–256, 1994.

6. Liu, W. K., Jun, S., and Zhang, Y. F., "Reproducing kernel particle methods," Int. J. Num. Methods in Fluids, 20, 1081–1106, 1995.

7. Chen, J. S., Pan, C., Wu, C. T., and Liu, W. K., "Reproducing kernel particle methods for large deformation analysis of nonlinear structures," Comput. Methods Appl. Mech. Eng., 139, 195–227, 1996.

8. Melenk, J. M., and Babuska, I., "The partition of unity finite element method: basic theory and applications," Computational Methods in Applied Mechanical Engineering, 139, 289–314, 1996.

9. Duarte, C. A. M., and Oden, J. T., "A H–P adaptive method using clouds," Comput. Methods Appl. Mech. Eng., 139, 237–262, 1996.

10. Wu, C. T., Botkin, Mark E. and Wang, Hui-Ping, "Development of A Coupled Finite Element and Mesh-free Method in LS-DYNA," Proceedings of 7th International LS-DYNA Users Conference, 2002.

11. Wu, C. T. and Guo, Y., "Development of a Coupled Finite Element/Mesh-free Method and Mesh-free Shell Formulation" GM R&D research report, 2002.

PVP-Vol. 458, Computer Technology and Applications

PVP2003-1904

BEHAVIOR OF EMBEDDED STEEL PLATES IN REINFORCED CONCRETE : NEW DESIGN BASIS

Sekhar K. Chakrabarti
Professor
Dept. of Civil Engineering
Indian Institute of Technology Kanpur
Kanpur, 208016, India

ABSTRACT

The behavior and capacity of embedded steel plates in reinforced concrete structures, are studied using a finite-element model developed for non-linear analysis. Strength interaction diagrams and moment-rotation charts useful for analysis and design of such plates, are developed for eccentric compressive and eccentric tensile loading, at failure and collapse. Capacities for a common class of embedded plates with variation in its thickness are computed for the cases of combined eccentric compression and shear.

INTRODUCTION

In many engineering applications, the cast-in-place embedded steel plates are used to support machine, equipment, pipe-lines etc., from the reinforced concrete members. Generally, the embedded steel plates are the assemblies consisting of steel plate, anchors and concrete base. An attachment, normally welded to the exposed face of the plate, is used to support pipes, equipment etc. and to transfer the loads to the reinforced concrete structures through plate-anchor assembly.

The analysis of embedded plates subjected to compressive load (at small eccentricity) is normally based on their behavior like the thin plates resting on tensionless elastic foundation (Timoshenko and Kreiger 1970), wherein transverse shear effects in the plate are neglected. The entire plate-anchor assembly becomes active when the embedded plate is subjected to compression (at large eccentricity) or tension, and/or shear forces. Generally, analysis and design of such plates is based on rigid plate assumption.

DeWolf (1978) experimentally investigated the behavior of the base plates subjected to axial load resulting in plate bending. Celep (1988) studied the problem of rectangular plates on tensionless foundations subjected to distributed load, concentrated loads, and moments. Weitshman (1970) and Celep (1988) demonstrated that regions of no contact can develop under beams and plates that are supported on tensionless foundation. Mishra and Chakrabarti (1996, 1997) analytically studied the effects of transverse shear and attachment on freely supported plates on tensionless elastic base, subjected to different loading conditions. The investigators concluded that vertical downward plate displacement and base pressure were underestimated when analyzed using thin-plate theory.

The PCI Design Handbook (Precast 1970) has suggested empirical expressions to assess the design capacity of welded, headed studs governed by steel fracture failure as well as concrete wedge failure or bond failure for anchorage loaded in direct tension, direct shear, and combined tension and shear. Cannon et.al. (1981) prepared a guide to the design of anchor bolts and other steel embedments. Klingner and Mendonca (1982) reviewed available procedures for computing the capacity of short anchor bolts and welded studs loaded monotinically in direct tension and direct shear, and compared them with available experimental results. Load-deflection behavior of liner enchorage system used in steel-lined concrete containment structures of nuclear power plants was examined by Armentrout and Burdette (1982). Brown and Whitlock (1983) made experimental investigation on the behavior of J-shaped bolts embedded in concrete masonry in direct tension, direct shear, and combined tension and shear. Ueda et.al. (1990) experimentally studied the anchor bolts in unreinforced concrete, monotonically loaded in shear.

Salmon et.al. (1957) analytically studied the moment-resisting capabilities and moment-rotation characteristics of column anchorage, keeping the axial load constant. Behavior of the moment-resistant base plate was studied by Diluna and Flaherty (1979) for examining the effect of plate flexibility. The investigation concluded that the anchor-bolt tension and the bearing pressure distribution on the concrete base are dependent on the plate flexibility. A state-of-the art report on steel embedments was prepared by the ASCE Nuclear Structures and Materials Committee, and Materials and Structural Design Committee (1984).

Cannon (1992) studied the effect of plate stress on the location of the compressive reaction and redistribution of load to anchor, and the effect of preloading of anchors on performance and capacity of anchorage. Chakrabarti and Tripathi (1992) analytically studied the impact of plate flexibility, base stiffness, and anchor stiffness on the behavior of an embedded plate-anchor assembly for thin plate behavior. The analytical results were compared with those obtained from the modified "concrete beam" analogy method suggested by the investigators. Finally, the investigators recommended a set of design guidelines.

Experimental studies were conducted by DeWolf and Sarislly (1980), Thambiratnam and Paramsivam (1986), and Cook and Klingner (1992) on the behavior of plate-anchor assemblies. It was observed that the primary mode of failure (yielding of any component) and the load-capacity at failure/collapse depend on factors that are ignored in the ultimate load design method. For example, the ultimate load design method predicts the capacity of the plate-anchor assembly, disregarding plate thickness.

Klingner et.al. (1998) have investigated the relative influences of the plate-characteristics and the load-displacement behavior of anchors on the load-displacement behavior of plate-anchor assembly. The investigators conducted tests in which the tension-shear interaction of anchors was evaluated using combination of tension and shear loads, and also tension and shear displacements. Load-displacement curves for the anchors under different orientations of tensile and shear loading were developed from the test results. These curves were used as input for the computer program developed at the University of Stuttgart (Li 1994); this program considers the plate as rigid to finally produce a load-displacement relationship for the assembly.

Mishra and Charabarti (2000) have developed a finite-element model to predict the behavior of embedded plates (with and without anchors) considering plate flexibility, effects of transverse shear in plate, base stiffness (for tensionless base), anchor parameters and attachment size. The model includes material nonlinearity for prediction of capacity (both at failure and at collapse) of the embedded plate-anchor assembly while providing the load-deformation behavior, along with indication of failure modes during the entire loading process. The model has been validated with the available experimental and analytical results.

From the review of the past studies it emerges that a definite need existed in undertaking a systematic investigation for knowing the pattern of variation of strength and serviceability characteristics of embedded plate-anchor assemblies based upon a realistic model like the one developed by Mishra and Chakrabarti (2000). This paper presents the salient features of the development (using Mishra and Chakrabarti Model) (Panisetty 1997) of strength interaction diagrams and moment-rotation charts useful for analysis and design of such assemblies subjected to eccentric compressive and eccentric tensile loading at failure, and collapse. Typical computation (Panisetty 1997) of the plate capacities when the plate is subjected to combined eccentric compression and shear, is also presented.

DEVELOPMENT OF INTERACTION DIAGRAMS AND MOMENT-ROTATION CHARTS

Strength interaction diagrams and moment-rotation charts have been developed for a common type of plate-anchor assembly (Fig.1) considering the possible variations in the different design parameters. Variations in plate thickness; plan dimensions of plate; area of attachment; grade of base concrete; and size, location and number of anchors have been taken into account. To facilitate the availability of information regarding reserve strength of the assemblies in post-failure phase, failure and collapse interaction diagrams have been developed.

Interaction Diagrams and Moment-Rotation Charts for Assemblies subjected to Eccentric Compressive and Tensile Loading

Strength interaction diagrams and moment-rotation charts have been developed for different plate-thicknesses with anchors placed as per normal practice. The dimensionless terms used in the development and presentation of interaction diagrams and moment-rotation charts are :
t/B; A_{anc}/s_1s_2; f_y/f_{ck}; $M/f_{ck}\ BD^2$; $P/f_{ck}\ BD$; A_{atch}/A_{plate}; d'/D; f_y'/f_{ckl}; A_{base}/BD.

Where, B = Shorter dimension of plate,
A_{anc}=Cross-sectional area of a single anchor,
A_{atch}= Area of attachment,
A_{plate}= Area of plate,
A_{base}= Area of concrete base,
s_1 , s_2 = Spacing of anchors in long and short directions of plate, respectively,
f_y = Yield strength of plate,
f_{ck} = Characteristic compressive strength of concrete,
M = Moment on the plate,
D = Longer dimension of plate,
P = Point load on plate,
d' =Distance of anchors from plate edge,
f_y' = Yield strength of anchor.

For the cases of eccentric compressive loading, the failure interaction diagrams were developed for conditions giving cracking of concrete, yielding of plate or yielding of anchors; and, the collapse interaction diagrams were developed for conditions giving crushing of concrete, failure of anchors or complete yielding of plate along with failure of anchors. The collapse of the assembly at/around zero compressive load-eccentricity has been found to occur due to crushing of concrete, and, with increase in compressive load-eccentricity, the collapse has been found to occur due to failure of anchors or complete yielding of plate along with failure of anchors. Typical interaction diagrams developed for eccentric compression at failure and collapse are illustrated in Figures 2 and 3, respectively. For the cases of eccentric tensile loading, the failure and collapse interaction diagrams were developed on the same basis as that for eccentric compressive loading excepting that collapse due to crushing of base concrete has been appeared practically impossible, as the moment due to eccentric tension has been found to be never capable of inducing concrete crushing. Typical interaction diagrams for eccentric tension at failure and collapse are presented in Figures 4 and 5, respectively.

Typical moment-rotation charts developed for eccentric compression cases, are presented in Fig.6. The rotation due to eccentric compression at every stage of incremental loading till the assembly collapses, was computed. It is observed that the assembly-rotation increases with the increase in load-eccentricity, while keeping the moment same. This is possibly due to the diminishing effect of axial compressive load against rotation with increase in load-eccentricity. The assembly rotation gets reduced with increase in plate-thickness for the same level of eccentric compressive load due to increase in plate stiffness. It is noted that for a given plate-anchor assembly the rate of increase of rotation with increase in moment, is relatively much lower initially (till failure of plate/anchor) than that observed in the post-failure phase of plate/anchor. Also, it is evident that the shape of the moment-rotation curves becomes increasingly flatter with increase in load-eccentricities.

Typical moment-rotation charts for eccentric tension cases, are presented in Fig.7. The assembly rotation decreases with the increase in load-eccentricity while keeping the moment same, which is reverse to the case of eccentric compression. As the load-eccentricity increases, the effect of axial tensile force in causing rotation decreases as compared to the effect of moment.

ASSEMBLY CAPACITY FOR COMBINED ECCENTRIC COMPRESSION AND SHEAR

An embedded plate of plan-dimensions 400mmx250mm with 6 mild steel anchors (yield strength = 330 N/mm^2) of 12mm diameter, was considered for computation of its capacities for its different plate-thicknesses, when subjected to combined eccentric compression and shear through an attachment of size 100mm square. A base concrete of characteristic strength of 25 N/mm^2 was adopted. For each plate-thickness, the eccentricity of the compressive load (P) was varied while the shear force (H) was applied at the plate level. The ratio, H/P was varied for each load-eccentricity for a particular plate-thickness. Assembly capacity (for eccentric compressive load) at collapse for each case was computed. Typical results are presented in Fig.8.

CONCLUSIONS

Strength interaction diagrams and moment-rotation charts are developed for a common type of embedded plates subjected to eccentric compressive/tensile loading. A possible variation in common design parameters are included to demonstrate the variation pattern for behavior and capacity with due regard for reserve capacity in the post-failure phase. Typical computation for plate capacity (at collapse) when subjected to eccentric compression and shear, is also presented.

REFERENCES

1. Armentrout, D.R., and Burdette, E.G. (1982), "Analysis of behavior of steel linear anchorage," *J. Struct. Div.*, ASCE, 108(7), 1451-1463.
2. ASCE, (1984), "*State-of-the-art report on steel embedments*," ASCE Nuclear Structures and Materials Committee, Materials and Structural Design Committee, New York.
3. Brown, R.H. and Whitlock, A.R. (1983), "Strength of anchor bolts in grouted concrete masonry," *J.Struct.Engrg., ASCE*, 109(6), 1362-1374.
4. Cannon, R.W. (1992), "Flexible base plates : Effects of plate flexibility and preload on anchor loading and capacity," *ACI Struct.J.*, 89(3).
5. Cannon, R.W., Godfrey, D.A., and Moreadith, F.L. (1981), "Guide to the design of anchor bolts and other steel embedments," *Concrete Int.*, July.
6. Celep, Z. (1988), "Rectangular plates resting on tensionless elastic foundation," *J. Engrg. Mech. Div.*, ASCE, 114(12), 2083-1092.
7. Chakrabarti, S.K. and Tripathi, R.P. (1992), "Design of embedded steel plates in reinforced concrete structures," *Struct. Engrg. Rev.*, 4(1), 81-91.
8. Cook, R.A. and Klingner, R.E. (1992), "Ductile multiple anchor steel to concrete connections," *J.Struct.Engrg.*, ASCE, 118(6), 1645-1655.
9. DeWolf, J.T. (1978), "Axially loaded column base plates," *J. Struct. Div.*, ASCE, 104(5), 781-794.
10. DeWolf, J.T. and Sarislly, E.F. (1978), "Axially loaded base plates : Effect of concrete base depth," Rep. No.78-119, Civil Engineering Dept., Univ. of Connecticut, Storrs, Conn.
11. DeWolf, J.T., and Sarislly, E.F. (1980), "Column base plates with axial loads and moments," *J. Struct. Engrg. Div.*, ASCE, 106(11).
12. Diluna, L.J., and Flaherty, J.A. (1979), "An assessment of the effect of plate flexibility on the design of moment resisting base plates," Pressure Vessels and Piping Div., American Society of Mechanical Engineers, New York.
13. Klingner, R.E., Hallowell, J.M., Lotze, D., Park, H-G, Rodriguez, M. and Zhang, Y-G (1998), "Anchor bolt behavior and strength during earthquakes," Report prepared for the U.S. Nuclear Regulatory Commission (NUREG/CR-5434).
14. Klingner, R.E. and Mendonca, J.A. (1982a)," Tensile capacity of short anchor bolts and welded studs : A literature review," *ACI J.*, 79(4).
15. Klingner, R.E. and Mendonca, J.A. (1982b), "Shear capacity of short anchor bolts and welded studs: A literature review," *ACI J.*, 79(5).
16. Li, L. (1994), "BDA : Programm zur Berechnung des Trag – und Verformungsverhaltens von Gruppenbefestigungen unter Kombinierter Schragzug-und Momentenbean spruchung (Programmbeschreibung)," The University of Stuttgart.
17. Mishra, R.C., and Chakrabarti, S.K. (1996), "Rectangular plates resting on tensionless elastic foundation : Some new results," *J. Engrg. Mech.*, ASCE, 122(4), 385-387.
18. Mishra, R.C. and Chakrabarti, S.K. (1997), "Shear and attachment effects on the behavior of rectangular plates resting on tensionless elastic foundation," *Engrg. Struct.*, 19(7), 551-567.
19. Mishra, R.C. and Chakrabarti, S.K. (2000), "Comprehensive model for embedded plates," *J. Struct. Engrg.*, ASCE, 126(5), 560-572.
20. Panisetty, R. (1997), "Behaviour of Steel Embedded Plates in R.C.C. Structures : A Parametric Study," M.Tech. thesis, Indian Institute of Technology, Kanpur, India.

21. *Precast and prestressed concrete*, 2nd Ed. (1970), Prestressed Concrete Institute, Chicago, III.
22. Salmon, C.G., Schenker, L., and Johnston, B.G. (1957), "Moment-rotation characteristics of column anchorage," *Trans. ASCE*, 122, 132-254.
23. Thambiratnam, D.P. and Paramsivam, P. (1986), "Base plates under axial loads and moments," *J. Struct. Engrg.*, ASCE, 112(5), 1166-1181.
24. Timoshenko, S.P., and Krieger, W. (1970), *Theory of plates and shells*, 2nd Ed., Mc-Graw Hill, New York.
25. Ueda, T., Kitipornchai, S., and Ling, K. (1990), "Experimental investigation of anchor bolts under shear," *J. Strut. Engrg.*, ASCE, 116(4), 910-924.
26. Weitsman, Y. (1970), "On foundations that react in compression only," *J. Appl. Mech. Div.*, Trans. ASME, 37(4), 1019-1030.

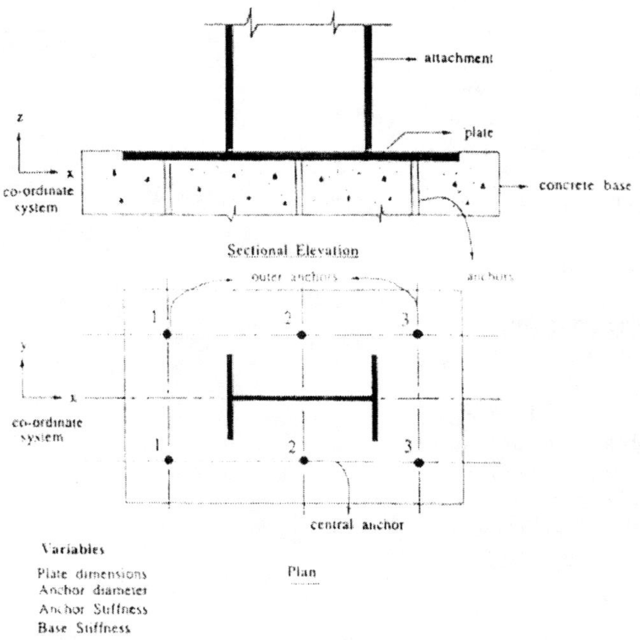

Fig. 1 General arrangement of plate - anchor assemblies

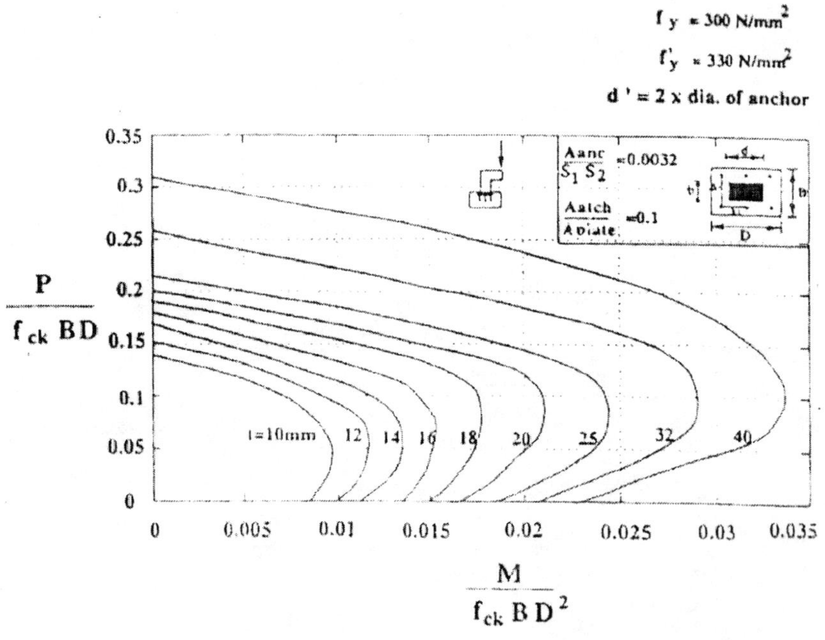

Fig. 2 Typical interaction diagram for eccentric compression at failure.

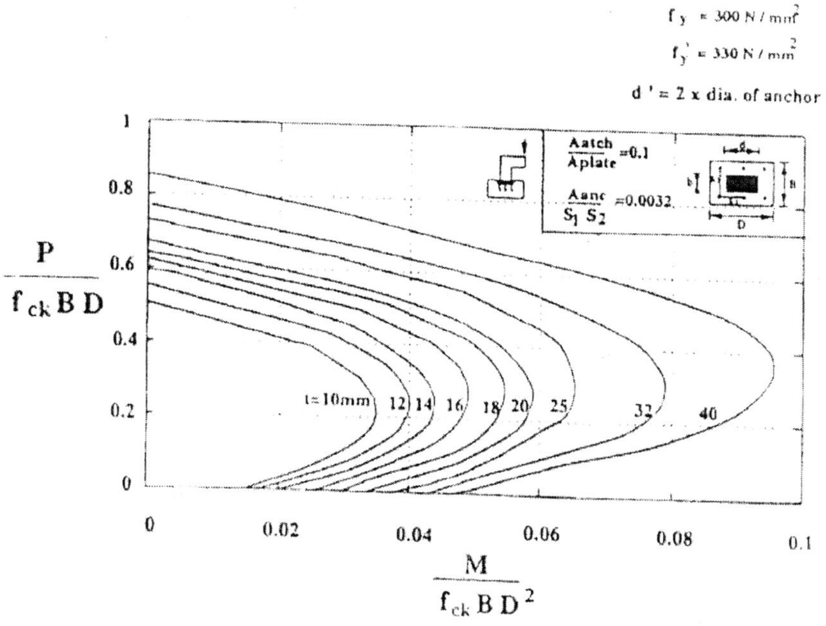

Fig. 3 Typical interaction diagram for eccentric compression at collapse.

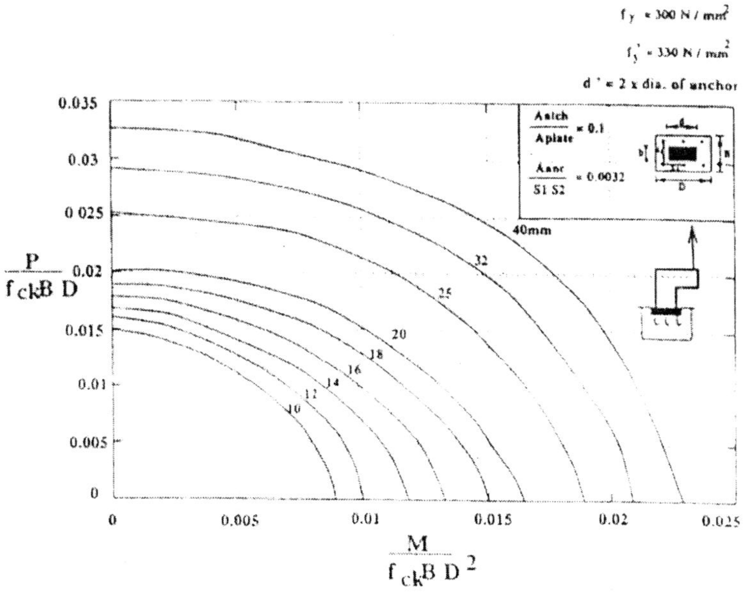

Fig. 4 Typical interaction diagram for eccentric tension at failure

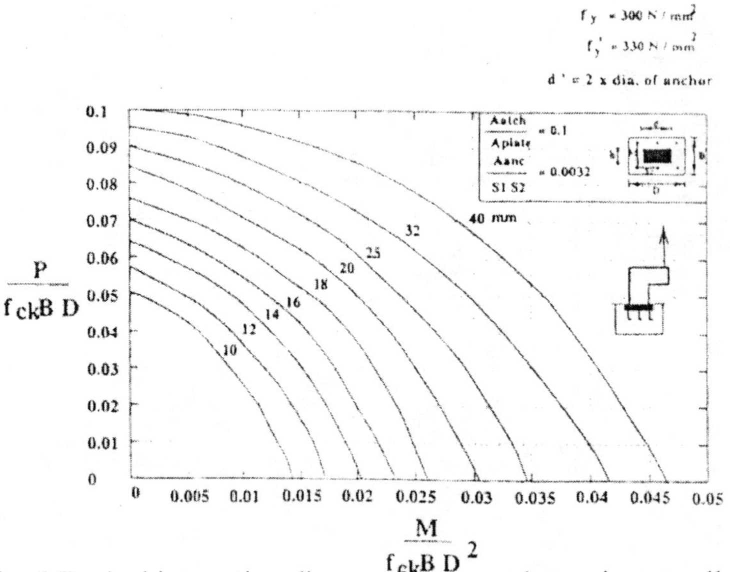

Fig. 5 Typical interaction diagram for eccentric tension at collapse

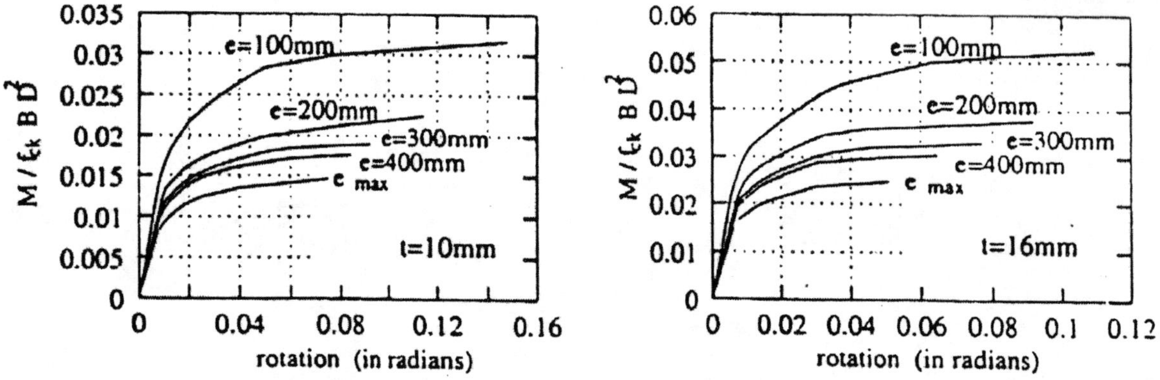

Fig. 6 Typical moment - rotation charts for eccentric compression cases.

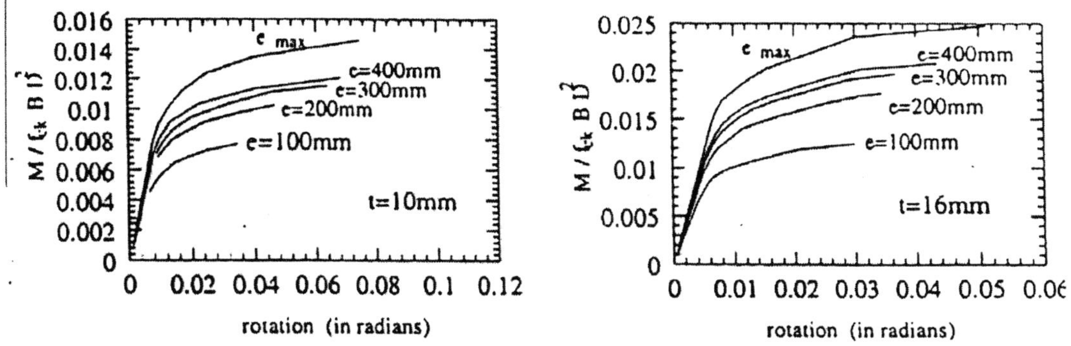

Fig. 7 Typical moment - rotation charts for eccentric tension cases.

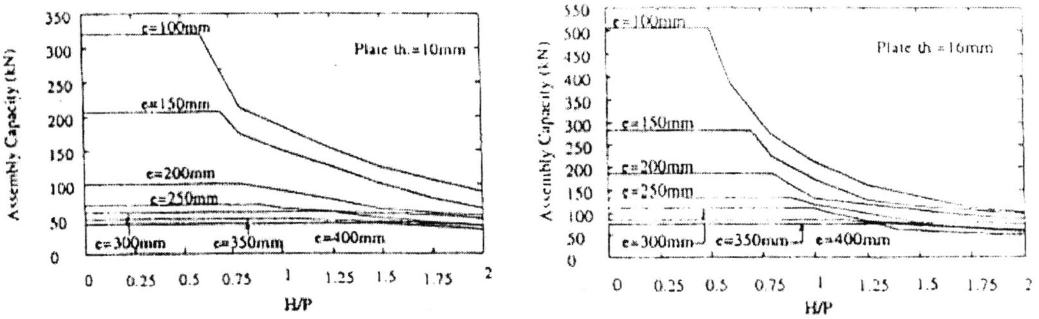

Fig. 8 Assembly capacity at collapse Vs. H/P Ratio

PVP-Vol. 458, Computer Technology and Applications

PVP2003-1905

GENERALIZED DIFFERENTIAL QUADRATURE METHOD FOR BURGERS EQUATION

Mladen Meštrović
Faculty of Civil Engineering
University of Zagreb
Zagreb, Croatia
Email: mestar@grad.hr

ABSTRACT

The generalized differential quadrature method as an accurate and efficient numerical method is developed for the Burgers equation. The numerical algorithm for this class of problem is presented. Differential quadrature approximation of needed derivatives is given by a weighted linear sum of the function values at grid points. Recurrence relationship is used for calculation of weighting coefficients. The calculated numerical results are compared with exact solutions to show the quality of the generalized differential quadrature solutions for each example. Numerical examples have shown accuracy of the GDQ method with relatively small computational effort.

INTRODUCTION

According the great advance in computer technology, the numerical methods have been applied on many problems in modern engineering and computational mechanics. The finite element method is the most developed numerical method and involved in almost all kind of engineering problems. This method gives very good results by using a large number of elements. If we need results only for few grid points, we have often to compute the solutions for large number of the points. In order to find an alternate numerical method to obtain results in specific points in the domain with acceptable accuracy and without large number of grid points, the differential quadrature method was introduced to solve some problems in engineering (Bellman, Casti, 1971). The differential quadrature method (DQM) is numerical method which approximates a derivative of the function in the governing equation at a discrete point as a weighted linear sum of the function values at all discrete points in domain used for approximation. The drawback of this method is the method to determine the weighting coefficients for needed partial derivatives. We have to solve a set of linear algebraic equations or to compute the weighting coefficients by an algebraic formula with grid points chosen as the roots of N-th order Legendre polynomial. The generalized differential quadrature method (GDQM) was introduced to overcome this drawback (Shu, Richards, 1992). The weighting coefficients for derivatives are given by a simple algebraic expression and the recurrence relation with arbitrary choice of grid points. In this paper, the GDQM is used to solve Burgers equation, nonlinear partial differential equation important in fluid dynamics, to show the quality of solution calculated by GDQ method for this kind of problem. A simple explicit Euler scheme is used for time step solution. Numerical examples have shown accuracy of the method with relatively small computational effort.

BURGERS EQUATION

The Burgers equation is nonlinear parabolic partial differential equation of second order as follows

$$\frac{\partial u}{\partial t} + u\frac{\partial u}{\partial x} = \lambda\frac{\partial^2 u}{\partial x^2} \quad , \tag{1}$$

with according initial condition

$$u(x,0) = u_0(x) \quad . \tag{2}$$

The equation includes the nonlinear convection term, uu_x, and the second order viscous diffusion term, u_{xx}, where the coefficient λ is viscosity, the reciprocal of the flow Reynolds number ($\lambda = 1/\text{Re}$). It is used in fluid dynamics and engineering as a simplified model for turbulence, viscous fluid flow, boundary layer behavior, shock wave formation and mass transport. It is a very simple form of Navier-Stokes equation in one dimension. The unknown function u is interpreted as velocity of fluid. The Burgers model of turbulence is very important fluid dynamical model for the conceptual understanding of physical flows and for testing various numerical algorithms. It is a nonlinear equation with known analytical solution. The analytical solution can be constructed from a linear partial differential equation. With change of variables, the Burgers equation can be reduced to the heat equation. The exact solution can be used to evaluate the numerical solutions.

GENERALIZED DIFFERENTIAL QUADRATURE METHOD

The differential quadrature approximation of nth derivative of the function $u(x)$ at ith discrete point on a grid is given by a weighted linear sum of the function values at all discrete points (N points) along that direction (direction x) as

$$\frac{d^n u(x)}{dx^n} = \sum_{j=1}^{N} c_{ij}^{(n)} u(x_j) \quad , \tag{3}$$

where $c_{ij}^{(n)}$ are weighting coefficients for needed nth derivative and $u(x_j)$ are function values at grid points $x_j, j = 1, 2, \ldots N$. The goal of the generalized differential quadrature method is to find a simple algebraic expression for calculating the weighting coefficients $c_{ij}^{(n)}$ for arbitrary choice of grid points. The weighting coefficients for mth order derivative are given by recurrence relations in general form as (Shu, Richards, 1992)

$$c_{ij}^{(1)} = \frac{M^{(1)}(x_i)}{(x_i - x_j)M^{(1)}(x_j)}$$
$$\text{for} \quad i \neq j, \quad i,j = 1,2,\ldots N, \tag{4}$$

$$c_{ij}^{(m)} = m\left(c_{ii}^{(m-1)} c_{ij}^{(1)} - \frac{c_{ij}^{(m-1)}}{x_i - x_j}\right)$$
$$\text{for} \quad i \neq j, \quad m = 2,3,\ldots,N-1, \quad i,j = 1,\ldots,N, \tag{5}$$

$$c_{ii}^{(m)} = -\sum_{j=1, j\neq i}^{N} c_{ij}^{(m)}$$
$$\text{for} \quad m = 1,2,\ldots N-1, \quad i = 1,2,\ldots,N, \tag{6}$$

where

$$M(x_i) = \prod_{j=1}^{N} (x - x_j). \tag{7}$$

These recurrence expressions are very useful for implementation in computational programming. There is no need to solve a set of algebraic equations to find the values of the weighting coefficients and we have no restriction on the chosen grid points. According this features we have a great convenience for solving engineering problems by using generalized differential quadrature method. The GDQ method will be applied to one-dimensional Burgers equation.

TIME STEP SOLUTION

Standard Euler method is used for time step solution. The explicit calculation of the values at $t+\Delta t$ is taken as the simplest method for calculation without a set of algebraic equations. Euler method generates approximation simply by following the direction field. The slope at the point (t, x_i^t) is

$$\left(\frac{\partial u}{\partial t}\right)_i^t = f(t, x_i^t) \quad , \tag{8}$$

where the function on the right side of equation is given by

$$f(t, x_i^t) = (\lambda u_{xx} - uu_x)_i^t \quad . \tag{9}$$

The ordinate of the line passing through (t, x_i^t) having this slope at $t + \Delta t$ is

$$u_i^{t+\Delta t} = u_i^t + \Delta t f(t, x_i^t) \quad . \tag{10}$$

The numerical value at time $t + \Delta t$ at each grid point is calculated from the value at according grid point at time t in following form

$$u_i^{t+\Delta t} = u_i^t + \Delta t \left(\lambda u_{xx} - u u_x\right)_i^t$$
$$= u_i^t + \Delta t \left[\lambda \left(u_{xx}\right)_i^t - u_i^t \left(u_x\right)_i^t\right] \quad . \qquad (11)$$

APPLICATION OF GDQ METHOD TO BURGERS EQUATION

Generalized differential quadrature method could be applied to Burgers equation. The governing Burgers equation is

$$u_t + u u_x = \lambda u_{xx} \quad , \qquad (12)$$

with according initial condition

$$u(x,0) = u_0(x) \quad . \qquad (13)$$

All space derivatives, u_x, u_{xx}, needed in governing Burgers equation are expressed in general differential quadrature form as linear combination of function values at chosen grid points with according, by formerly presented recurrence calculated, weighting coefficients $c_{ij}^{(m)}$. At each discrete point on the grid, x_i, at time t , we have now following equation for needed derivatives

$$\left(u_x\right)_i^t = \sum_{j=1}^{N} c_{ij}^{(1)} u_j^t \quad , \qquad (14)$$

$$\left(u_{xx}\right)_i^t = \sum_{j=1}^{N} c_{ij}^{(2)} u_j^t \quad . \qquad (15)$$

At each grid point, x_i, explicit Euler method, equation (11), for time derivative can be expressed, with involved GDQ terms for needed derivatives, in the following form

$$u_i^{t+\Delta t} = u_i^t + \Delta t \left(\lambda \sum_{j=1}^{N} c_{ij}^{(2)} u_j^t - u_i^t \sum_{j=1}^{N} c_{ij}^{(1)} u_j^t\right)$$
$$\text{for} \quad i = 1, 2, \ldots N \quad . \qquad (16)$$

The application of GDQ method to differential operator in explicit Euler method will lead to direct solution of N unknown function values at the grid points at time $t + \Delta t$, $u_i^{t+\Delta t}, i = 1, 2, \ldots N$. After the determination the numerical values for unknown function at the grid points, $u_i^{t+\Delta t}$, it is easy to obtain the values for unknown function at any point x of the domain in terms of polynomial approximation

$$u^{t+\Delta t}(x) = \sum_{i=1}^{N} u_i^{t+\Delta t} p_i(x) \quad , \qquad (17)$$

where $p_i(x), i = 1, 2, \ldots N$ are the Lagrange interpolated polynomials.

NUMERICAL EXAMPLES

The first simple example with analytical solution is taken for numerical computation to show the application of GDQ method to Burgers equation. The governing Burgers equation with according initial condition is

$$u_t = u_{xx} - u u_x \quad \text{for} \quad x \in [0,1]$$
$$u(x,0) = 1 - 2 \tanh x \quad . \qquad (18)$$

The analytical solution of this problem is

$$u(x,t) = 1 - 2 \tanh(x - t) \quad . \qquad (19)$$

Numerical solutions are calculated for this problem with 11 equidistant grid points and for two different time steps, $\Delta = 0.001$ and $\Delta t = 0.01$, (Table 1).

Table 1. NUMERICAL RESULTS FOR THE GIVEN PROBLEM (18)

11 grid points	u(1,0.001)	u(1,0.01)	u(1,0.1)
$\Delta t = 0.001$	-0.5223482	-0.5147204	-0.4128528
$\Delta t = 0.01$		-0.5147876	-0.4222816
exact solution	-0.5223477	-0.5147246	-0.4325957

The evaluated numerical results show very good accuracy of presented generalized differential quadrature method applied to Burgers equation. Extreme error is less than 3%. We can see the influence of the number of time

steps on the quality of numerical solution. There is no need to take very large number of small time steps. For the values after small time interval, at $t = 0.01$, small time step is necessary. But, we have calculated better solution for larger time interval, $t = 0.1$ with larger time step but with much less numerical computation.
In the second example, Burgers equation with given viscosity coefficient, λ, has been calculated. The governing equation with according initial condition is

$$\begin{aligned} u_t &= \lambda u_{xx} - u u_x \quad \text{for} \quad x \in [0,1] \\ u(x,0) &= -2\lambda \frac{\pi \cos \pi x}{\sin \pi x + 1.0} \quad . \end{aligned} \tag{20}$$

The analytical solution of this problem is

$$u(x,t) = -2\lambda \frac{\pi e^{-\lambda \pi^2 t} \cos \pi x}{e^{-\lambda \pi^2 t} \sin \pi x + 1.0} \quad . \tag{21}$$

Numerical solutions are calculated for 11 and 21 equidistant grid points, for two different time steps, $\Delta t = 0.001$ and $\Delta t = 0.01$. The generalized differential quadrature method has been provided for two different viscosity coefficients, $\lambda_1 = 10^{-2}$, (Table 2), and $\lambda_2 = 10^{-4}$, (Table 3).

Table 2. NUMERICAL RESULTS FOR THE GIVEN PROBLEM (20) WITH $\lambda = 10^{-2}$

	u(0,0.1)	u(0,1.0)
11 grid points, $\Delta t = 0.001$	-0.06221002	-0.05685703
11 grid points, $\Delta t = 0.01$	-0.06220981	-0.05685767
21 grid points, $\Delta t = 0.001$	-0.06221474	-0.05692484
21 grid points, $\Delta t = 0.01$	-0.06221446	-0.05692068
exact solution	-0.06221478	-0.05692679

The evaluated results show good accuracy for both viscosity values. The error depends on the value of viscosity. For smaller viscosity, $\lambda_2 = 10^{-4}$, the error is less than for the other value, $\lambda = 10^{-2}$. Even the larger error is much less than 1%. The numerical solution with larger number of grid points has improved the solution former calculated with less grid points, but numerical solutions with less grid points are already very good approximation of the exact values, specially for smaller viscosity.

Table 3. NUMERICAL RESULTS FOR THE GIVEN PROBLEM (20) WITH $\lambda = 10^{-4}$

	u(0,0.1)	u(0,1.0)
11 grid points, $\Delta t = 0.001$	-0.0006282561	-0.0006276947
11 grid points, $\Delta t = 0.01$	-0.0006280066	-0.0006276947
21 grid points, $\Delta t = 0.001$	-0.0006282565	-0.0006276987
21 grid points, $\Delta t = 0.01$	-0.0006282565	-0.0006276987
exact solution	-0.0006282565	-0.0006276987

CONCLUSION

The generalized differential quadrature method is applied to Burgers equation. It has been shown good quality of solution with relatively small computational effort. The needed derivatives are calculated by recurrence relation. Time step solution is calculated by simple Euler method, without a set of linear equation. The influence of time step size and number of grid points on the quality of solution has been shown. It has been shown the influence of the viscosity coefficient on the quality of numerical solution. In further, the presented GDQ method could be applied to the other nonlinear equations, specially to Korteveg-deVries equation or to two-dimensional fluid dynamical problems.

REFERENCES

Bellman, R.E., Casti, J., *Differential Quadrature and Long-term Integration*, J. Math. Anal. Appl., 34, (1971), 235–238.

Shu, C., Richards, B. E., *Application of Generalized Differential Quadrature to Solve Two-dimensional incompressible Navier-Stokes equation*, Int. J. Num. Meth. Fluids, 15, (1992), 791–798.

PVP-Vol. 458, Computer Technology and Applications

PVP2003-1906

A NEW INTERACTION INTEGRAL METHOD FOR ANALYSIS OF CRACKS IN ORTHOTROPIC FUNCTIONALLY GRADED MATERIALS

B. N. Rao and S. Rahman
Center for Computer-Aided Design
The University of Iowa
Iowa City, IA 52242

ABSTRACT

This paper presents two new interaction integrals for calculating stress-intensity factors (SIFs) for a stationary crack in two-dimensional orthotropic functionally graded materials of arbitrary geometry. The method involves the finite element discretization, where the material properties are smooth functions of spatial co-ordinates and two newly developed interaction integrals for mixed-mode fracture analysis. These integrals can also be implemented in conjunction with other numerical methods, such as meshless method, boundary element method, and others. Three numerical examples including both mode-I and mixed-mode problems are presented to evaluate the accuracy of SIFs calculated by the proposed interaction integrals. Comparisons have been made between the SIFs predicted by the proposed interaction integrals and available reference solutions in the literature, generated either analytically or by finite element method using various other fracture integrals or analyses. An excellent agreement is obtained between the results of the proposed interaction integrals and the reference solutions.

INTRODUCTION

In recent years, functionally graded materials (FGMs) have been introduced and applied in the development of structural components subject to non-uniform service requirements. FGMs, which possess continuously varying microstructure and mechanical and/or thermal properties, are essentially two-phase particulate composites, such as ceramic and metal alloy phases, synthesized such that the composition of each constituent changes continuously in one direction, to yield a predetermined composition profile [1]. Even though the initial developmental emphasis of FGMs was to synthesize thermal barrier coating for space applications [2], later investigations uncovered a wide variety of potential applications, including nuclear fast breeder reactors [3], piezoelectric and thermoelectric devices [4-6], graded refractive index materials in audio-video disks [7], thermionic converters [8], dental and medical implants [9], and others [10]. The absence of sharp interfaces in FGM largely reduces material property mismatch, which has been found to improve resistance to interfacial delamination and fatigue crack propagation [11]. However, the microstructure of FGM is generally heterogeneous, and the dominant type of failure in FGM is crack initiation and growth from inclusions. The extent to which constituent material properties and microstructure can be tailored to guard against potential fracture and failure patterns is relatively unknown. Such issues have motivated much of the current research into the numerical computation of crack-driving forces and the simulation of crack growth in FGMs.

Analytical work on functionally graded materials begins as early as 1960 when soil was modeled as a non-homogeneous material by Gibson [12]. Thereafter, extensive research on various aspects of isotropic FGMs fracture under mechanical [13–15] or thermal [16–22] loads has been carried out by various investigators. Crack problems under both mode-I [23,24] and mixed mode [25,26] loading conditions were studied using finite element method (FEM) and J_k^*-integral method. Gu *et al.* [27] presented a simplified method for calculating the crack-tip field of FGMs using the equivalent domain integral technique. Bao and Wang [28] studied multi-cracking in an FGM coating. Bao and Cai [29] studied delamination cracking in a functionally graded ceramic/metal substrate. Kim and Paulino [26] evaluated the mixed-mode fracture parameters in FGMs using FEM analysis with three different approaches: the path-independent J_k^*-integral method, the modified crack-closure integral method, and the displacement correlation technique. In the J_k^*-integral method [16,26], there is need to perform integration along the crack face of the discontinuity (e.g., in calculating J_2^*). Recently, Rao and Rahman [30] developed two new interaction integrals in conjunction with the element-free Galerkin method for mixed-mode fracture analysis in isotropic FGM. However, most analytical methods reviewed above are developed to quantify crack-driving force in isotropic FGMs.

This paper presents two new interaction integrals for calculating the fracture parameters of a stationary crack in orthotropic FGM of arbitrary geometry. This method involves FEM, where the material properties are smooth functions of spatial co-ordinates and two newly developed interaction integrals for mixed-mode fracture analysis. In conjunction with the proposed method, both mode-I and mixed-mode two-dimensional problems have been solved. Three numerical examples are presented to evaluate the accuracy of SIFs calculated by the proposed method. Comparisons have been made between the SIFs predicted by the proposed method and the existing results available in the current literature.

CRACK TIP FIELDS IN ORTHOTROPIC FGM

Consider a plane problem in rectilinear anisotropic elasticity. The basic equations that describe the deformation of anisotropic materials are the same as those for isotropic materials except for the adoption of a generalized Hooke's law. The most general anisotropic form of linear elastic stress-strain relation is given by

$$\varepsilon_{ij} = S_{ijkl}\sigma_{kl} \quad (i,j,k,l = 1,2,3) \tag{1}$$

where σ_{kl} is the stress tensor, ε_{ij} is the strain tensor, and S_{ijkl} is the fourth-order compliance tensor. Due to the symmetry of σ_{kl} and ε_{ij}, 81 independent components of the compliance tensor reduces to 36 independent components. The existence of a strain energy function provides a further reduction in the number of independent components to 21 $(S_{ijkl} = S_{klij})$. In order to represent S_{ijkl} in compact form, a contracted notation a_{ij} is introduced as follows:

$$\varepsilon_i = \sum_{j=1}^{6} a_{ij}\sigma_j \quad i,j = 1,2,\ldots 6, \tag{2}$$

where a_{ij} are compliance coefficients with a_{ij}=a_{ji} and

$$\begin{aligned}&\varepsilon_1 = \varepsilon_{11},\ \varepsilon_2 = \varepsilon_{22},\ \varepsilon_3 = \varepsilon_{33},\ \varepsilon_4 = 2\varepsilon_{23},\ \varepsilon_5 = 2\varepsilon_{13},\ \varepsilon_6 = 2\varepsilon_{12}\\ &\sigma_1 = \sigma_{11},\ \sigma_2 = \sigma_{22},\ \sigma_3 = \sigma_{33},\ \sigma_4 = \sigma_{23},\ \sigma_5 = \sigma_{13},\ \sigma_6 = \sigma_{12}\end{aligned} \quad .(3)$$

At each point through the thickness of transversely isotropic materials there is a plane of material symmetry that runs parallel to the plane of the problem. For this special case, the compliance coefficients in Equation 2 can be reduces to depend upon six independent elastic constants, a_{ij} $i,j = 1,2,6$ for plane stress conditions and $b_{ij} = a_{ij} - \dfrac{a_{i2}a_{j3}}{a_{33}}$ $i,j = 1,2,6$ for plane strain conditions.

Figure 1 shows a crack tip that is referred to the Cartesian coordinate system in orthotropic FGMs. Two dimensional anisotropic elasticity problems can be formulated in terms of the analytic functions $\phi_j(z_j)$ of the complex variable $z_j = x_j + iy_j$ $(j = 1,2)$, where

$$x_j = x + \alpha_j y,\ y_j = \beta_j y \quad (j = 1,2) \tag{4}$$

The parameters α_j and β_j are the real and imaginary parts of $\mu_j = \alpha_j + i\beta_j$, which can be determined from [37]

$$a_{11}\mu^4 - 2a_{16}\mu^3 + (2a_{12} + a_{66})\mu^2 - 2a_{26}\mu + a_{22} = 0 \tag{5}$$

The roots μ_j are always either complex or purely imaginary in conjugate pairs as $\mu_1, \bar{\mu}_1, \mu_2$, and $\bar{\mu}_2$. Hence, the linear-elastic singular stress field near the crack tip can be obtained as [38]

$$\sigma_{11} = \frac{1}{\sqrt{2\pi r}}\left[K_I f_{11}^I(\mu_1,\mu_2,\theta) + K_{II} f_{11}^{II}(\mu_1,\mu_2,\theta)\right], \tag{6}$$

$$\sigma_{22} = \frac{1}{\sqrt{2\pi r}}\left[K_I f_{22}^I(\mu_1,\mu_2,\theta) + K_{II} f_{22}^{II}(\mu_1,\mu_2,\theta)\right], \tag{7}$$

$$\sigma_{12} = \frac{1}{\sqrt{2\pi r}}\left[K_I f_{12}^I(\mu_1,\mu_2,\theta) + K_{II} f_{12}^{II}(\mu_1,\mu_2,\theta)\right], \tag{8}$$

where $f_{ij}^I(\mu_1,\mu_2,\theta)$ and $f_{ij}^{II}(\mu_1,\mu_2,\theta)$ $(i,\ j = 1,\ 2)$ are the standard angular functions for a crack in an orthotropic elastic medium, given by

$$f_{11}^I(\mu_1,\mu_2,\theta) = \mathrm{Re}\left[\frac{\mu_1\mu_2}{\mu_1-\mu_2}\left(\frac{\mu_2}{\sqrt{\cos\theta+\mu_2\sin\theta}} - \frac{\mu_1}{\sqrt{\cos\theta+\mu_1\sin\theta}}\right)\right], \tag{9}$$

$$f_{11}^{II}(\mu_1,\mu_2,\theta) = \mathrm{Re}\left[\frac{1}{\mu_1-\mu_2}\left(\frac{\mu_2^2}{\sqrt{\cos\theta+\mu_2\sin\theta}} - \frac{\mu_1^2}{\sqrt{\cos\theta+\mu_1\sin\theta}}\right)\right], \tag{10}$$

$$f_{22}^I(\mu_1,\mu_2,\theta) = \mathrm{Re}\left[\frac{1}{\mu_1-\mu_2}\left(\frac{\mu_1}{\sqrt{\cos\theta+\mu_2\sin\theta}} - \frac{\mu_2}{\sqrt{\cos\theta+\mu_1\sin\theta}}\right)\right], \tag{11}$$

$$f_{22}^{II}(\mu_1,\mu_2,\theta) = \mathrm{Re}\left[\frac{1}{\mu_1-\mu_2}\left(\frac{1}{\sqrt{\cos\theta+\mu_2\sin\theta}} - \frac{1}{\sqrt{\cos\theta+\mu_1\sin\theta}}\right)\right], \tag{12}$$

$$f_{12}^I(\mu_1,\mu_2,\theta) = \mathrm{Re}\left[\frac{\mu_1\mu_2}{\mu_1-\mu_2}\left(\frac{1}{\sqrt{\cos\theta+\mu_1\sin\theta}} - \frac{1}{\sqrt{\cos\theta+\mu_2\sin\theta}}\right)\right], \tag{13}$$

$$f_{12}^I(\mu_1,\mu_2,\theta) = \mathrm{Re}\left[\frac{1}{\mu_1-\mu_2}\left(\frac{\mu_1}{\sqrt{\cos\theta+\mu_1\sin\theta}} - \frac{\mu_2}{\sqrt{\cos\theta+\mu_2\sin\theta}}\right)\right], \tag{14}$$

The near tip displacement field $\boldsymbol{u} = \{u_1, u_2\}^T$ can be obtained as [38]

$$u_1 = \sqrt{\frac{2r}{\pi}}\left[K_I g_1^I(\mu_1,\mu_2,\theta) + K_{II} g_1^{II}(\mu_1,\mu_2,\theta)\right], \tag{15}$$

and

$$u_2 = \sqrt{\frac{2r}{\pi}}\left[K_I g_2^I(\mu_1,\mu_2,\theta) + K_{II} g_2^{II}(\mu_1,\mu_2,\theta)\right], \tag{16}$$

where $g_i^I(\mu_1,\mu_2,\theta)$ and $g_i^{II}(\mu_1,\mu_2,\theta)$, $i = 1, 2$ are the standard angular functions for a crack in an orthotropic elastic medium, given by

$$g_1^I(\mu_1,\mu_2,\theta) = \mathrm{Re}\left[\frac{\left(\mu_1 p_2\sqrt{\cos\theta+\mu_2\sin\theta}-\mu_2 p_1\sqrt{\cos\theta+\mu_1\sin\theta}\right)}{\mu_1-\mu_2}\right], \quad (17)$$

$$g_1^{II}(\mu_1,\mu_2,\theta) = \mathrm{Re}\left[\frac{\left(p_2\sqrt{\cos\theta+\mu_2\sin\theta}-p_1\sqrt{\cos\theta+\mu_1\sin\theta}\right)}{\mu_1-\mu_2}\right], \quad (18)$$

$$g_2^I(\mu_1,\mu_2,\theta) = \mathrm{Re}\left[\frac{\left(\mu_1 q_2\sqrt{\cos\theta+\mu_2\sin\theta}-\mu_2 q_1\sqrt{\cos\theta+\mu_1\sin\theta}\right)}{\mu_1-\mu_2}\right], \quad (19)$$

$$g_2^{II}(\mu_1,\mu_2,\theta) = \mathrm{Re}\left[\frac{\left(q_2\sqrt{\cos\theta+\mu_2\sin\theta}-q_1\sqrt{\cos\theta+\mu_2\sin\theta}\right)}{\mu_1-\mu_2}\right]. \quad (20)$$

In the above equations, μ_1 and μ_2 denote the crack tip parameters calculated as the roots of Equation 5, which are taken such that $\beta_j > 0$ $(j = 1,2)$, and p_j and q_j are given by

$$p_j = a_{11}\mu_j^2 + a_{12} - a_{16}\mu_j, \quad (21)$$

$$q_j = a_{12}\mu_j + \frac{a_{22}}{\mu_j} - a_{26}. \quad (22)$$

Even though the material gradient does not influence the square-root singularity or the singular stress distribution, the material gradient does affect the SIFs. Hence, the fracture parameters are functions of the material gradients, external loading, and geometry.

THE INTERACTION INTEGRAL METHOD

The interaction integral method is an effective tool for calculating mixed-mode fracture parameters in homogeneous orthotropic materials [39,40]. In this section the interaction integral method for homogeneous orthotropic materials is first briefly summarized, then extended for cracks in orthotropic FGM. In fact, the study of orthotropic FGM would enhance the understanding of a fracture in a generic material, since upon shrinking the gradient layer in FGM is expected to behave like a sharp interface, and upon expansion, the fracture behavior would be analogous to that of an orthotropic homogeneous material.

Homogeneous Materials

The path independent J-integral for a cracked body is given by [41]

$$J = \int_\Gamma \left(W\delta_{1j} - \sigma_{ij}\frac{\partial u_i}{\partial x_1}\right) n_j \, d\Gamma, \quad (23)$$

where $W = \int \sigma_{ij} d\varepsilon_{ij}$ is the strain energy density and n_j is the jth component of the outward unit vector normal to an arbitrary contour Γ enclosing the crack tip. For linear elastic material models it can shown that $W = \sigma_{ij}\varepsilon_{ij}/2 = \varepsilon_{ij}D_{ijkl}\varepsilon_{kl}/2$, where D_{ijkl} is a component of constitutive tensor. Applying the divergence theorem, the contour integral in Equation 23 can be converted into an equivalent domain form, given by [42]

$$J = \int_A \left(\sigma_{ij}\frac{\partial u_i}{\partial x_1} - W\delta_{1j}\right)\frac{\partial q}{\partial x_j} dA + \int_A \frac{\partial}{\partial x_j}\left(\sigma_{ij}\frac{\partial u_i}{\partial x_1} - W\delta_{1j}\right) q dA, \quad (24)$$

where A is the area inside the contour and q is a weight function chosen such that it has a value of *unity* at the crack tip, *zero* along the boundary of the domain, and arbitrary elsewhere. By expanding the second integrand, Equation 24 leads to

$$J = \int_A \left(\sigma_{ij}\frac{\partial u_i}{\partial x_1} - W\delta_{1j}\right)\frac{\partial q}{\partial x_j} dA + \int_A \left(\frac{\partial \sigma_{ij}}{\partial x_j}\frac{\partial u_i}{\partial x_1} + \sigma_{ij}\frac{\partial^2 u_i}{\partial x_j \partial x_1} - \sigma_{ij}\frac{\partial \varepsilon_{ij}}{\partial x_1} - \frac{1}{2}\varepsilon_{ij}\frac{\partial D_{ijkl}}{\partial x_1}\varepsilon_{kl}\right) q dA \quad (25)$$

Using equilibrium $(\partial\sigma_{ij}/\partial x_j = 0)$ and strain-displacement $(\varepsilon_{ij} = \partial u_i/\partial x_j)$ conditions and noting that $\partial D_{ijkl}/\partial x_1 = 0$ in homogenous orthotropic materials, the second integrand of Equation 25 vanishes, yielding

$$J = \int_A \left(\sigma_{ij}\frac{\partial u_i}{\partial x_1} - W\delta_{1j}\right)\frac{\partial q}{\partial x_j} dA, \quad (26)$$

which is the classical domain form of the J-integral in homogenous orthotropic materials.

Consider two independent equilibrium states of the cracked body. Let state 1 correspond to the *actual* state for the given boundary conditions, and let state 2 correspond to an *auxiliary* state, which can be either mode-I or mode-II near tip displacement and stress fields in an orthotropic elastic medium. Superposition of these two states leads to another equilibrium state (state S) for which the domain form of the J-integral is

$$J^{(S)} = \int_A \left[\left(\sigma_{ij}^{(1)} + \sigma_{ij}^{(2)}\right)\frac{\partial\left(u_i^{(1)} + u_i^{(2)}\right)}{\partial x_1} - W^{(S)}\delta_{1j}\right]\frac{\partial q}{\partial x_j} dA, \quad (27)$$

where superscript i = 1, 2, and S indicate fields and quantities associated with state i and

$$W^{(S)} = \frac{1}{2}\left(\sigma_{ij}^{(1)} + \sigma_{ij}^{(2)}\right)\left(\varepsilon_{ij}^{(1)} + \varepsilon_{ij}^{(2)}\right). \quad (28)$$

By expanding Equations 27,

$$J^{(S)} = J^{(1)} + J^{(2)} + M^{(1,2)}, \quad (29)$$

where

$$J^{(1)} = \int_A \left[\sigma_{ij}^{(1)}\frac{\partial u_i^{(1)}}{\partial x_1} - W^{(1)}\delta_{1j}\right]\frac{\partial q}{\partial x_j} dA, \quad (30)$$

and

$$J^{(2)} = \int_A \left[\sigma_{ij}^{(2)}\frac{\partial u_i^{(2)}}{\partial x_1} - W^{(2)}\delta_{1j}\right]\frac{\partial q}{\partial x_j} dA \quad (31)$$

are the J-integrals for states 1 and 2, respectively, and

$$M^{(1,2)} = \int_A \left[\sigma_{ij}^{(1)} \frac{\partial u_i^{(2)}}{\partial x_1} + \sigma_{ij}^{(2)} \frac{\partial u_i^{(1)}}{\partial x_1} - W^{(1,2)} \delta_{1j} \right] \frac{\partial q}{\partial x_j} dA, \quad (32)$$

is an interaction integral. In Equations 30-32, $W^{(1)} = \frac{1}{2}\sigma_{ij}^{(1)}\varepsilon_{ij}^{(1)}$, $W^{(2)} = \frac{1}{2}\sigma_{ij}^{(2)}\varepsilon_{ij}^{(2)}$, and $W^{(1,2)} = \frac{1}{2}\left(\sigma_{ij}^{(1)}\varepsilon_{ij}^{(2)} + \sigma_{ij}^{(2)}\varepsilon_{ij}^{(1)}\right)$ represent various strain energy densities, which satisfy

$$W^{(S)} = W^{(1)} + W^{(2)} + W^{(1,2)}. \quad (33)$$

For linear-elastic homogeneous orthotropic solids under mixed-mode loading conditions, the J-integral is also related to the stress intensity factors as

$$J = \alpha_{11} K_I^2 + \alpha_{12} K_I K_{II} + \alpha_{22} K_{II}^2, \quad (34)$$

where

$$\alpha_{11} = -\frac{a_{22}}{2} \mathrm{Im}\left(\frac{\mu_1 + \mu_2}{\mu_1 \mu_2}\right), \quad (35)$$

$$\alpha_{22} = \frac{a_{11}}{2} \mathrm{Im}(\mu_1 + \mu_2), \quad (36)$$

and

$$\alpha_{12} = -\frac{a_{22}}{2} \mathrm{Im}\left(\frac{1}{\mu_1 \mu_2}\right) + \frac{a_{11}}{2} \mathrm{Im}(\mu_1 \mu_2). \quad (37)$$

Applying Equation 34 to states 1, 2, and the superimposed state S gives

$$J^{(1)} = \alpha_{11} K_I^{(1)2} + \alpha_{12} K_I^{(1)} K_{II}^{(1)} + \alpha_{22} K_{II}^{(1)2}, \quad (38)$$

$$J^{(2)} = \alpha_{11} K_I^{(2)2} + \alpha_{12} K_I^{(2)} K_{II}^{(2)} + \alpha_{22} K_{II}^{(2)2}, \quad (39)$$

and

$$\begin{aligned} J^{(S)} &= \alpha_{11}\left(K_I^{(1)} + K_I^{(2)}\right)^2 + \alpha_{12}\left(K_I^{(1)} + K_I^{(2)}\right)\left(K_{II}^{(1)} + K_{II}^{(2)}\right) + \alpha_{22}\left(K_{II}^{(1)} + K_{II}^{(2)}\right)^2 \\ &= \alpha_{11} K_I^{(1)2} + \alpha_{12} K_I^{(1)} K_{II}^{(1)} + \alpha_{22} K_{II}^{(1)2} + \alpha_{11} K_I^{(2)2} + \alpha_{12} K_I^{(2)} K_{II}^{(2)} + \alpha_{22} K_{II}^{(2)2} \\ &\quad + 2\alpha_{11} K_I^{(1)} K_I^{(2)} + \alpha_{12}\left(K_I^{(1)} K_{II}^{(2)} + K_I^{(2)} K_{II}^{(1)}\right) + 2\alpha_{22} K_{II}^{(1)} K_{II}^{(2)} \\ &= J^{(1)} + J^{(2)} + 2\alpha_{11} K_I^{(1)} K_I^{(2)} + \alpha_{12}\left(K_I^{(1)} K_{II}^{(2)} + K_I^{(2)} K_{II}^{(1)}\right) + 2\alpha_{22} K_{II}^{(1)} K_{II}^{(2)} \end{aligned} . \quad (40)$$

Comparing Equations 29 and 40,

$$M^{(1,2)} = 2\alpha_{11} K_I^{(1)} K_I^{(2)} + \alpha_{12}\left(K_I^{(1)} K_{II}^{(2)} + K_I^{(2)} K_{II}^{(1)}\right) + 2\alpha_{22} K_{II}^{(1)} K_{II}^{(2)}. \quad (41)$$

The individual SIFs for the actual state can obtained by judiciously choosing the auxiliary state (state 2). For example, if state 2 is chosen to be state I, *i.e.*, the mode-I near tip displacement and stress field is chosen as the auxiliary state, then $K_I^{(2)}$ = 1 and $K_{II}^{(2)}$ = 0. Hence, Equation 41 can be reduced to

$$M^{(1,I)} = 2\alpha_{11} K_I^{(1)} + \alpha_{12} K_{II}^{(1)}, \quad (42)$$

Similarly, if state 2 is chosen to be state II, *i.e.*, the mode-II near tip displacement and stress field is chosen as the auxiliary state, then $K_I^{(2)} = 0$ and $K_{II}^{(2)}$ = 1. Following similar considerations,

$$M^{(1,II)} = \alpha_{12} K_I^{(1)} + 2\alpha_{22} K_{II}^{(1)}, \quad (43)$$

The interaction integrals $M^{(1,I)}$ and $M^{(1,II)}$ can be evaluated from Equation 32. Equations 42 and 43 provide a system of linear algebraic equations which can be solved for SIFs, $K_I^{(1)}$ and $K_{II}^{(1)}$ under various mixed-mode loading conditions.

Functionally Graded Materials

For non-homogeneous materials, even though the equilibrium and strain-displacement conditions are satisfied, the material gradient term of the second integrand of Equation 25 does not vanish. So Equation 25 leads to a more general integral, henceforth referred to as the $\tilde{J}$-integral [27], which is

$$\tilde{J} = \int_A \left(\sigma_{ij} \frac{\partial u_i}{\partial x_1} - W \delta_{1j} \right) \frac{\partial q}{\partial x_j} dA - \int_A \frac{1}{2} \varepsilon_{ij} \frac{\partial D_{ijkl}}{\partial x_1} \varepsilon_{kl} q dA. \quad (44)$$

By comparing Equation 44 to the classical J-integral (see Equation 26), the presence of material non-homogeneity results in the addition of the second domain integral. Although this integral is negligible for a path very close to the crack tip, it must be accounted for with relatively large integral domains, so that the $\tilde{J}$-integral can be accurately calculated. The $\tilde{J}$-integral in Equation 44 is actually the first component of the $J^* = \{J_1^*, J_2^*\}^T$ vector integral (i.e., J_1^*) proposed by Eischen [25]. Hence, $\tilde{J}$ also represents the energy release rate of an elastic body.

In order to derive interaction integral for orthotropic FGMs, consider again actual (state 1), auxiliary (state 2), and superimposed (state S) equilibrium states. For the actual state, Equation 44 can be directly invoked to represent the $\tilde{J}$-integral. However, a more general form, such as Equation 24, must be used for auxiliary and superimposed states. For example, the $\tilde{J}$-integral for the superimposed state S can written as

$$\begin{aligned} \tilde{J}^{(S)} &= \int_A \left(\left(\sigma_{ij}^{(1)} + \sigma_{ij}^{(2)}\right) \frac{\partial\left(u_i^{(1)} + u_i^{(2)}\right)}{\partial x_1} - W^{(S)} \delta_{1j} \right) \frac{\partial q}{\partial x_j} dA \\ &+ \int_A \frac{\partial}{\partial x_j} \left(\left(\sigma_{ij}^{(1)} + \sigma_{ij}^{(2)}\right) \frac{\partial\left(u_i^{(1)} + u_i^{(2)}\right)}{\partial x_1} - W^{(S)} \delta_{1j} \right) q dA \end{aligned}, \quad (45)$$

Clearly, the evaluations of $\tilde{J}^{(S)}$ and the resulting interaction integral depend on how the auxiliary field is defined. There are several options in choosing the auxiliary field. Two methods, developed in this study, are described in the following.

Method I – Homogeneous Auxiliary Field

The method I involves selecting the auxiliary stress and displacement fields in an orthotropic elastic medium given by Equations 6-8 and Equations 15-16 and calculating the auxiliary strain field from the symmetric gradient of the auxiliary displacement field. In this approach, the auxiliary stress and strain fields are related through a constant constitutive tensor evaluated at the crack tip. Hence, both equilibrium ($\partial\sigma_{ij}^{(2)}/\partial x_j = 0$) and strain-displacement ($\varepsilon_{ij}^{(2)} = \partial u_i^{(2)}/\partial x_j$) conditions are satisfied in the auxiliary state. However, the non-homogeneous constitutive relation of FGM is not strictly satisfied in the auxiliary state, which would introduce gradients of stress fields as extra terms in the interaction integral.

Using Equation 33 and invoking both equilibrium and strain-displacement conditions, Equation 45 can be further simplified to

$$\tilde{J}^{(S)} = \int_A \left(\left(\sigma_{ij}^{(1)} + \sigma_{ij}^{(2)}\right) \frac{\partial\left(u_i^{(1)} + u_i^{(2)}\right)}{\partial x_1} - \left(W^{(1)} + W^{(2)} + W^{(1,2)}\right)\delta_{1j} \right) \frac{\partial q}{\partial x_j}\, dA$$
$$+ \int_A \frac{1}{2}\left[-\varepsilon_{ij}^{(1)} \frac{\partial D_{ijkl}}{\partial x_1} \varepsilon_{kl}^{(1)} + \sigma_{ij}^{(1)} \frac{\partial \varepsilon_{ij}^{(2)}}{\partial x_1} - \frac{\partial \sigma_{ij}^{(2)}}{\partial x_1} \varepsilon_{ij}^{(1)} + \sigma_{ij}^{(2)} \frac{\partial \varepsilon_{ij}^{(1)}}{\partial x_1} - \frac{\partial \sigma_{ij}^{(1)}}{\partial x_1} \varepsilon_{ij}^{(2)} \right] q\, dA \quad .(46)$$

By expanding Equation 46,

$$\tilde{J}^{(S)} = \tilde{J}^{(1)} + \tilde{J}^{(2)} + \tilde{M}^{(1,2)}, \quad (47)$$

where

$$\tilde{J}^{(1)} = \int_A \left[\sigma_{ij}^{(1)} \frac{\partial u_i^{(1)}}{\partial x_1} - W^{(1)}\delta_{1j} \right] \frac{\partial q}{\partial x_j}\, dA - \int_A \frac{1}{2} \varepsilon_{ij}^{(1)} \frac{\partial D_{ijkl}}{\partial x_1} \varepsilon_{kl}^{(1)} q\, dA \quad ,(48)$$

$$\tilde{J}^{(2)} = \int_A \left[\sigma_{ij}^{(2)} \frac{\partial u_i^{(2)}}{\partial x_1} - W^{(2)}\delta_{1j} \right] \frac{\partial q}{\partial x_j}\, dA, \quad (49)$$

are the $\tilde{J}$-integrals for states 1 and 2, respectively, and

$$\tilde{M}^{(1,2)} = \int_A \left[\sigma_{ij}^{(1)} \frac{\partial u_i^{(2)}}{\partial x_1} + \sigma_{ij}^{(2)} \frac{\partial u_i^{(1)}}{\partial x_1} - W^{(1,2)}\delta_{1j} \right] \frac{\partial q}{\partial x_j}\, dA$$
$$+ \int_A \frac{1}{2}\left[\sigma_{ij}^{(1)} \frac{\partial \varepsilon_{ij}^{(2)}}{\partial x_1} - \frac{\partial \sigma_{ij}^{(2)}}{\partial x_1} \varepsilon_{ij}^{(1)} + \sigma_{ij}^{(2)} \frac{\partial \varepsilon_{ij}^{(1)}}{\partial x_1} - \frac{\partial \sigma_{ij}^{(1)}}{\partial x_1} \varepsilon_{ij}^{(2)} \right] q\, dA \quad ,(50)$$

is the modified interaction integral for non-homogeneous materials.

Method II – Non-homogeneous Auxiliary Field

The method II entails selecting the auxiliary stress and displacement fields in an orthotropic elastic medium given by Equations 6-8 and Equations 15-16 and calculating the auxiliary strain field using the same spatially varying constitutive tensor of FGM. In this approach, the auxiliary stress field satisfies equilibrium ($\partial \sigma_{ij}^{(2)}/\partial x_j = 0$); however, the auxiliary strain field is not compatible with the auxiliary displacement field ($\varepsilon_{ij}^{(2)} \neq \partial u_i^{(2)}/\partial x_j$). If the auxiliary fields are not compatible, extra terms that will arise due to lack of compatibility should be taken into account while evaluating the interaction integral, even though they may not be sufficiently singular in the asymptotic limit to contribute to the value of the integral [30]. Hence, this method also introduces additional terms to the resulting interaction integral.

Following similar considerations, but using only equilibrium condition in the auxiliary state, Equation 45 can also be simplified to

$$\tilde{J}^{(S)} = \int_A \left(\left(\sigma_{ij}^{(1)} + \sigma_{ij}^{(2)}\right) \frac{\partial\left(u_i^{(1)} + u_i^{(2)}\right)}{\partial x_1} - \left(W^{(1)} + W^{(2)} + W^{(1,2)}\right)\delta_{1j} \right) \frac{\partial q}{\partial x_j}\, dA$$
$$+ \int_A \left(\left(\sigma_{ij}^{(1)} + \sigma_{ij}^{(2)}\right) \left(\frac{\partial^2 u_i^{(2)}}{\partial x_j \partial x_1} - \frac{\partial \varepsilon_{ij}^{(2)}}{\partial x_1} \right) - \frac{1}{2}\left(\varepsilon_{ij}^{(1)} + \varepsilon_{ij}^{(2)}\right) \frac{\partial D_{ijkl}}{\partial x_1} \left(\varepsilon_{kl}^{(1)} + \varepsilon_{kl}^{(2)}\right) \right) q\, dA \quad .(51)$$

Comparing Equations 47 and 51,

$$\tilde{J}^{(1)} = \int_A \left[\sigma_{ij}^{(1)} \frac{\partial u_i^{(1)}}{\partial x_1} - W^{(1)}\delta_{1j} \right] \frac{\partial q}{\partial x_j}\, dA - \int_A \frac{1}{2} \varepsilon_{ij}^{(1)} \frac{\partial D_{ijkl}}{\partial x_1} \varepsilon_{kl}^{(1)} q\, dA \quad (52)$$

$$\tilde{J}^{(2)} = \int_A \left[\sigma_{ij}^{(2)} \frac{\partial u_i^{(2)}}{\partial x_1} - W^{(2)}\delta_{1j} \right] \frac{\partial q}{\partial x_j}\, dA$$
$$+ \int_A \left[\sigma_{ij}^{(2)} \left(\frac{\partial^2 u_i^{(2)}}{\partial x_j \partial x_1} - \frac{\partial \varepsilon_{ij}^{(2)}}{\partial x_1} \right) - \frac{1}{2} \varepsilon_{ij}^{(2)} \frac{\partial D_{ijkl}}{\partial x_1} \varepsilon_{kl}^{(2)} \right] q\, dA \quad (53)$$

are the $\tilde{J}$-integrals for states 1 and 2, respectively, and

$$\tilde{M}^{(1,2)} = \int_A \left[\sigma_{ij}^{(1)} \frac{\partial u_i^{(2)}}{\partial x_1} + \sigma_{ij}^{(2)} \frac{\partial u_i^{(1)}}{\partial x_1} - W^{(1,2)}\delta_{1j} \right] \frac{\partial q}{\partial x_j}\, dA$$
$$+ \int_A \left[\sigma_{ij}^{(1)} \left(\frac{\partial^2 u_i^{(2)}}{\partial x_j \partial x_1} - \frac{\partial \varepsilon_{ij}^{(2)}}{\partial x_1} \right) - \varepsilon_{ij}^{(1)} \frac{\partial D_{ijkl}}{\partial x_1} \varepsilon_{kl}^{(2)} \right] q\, dA \quad (54)$$

is another modified interaction integral for non-homogeneous materials.

Note, for homogeneous materials, $\partial D_{ijkl}/\partial x_1 = 0$, $\varepsilon_{ij}^{(2)} = \partial u_i^{(2)}/\partial x_j$, $\sigma_{ij}^{(1)} \partial \varepsilon_{ij}^{(2)}/\partial x_1 = \partial \sigma_{ij}^{(2)}/\partial x_1\, \varepsilon_{ij}^{(1)}$ and $\sigma_{ij}^{(2)} \partial \varepsilon_{ij}^{(1)}/\partial x_1 = \partial \sigma_{ij}^{(1)}/\partial x_1\, \varepsilon_{ij}^{(2)}$, regardless of how the auxiliary field is defined. As a result, the $\tilde{J}^{(1)}$, $\tilde{J}^{(2)}$, and $\tilde{M}^{(1,2)}$ integrals in methods I and II degenerate to their corresponding homogeneous solutions, as expected.

Stress-Intensity Factors

For linear-elastic solids, the $\tilde{J}$-integral also represents the energy release rate and, hence,

$$\tilde{J} = \alpha_{11,tip} K_I^2 + \alpha_{12,tip} K_I K_{II} + \alpha_{22,tip} K_{II}^2, \quad (55)$$

where $\alpha_{11,tip}$, $\alpha_{12,tip}$, and $\alpha_{22,tip}$ are evaluated at the crack tip. Regardless of how the auxiliary fields are defined, Equation 55 applied to states 1, 2, and S yields

$$\tilde{J}^{(1)} = \alpha_{11,tip} K_I^{(1)2} + \alpha_{12,tip} K_I^{(1)} K_{II}^{(1)} + \alpha_{22,tip} K_{II}^{(1)2}, \quad (56)$$

$$\tilde{J}^{(2)} = \alpha_{11,tip} K_I^{(2)2} + \alpha_{12,tip} K_I^{(2)} K_{II}^{(2)} + \alpha_{22,tip} K_{II}^{(2)2}, \quad (57)$$

and

$$
\begin{aligned}
\bar{J}^{(S)} &= \alpha_{11,tip}\left(K_I^{(1)}+K_I^{(2)}\right)^2+\alpha_{12,tip}\left(K_I^{(1)}+K_I^{(2)}\right)\left(K_{II}^{(1)}+K_{II}^{(2)}\right) \\
&\quad +\alpha_{22,tip}\left(K_{II}^{(1)}+K_{II}^{(2)}\right)^2 \\
&= \alpha_{11,tip}K_I^{(1)2}+\alpha_{12,tip}K_I^{(1)}K_{II}^{(1)}+\alpha_{22,tip}K_{II}^{(1)2}+\alpha_{11,tip}K_I^{(2)2}+\alpha_{12,tip}K_I^{(2)}K_{II}^{(2)} \\
&\quad +\alpha_{22,tip}K_{II}^{(2)2}+2\alpha_{11,tip}K_I^{(1)}K_I^{(2)}+\alpha_{12,tip}\left(K_I^{(1)}K_{II}^{(2)}+K_I^{(2)}K_{II}^{(1)}\right) \\
&\quad +2\alpha_{22,tip}K_{II}^{(1)}K_{II}^{(2)} \\
&= J^{(1)}+J^{(2)}+2\alpha_{11,tip}K_I^{(1)}K_I^{(2)}+\alpha_{12,tip}\left(K_I^{(1)}K_{II}^{(2)}+K_I^{(2)}K_{II}^{(1)}\right) \\
&\quad +2\alpha_{22,tip}K_{II}^{(1)}K_{II}^{(2)}
\end{aligned}
\tag{58}
$$

Comparing Equations 47 and 58,

$$
\tilde{M}^{(1,2)} = 2\alpha_{11,tip}K_I^{(1)}K_I^{(2)}+\alpha_{12,tip}\left(K_I^{(1)}K_{II}^{(2)}+K_I^{(2)}K_{II}^{(1)}\right)+2\alpha_{22,tip}K_{II}^{(1)}K_{II}^{(2)} \tag{59}
$$

Following a similar procedure and judiciously choosing the intensity of the auxiliary state as described earlier, the SIFs for non-homogenous materials can also be derived as

$$
\tilde{M}^{(1,I)} = 2\alpha_{11,tip}K_I^{(1)}+\alpha_{12,tip}K_{II}^{(1)}, \tag{60}
$$

and

$$
\tilde{M}^{(1,II)} = \alpha_{12,tip}K_I^{(1)}+2\alpha_{22,tip}K_{II}^{(1)}, \tag{61}
$$

where $\tilde{M}^{(1,I)}$ and $\tilde{M}^{(1,II)}$ are two modified interaction integrals for modes I and II, respectively. The interaction integrals $\tilde{M}^{(1,I)}$ and $\tilde{M}^{(1,II)}$ can be evaluated using either Equation 50 or Equation 54. Equations 60 and 61 provide a system of linear algebraic equations which can be solved for SIFs, $K_I^{(1)}$ and $K_{II}^{(1)}$ under various mixed-mode loading conditions. Both methods developed in this study were used in performing numerical calculations, to be presented in a forthcoming section.

Note, Equations 60 and 61 are the result of a simple generalization of the interaction integral method for calculating fracture parameters in linear-elastic non-homogenous materials. When both the elastic modulus and the Poisson's ratio have no spatial variation, $\tilde{M}^{(1,2)} = M^{(1,2)}$. Consequently, Equations 60 and 61 degenerate into Equations 42 and 43, as expected.

NUMERICAL EXAMPLES

The newly modified interaction integrals (methods I and II) developed in this study were applied to evaluate the SIFs of cracks in othrotropic FGMs. Both single- (mode I) and mixed-mode (modes I and II) conditions were considered and three examples are presented here. The results obtained in the current study were compared with the semi-analytical solutions by Ozturk and Erdogan [34,35]. For the sake of comparison the independent engineering constants, $E_{11}, E_{22}, G_{12}, \nu_{12}$, and ν_{21} are replaced by a stiffness parameter E, a stiffness ratio δ^4, a Poisson's ratio ν, and a shear parameter κ [34,35], which are defined as

$$
E=\sqrt{E_{11}E_{22}}, \quad \delta^4=\frac{E_{11}}{E_{22}}=\frac{\nu_{12}}{\nu_{21}}, \quad \nu=\sqrt{\nu_{12}\nu_{21}}, \quad \kappa=\frac{E}{2G_{12}}-\nu, \tag{62}
$$

for plane stress, and

$$
\begin{aligned}
E &= \sqrt{\frac{E_{11}E_{22}}{(1-\nu_{13}\nu_{31})(1-\nu_{23}\nu_{32})}}, \quad \delta^4=\frac{E_{11}}{E_{22}}\frac{(1-\nu_{23}\nu_{32})}{(1-\nu_{13}\nu_{31})}, \\
\nu &= \sqrt{\frac{(\nu_{12}+\nu_{13}\nu_{32})(\nu_{21}+\nu_{23}\nu_{31})}{(1-\nu_{13}\nu_{31})(1-\nu_{23}\nu_{32})}}, \quad \kappa=\frac{E}{2G_{12}}-\nu
\end{aligned}
\tag{63}
$$

for plane strain.

Example 1: Plate with an Interior Crack Parallel to Material Gradation under Mode-I

Consider an orthotropic square plate of dimensions $2L = 2W = 20$ *units* ($L/W = 1$) with a central crack of length $2a = 0.2$ *units*, subjected to crack-face pressure loading, as shown in Figure 2(a). A plane stress condition was assumed. The crack is parallel to the material gradation and the following material property data were employed for FEM analysis: $E_{11}(x_1) = E_{11}^0 e^{\beta x_1}$, $E_{22}(x_1) = E_{22}^0 e^{\beta x_1}$, $G_{12}(x_1) = G_{12}^0 e^{\beta x_1}$, where the average modulus of elasticity $E(x_1) = E^0 e^{\beta a\left(\frac{x_1}{a}\right)}$, with $E^0 = \sqrt{E_{11}^0 E_{22}^0}$. The non-homogeneity parameter βa is varied from 0.0 to 1.0. Two different values of the shear parameter $\kappa = -0.25$ and 5.0 were employed. The stiffness ratio $\delta^4 = 0.25$ and the average Poisson's ratio $\nu = 0.3$ were used in the FEM analysis.

The applied load corresponds to $\sigma_{22}(-1 \le x_1 \le 1, \pm 0) = \pm\sigma_0 = \pm 1.0$ along top and bottom crack faces. The displacement boundary condition is prescribed such that $u_1 = u_2 = 0$ for the node in the middle of the left edge, and $u_2 = 0$ for the node in the middle of the right edge. The FEM discretization involves 5756 nodes, 1696 8-noded quadrilateral elements, 214 6-noded triangular elements, and 8 focused quarter-point 6-noded triangular elements in the vicinity of each crack tip, as shown in Figure 2(b). Figure 2(c) depicts the enlarged view of discretization around the crack tips. A 2 × 2 Gaussian integration was employed.

Ozturk and Erdogan [34] investigated an infinite plate with the same configuration. Obviously, a FEM model cannot represent the infinite domains addressed in the analysis of Ozturk and Erdogan [34], but as long as the ratios a/W and a/L are kept relatively small (e.g. $a/W = a/L \le 1/10$), the approximation is acceptable. Tables 1 and 2 provide a comparison between predicted normalized stress intensity factors $K_I(a)/\sigma_0\sqrt{\pi a}$ and $K_I(-a)/\sigma_0\sqrt{\pi a}$ at both crack tips for several values of non-homogeneous parameter βa, obtained by proposed interaction integral methods I and II, and those of Ozturk and Erdogan [34] for shear parameters $\kappa = -0.25$ and 5.0, respectively. The numerical results from proposed methods I and II shows that effect of κ on normalized stress intensity factors is less significant than that of βa. Also the stress intensity factor on the stiffer side of the medium is always greater than that on the less stiff side. The agreement between the present results of proposed methods and Ozturk and Erdogan's [34] analytical solution is excellent, regardless of methods I and II.

Example 2: Plate with an Interior Crack Perpendicular to Material Gradation (Mixed Mode)

Consider an orthotropic square plate of dimensions $2L = 2W = 20$ *units* ($L/W = 1$) with a central crack of length $2a = 0.2$ *units*, as shown in Figure 3. Except for material properties and loading conditions, all other conditions including FEM discretization were same as in Example 1. However, the crack is perpendicular to the material gradation. In Example 2, both crack-face pressure loading and crack-face shear loading were considered separately. The following material property data were employed for FEM analysis: $E_{11}(x_2) = E_{11}^0 e^{\beta x_2}$, $E_{22}(x_2) = E_{22}^0 e^{\beta x_2}$, $G_{12}(x_2) = G_{12}^0 e^{\beta x_2}$, where the average modulus of elasticity $E(x_2) = E^0 e^{\beta a\left(\frac{x_2}{a}\right)}$, with $E^0 = \sqrt{E_{11}^0 E_{22}^0}$. The non-homogeneity parameter βa is varied from 0.0 to 2.0. Two different values of the shear parameter $\kappa = 0.5$ and 5.0 and two different stiffness ratio δ^4 = 0.25, and 10 were employed in the FEM analysis. Three different values of the average Poisson's ratio $\nu = 0.15, 0.30$, and 0.45 were used in the computations.

For crack-face pressure loading the applied load corresponds to $\sigma_{22}(-1 \le x_1 \le 1, \pm 0) = \pm\sigma_0 = \pm 1.0$ and for crack-face shear loading the applied load corresponds to $\sigma_{12}(-1 \le x_1 \le 1, \pm 0) = \pm\tau_0 = \pm 1.0$ along top and bottom crack faces. A 2 x 2 Gaussian integration was adopted.

The effect of the non-homogeneity parameter βa and effect of the average Poisson's ratio ν on the normalized stress intensity factors for both crack-face pressure loading and crack-face shear loading were studied. The present results obtained by proposed interaction integral methods I and II are compared with those reported by Ozturk and Erdogan [35], who investigated an infinite plate with the same configuration. Tables 3 and 4 provide a comparison between predicted normalized stress intensity factors $K_I(a)/\sigma_0\sqrt{\pi a}$ and $K_I(-a)/\sigma_0\sqrt{\pi a}$ under uniform crack-face pressure loading and uniform crack-face shear loading, respectively, obtained by proposed interaction integral methods I and II, and those of Ozturk and Erdogan [35] for various values of non-homogeneity parameter βa, and $\delta^4 = 0.25$; ν=0.3; and $\kappa = 0.5$. Tables 5 and 6 provide a comparison between predicted normalized stress intensity factors under uniform crack face pressure loading for $\delta^4 = 0.25$, and $\delta^4 = 10$ respectively, obtained by proposed interaction integral methods I and II, and those of Ozturk and Erdogan [35] for ν = 0.15, 0.3, and 0.45 and βa=0.5, and 1.0. Tables 7 and 8 provide a similar comparison for an orthotropic plate under uniform crack face shear loading. The agreement between the results of the proposed methods and Ozturk and Erdogan's [35] analytical solution is excellent, irrespective of methods I and II.

<u>**Example 3: Slanted Crack in a Plate under Mixed-Mode**</u>

Consider a centrally located, inclined crack of length $2a$ = $2\sqrt{2}$ units in a finite two-dimensional orthotropic rectangular plate of size $2L$ = 40 units and $2W$ = 20 units, as shown in Figure 4(a). A plane stress condition was assumed. This example is investigated with material variation following exponential functions and the following material property data were used for FEM analysis: $E_{11}(x_1) = E_{11}^0 e^{\alpha x_1}$, $E_{22}(x_1) = E_{22}^0 e^{\beta x_1}$, $G_{12}(x_1) = G_{12}^0 e^{\gamma x_1}$, $E_{11}^0 = 3.5 \times 10^6$, $E_{22}^0 = 12 \times 10^6$, $G_{12}^0 = 3 \times 10^6$, $\nu_{12} = 0.204$, and $E_{11}/E_{22} = \nu_{12}/\nu_{21}$. The following two cases were examined: (1) an orthotropic FGM with proportional material variation $(\alpha, \beta, \gamma) = (0.2, 0.2, 0.2)$, and (2) an orthotropic FGM with non-proportional material variation $(\alpha, \beta, \gamma) = (0.5, 0.4, 0.3)$.

The applied load corresponds to $\sigma_{22}(-10 \le x_1 \le 10, \pm 20) = \pm\sigma^{\infty} = \pm 1.0$ along top and bottom edges. The displacement boundary condition is prescribed such that $u_1 = u_2 = 0$ for the node in the middle of the left edge, and $u_2 = 0$ for the node in the middle of the right edge. The FEM discretization involves 3606 Nodes, 898 8-noded quadrilateral elements, 354 6-noded triangular elements, and 8 focused quarter-point 6-noded triangular elements in the vicinity of each crack tip, as shown in Figure 4(b). Figure 4(c) depicts the enlarged view of discretization around the crack tips.

Table 9 provides a comparison between predicted SIFs $K_I^+(a)$, $K_{II}^+(a)$, $K_I^-(a)$, and $K_{II}^-(a)$ at both crack tips, obtained by proposed interaction integral methods I and II, and existing results obtained by Kim and Paulino from the literature [36]. A good agreement is obtained between the results of the proposed methods and Kim and Paulino's [36] numerical results by using the MCC and DCT.

SUMMARY AND CONCLUSIONS

Two new interaction integrals have been developed for calculating stress-intensity factors for a stationary crack in two-dimensional orthotropic functionally graded materials of arbitrary geometry. The method involves finite element discretization, where the material properties are smooth functions of spatial co-ordinates and two newly developed interaction integrals for mixed-mode fracture analysis. The proposed interaction integral can also be implemented in conjunction with other numerical methods, such as meshless method, boundary element method, and other. Three numerical examples, including both mode-I and mixed-mode problems, are presented to evaluate the accuracy of fracture parameters calculated by the proposed interaction integrals. Comparisons have been made between the stress-intensity factors predicted by the proposed interaction integrals and available reference solutions in the literature, generated either analytically or numerically using various other fracture integrals or analyses. An excellent agreement is obtained between the results of proposed interaction integrals and the previously obtained solutions.

ACKNOWLEDGMENTS

The authors would like to acknowledge the financial support of the U.S. National Science Foundation (NSF) under Award No. CMS-9900196. The NSF program director was Dr. Ken Chong.

REFERENCES

1. Suresh, S. and Mortensen, A., "Fundamentals of Functionally Graded Materials," *IOM Communications Ltd.*, London, 1998.

2. Hirano, T., Teraki, J., and Yamada, T., "On the Design of Functionally Gradient Materials," In: Yamanouochi, M., Koizumi, M., Hirai, T., and Shiota, I., (eds.), *Proceedings of the First International Symposium on Functionally Gradient Materials*, Sendai, Japan, pp. 5-10, 1990.

3. Igari, T., Notomi, A., Tsunoda, H., Hida, K., Kotoh, T., and Kunishima, S., "Material Properties of Functionally Gradient Material for Fast Breeder Reactor," In: Yamanouochi, M., Koizumi, M., Hirai, T., and Shiota, I., (eds.), *Proceedings of the First International Symposium on Functionally Gradient Materials*, Sendai, Japan, pp. 209-214, 1990.

4. Tani, J. and Liu, G, R., "Surface Waves in Functionally Gradient Piezoelectric Plates," *JSME International Journal Series A (Mechanics and Material Engineering)*, Vol. 36, No. 2, pp. 152-155, 1993.

5. Osaka, T., Matsubara, H., Homma, T., Mitamura, S., and Noda, K., "Microstructural Study of Electroless-Plated CoNiReP/NiMoP Double-Layered Media for Perpendicular Magnetic Recording," *Japanese Journal of Applied Physics*, Vol. 29, No. 10, pp. 1939-1943, 1990.

6. Watanabe, Y., Nakamura, Y., Fukui, Y., and Nakanishi, K., "A Magnetic-Functionally Graded Material Manufactured with Deformation-Induced Martensitic Transformation," *Journal of Materials Science Letters*, Vol. 12, No. 5, pp. 326-328, 1993.

7. Koike, Y., "Graded-Index and Single Mode Polymer Optical Fibers," In: Chiang, L, Y., Garito, A, G., and Sandman, D, J., (eds.), *Electrical, Optical, and Magnetic Properties of Organic Solid State Materials,* Materials Research Proceedings, Pittsburgh, PA, MRS, Vol. 247, pp. 817, 1992.

8. Desplat, J. L., "Recent Developments on Oxygenated Thermionic Energy Converter-Overview," *Proceedings of the Fourth International Symposium on Functionally Graded Materials*, Tsukuba City, Japan, 1996.

9. Oonishi, H., Noda, T., Ito, S., Kohda, A., Yamamoto, H., and Tsuji, E., "Effect of Hydroxyapatite Coating on Bone Growth into Porous Titanium Alloy Implants Under Loaded Conditions," *Journal of Applied Biomaterials,* Vol. 5, No. 1, pp. 23-27, 1994.

10. Getto, H. and Ishihara, S., "Development of the Fire Retardant Door with Functionally Gradient Wood," *Proceedings of the Fourth International Symposium on Functionally Graded Materials*, Tsukuba City, Japan, 1996.

11. Erdogan, F., "Fracture Mechanics of Functionally Graded Materials," *Composites Engineering*, Vol. 5, No. 7, pp. 753-770, 1995.

12. Gibson, R. E., "Some Results Concerning Displacements and Stresses in a Nonhomogeneous Elastic Half Space," *Geotechnique*, Vol. 17, pp. 58-67, 1967.

13. Delale, F. and Erdogan, F., "The Crack Problem for a Nonhomogeneous Plane," *Journal of Applied Mechanics*, Vol. 50, pp. 609-614, 1983.

14. Erdogan, F. and Wu, B, H., "Analysis of FGM Specimens for Fracture Toughness Testing," *Proceedings of the Second International Symposium on Functionally Graded Materials, Ceramic Transactions*, Westerville, Ohio, The American Ceramic Society, Vol. 34, pp. 39-46, 1993.

15. Marur, P. R. and Tippur, H, V., "Numerical Analysis of Crack Tip Fields in Functionally Graded Materials with a Crack Normal to the Elastic Gradient," *International Journal of Solids and Structures*, Vol. 37, pp. 5353-5370, 2000.

16. Noda, N. and Jin, Z. H., "Thermal Stress Intensity Factors for a Crack in a Strip of a Functionally Graded Materials," *International Journal of Solids and Structures*, Vol. 30, pp. 1039-1056, 1993.

17. Jin, Z. H. and Noda, N., "An Internal Crack Parallel to the Boundary of a Nonhomogeneous Half Plane Under Thermal Loading," *International Journal of Engineering Science*, Vol. 31, pp. 793-806, 1993.

18. Erdogan, F., and Wu, B. H., "Crack Problems in FGM layers Under Thermal Stresses," *Journal of Thermal Stresses*, Vol. 19, pp. 237-265, 1996.

19. Jin, Z. H. and Batra, R. C., "Stress Intensity Relaxation at the Tip of an Edge Crack in a Functionally Graded Material Subjected to a Thermal Shock," *Journal of Thermal Stresses*, Vol. 19, pp. 317-339, 1996.

20. Wang, B. L., Han, J. C., and Du, S. Y., "Crack Problems for Functionally Graded Materials Under Transient Thermal Loading," *Journal of Thermal Stresses*, Vol. 23(2), pp. 143-168, 2000.

21. Jin, Z. H. and Paulino, G. H, "Transient Thermal Stress Analysis of an Edge Crack in a Functionally Graded Material," *International Journal of Fracture*, Vol. 107(1), pp. 73-98, 2001.

22. Wang, B. L., and Noda, N., "Thermal Induced Fracture of a Smart Functionally Graded Composite Structure," *Theoretical and Applied Fracture Mechanics*, Vol. 35(2), pp. 93-109, 2001.

23. Anlas, G., Santare, M, H., and Lambros, J., "Numerical Calculation of Stress Intensity Factors in Functionally Graded Materials," *International Journal of Fracture*, Vol. 104, pp. 131-143, 2000.

24. Chen, J., Wu, L., and Du, S., "Element Free Galerkin Methods for Fracture of Functionally Graded Materials," *Key Engineering Materials*, Vol. 183-187, pp. 487-492, 2000.

25. Eischen, J, W., "Fracture of Nonhomogeneous Materials," *International Journal of Fracture*, Vol. 34, pp. 3-22, 1987.

26. Kim, J. H. and Paulino, G, H., "Finite Element Evaluation of Mixed Mode Stress Intensity Factors in Functionally Graded Materials," *International Journal for Numerical Methods in Engineering*, Vol. 53, No. 8, pp. 1903-1935, 2002.

27. Gu, P., Dao, M., and Asaro, R, J., "A Simplified Method for Calculating the Crack Tip Field of functionally Graded Materials Using the Domain Integral," *Journal of Applied Mechanics*, Vol. 66, pp. 101-108, 1999.

28. Bao, G. and Wang, L., "Multiple Cracking in Functionally Graded Ceramic/Metal Coatings," *International Journal of Solids and Structures*, Vol. 32, No. 19, pp. 2853-2871, 1995.

29. Bao, G. and Cai, H., "Delamination Cracking in Functionally Graded Coating/Metal Substrate systems," *Acta Mechanica*, Vol. 45, No. 3, pp. 1055-1066, 1997.

30. Rao, B. N., and Rahman, S., "Meshfree Analysis of Cracks in Isotropic Functionally Graded Materials," *Engineering Fracture Mechanics*, Vol. 70, pp. 1-27, 2003.

31. Sampath, S., Herman, H., Shimoda, N., and Saito, T., "Thermal Spray Processing of FGMs," *M.R.S. Bulletin*, Vol. 20(1), pp. 27-31, 1995.

32. Kaysser, W. A., Ilschner, B., "FGM Research Activities in Europe," *M.R.S. Bulletin*, Vol. 20(1), pp. 22-26, 1995.

33. Gu, P. and Asaro, R, J., "Cracks in Functionally Graded Materials," *International Journal of Solids and Structures*, Vol. 34, pp. 1-17, 1997.

34. Oztuk, M., and Erdogan, F., "Mode I Crack Problem in an Inhomogeneous Orthotropic Medium," *International Journal of Engineering Science*, Vol. 35(9), pp. 869-883, 1997.

35. Oztuk, M., and Erdogan, F., "The Mixed Mode Crack Problem in an Inhomogeneous Orthotropic Medium," *International Journal of Fracture*, Vol. 98, pp. 243-261, 1999.

36. Kim, J. H. and Paulino, G, H., "Mixed-Mode Fracture of Orthotropic Functionally Graded Materials using Finite Elements and the Modified Crack Closure Method," *Engineering Fracture Mechanics*, Vol. 69, pp. 1557-1586, 2002.

37. Lekhnitskii, S. G., Tsai, S. W., and Cheron, T., "Anisotropic Plates," *Gorden and Breach Science Publishers*, New York, 1986.

38. Shih, G. C., Paris, P. C., and Irwin, G. R., "On Cracks in Rectilinearly Anisotropic Bodies," *International Journal of Fracture*, Vol. 1, pp. 189-203, 1965.

39. Yau, J. F., Wang, S. S., and Corten, H. T., "A Mixed-Mode Crack Analysis of Isotropic Solids using Conservation Laws of Elasticity," *Journal of Applied Mechanics*, Vol. 47, pp. 335-341, 1980.

40. Wang, S. S., Yau, J. F., and Corten, H. T., "A Mixed-Mode Crack Analysis of Rectilinear Anisotropic Solids using Conservation Laws of Elasticity," *International Journal of Fracture*, Vol. 16(3), pp. 247-259, 1980.

41. Rice, J. R., "A Path independent Integral and the Approximate Analysis of Strain Concentration by Notches and Cracks," *Journal of Applied Mechanics*, Vol. 35, pp. 379-386, 1968.

42. Moran, B. and Shih, F., "Crack Tip and Associated Domain Integrals from Momentum and Energy Balance," *Engineering Fracture Mechanics*, Vol. 27, pp. 615-642, 1987.

Table 1. Normalized stress intensity factors for an orthotropic plate under uniform crack face pressure loading ($\kappa = -0.25$)

βa	Proposed Method-I $\left[\tilde{M}^{(1,2)}\right]$ $\frac{K_I(a)}{\sigma_0\sqrt{\pi a}}$	Proposed Method-I $\left[\tilde{M}^{(1,2)}\right]$ $\frac{K_I(-a)}{\sigma_0\sqrt{\pi a}}$	Proposed Method-II $\left[\tilde{M}^{(1,2)}\right]$ $\frac{K_I(a)}{\sigma_0\sqrt{\pi a}}$	Proposed Method-II $\left[\tilde{M}^{(1,2)}\right]$ $\frac{K_I(-a)}{\sigma_0\sqrt{\pi a}}$	Ozturk & Erdogan [34] $\frac{K_I(a)}{\sigma_0\sqrt{\pi a}}$	Ozturk & Erdogan [34] $\frac{K_I(-a)}{\sigma_0\sqrt{\pi a}}$
0.0	1.0003	1.0003	1.0009	1.0008	1.0	1.0
0.01	1.0013	0.9997	1.0016	1.0005	1.0025	0.9975
0.1	1.0176	0.9689	1.0354	0.9844	1.0246	0.9747
0.25	1.0587	0.9335	1.0770	0.9497	1.0604	0.9364
0.50	1.1099	0.8699	1.1297	0.8823	1.1177	0.8740
0.75	1.1709	0.8099	1.1695	0.8169	1.1720	0.8154
1.00	1.2287	0.7703	1.2137	0.7621	1.2235	0.7616
1.50	1.2999	0.6698	1.3065	0.6722	1.3184	0.6701
2.00	1.3899	0.5909	1.3908	0.6016	1.4043	0.5979

Table 2. Normalized stress intensity factors for an orthotropic plate under uniform crack face pressure loading ($\kappa = 5.0$)

βa	Proposed Method-I $\left[\tilde{M}^{(1,2)}\right]$ $\frac{K_I(a)}{\sigma_0\sqrt{\pi a}}$	Proposed Method-I $\left[\tilde{M}^{(1,2)}\right]$ $\frac{K_I(-a)}{\sigma_0\sqrt{\pi a}}$	Proposed Method-II $\left[\tilde{M}^{(1,2)}\right]$ $\frac{K_I(a)}{\sigma_0\sqrt{\pi a}}$	Proposed Method-II $\left[\tilde{M}^{(1,2)}\right]$ $\frac{K_I(-a)}{\sigma_0\sqrt{\pi a}}$	Ozturk & Erdogan [34] $\frac{K_I(a)}{\sigma_0\sqrt{\pi a}}$	Ozturk & Erdogan [34] $\frac{K_I(-a)}{\sigma_0\sqrt{\pi a}}$
0.0	0.9967	0.9967	1.0005	1.0005	1.0	1.0
0.01	1.0014	1.0002	1.0052	1.0049	1.0025	0.9975
0.1	1.0199	0.9833	1.0247	0.9867	1.0231	0.9733
0.25	1.0499	0.9289	1.0587	0.9289	1.0531	0.9306
0.50	1.0961	0.8543	1.0887	0.8602	1.0946	0.8594
0.75	1.1203	0.7878	1.1301	0.8069	1.1281	0.7932
1.00	1.1499	0.7298	1.1603	0.7363	1.1556	0.7339
1.50	1.2007	0.6295	1.2015	0.6401	1.1979	0.6367
2.00	1.2305	0.5586	1.2307	0.5598	1.2290	0.5636

Table 3. The effect of the inhomogeneity parameter on the normalized stress intensity factors for an orthotropic plate under uniform crack face pressure loading ($\delta^4 = 0.25$, $\nu=0.3$, $\kappa = 0.5$)

βa	Proposed Method-I $\left[\tilde{M}^{(1,2)}\right]$ $\frac{K_I(a)}{\sigma_0\sqrt{\pi a}}$	Proposed Method-I $\left[\tilde{M}^{(1,2)}\right]$ $\frac{K_{II}(a)}{\sigma_0\sqrt{\pi a}}$	Proposed Method-II $\left[\tilde{M}^{(1,2)}\right]$ $\frac{K_I(a)}{\sigma_0\sqrt{\pi a}}$	Proposed Method-II $\left[\tilde{M}^{(1,2)}\right]$ $\frac{K_{II}(a)}{\sigma_0\sqrt{\pi a}}$	Ozturk & Erdogan [35] $\frac{K_I(a)}{\sigma_0\sqrt{\pi a}}$	Ozturk & Erdogan [35] $\frac{K_{II}(a)}{\sigma_0\sqrt{\pi a}}$
0.0	1.0067	0.0	1.0059	0.0	1.0	0.0
0.1	1.0206	0.0248	1.0193	0.0246	1.0115	0.0250
0.25	1.0502	0.0599	1.0513	0.0609	1.0489	0.0627
0.50	1.1373	0.1303	1.1355	1.1299	1.1351	0.1263
1.00	1.3398	0.2497	1.3406	0.2501	1.3494	0.2587
2.00	1.8623	0.5493	1.8598	0.5564	1.8580	0.5529

Table 4. The effect of the inhomogeneity parameter on the normalized stress intensity factors for an orthotropic plate under uniform crack face shear loading ($\delta^4 = 0.25, \nu=0.3, \kappa = 0.5$)

βa	Proposed Method-I $\left[\tilde{M}^{(1,2)}\right]$		Proposed Method-II $\left[\tilde{M}^{(1,2)}\right]$		Ozturk & Erdogan [35]	
	$\frac{K_I(a)}{\tau_0\sqrt{\pi a}}$	$\frac{K_{II}(a)}{\tau_0\sqrt{\pi a}}$	$\frac{K_I(a)}{\tau_0\sqrt{\pi a}}$	$\frac{K_{II}(a)}{\tau_0\sqrt{\pi a}}$	$\frac{K_I(a)}{\tau_0\sqrt{\pi a}}$	$\frac{K_{II}(a)}{\tau_0\sqrt{\pi a}}$
0.0	0.0	0.9998	0.0	1.0007	0.0	1.0
0.1	−0.0487	1.0005	−0.0496	0.9999	−0.0494	0.9989
0.25	−0.1156	0.9971	−0.1167	0.9973	−0.1191	0.9968
0.50	−0.2196	0.9964	−0.2203	0.9969	−0.2217	0.9965
1.00	−0.3874	0.9999	−0.3799	1.0006	−0.3862	1.0071
2.00	−0.5698	1.0399	−0.5708	1.0453	−0.5725	1.0499

Table 5. The effect of the Poisson's ratio on the normalized stress intensity factors for an orthotropic plate under uniform crack face pressure loading ($\kappa = 5.0$)

βa	$\delta^4 = 0.25$	Proposed Method-I $\left[\tilde{M}^{(1,2)}\right]$			Proposed Method-II $\left[\tilde{M}^{(1,2)}\right]$			Ozturk & Erdogan [35]		
	ν	0.15	0.30	0.45	0.15	0.30	0.45	0.15	0.30	0.45
0.5	$K_I(a)/\sigma_0\sqrt{\pi a}$	1.2531	1.2578	1.2599	1.2528	1.2583	1.2608	1.2516	1.2596	1.2674
	$K_{II}(a)/\sigma_0\sqrt{\pi a}$	0.1217	0.1219	0.1219	0.1267	0.1270	0.1271	0.1259	0.1259	0.1259
1.0	$K_I(a)/\sigma_0\sqrt{\pi a}$	1.5603	1.5729	1.5845	1.5599	1.5748	1.5836	1.5589	1.5739	1.5884
	$K_{II}(a)/\sigma_0\sqrt{\pi a}$	0.2600	0.2599	0.2516	0.2583	0.2581	0.2499	0.2555	0.2557	0.2558

Table 6. The effect of the Poisson's ratio on the normalized stress intensity factors for an orthotropic plate under uniform crack face pressure loading ($\kappa = 5.0$)

βa	$\delta^4 = 10$	Proposed Method-I $\left[\tilde{M}^{(1,2)}\right]$			Proposed Method-II $\left[\tilde{M}^{(1,2)}\right]$			Ozturk & Erdogan [35]		
	ν	0.15	0.30	0.45	0.15	0.30	0.45	0.15	0.30	0.45
0.5	$K_I(a)/\sigma_0\sqrt{\pi a}$	1.0699	1.0720	1.0799	1.0710	1.0716	1.0810	1.0748	1.0776	1.0804
	$K_{II}(a)/\sigma_0\sqrt{\pi a}$	0.1239	0.1252	0.1259	0.1227	0.1248	0.1256	0.1252	0.1252	0.1251
1.0	$K_I(a)/\sigma_0\sqrt{\pi a}$	1.1889	1.1955	1.2009	1.1897	1.1993	1.2056	1.1892	1.1955	1.2017
	$K_{II}(a)/\sigma_0\sqrt{\pi a}$	0.2499	0.2503	0.2536	0.2538	0.2543	0.2537	0.2511	0.2512	0.2512

Table 7. The effect of the Poisson's ratio on the normalized stress intensity factors for an orthotropic plate under uniform crack face shear loading ($\kappa = 5.0$)

βa	$\delta^4 = 0.25$	Proposed Method-I $\left[\tilde{M}^{(1,2)}\right]$			Proposed Method-II $\left[\tilde{M}^{(1,2)}\right]$			Ozturk & Erdogan [35]		
	ν	0.15	0.30	0.45	0.15	0.30	0.45	0.15	0.30	0.45
0.5	$K_I(a)/\sigma_0\sqrt{\pi a}$	−0.1926	−0.1909	−0.1900	−0.1956	−0.1945	−0.1923	−0.1980	−0.1971	−0.1963
	$K_{II}(a)/\sigma_0\sqrt{\pi a}$	0.9899	0.9907	0.9960	0.9888	0.9999	0.9919	0.9898	0.9915	0.9931
1.0	$K_I(a)/\sigma_0\sqrt{\pi a}$	−0.3199	−0.3099	−0.3067	−0.3199	−0.3056	−0.3048	−0.3203	−0.3186	−0.3169
	$K_{II}(a)/\sigma_0\sqrt{\pi a}$	0.9899	0.9901	0.9968	0.9901	0.9953	0.9968	0.9888	0.9921	0.9953

Table 8. The effect of the Poisson's ratio on the normalized stress intensity factors for an orthotropic plate under uniform crack face shear loading ($\kappa = 5.0$)

βa	$\delta^4 = 10$	Proposed Method-I $\left[\tilde{M}^{(1,2)}\right]$			Proposed Method-II $\left[\tilde{M}^{(1,2)}\right]$			Ozturk & Erdogan [35]		
	ν	0.15	0.30	0.45	0.15	0.30	0.45	0.15	0.30	0.45
0.5	$K_I(a)/\sigma_0\sqrt{\pi a}$	−0.0359	−0.0379	−0.0380	−0.0360	−0.0383	−0.0383	−0.0366	−0.0365	−0.0365
	$K_{II}(a)/\sigma_0\sqrt{\pi a}$	0.9909	0.9911	0.9925	0.9911	0.9917	0.9977	0.9956	0.9961	0.9965
1.0	$K_I(a)/\sigma_0\sqrt{\pi a}$	−0.0659	−0.0649	−0.0627	−0.0655	−0.0648	−0.0631	−0.0660	−0.0657	−0.0654
	$K_{II}(a)/\sigma_0\sqrt{\pi a}$	0.9899	0.9905	0.9917	0.9898	0.9913	0.9949	0.9913	0.9925	0.9938

Table 9. Mixed-mode stress-intensity factor in an orthotropic functionally graded plate with a slant crack

Material	Method	$K_I^*(a)$	$K_{II}^*(a)$	$K_I(a)$	$K_{II}(a)$
FGM (proportional) $(\alpha,\beta,\gamma) = (0.2, 0.2, 0.2)$	Kim and Paulino (MCC) [36]	1.762	1.439	1.403	1.288
	Kim and Paulino (DCT) [36]	1.769	1.419	1.419	1.284
	Proposed Method-I $\left[\tilde{M}^{(1,2)}\right]$	1.728	1.429	1.411	1.299
	Proposed Method-II $\left[\tilde{M}^{(1,2)}\right]$	1.725	1.432	1.409	1.300
FGM (non-proportional) $(\alpha,\beta,\gamma) = (0.5, 0.4, 0.3)$	Kim and Paulino (MCC) [36]	2.384	1.581	1.437	1.225
	Kim and Paulino (DCT) [36]	2.387	1.553	1.456	1.229
	Proposed Method-I $\left[\tilde{M}^{(1,2)}\right]$	2.393	1.603	1.399	1.217
	Proposed Method-II $\left[\tilde{M}^{(1,2)}\right]$	2.399	1.611	1.407	1.204

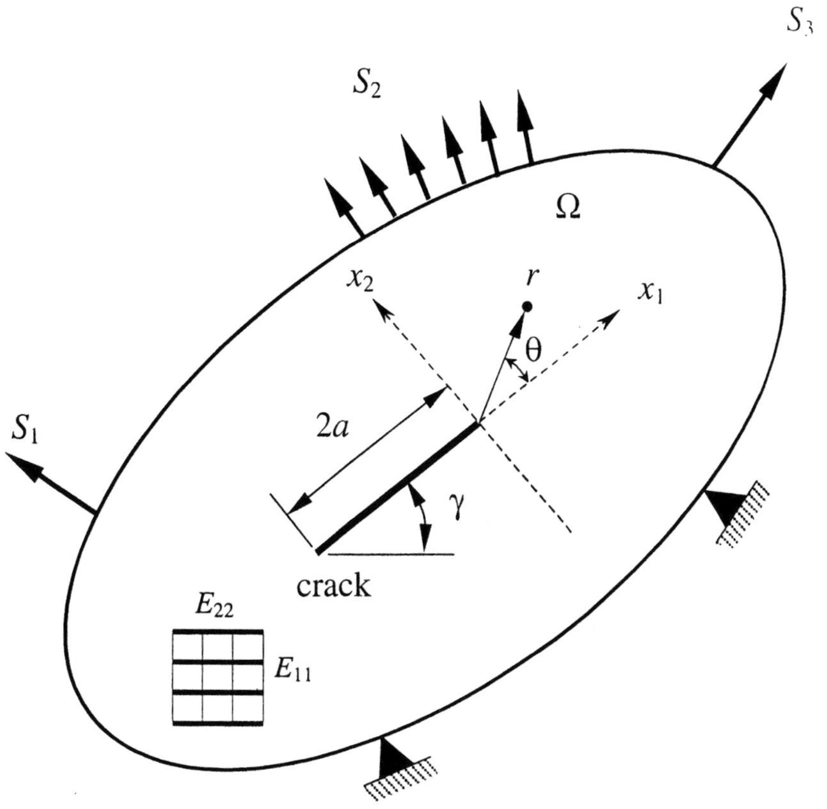

Figure 1. A crack in an orthotropic functionally graded material

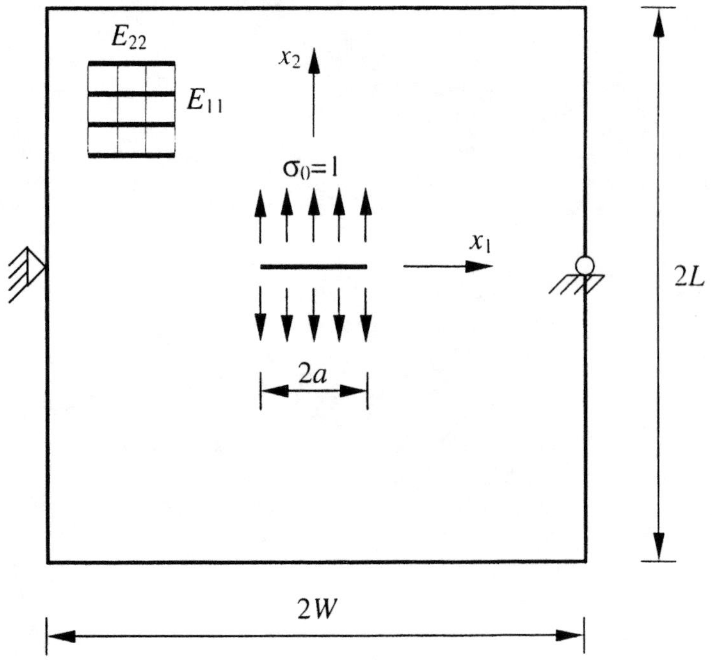

(a)

(b)

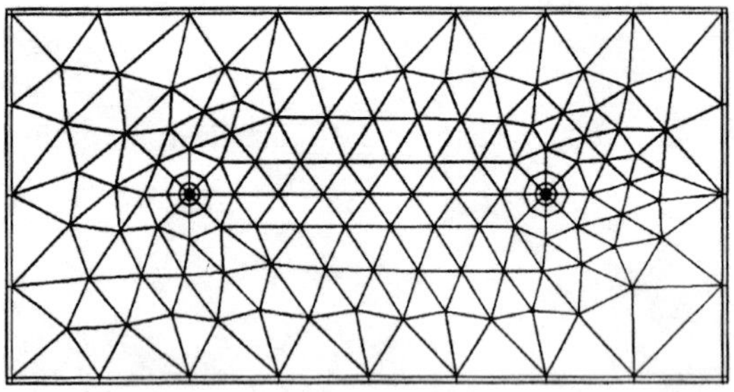

(c)

Figure 2. Plate with an Interior Crack Parallel to Material Gradation under Mode-I: (a) Geometry and loads; (b) FEM discretization (5756 nodes, 1696 8-noded quadrilateral elements, 214 6-noded triangular elements, and 16 focused quarter-point 6-noded triangular elements); and (c) Enlarged view of discretization around the crack-tips

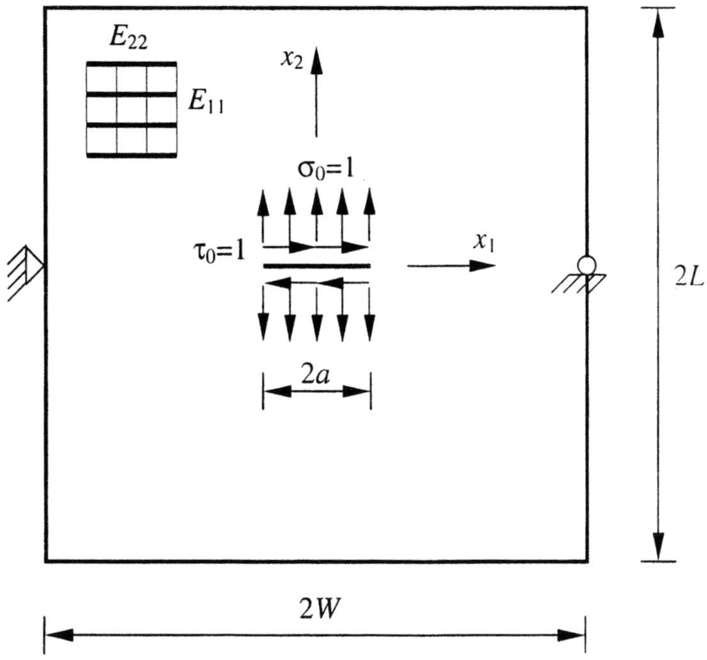

Figure 3. Plate with an Interior Crack Perpendicular to Material Gradation: Geometry and loads

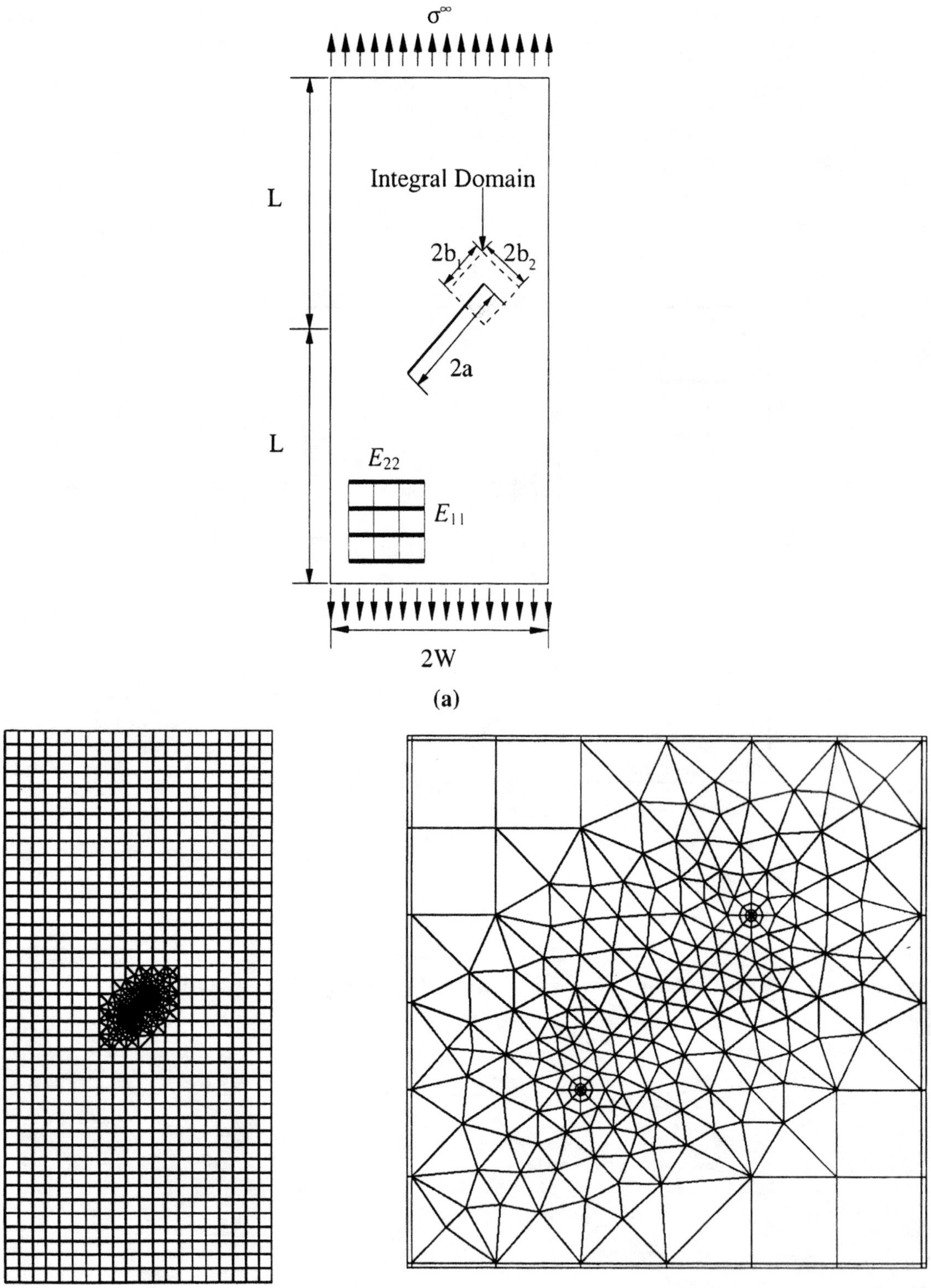

(b) (c)

Figure 4. Slanted Crack in a Plate under Mixed-Mode: (a) Geometry and loads; (b) FEM **discretization (3606 Nodes, 898 8-noded quadrilateral elements, 354 6-noded triangular elements, and 16 focused quarter-point 6-noded triangular elements);** and (c) Enlarged view of **discretization around the crack-tips**

PVP-Vol. 458, Computer Technology and Applications

PVP2003-1907

RIGID-PLASTIC MESHFREE METHOD FOR METAL FORMING SIMULATION

Young H. Park
Assistant Professor
Mechanical Engineering Department
New Mexico State University
Las Cruces, NM 88003-8001
e-mail: ypark@nmsu.edu

ABSTRACT

In this paper, material processing simulation is carried out using a meshfree method. With the use of a meshfree method, the domain of the workpiece is discretized by a set of particles without using a structured mesh to avoid mesh distortion difficulties which occurred during the course of large plastic deformation. The proposed meshfree method is formulated for rigid-plastic material. This approach uses the flow formulation based on the assumption that elastic effects are insignificant in the metal forming operation. In the rigid-plastic analysis, the main variable of the problem becomes flow velocity rather than displacement. A numerical example is solved to validate the proposed method.

1. INTRODUCTION

Manufacturing processes such as metal forming are important in industry. Since these operations occur within high-volume industries, even small improvements in material properties, simulations, and manufacturing costs will translate into substantial savings. The high cost of manufacturing critical structural components can be significantly reduced with the development of mathematically and physically sound computational methodologies for process design and control. In the forming process simulation, the rigid-plastic material model is widely used since its behavior is numerically robust. This approach uses the flow formulation based on the assumption that elastic effects are insignificant in the metal forming operation. In large deformation process involving plastic (or viscoplastic) materials the elastic deformations can be considered as negligible. In the rigid-plastic analysis, the main variable of the problem becomes flow velocity rather than displacement. Zienkiewicz and Godbole (1974) proposed an iterative procedure to solve the nonlinear response analysis, whereas Lee and Kobayashi (1973) suggested a Newton-Raphson method by computing a tangent stiffness matrix, that guarantees a quadratic convergence in the vicinity of the solution. The rigid-plastic and viscoplastic finite element methods have been applied to various steady and non-steady metal forming processes such as rolling, extrusion, deep drawing, and sheet metal forming (Oh and Kobayashi, 1980; Mori et al., 1982; Osakada, 1982; Chandra, 1988; Balaji et al., 1991; Antunez and Kleiber, 1996).

When a material is shaped by metal forming processes, it experiences large plastic deformation. The deformation which occurs in the forming process exhibits characteristics of both solid and fluid flow, and conventional finite element methods often fail due to severe mesh distortion. An effective numerical method, which can handle severe mesh distortion problems in conventional finite element analysis, is highly desirable in analyzing large deformation problems. Recently, in order to alleviate these difficulties, a number of meshfree methods have been proposed that do not require explicit meshes in domain discretization (Belytschko et al., 1994; Liu et al., 1995; Chen et al., 1996; Chen et al., 1998). The shape function is not a function of the reference domain but a function of the material points and the order of shape function can be easily changed.

Insensitiveness to the mesh distortion is a very important feature in nonlinear analysis. Higher accuracy can be achieved by simply adding more nodes to the structure without re-modeling the total structure. In this research, a computer simulation method for unsteady metal forming processes will be developed based on the flow formulation using a meshfree method. With the use of a meshfree method, the domain of the workpiece will be discretized by a set of particles without using a structured mesh to avoid mesh distortion difficulties which occurred during the course of large plastic deformation. This research will employ the Reproducing Kernel Particle Method (RKPM) (Liu et al., 1995) where a modified kernel function is constructed based on the enforcement of reproducing conditions such that the kernel estimate of displacement exactly reproduces polynomials. Meshfree methods developed in recent discretization (Belytschko et al., 1994; Liu et al., 1995; Chen et al., 1996; Chen et al., 1998) allow the shape functions to be constructed entirely in terms of arbitrarily placed nodes without the use of an element connectivity in the classical sense. Several authors applied meshfree methods to metal forming process (Chen et al., 1996; Chen et al., 1998). Their approaches, however, were based on the elastoplasticity where deformation is decomposed into elastic and plastic parts. In this paper, a rigid-plastic meshfree method is developed and utilized for metal forming simulation. For a contact condition, a variational formulation is developed in the case where the frictional stress is dependent on the relative velocity at the die-workpiece interface. In validating the proposed method, a ring compression problem is solved.

2. FORMULATION FOR RIGID-PLASTIC MATERIALS

The basis of meshfree modeling using the variational approach is to formulate proper functionals depending upon specific constitutive relations. The meshfree equations are then constructed by the condition that the first-order variation of the functional vanishes.

2.1 Variational Formulation

The variational approach for rigid-plastic materials requires that among admissible velocities $\boldsymbol{v}$ that satisfy the conditions of compatibility and incompressibility, as well as the velocity boundary conditions, the actual solution gives the following functional (Hill, 1956) a stationary value:

$$\pi=\int_{\Omega}\boldsymbol{\sigma}'\cdot\dot{\boldsymbol{\varepsilon}}\,d\Omega-\int_{\Gamma_f}\mathbf{f}\cdot\mathbf{v}\,d\Gamma \tag{1}$$

where $\dot{\boldsymbol{\varepsilon}}$ is the strain-rate vector, and $\boldsymbol{\sigma}'$ is the deviatoric stress vector. In Eq. (1), traction $\mathbf{f}$ is prescribed on the surface, Γ_f. The incompressibility constraint on admissible velocity fields in Eq. (1) may be removed by introducing a Lagrange multiplier λ and modifying the functional (1) by adding the term $\int_{\Omega}\lambda\dot{\varepsilon}_v\,d\Omega$, where $\dot{\varepsilon}_v=\dot{\varepsilon}_{ii}$, is the volumetric strain-rate. The functional given by Eq. (1) can then be written as

$$\begin{aligned}\pi&=\int_{\Omega}\boldsymbol{\sigma}'\cdot\dot{\boldsymbol{\varepsilon}}\,d\Omega-\int_{\Gamma_f}\mathbf{f}\cdot\mathbf{v}\,d\Gamma+\int_{\Omega}\lambda\dot{\varepsilon}_v\,d\Omega\\&=\pi_1+\pi_F+\pi_v\end{aligned} \tag{2}$$

For the materials obeying

$$\sigma'_{ij}=\frac{2}{3}\frac{\bar{\sigma}}{\dot{\bar{\varepsilon}}}\dot{\varepsilon}_{ij} \tag{3}$$

the first term of the functional (2) can be rewritten as

$$\int\boldsymbol{\sigma}'\cdot\dot{\boldsymbol{\varepsilon}}\,d\Omega=\int\bar{\sigma}\dot{\bar{\varepsilon}}\,d\Omega \tag{4}$$

where $\bar{\sigma}=\sqrt{[(3/2)\sigma'_{ij}\sigma'_{ij}]}$ and $\dot{\bar{\varepsilon}}=\sqrt{[(2/3)\dot{\varepsilon}_{ij}\dot{\varepsilon}_{ij}]}$.

The Lagrangian multiplier, λ, in Eq. (2) can be shown to be equal to the hydrostatic stress component σ_m. Another approach to removing the incompressibility constraint is to use the penalty function method. This method introduces a very large positive constraint, k_v, to penalize the dilatation strain as follows;

$$\begin{aligned}\pi&=\int_{\Omega}\boldsymbol{\sigma}'\cdot\dot{\boldsymbol{\varepsilon}}\,d\Omega-\int_{\Gamma_f}\mathbf{f}\cdot\mathbf{v}\,d\Gamma+\int_{\Omega}\frac{1}{2}k_v(\dot{\varepsilon}_{ii})^2\,d\Omega\\&=\pi_1+\pi_F+\pi_v\end{aligned} \tag{5}$$

The solution of the mechanics of plastic deformation for rigid-plastic materials can be obtained from the solution of the dual variational problem, where the first-order variation of the functional vanishes, namely,

$$\delta\pi=\int_{\Omega}\bar{\boldsymbol{\sigma}}\cdot\delta\dot{\bar{\boldsymbol{\varepsilon}}}\,d\Omega-\int_{\Gamma_f}\mathbf{f}\cdot\delta\mathbf{v}\,d\Gamma+\int_{\Omega}k_v\dot{\varepsilon}_v\delta\dot{\varepsilon}_v\,d\Omega=0 \tag{6}$$

with the incompressibility constraint given by $\dot{\varepsilon}_{ii}=\dot{\varepsilon}_v=0$ in Ω.

2.2 Boundary Conditions

The traction boundary condition on Γ_f is either zero-traction or ordinarily a uniform hydrostatic pressure. However, the boundary conditions along the die-workpiece interface are mixed. Further, in general, neither velocity nor force (including magnitude and direction) can be prescribed completely along the interface, because the direction of the frictional stress is opposite to the direction of the relative velocity between the deforming workpiece and the die, and this relative velocity is not known a priori. Situations exist, e.g., in extrusion and drawing, in which the direction of metal flow relative to the die is known. This class of problems can be solved if the magnitude of the frictional stress f_s is given according to the well-known Coulomb law, $f_s=\mu p$, or the friction law of constant factor m, expressed by $f_s=mk$ (where $k=Y/\sqrt{3}$); here, p is the die pressure and k is the shear yield stress.

For problems such as ring compression, rolling, and forging, the unknown direction of the relative velocity between the die-workpiece interface makes it difficult to handle the boundary condition in a straightforward manner. A unique feature of this type of problem is that there exists a point (or a region) along the die-workpiece interface where the velocity of the deforming material relative to the die becomes zero, and the location of this point (or region) depends on the magnitude of the frictional stress itself. In order to deal with these situations, a velocity-dependent frictional stress is used as an approximation to the condition of constraint frictional stress. At the interface Γ_c the velocity boundary condition is given in the direction normal to the interface by the die velocity, and the traction boundary condition is expressed by

$$\mathbf{f}_s=mk\boldsymbol{\ell}\cong mk\left\{\frac{2}{\pi}\tan^{-1}\left(\frac{|v_s|}{v_0}\right)\right\}\boldsymbol{\ell} \tag{7}$$

where $\mathbf{f}_s$ is the frictional stress, $\boldsymbol{\ell}$ is the unit vector in the opposite direction of relative sliding, v_s is the sliding velocity of a material relative to the die velocity (relative sliding velocity), and v_0 is a small positive number compared to v_s. The approximate expression (7) for a constant frictional stress has been used for the smooth transition of the frictional stress in the range near the neutral point.

3. REPRODUCING KERNEL PARTICLE METHOD (RKPM)

3.1 Kernel Approximation using RKPM

Liu et al. (1995) developed the Reproducing Kernel Particle Method (RKPM) by introducing a modified kernel function that is constructed based on the enforcement of reproducing conditions such that the kernel estimate of displacement exactly reproduces polynomials. In RKPM, a displacement function $u^h(x)$ is approximated using a kernel estimate,

$$u^h(x)=\int_{\Omega}C(x;y-x)\phi_a(y-x)u(y)\,dy \tag{8}$$

where $u^h(x)$ is the *reproduced displacement function* of u(x) and $\phi_a(y-x)$ is the *kernel function* (or weight function) with a support measure of "a". The kernel function $\phi_a(y-x)$ has the following properties:

$$\int_\Omega \phi_a(y-x)dy = 1 \tag{9}$$

$$u^h(x) \rightarrow u(x) \quad \text{as} \quad a \rightarrow 0 \tag{10}$$

In Eq. (8), C(x;y–x) is the correction function defined by

$$C(x;y-x) = \mathbf{q}(x)^T \mathbf{H}(y-x) \tag{11}$$

where $\mathbf{H}(y-x)^T = [1,(y-x),(y-x)^2,\cdots,(y-x)^N]$, and $\mathbf{q}(x)^T = [q_0(x),q_1(x),\cdots,q_N(x)]$ are the interpolation function and unknown coefficient vector, respectively. In Eq. (11), $\mathbf{q}(x)$ is determined by imposing the N-th order completeness requirement. Expanding u(y) in Eq. (8) using Taylor series expansion and imposing the N-th order completeness condition yield

$$\begin{aligned} u^h(x) &= \int_\Omega C(x;y-x)\phi_a(y-x)u(y)dy \\ &= \bar{m}_0(x)u(x) + \sum_{n=1}^{\infty}\frac{(-1)^n}{n!}\bar{m}_n(x)\frac{d^n u(x)}{dx^n} \end{aligned} \tag{12}$$

where $m_n(x) = \int_\Omega (y-x)^n \phi_a(y-x)dy$ and $\bar{m}_n(x) = \sum_{k=1}^{n} q_k(x)m_{n+k}(x)$. For the reproduced function to completely represent original function up to N-th order which is called the *reproducing condition*, the following condition should be satisfied,

$$\bar{m}_0(x) = 1 \qquad \bar{m}_k(x) = 0 \qquad k = 1,\ldots,N \tag{13}$$

which means $u^h(x)$ in Eq. (12) represents z(x) completely up to the N-th order derivative. Equation (13) represents the following set of equations (reproducing condition)

$$\mathbf{M}(x)\mathbf{q}(x) = \mathbf{H}(0) \tag{14}$$

$$\mathbf{M}(x) = \begin{bmatrix} m_0(x) & m_1(x) & \ldots & m_N(x) \\ m_1(x) & m_2(x) & \ldots & m_{N+1}(x) \\ \cdot & \cdot & \ldots & \cdot \\ m_N(x) & m_{N+1}(x) & \ldots & m_{N+N}(x) \end{bmatrix} \tag{15}$$

$$\mathbf{H}(0)^T = [1,0,\ldots,0] \tag{16}$$

Thus, the coefficient of reproducing condition $\mathbf{q}(x)$ can be obtained by solving Eq. (14). The correction function C(x;x–y) can be computed from Eq. (11),

$$C(x;y-x) = \mathbf{H}(0)^T \mathbf{M}(x)^{-1} \mathbf{H}(y-x) \tag{17}$$

Introducing Eq. (17) into Eq. (8) leads to the following Reproducing Kernel approximation:

$$u^h(x) = \mathbf{H}(0)^T \mathbf{M}(x)^{-1} \int_\Omega \mathbf{H}(y-x)\phi_a(y-x)u(y)\,dy \tag{18}$$

To develop a multidimensional shape function for discrete approximation, Eq. (18) must be discretized. Suppose that the domain Ω is discretized by a set of nodes $[\mathbf{x}_1,\ldots,\mathbf{x}_{NP}]$, where $\mathbf{x}_I$ is the position vector of node I, and NP is the total number of nodes. Using a simple trapezoidal rule, Eq. (18) is discretized into

$$u^h(\mathbf{x}) = \sum_{I=1}^{NP} C(\mathbf{x};\mathbf{x}_I-\mathbf{x})\phi_a(\mathbf{x}_I-\mathbf{x})\mathbf{u}_I\Delta\mathbf{x}_I \tag{19}$$

where Δx_I is a measure of length associate with node I. It is difficult to determine Δx_I in a multi-dimensional case but it can be treated as a weight of the nodal value. However, its effect will be canceled if we compute moment $\mathbf{M}(x)$ in Eq. (15) consistently with Eq. (9) since its inverse is multiplied. The property of Eq. (9) is not required by the same reason. Equation (19) can be rewritten, using generalized displacement d_I, as

$$u^h(\mathbf{x}) = \sum_{I=1}^{NP} \Phi_I(\mathbf{x})d_I \tag{20}$$

where $\Phi_I(\mathbf{x}) = C(\mathbf{x};\mathbf{x}_I-\mathbf{x})\phi_a(\mathbf{x}_I-\mathbf{x})$. The function $\Phi_I(\mathbf{x})$ is interpreted as the *particle* or *meshfree shape function* of node I, and d_I is the associated coefficient of approximation, often called the generalized displacement. The shape function $\Phi_I(\mathbf{x}_J)$ depends on the current coordinate x_J, whereas the shape function of the finite element method depends only on a coordinate of reference geometry. It should also be noted that, in general, the shape function does not bear the Kronecker delta properties, i.e., $\Phi_I(x_J) \neq \delta_{IJ}$. Therefore, for a general function z(x) which is not a polynomial, d_I in Eq. (20) is not the nodal value of z(x). To specify essential boundary conditions, i.e., prescribed value at node points, the Lagrange multiplier method must be used to impose those conditions.

3. MESHFREE FORMULATION WITH RIGID-PLASTIC MATERIAL

3.1 Solution Procedures

Define a set of nodal point velocities $[\boldsymbol{v}_1, \boldsymbol{v}_2, \ldots, \boldsymbol{v}_{NP}]$, where $\boldsymbol{v}_I$ is the velocity vector at node I, and NP is the total number of nodes. An admissibility requirement for the velocity field is that the velocity boundary condition prescribed on surface Γ_u (essential boundary condition) must be satisfied. This condition can be imposed at nodes on Γ_u by assigning known values to the corresponding variables. Equation (6) is now expressed in items of nodal point velocities $\boldsymbol{v}$ and their variations $\delta\boldsymbol{v}$. From arbitrariness of $\delta\boldsymbol{v}_I$, a set of algebraic equations (stiffness equations) are obtained as

$$\frac{\partial\pi}{\partial\boldsymbol{v}_I} = \sum_j \left(\frac{\partial\pi}{\partial\boldsymbol{v}_I}\right)_{(j)} = 0 \tag{21}$$

where "j" indicates the quantity at the j-th element. The capital letter suffix signifies that it refers to the nodal point number.

Equation (21) is obtained by evaluating the $(\partial\pi/\partial v_I)$ at the elemental level and assembling them into the global equation under appropriate constraints.

In metal-forming, the stiffness equation (21) is nonlinear and the solution is obtained iteratively by using the Newton-Raphson method. The method consists of linearization and application of convergence criteria to obtain the final solution. Linearization is achieved by Taylor expansion near an assumed solution point $v=v_0$ (initial guess), namely,

$$\left[\frac{\partial\pi}{\partial v_I}\right]_{v=v_0}+\left[\frac{\partial^2\pi}{\partial v_I\partial v_J}\right]_{v=v_0}\Delta v_J=0 \tag{22}$$

where Δv_J is the first-order correction of the velocity v_0. Equation (22) can be written in the form

$$\mathbf{K}\Delta \boldsymbol{v}=\mathbf{f} \tag{23}$$

where **K** is called the stiffness matrix and **f** is the residual of the nodal point force vector.

3.2 Reproducing Kernel Discretization

Using the meshfree shape function given in Eq. (20), velocity components at any arbitrary point inside the element can be determined in terms of the surrounding nodal velocities. This is expressed by

$$v^h(\mathbf{x})=\sum_{I=1}^{NP}\Phi_I(\mathbf{x})g_I\text{ , I=1,2} \tag{24}$$

where g_I is the generalized velocity. The stress and strain components can be represented using Eq. (24) as

$$\dot{e}=\sum_{i=1}^{NP}\mathbf{B}_I\,\mathbf{g}_I \tag{25}$$

and

$$\boldsymbol{\sigma}'=\frac{3}{2}\frac{\bar{\sigma}}{\dot{\bar{\varepsilon}}}\dot{\boldsymbol{\varepsilon}} \tag{26}$$

where **B** is the strain-rate matrix. The effective strain-rate is defined in terms of strain-rate components as

$$\dot{\bar{\varepsilon}}=\sqrt{\frac{2}{3}}\{\dot{\varepsilon}_{ij}\,\dot{\varepsilon}_{ij}\}^{1/2} \tag{27}$$

or, in the matrix form

$$(\dot{\bar{\varepsilon}})^2=\dot{\boldsymbol{\varepsilon}}^T\mathbf{D}\dot{\boldsymbol{\varepsilon}} \tag{28}$$

where the diagonal matrix **D** has $\frac{2}{3}$ and $\frac{1}{3}$ as its components; corresponding to normal strain-rate and engineering shear strain-rate, respectively.

The volumetric strain-rate $\dot{\varepsilon}_v$ is given by

$$\dot{\varepsilon}_v=\dot{\varepsilon}_{kk} \tag{29}$$

and expressed by

$$\dot{\varepsilon}_v=\mathbf{C}^T\boldsymbol{v} \tag{30}$$

where **C** includes sum of components of matrix **B** corresponding to normal strain rate.

According to Eqs. (25), (27), and (30), the effective strain rate and volumetric strain rate can be expressed in terms of the nodal velocities as follows:

$$(\dot{\bar{\varepsilon}})^2=\mathbf{g}^T\mathbf{P}\mathbf{g} \tag{31}$$

where

$$\mathbf{P}_{IJ}=\mathbf{B}_I^{\,T}\mathbf{D}\mathbf{B}_J \tag{32}$$

By inserting all the corresponding values into Eq. (22), the meshfree discretization formulation is obtained. These are given by

$$\begin{aligned}
&\frac{\partial\pi_1}{\partial v_I}=\int_\Omega\frac{\bar{\sigma}}{\dot{\bar{\varepsilon}}}\mathbf{P}_{Im}\mathbf{g}_m d\Omega\\
&\frac{\partial^2\pi_1}{\partial v_I\partial v_J}=\int_\Omega\left[\frac{\bar{\sigma}}{\dot{\bar{\varepsilon}}}\mathbf{P}_{Im}+\frac{1}{\dot{\bar{\varepsilon}}}\left(\frac{1}{\dot{\bar{\varepsilon}}}\frac{\partial\bar{\sigma}}{\partial\dot{\bar{\varepsilon}}}-\frac{\bar{\sigma}}{\dot{\bar{\varepsilon}}^2}\right)(\mathbf{P}_{Im}\mathbf{P}_{Jn}\mathbf{g}_m\mathbf{g}_n)\right]d\Omega\\
&\frac{\partial\pi_v}{\partial v_I}=\int_\Omega k_v\mathbf{C}_J\mathbf{C}_I\,\mathbf{g}_I d\Omega\\
&\frac{\partial^2\pi_v}{\partial v_I\partial v_J}=\int_\Omega k_v\mathbf{C}_J\mathbf{C}_I\,d\Omega
\end{aligned} \tag{33}$$

3.3 Boundary Treatment

Equation (7) expresses that the magnitude of the frictional stress is dependent on the magnitude of the relative sliding and that their directions are opposite to each other.

$$\mathbf{f}_s=mk\ell\cong mk\left\{\frac{2}{\pi}\tan^{-1}\left(\frac{|v_s|}{v_0}\right)\right\}\ell \tag{34}$$

The approximation of the frictional stress by the arctangent function of the relative sliding velocity eliminates the sudden change of direction of the frictional stress *mk* at the neutral point. It is known that the ratio v_s/v_0 should be equal to or larger than 10 in order to attain the friction value within 9% of the one originally intended. On the other hand, if we choose the ratio too large, then the sudden change of the frictional stress near the neutral point can cause difficulties in numerical calculation.

At the contact surface, the relative sliding velocity at the nodes $\mathbf{v}_s$ can be evaluated. It may be assumed that the relative sliding velocity v_s can be approximated in terms of nodal-point values $v_{s\alpha}$ by using a shape function of the elements as

$$v_s=\sum_\alpha q_\alpha v_{s\alpha} \tag{35}$$

where the subscript α denotes the value at α-th node.

In deriving the stiffness equation, $\delta\pi$ should include the term $\delta\pi_{s_c}$, and the final form of the stiffness equation should contain the terms

$$\frac{\partial \pi_{s_c}}{\partial v_\alpha} = \int_{s_c} \mathrm{mk}\frac{2}{\pi} q_\alpha \tan^{-1}\left[{q_\beta v_{s_\beta}}\big/{v_0} \right] dS \qquad (36)$$

and

$$\frac{\partial^2 \pi_{s_c}}{\partial v_\alpha \partial v_\beta} = \int_{s_c} \mathrm{mk}\frac{2}{\pi} q_\alpha q_\beta \left[\frac{v_0}{v_0^2 + (q_k v_{s_k})^2} \right] dS \qquad (37)$$

in addition to those given in Eq. (33). The surface integration in Eqs. (36) and (37) is carried out over the contact. When linear elements are used with a curved die, the interface area represented by elements is always smaller than the actual interface area, and the effect of friction in the analysis is always smaller than the actual.

3.4 Transformation Method

Due to the lack of Kronecker delta properties in the reproducing kernel shape functions, the essential boundary conditions need to be introduced with additional effort. Using the Lagrangian reproducing kernel shape functions, the kinematic admissible velocity $v_i^h(\mathbf{X})$ of the variational equation needs to satisfy the following conditions:

$$\left.\begin{aligned} v_i^h(\mathbf{X}_J) &= \sum_{I=1}^{NP} \Phi_I(\mathbf{X}_J) g_{iI} = h_i(\mathbf{X}_J) \\ \delta v_i^h(\mathbf{X}_J) &= \sum_{I=1}^{NP} \Phi_I(\mathbf{X}_J) \delta g_{iI} \end{aligned}\right\} \forall J \in \eta_{h_i} \qquad (38)$$

where η_{hi} denotes a set of particle numbers in which the associated particles are located on $\Gamma_x^{h_i}$. Equation (38) represents two sets of constraint equations that are needed to be solved simultaneously, with the equilibrium equation. Using the transformation method, the transformation matrix is formed by establishing the relationship between the nodal value $v_i^h(\mathbf{X}_J) \equiv \hat{g}_{iJ}$ and the "generalized" displacement g_{iJ} by

$$\hat{g}_{iJ} = \sum_{I=1}^{NP} \Phi_I(\mathbf{X}_J) g_{iI} \qquad (39)$$

$$\begin{bmatrix} \hat{\mathbf{g}}_1 \\ \hat{\mathbf{g}}_2 \\ \vdots \\ \hat{\mathbf{g}}_N \end{bmatrix} = \begin{bmatrix} \Phi_1(\mathbf{X}_1)\mathbf{I} & \Phi_1(\mathbf{X}_2)\mathbf{I} & \cdots & \Phi_1(\mathbf{X}_N)\mathbf{I} \\ \Phi_2(\mathbf{X}_1)\mathbf{I} & \Phi_2(\mathbf{X}_2)\mathbf{I} & \cdots & \Phi_2(\mathbf{X}_N)\mathbf{I} \\ \vdots & \vdots & \ddots & \vdots \\ \Phi_N(\mathbf{X}_1)\mathbf{I} & \Phi_N(\mathbf{X}_2)\mathbf{I} & \cdots & \Phi_N(\mathbf{X}_N)\mathbf{I} \end{bmatrix} \begin{bmatrix} \mathbf{g}_1 \\ \mathbf{g}_2 \\ \vdots \\ \mathbf{g}_N \end{bmatrix} (N = NP)$$

$$= \Lambda \mathbf{g} \qquad (40)$$

where Λ is the coordinate transformation matrix. The shape functions can be also be transformed by

$$\hat{\Phi}_I(\mathbf{X}) = \sum_{I=1}^{NP} \Lambda_{IJ}^{-1} \Phi_I(\mathbf{X}) \qquad (41)$$

$$v_i^h(\mathbf{X}) = \sum_{I=1}^{NP} \Phi_I(\mathbf{X}) g_{iI} = \sum_{I=1}^{NP} \hat{\Phi}_I(\mathbf{X}) \hat{g}_{iI} \qquad (42)$$

Note that $\hat{\Phi}_I(\mathbf{X}_J) = \delta_{IJ}$, and with this transformation, the essential boundary conditions of Eq. (38) are imposed by

$$\left.\begin{aligned} \hat{g}_{iI} &= h_i(\mathbf{X}_I) \\ \delta \hat{g}_{iI} &= 0 \end{aligned}\right\} \forall I \in \eta_{h_i} \qquad (43)$$

Using the Lagrangian reproducing kernel shape functions, the transformation matrix is formed only once and that can be performed at the undeformed configuration or even at a pre-processing stage.

The matrix equation is obtained by introducing kernel shape function and coordinate transformation to the variational equation as

$$\hat{\mathbf{K}} \Delta \hat{\mathbf{g}} = \hat{\mathbf{f}};\ \hat{\mathbf{K}} = \Lambda^{-T} \mathbf{K} \Lambda^{-1} + \hat{\mathbf{K}}^C;\ \hat{\mathbf{f}} = \Lambda^{-T} \mathbf{f} - \hat{\mathbf{f}}^C \qquad (44)$$

where **K**, and **f** are stiffness and external force in a generalized coordinate, $\hat{\mathbf{K}}^C$ and $\hat{\mathbf{f}}^C$ are the stiffness and force associated with the contact penalty terms, respectively, and they can be obtained directly using a nodal coordinate.

4. NUMERICAL EXAMPLE

Figure 1 shows the initial geometry of a cylinder compression problem. The initial billet geometry is: diameter=1.0 in., height=1.0 in.. The lower rigid wall is fixed and the upper die moves downward. The domain is discretized using 81 RKPM particles. The die velocity is -1.0 (in./s) with friction factor of 0.5, and the flow stress is $10.0(\bar{\varepsilon})^{0.1}$ (Ksi). In this example, a simple contact condition is imposed by an external force for the demonstration purpose of the proposed method.

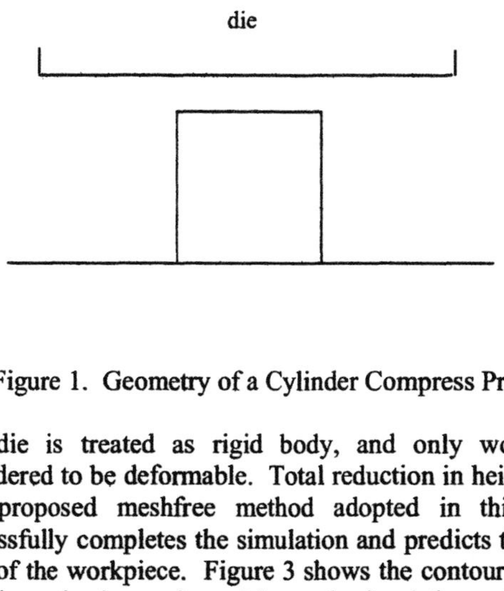

Figure 1. Geometry of a Cylinder Compress Problem

The die is treated as rigid body, and only workpiece is considered to be deformable. Total reduction in height is 40%. The proposed meshfree method adopted in this research successfully completes the simulation and predicts the material flow of the workpiece. Figure 3 shows the contour plot of the effective plastic strain. More plastic deformation of the workpiece is observed near the inside of the cylinder.

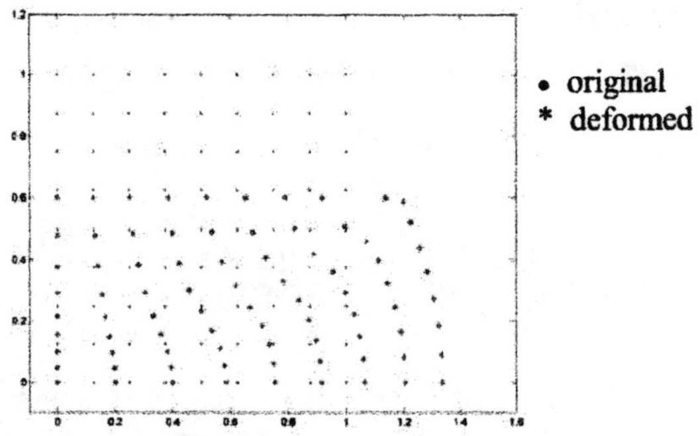

Figure 2. Undeformed and Deformed Configuration of the Symmetric Half-Model

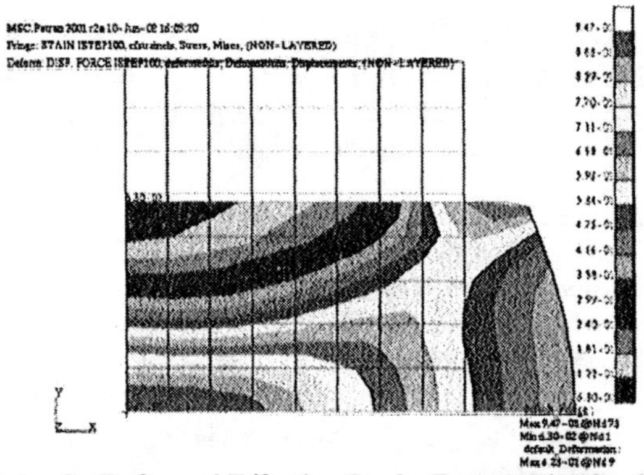

Figure 3. Deformed Effective Strain Contour Plot for the Symmetric Half-Model

CONCLUSION

A meshfree formulation for rigid-plastic material has been presented. The domain of the workpiece is discretized using the Reproducing Kernel Particle Method (RKPM) where no external meshes are used. In the large deformation process involving plastic materials, the elastic deformations can be considered as negligible. For a contact condition, a variational formulation is developed in the case where the frictional stress is dependent on the relative velocity at the die-workpiece interface. A numerical example is presented to validate the proposed method.

5. ACKNOWLEDGEMENT

This research is supported by Sandia National Laboratories.

REFERENCES

Antunez, H,J, and Kleiber, M., Sensitivity of forming processes to shape parameters, *Comput. Meth. Appl. Mech. Engrg.*, Vol. 137, pp. 189-206, 1996.

Balaji, P.A., Sundararajan T., and Lal, G.K., Viscoplastic deformation analysis and extrusion die design by FEM, *J. Appl. Mech., Trans. ASME*, Vol. 58, pp. 644-650, 1991.

Belytschko, T. Lea Y., and Gu, L.,"Element-free galerkin methods," *Int. J. Numer. Meth. Eng.*, Vol. 37, pp. 229-256, 1994.

Chandra, N., Analysis of superplastic metal forming by a finite element method, *Int. J. for Numer. Meth. Engr.*, Vol. 26, No. 9, pp. 1925-1944, 1988.

Chen, J.S., Pan, C., Wu, C.T., and Liu, W.K., "Reproducing kernel particle methods for large deformation analysis of nonlinear structures," *Comput. Meth. Appl. Mech. Engrg.*, Vol. 139, pp. 195-229, 1996.

Chen, J.S., Roque, C.M.O.L., Pan, C., and Button, S.T., "Analysis of Metal Forming Process Based on Meshfree Method," *J. Mat. Process. Tech.*, Vol. 80-81, pp. 642-646, 1998.

Hill, R., New horizons in the mechanics of solids, *J. Mech. Phys. Solids*, Vol. 5, No. 66., 1956.

Lee C.H. and Kobayashi S., New solution of rigid-plastic deformation problems using a matrix method, *Trans. ASME J. Engr. Ind.* Vol. 95, pp. 865-873, 1973.

Liu, W., Jun, S., Zhang, Y., "Reproducing kernel particle methods," *Int. J. Numer. Meth. Fluids*, Vol. 20, pp. 1081-1106, 1995.

Mori, K., Osakada K., and Oda, T., Simulation of plane-strain rolling by the rigid-plastic finite element method, *Int. J. Mech. Sci.*, Vol. 24, No. 9, pp. 519-527, 1982.

Oh, S.I. and Kobayashi, S., Finite element analysis of plane-strain sheet bending, *Int. J. Mech. Sci.*, Vol. 22, pp. 583-594, 1980.

Osakada K., Nakano, J., and Mori, K., Finite element method for rigid-plastic analysis of metal forming-formulation for finite deformation, *Int. J. Mech. Sci.*, Vol. 24, No. 8, pp. 459-468, 1982.

Zienkiewicz, O.C. and Godbole P.N., Flow of plastic and visco-plastic solids with special reference to extrusion and forming processes, *Int. J. Num. Meth. Engr.* Vol. 8, pp 3-16, 1974.

PVP-Vol. 458, Computer Technology and Applications

PVP2003-1908

STUDY ABOUT REAL-TIME FINITE ELEMENT METHOD USING CNN

Gao JingBo , Xu MinQiang and Wang RiXin
Dept. of Astronautic Engineering and Mechanics
Harbin Institute of Technology
Harbin CHINA

ABSTRACT

The paper presents cellular neural network (CNN) for real-time finite element method, useful as calculating temperature field and thermal stress field for rotor of turbine and so on. The comparability between template of CNN and the stiffness matrix of finite element is analyzed, and the conception of finite element template (FMT) of CNN is discussed. The FMT can be suitable for finite element grid with arbitrary shape. In this paper, the FMT is simulated by temperature field of rotor of turbine, the result is right.

INTRODUCTION

The finite element method is widely applied to many domains, such as engineering, thermokinetics, oceanography, biology, etc. The major drawback of the finite element method is that execution takes a lot of time and memory spaces[1]. To solve problems using FEM on a distributed memory parallel computer, a large number of parallel algorithms have been proposed in the literature, for example the domain decomposition method [2] and the EBE (Element-by-Element) method [3], inverse electromagnetic problems [4], and these methods efficiently execute a finite element application program on a distributed memory multicomputer□

In recent years , the parallel finite element algorithm based on neural network has been used in image reconstruction in electrical impedance imaging[5], electromagnetic fields[4] and finite element mesh generating [6-7] etc. In the literature[8,9], the Hopfield Neural Network(HNN) is directly used for parallel finite element analysis. The main property of the HNN, constructed of interconnected neurons, is to decrease the energy of the network until it reaches minimum with the time evolution of the system. This process is very similar to the minimization of the energy functional defined by ordinary finite element analysis[8]. This similarity makes the usage of the HNN in ordinary FEA relatively easy[8].

Cellular neural networks that develop the interconnecting mode on the basis of the Hopfield neural network were introduced by Chua and Yang [10] for image processing problem. They have been applied in such as image and signal processing, vision, and pattern recognition, optimization. Recently, many authors have paid much attention to the research on the theory and application of the CNNs, e.g.,[11-13], and the CNN chips have been designed and fabricated, and the chip of CNN not only have small dimensions but also have high cell density , for example the chip of CNN_UM that only have 9.27*8.45 sq.mm about dimension and 29.24 cells/mm^2 about density[14]. The realization of chip of CNN provide a sufficient condition of the micro-machine of realizing FEM.

The work concentrates on solution of the system of equation(1) and equation(2) generated in the FEM processing using cellular neural network(CNN).

$$[\mathbf{C}]\{\dot{\mathbf{x}}\} + [\mathbf{K}]\{\mathbf{x}\} = \mathbf{P} \qquad (1)$$

$$[\mathbf{K}]\{\mathbf{x}\} = \mathbf{P} \qquad (2)$$

Where: {x} is the displacement vector at time t, or it can also stand for other field variables in field problems, [K] is the stiffness matrix, and P is the factor vector. It is assumed that the initial conditions and boundary conditions have already been known. In the paper, the comparability between template of CNN and the stiffness matrix of finite element method is analyzed, and the conception of finite element template (FMT) of CNN is discussed. Meanwhile, an example is given to simulate temperature field of rotor of turbine using FMT.

CELLULAR NEURAL NETWORK

As in the well-known case of cellular automata, a CNN is defined as an *n*-dimensional array of dynamic processing elements called cells with local iterations between them. However, they are different in the sense that all state variables take continuous values. CNN may present any dimension according to the dimension of structure of the finite element analysis. In the paper, the bidimensional CNN is applied to analyze temperature field of rotor of turbine.

An example of a two-dimensional CNN is shown in Figure 1 with $C(i,j)$ the cell located in the *i*th row and the *j*th column of the array.

The links in Figure 1 indicate the connections between neighbouring cells. Let $C(i,j)$ denote the cell at the (i,j)position of the lattice, the *r*-neighborhood $N_r(i,j)$ of $C(i,j)$ is:

$$N_r(i£j) = \{C(k£l) \mid \max[|k-i|, \ |j-l|] \le r\}$$
$$1 \le k \le m£ 1 \le l \le n$$

(3)

The order *r* is defined as *r*-neighborhood radius.

Each cell is formed with circuit elements. A cell of neural is shown in Figure 2. Observe from Figure 2 that each cell $C(i,j)$ contain current source I, capacitor, resistor, independent and dependent sources.

Applying the Kirckoff laws for solving the circuit, a classical state equation for a cell is obtained:

$$C\frac{dx_{ij}(t)}{dt} = -\frac{1}{R_x}x_{ij}(t) + \sum_{C(k,l)\in N_r(i,j)} I_{xy}(i,j;k,l) + \sum_{C(k,l)\in N_r(i,j)} I_{xu}(i,j;k,l) + I \quad (4)$$

The cell external input is given as an independent voltage source u_{ij}, x_{ij} and y_{ij} are the internal state and output voltages respectively. The offset term is represented as an independent current source I. The influence of external inputs and outputs from neighbouring cells $C(k, l)$ is represented as voltage-controlled current sources $I_{xy}(i,j,k,l)$ and $I_{xu}(i,j,k,l)$. In the particular case of a linear relation, I_{xy}□ I_{xu} can be described as:

$$I_{xy}(i,j;k,l) = A(i,j;k,l)y_{kl}$$
$$I_{xu}(i,j;k,l) = B(i,j;k,l)u_{kl}$$

(5)

Finally the output of voltage-controlled current source I_{yx} describes as:

$$I_{yx} = \left(\frac{1}{R_y}\right)f(x_{ij})$$

The set (A, B, I) is called the cloning template. For solving each specified problem it is necessary to obtain appropriate cloning templates for the network[15].

THE FINITE ELEMENT TEMPLATE (FMT) OF CNN

In order to solve the equation (1) and equation (2) on real-time using CNN, the characteristic of the global stiffness matrix [K] will be discussed.

The global stiffness matrix [K] generated in FEM processing is sparse and symmetric. A simple grid of FEM is shown in Figure 3. The distribution of nonzero elements of the global stiffness matrix [K] is described in Figure 4.

Take the 6th node as example. Notice that the 6th row of the global matrix [K] in Figure 4 have zero elements of node 4, 8 and 12 since nodes 4, 8 and 12 have no direct connection to node 6. Meanwhile the other nodes have the same connection characteristic as the 6th node. Therefore the changer of one node associated with only the changers of its neighbor nodes connecting the node but the changer of neighborless nodes.

As analyzed as above, the conception of FMT of CNN discuss as follows.

The conception of the finite element template (FMT) of CNN is described: any one of the nodes in the FEM grids is regarded as a central cell of CNN, and the neighbor nodes connecting the same elements with the central cell is the neighbor cells of radius $r=1$ of CNN. The template (A,B,I) of CNN is decided by the global stiffness matrix in the FEM processing.

The finite element template (FMT) of CNN can be suitable for finite element grid with arbitrary shape. Whatever the FEM grid it is, the distribution of nonzero elements of the stiffness matrix generated in the FEM processing only related with the neighborhood nodes that connected the same element with the central nodes, and not related with the element shapes.

Therefore, there are two characteristics of the FMT in this article.

1. The radius of FMT of CNN is equal to 1; The dimension of matrix of FMT is related to the number of neighborhood cells of the central cell.

2. Different cells have different FMT, and the matrix of FMT of the central cell is made of non-zero elements of the global stiffness matrix of the neighborhood cells.

THE IMPLEMENTATION USING FMT OF CNN

Take the temperature field of rotor of turbine as example: the equation of temperature stiffness matrix of rotor's temperature field based on FEM is:

$$[\mathbf{C}]\{\dot{\mathbf{T}}\} + [\mathbf{K}]\{\mathbf{T}\} = \mathbf{P} \quad (6)$$

The state equation of a certain node i inside the temperature field is described by the equation as follow:

$$C^{(i)}\frac{dT^{(i)}}{dt}=-\sum_{j=1}^{m}K_{ij}T_j+P^{(i)} \tag{7}$$

Where $C^{(i)}$ stands for the thermal capacity matrix value of the jth node at the ith line.

Compare with equation (4) and equation (7), the result can be got that the state equation of the temperature field is similar to the cell circuit of CNN.

So the process of calculating the temperature field can be realized by CNN.

But when we calculate the thermal stress of rotor turbine using FEM, the state equation of stiffness matrix of thermal stress field using FEM method can be described as $\mathbf{AX}=\mathbf{b}$. When we solute the equation $\mathbf{AX}=\mathbf{b}$ using neural network, the expected value x of the equation $\mathbf{AX}=\mathbf{b}$ is described by the minimum energy state of the energy function *E(x)* that will be constructed.

Through the transform of $\mathbf{AX}=\mathbf{b}$,the differential function equations of the minimum energy function *E(x)* are:

$$\begin{cases}\dfrac{dx}{dt}=-\mu(t)\nabla E(x)\\ \nabla E(x)=\mathbf{A}^{\mathrm{T}}(\mathbf{Ax}-\mathbf{b})\\ \mathbf{x}(0)=\mathbf{x}^{(0)}\end{cases} \tag{8}$$

where $\mu(t)=|\mu_{ij}(t)|$ is a $n\times n$ positive diagonal matrix, each columns of $\mu(t)$ depends on time *t* and vector x□ that is $\mu_{ij}=\mu_j\delta_{ij},\mu_j>0$

$\nabla E(x)$ is the energy gradient function, that is

$$\nabla E(x)=\left[\frac{\partial E(x)}{\partial x_1},\frac{\partial E(x)}{\partial x_2},\ldots,\frac{\partial E(x)}{\partial x_n}\right]^{\mathrm{T}} \tag{9}$$

In equation (8), matrix A is the global stiffness matrix based on the FEM analysis.

The equations (8) is similar to equation (7), the transformation equations (8) can realized by the FMT of CNN.

SIMULATION

The temperature field of turbine rotor can be simulated by the FMT of CNN that have been described above.

The analysis and calculation of rotor of 200MW turbine based on FEM can refer to literature [16]. This article compares the result of simulation of FMT with the result of FE analysis in the literature[16] . The conditions of simulation temperature field using FMT of CNN are same as those used in literature [16]. The shape of 200MW turbine rotor and the mapping of FEM grids are as described in the Figure 5.

The change of temperature is list in table 1, that is the curve of temperature change of cold start-up of turbine.

Simulation result:

The Figure 6 shows the isothermal diagram of turbine rotor through simulation of FMT. The simulation result of FMT is approximate to the result in literature [16].

CONCLUSIONS

In this paper, the comparability between template of CNN and the stiffness matrix of finite element is analyzed, and the conception of finite element template (FMT) of CNN is discussed. The FMT can be suitable for finite element grid with arbitrary shape. The FMT is simulated by temperature field of rotor of turbine, the result is right.

It can be concluded from the discussion in this article that it is feasible to realize the parallel calculation of FE structure using hardware of CNN. And as the method through VLSA of CNN is fast, it provides a real-time measuring way in engineering application.

REFERENCES

1. Chao-Jen Lee; Yeh-Ching Chung. "A web-based parallel PDE solver generation system for distributed memory computing environments" *Computer Software and Applications Conference, 2000. COMPSAC 2000. The 24th Annual International* , 2000 Page(s): 54 –59

2. Zuping Qian; Lei Yin; Wei Hong; "Application of domain decomposition and finite element method to electromagnetic compatible analysis" *Antennas and Propagation Society, 2001 IEEE International Sym* , Volume: 4 , 2001, Page(s): 642 -645 vol.4

3. Chellamuthu, K.C.; Ida, N.;."'A posteriori' element by element local error estimation technique and 2D & 3D adaptive finite element mesh refinement" *Magnetics, IEEE Transactions on* , Volume: 30 Issue: 5 , 31 Oct-4 Nov 1993, Page(s): 3527 –3530

4. T.S.Low, Bi Chao. "The use of finite elements and neural networks for the solution of inverse electromagnetic problems" *IEEE Transactions on Magnetics* , VOL 28, No5, sep1992, 2811-2813.

5. Kilic, B.; Korurek, M.; "A Finite Element Method based Neural Network Technique for Image Reconstruction in Electrical Impedance Imaging" *Biomedical Engineering Days, 1998. Proceedings of the 1998 2nd International Conference* , 20-22 May 1998, Page(s): 100 -102.

6. S.Alfonzetti. "A finite element mesh generator based on an adaptive neural" *IEEE transactions on Magnetics* VOL 34,No.5,pp3363-3366, Sep1998.

7. R.Chedid. "Automatic finite-element mesh generation using artificial neural networks" *IEEE transactions on Magnetics* VOL 32,No.5,pp5173-5178, Sep1996.

8. Hideo Yamashita and Norio Kowato etc. "Direct solution method for finite element analysis using Hopfield neural network" *IEEE transactions on Magnetics* VOL 31,No.3,pp1964-1967, Sep1995.

9. Ioannis T. Rekanos and Theodoros D. Tsiboukis. "A parallel method based on the hopfield neural network for the solution of the finite elements method equations". FSTE16

10. L.O.Chua and L.Yang, "cellular networks:Theory", *IEEE Trans. On Circuits and Systems,* VOL.35,No.10, pp1257-1272, Oct 1988.

11. Anping Chen a , Jinde Cao, "Almost periodic solution of shunting inhibitory CNNs with delays" *Physics Letters* A 298 (2002) 161–170.

12. Crounse, K.R.; Chinling Wee; Chua, L.O."Linear spatial filter design for implementation on the CNN Universal Machine" *Cellular Neural Networks and Their Applications, 2000. (CNNA 2000). Proceedings of the 2000 6th IEEE International Workshop on* , 2000 Page(s): 357 –362

13. Arena, P.; Basile, A.; Fortuna, L.; Yalcin, M.E.; Vandewalle, J. " Watermarking for the authentication of video on CNN-UM " *Cellular Neural Networks and Their Applications, 2002. (CNNA 2002). Proceedings of the 2002 7th IEEE International Workshop on* , 22-24 Jul 2002 Page(s): 78 -83.

14. Carmona, R.; Jimenez-Garrido, F.; Dominguez-Castro, R.; Espejo, S.; Rodriguez-Vazquez, A. "CMOS realization of a 2-layer CNN universal machine chip" *Cellular Neural Networks and Their Applications, 2002. (CNNA 2002). Proceedings of the 2002 7th IEEE International Workshop on* , 22-24 Jul 2002, Page(s): 444 -51,xviii-x

15. D.L. Vilarino, V.M. Brea, D. Cabello and J.M. Pardo . "Discrete-time CNN for image segmentation by active contours", *Pattern Recognition Letters* 19 1998 721–734.

16. Song Chunting. "Study on Estimate Remaining Life of rotor of 200MW turbogenerator". *Dissertation for The Master Degree Harbin Institute of Technology* 2000:10,12-15.

Table 1. The temperature variation relationship of cold startup

Time □ s□	0	15600	21600	25400
Temperature(□)	120	500	535	535

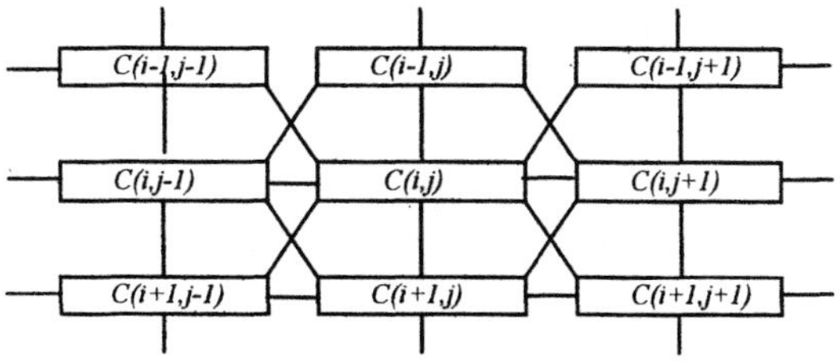

Figure 1. Structure for a bidimensional CNN.

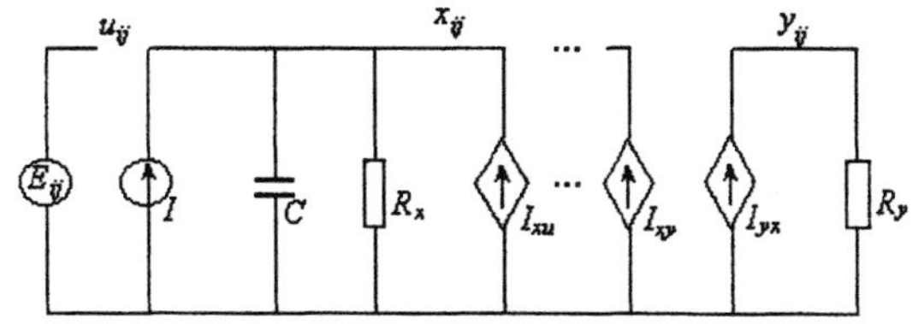

Figure 2. Example of a circuit representation of a cell

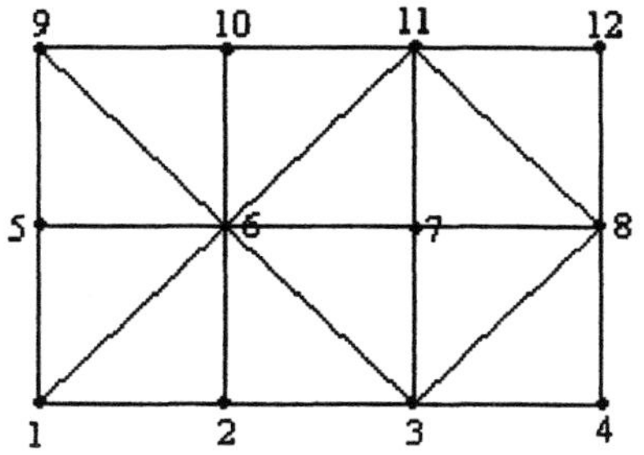

Figure 3. Example of grids

	1	2	3	4	5	6	7	8	9	10	11	12
1	Δ	Δ			Δ	Δ						
2	Δ	Δ	Δ			Δ						
3		Δ	Δ	Δ		Δ	Δ	Δ				
4			Δ	Δ				Δ				
5	Δ				Δ	Δ			Δ			
(6)	▲	▲	▲		▲	▲	▲		▲	▲	▲	
7			Δ			Δ	Δ	Δ			Δ	
8			Δ	Δ			Δ	Δ			Δ	Δ
9					Δ	Δ			Δ	Δ		
10						Δ			Δ	Δ	Δ	
11						Δ	Δ	Δ		Δ	Δ	Δ
12								Δ			Δ	Δ

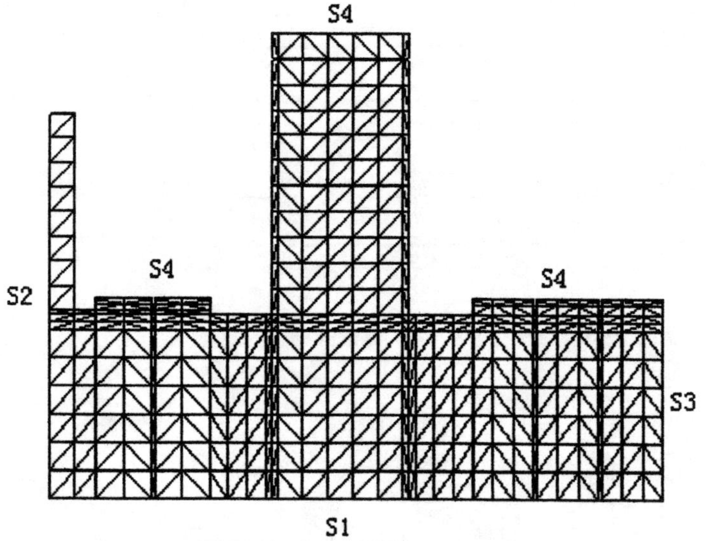

Figure 5. The FEM grids of high pressure rotor

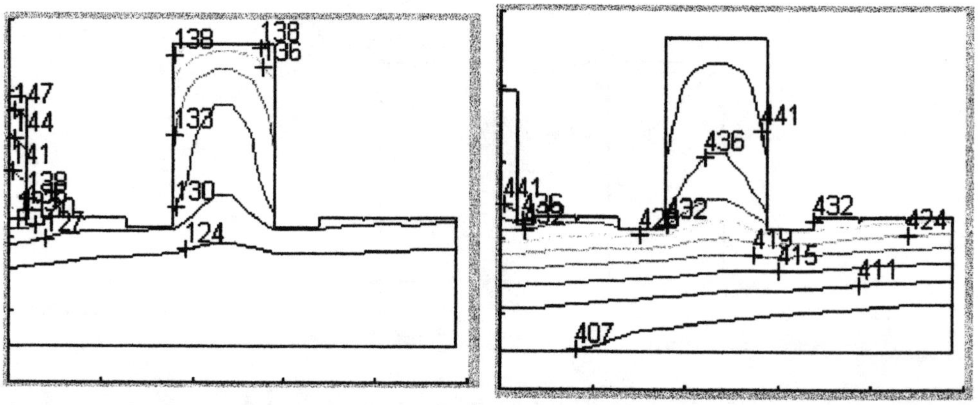

(a) the isothermal diagram at 300s **(b) the isothermal diagram of turbine rotor at 15600s**

Figure 6. The isothermal diagram of different time of rotor of 200MW turbine

PVP-Vol. 458, Computer Technology and Applications

PVP2003-1909

Crack Shape Development of Two Interacting Surface Elliptical Cracks

S. K. Patel*, B. Dattaguru+, K. Ramachandra*

+Department of Aerospace Engineering, Indian Institute of Science, Bangalore-560 012, INDIA, *Structural Integrity and Mechanical Analysis Group, Gas Turbine Research Establishment, Bangalore-560 093, INDIA.

Abstract

The issue of assessing residual life of an aged structure based on damage tolerance concepts attained significance in high technology fields such as aerospace, piping and pressure vessels and nuclear engineering. Computational fracture analysis of these structures in the presence of single or multi-site surface flaws is essential for life estimation and life extension. In this paper development of accurate post-processing technique (Modified Virtual Crack Closure Integral) to estimate strain energy release rates, and simple numerical method to simulate crack shape development in single and multiple interacting cracks (till they merge into single dominant crack) is presented.

Crack shape development in single surface elliptical cracks was carried out earlier in literature using 2 degree of freedom model wherein fatigue crack growth is estimated along the major and minor axis of the ellipse and new crack shape was derived by fitting an ellipse to these points. A special three-degree of freedom model is proposed and presented in this paper for interacting and coalescing cracks. The crack shape development was checked with experimental work on coupons with multi-site surface cracks tested under fatigue loading.

In safety critical aerospace and thick piping structures this work is significant in predicting the remaining life of aged components with multi-site damage.

1. INTRODUCTION

The issue of assessing residual life of an aged structure[1] based on damage tolerance concepts attained significance in high technology fields such as aerospace, piping and pressure vessels and nuclear engineering. Computational fracture analysis of these structures in the presence of single and multiple surface flaws is essential for life estimation and life extension. In this paper development of accurate post-processing technique (Modified Virtual Crack Closure Integral) to estimate strain energy release rates, and a simple numerical method to simulate crack shape development in multiple interacting cracks is presented.

Mainly three approaches are available for assessment of multiple surface cracks in fatigue viz. ASME XI code (Rules for in-service inspection of nuclear power plant components, 1977), BSI PD6493 code (Guidance for the derivation of acceptance levels for defects in fusion welded joints, 1980) and NIIT ('No interaction and immediate transition'). As per ASME XI and BSI PD6493 codes, the adjacent cracks should be re-characterized to a single crack if certain geometric conditions are satisfied. Clearly, this method excludes the stages of interaction and coalescence. The NIIT method assumes no interaction before the inner tips of the cracks touch and immediate formation to a single elliptical shaped crack after they touch. This actually means that the advance of each adjacent crack is individually computed until they come together, and the region of coalescence cracks is neglected. Here, a simplified semi-analytical/numerical approach is presented to simulate the growth of multiple surface cracks in fatigue considering interaction and coalescence of cracks.

The crack growth prediction is carried out using Paris's law with Elber correction for crack closure. Fatigue crack growth experiments were carried out on single and multiple surface elliptical cracks and crack closure values were measured from experiment. The stress intensity factor for numerical prediction of crack growth was obtained using modified virtual crack closure integral (MVCCI) technique extensively developed by authors in the past. Finite element analysis was conducted using eight-noded brick (HEXA8) elements in NASTRAN software. Post-processing FEM data for MVCCI estimation of strain energy release rate components in various modes of fracture is carried outside NASTRAN through special purpose software. This paper presents MVCCI expressions for Hexa8 element and also a special correction for unequal size elements on both sides of the crack front.

Fatigue crack growth shape development for single elliptical crack was presented earlier in literature using two-degree of freedom model [2,3,4]. In present paper, a three-degree of freedom model is presented for two surface elliptical cracks in interacting and coalescing phases. The crack shape development using 3 DOF model is carried out and the crack shape is compared with experimental measurements on fracture surface obtained in the experimental program. Excellent correlation is seen for both single and multiple cracks.

2. FATIGUE CRACK GROWTH OF SINGLE AND MULTIPLE SURFACE CRACKS

2.1 FATIGUE CRACK GROWTH OF SINGLE SURFACE CRACK

The change in shape of a single surface crack during fatigue loading was predicted earlier using semi-analytical/numerical and computational models. The "semi-analytical method" relies on empirical form of stress intensity factor solution and the shape of the crack at various stages of growth is determined by two degrees of freedom. Although "semi-analytical method" restricts the crack shape to be only elliptical or circular but it is computationally inexpensive.

Numerous studies [2,3,4] on the evolving shape of surface cracks in fatigue are reported in literature. All these studies are in general based on Paris equation. The Paris equation for prediction of crack growth rate for three-dimensional crack can be written in the following generalized form

$$\frac{da_i}{dN} = f\left(\Delta K_i, C_i, n_i\right)$$

where ΔK_i is discrete stress intensity factor range at "i^{th} degree of freedom", "N" is the number of load cycles, "C_i" and "n_i" are material constants of Paris equation, "i" is the degree of freedom, a_i is the characteristic dimension at i^{th} degree of freedom. Two degrees of freedom (DOF) model is based on crack propagation along two principal dimensions of an ellipse. The equations for the crack growth in 'a' and 'c' can be written as

$$\frac{da}{dN} = C_a\left(\Delta K_a\right)^{n_a}$$

$$\frac{dc}{dN} = C_c\left(\Delta K_c\right)^{n_c}$$

Where C_a, C_c and n are material properties, K_a and K_c are the values of stress intensity factor at points 'A' and 'C' respectively as shown in Fig.1.

Using eqs. (2) & (3) and adopting the plasticity induced closure concept, the conventional crack growth equation using stress intensity factor ranges may be modified to incorporate the effects of plasticity induced crack closure, the equation for crack shape development can be written as

$$\frac{dc}{da} = \left(\beta_R \frac{\Delta K_c}{\Delta K_a}\right)^n$$

where β_R is the ratio of crack closure factors at 'C' and 'A'.

Jolles and Tortoriello [5] assumed β_R=0.911. A similar value (an empirical factor) equal to 0.9 was suggested by Newman and Raju [3] in the Paris equation for surface crack.

Two degrees of freedom ("a" and "c" for a semi-elliptical crack model) model along with close form stress intensity factor solutions has been used. The empirical solutions (fitted equations for stress intensity factor from finite element solutions) provided by Newman-Raju [3] are used with proposed finite width correction as proposed in ref.[5]. The maximum and minimum values of crack closure at points 'C' and 'A' are determined experimentally in the present work .

2.2 FATIGUE CRACK GROWH OF MULTIPLE SURFACE CRACKS

As mentioned before the NIIT method assumes no interaction before the inner tips of the cracks touch and immediate formation to a single elliptical shaped crack after they touch. This actually means that the advance of each adjacent crack is individually computed until they come together, and the region of coalescence cracks is neglected. Here, a simplified semi-analytical/numerical approach is presented to simulate the growth of multiple surface cracks in fatigue considering interaction and coalescence of cracks. The growth of multiple surface cracks in fatigue can be divided in various phases

I. Multiple cracks without interaction
II. Multiple cracks with interaction
III. Coalescing cracks
IV. Formation and growth of single crack

Three degrees of freedom per crack is thought to be appropriate for simulation of multiple cracks in fatigue as shown in Fig.2. For clarity, the procedure is explained with respect to symmetric cracks where it is sufficient to consider one crack in simulation. The similar procedure can be applied to all the cracks in case of unsymmetric crack configuration. All the assumptions for single crack are also applicable here.

The crack extension is assumed to be in normal direction at the considered location. Until the interaction (phase-I) the simulation procedure is same as for single surface crack and stress intensity factor for single crack is assumed to be applicable. Once the cracks come in close vicinity, they start interacting. Here, the interaction effects are considered by multiplying the stress intensity factor of single crack with interaction factor. The interaction effect is considered at the location B only and its effect is considered negligible on points A and C. The similar procedure to account the effect of interaction was proposed by various authors [6, 7,8,9,10].

Here, a modified procedure for crack extension during the interaction phase is proposed although in principal it is same as proposed earlier [6, 7]. In reference [6, 7] the crack extensions at location A, B and C are assumed to occur in normal direction based on the stress intensity factors and Paris equation. But once the crack extensions occur at these three locations, the ellipse (semi-ellipse) can stay remain ellipse since the extension at locations A and C would be different due to interaction effects. This can be avoided if the center of ellipse is shifted. This is also logical since as cracks grow and form a single crack, it center would be shifted. Thus during interaction, if a_0 and c_0 represent the crack geometry at current step and a_1 and c_1 represent the geometry of the crack after the fatigue crack growth increments at A, B and C as shown in Fig.2.

Once the cracks approximately touch each other, coalesce phase begins. During coalesce the third degree of freedom is taken at coalesce plane i.e. at location 'B' as shown in Fig.3 and direction of extension is taken along the coalesce plane which is also bynormal to crack fronts at point 'B'.

The updated geometry after crack extension can be obtained using the generalised equation for ellipse having shifted center point by 'Δh' in coordinate system 'xy'. The formulation for interacting and coalescing cracks is given elsewhere [5]. During coalesce, the crack extensions Δc, Δa, Δy at locations A, B and C respectively are predicted using stress intensity factor and Paris equation. The stress intensity factor at B is multiplied by interaction factor eq.(12). Using Δc, Δa and Δy, new crack dimensions are predicted. The X_{coal} and Y_{coal} are updated as $X_{coal} = X_{coal} + \Delta h$ and $Y_{coal} = Y_{coal} + \Delta y$ respectively. The coalescence phase is assumed to be completed when Y_{coal} reaches 99% of a_1 and the length of the single crack would be $c+X_{coal}$. Subsequently, the crack growth follows the procedure same as that for a single crack. In the crack growth simulation of multiple crack Δa is taken equal to 0.0001 times of 'a_o'.

3. MVCCI for 3-D Cracks

Consider an arbitrary crack in a three dimensional continuum, as shown in Fig.4. Let the surface crack in the elastic system differ from the actual state of equilibrium by an infinitesimally small area shown by the area between virtually extended crack front and the original crack front. The new contour is a curve that encloses the point 'p' lying in the plane of the crack. An infinitesimally small area ΔA is shown at point 'p' behind and ahead of the original crack front.

The strain energy release rate due to virtual crack extension is an average value over the region of extension. The SERR in individual modes at point "p" (Fig. 4) with respect to local coordinate system n, t, y: where "n" is normal to the crack and lies in the plane of the crack surface, "t" is tangential to the crack front and "y" is orthogonal to n and t, are written as follows

$$G_I(p) = \lim_{\Delta A \to 0} \frac{1}{2\Delta A} \int_{\Delta A} \sigma_{yy}(n,0,t) U_y(\Delta a - n,0,t) dA \quad (5)$$

$$G_{II}(p) = \lim_{\Delta A \to 0} \frac{1}{2\Delta A} \int_{\Delta A} \sigma_{ny}(n,0,t) U_n(\Delta a - n,0,t) dA \quad (6)$$

$$G_{III}(p) = \lim_{\Delta A \to 0} \frac{1}{2\Delta A} \int_{\Delta A} \sigma_{ty}(n,0,t) U_t(\Delta a - n,0,t) dA \quad (7)$$

For effective utilisation of this concept in conjunction with the finite element method Badari Narayana et al [11] have proposed a general procedure for the derivation of MVCCI expressions in three-dimensional problems with cracks in terms of nodal forces and displacements in the elements forming the crack front. Consider the finite element idealization of surface crack configuration with the eight noded conventional brick elements along the crack front. Considering Fig.5, the MVCCI for $G_I(p)$ in eq.(5) can be written after geometric transformation in the natural coordinate system as follows (Part of derivation presented here is from Ref. 11, but is given here for the sack of completeness and to enable the present work to propose appropriate corrections):

$$G_I(p) = \frac{1}{2\Delta A_k} \int_{-1}^{1} \int_{-1}^{1} \sigma_y(\xi,\zeta) U_y |J| d\xi d\zeta \quad ...(8)$$

where stress σ_y and displacement U_y are assumed in polynomial form consistent with the shape functions of the element. For an eight noded element det|J| where J is Jacobian matrix is conveniently expressed as

$$|J(\xi,\zeta)| = J_0 + J_1\xi + J_2\zeta \quad ...(9)$$

where J_0, J_1 and J_2 are constants depending on the real coordinates of the nodes.

Substituting consistent stress and displacement, and eq.(9) in the eq.(8) and carrying out necessary integration, the expression [11] for $G_I(p)$ is given by

$$G_I(p) = \frac{1}{2\Delta A_k} [4J_0C_{11} + 4J_1C_{12} + 4J_2C_{13}] \quad ...(10)$$

where

$$C_{11} = a_0b_0 + a_1b_1/3 + a_2b_2/3 + a_3b_3/9$$
$$C_{12} = a_1b_1 + a_1b_0/3 + a_2b_3/3 + a_3b_2/9$$
$$C_{13} = a_2b_2 + a_1b_3/3 + a_2b_0/3 + a_3b_1/9$$

and coefficients a_i (i=0,1,2,3) and b_i (i=0,1,2,3) can be determined from the nodal values of displacements (Fig 5) and forces respectively.

For a straight crack front and a rectangular shape of the element J_1 and J_2 are zero and SERR reduces to a simple expression:

$$G_I(p) = \frac{1}{2\Delta A_k} \left(F_{y,j} U_{y,j-1} + F_{y,j+3} U_{y,j-2}\right) \quad ...(11)$$

This expression for $G_I(p)$ was used by Shivakumar et al [12] with nodal forces obtained by partitioning the common nodal reaction force. This was accomplished by assuming that the forces shared by adjacent strips are proportional to their respective lengths. No such assumption is required in the MVCCI expression (10) of $G_I(p)$ since the forces are extracted directly from the element based on consistent force approach.

4. Area Correction for Unequal Mesh

In the MVCCI method the integral is based on the assumption [11] that the size of elements are equal on both the sides of the crack front and the edges of the elements are normal to the crack front. However in practice it is difficult to satisfy the first assumption even in the case of simple configurations such as penny shaped or surface elliptical crack. In this paper a correction is proposed to account for the unequal size elements across the crack front and also to account for the local curvature.

The correction factor is obtained as the ratio of nodal forces between unequal elements across the crack front (Fig.6). The forces are expressed consistent with the stress field at crack tip over elements in front of crack and in wake of crack front. The "curvature correction factor" to be multiplied in MVCCI force coefficient matrix is as follows:

$$\text{Curvature Correction Factor} = \left[\sqrt{\frac{\Delta a_2}{\Delta a_1}} \left\{ \frac{1 - \frac{1}{3}\frac{\Delta a_2}{R}}{1 + \frac{1}{3}\frac{\Delta a_1}{R}} \right\} \right] \quad ...(12)$$

Where "R" is radius of curvature at point "p", and "a" and "c" are minor and major axes of ellipse respectively.

5. Testing procedure

The surface crack specimen with multiple cracks is considered for testing. The specimen is rectangular in shape in the gauge section having thickness, t, and width, 2W. The major dimensions of the specimen are shown in Fig.7. The specimen is approved by AGARD committee [13]. The specimen material selected for present study is superalloy Inconel-718 which is extensively used for components of gas turbine engines. The specimens are made out of rolled bar of 20mm diameter. The rolled bar was received in hot rolled, solution treated and machined condition. Subsequently the product was heat treated as 900°C/1-1/2 hr/AC. The specimens are made using wire cutting EDM (Electrical Discharge Machining). The specimens were aged after machining. The ageing heat treatment cycle was 720°C/8hrs/furnace cooled at 620°C @ 50°C/hr/8hrs/aircooled.

The initial flaw was created by EDM method. The test on specimen has been carried out under constant amplitude uni-axial loading at frequency 2Hz. The crack growth fatigue test was carried at room temperature in air.

6. Experimental and numerical study of multiple surface cracks

The details of crack configuration of multiple cracks (twin semi-elliptical cracks) have been given in Table 1. The geometry of initial notch is measured from the fracture surface of the specimens. The photographs of fracture surface of specimens are shown in Fig. 8.

The measured points on crack fronts are shown in Fig. 9. The points on each front are fitted by an ellipse using least square fit technique. The points near the free surface are not considered in curve fitting. It is observed that during coalescence the crack front fits well into elliptical shape if the shift in center of ellipse is considered. After the coalescence of cracks, the single crack is found to have less curvature at center whereas higher curvature near the free surface compared to an ellipse, still it can be reasonably represented by an elliptical shape.

The crack shape development (aspect ratio v/s thickness ratio) obtained by experiment are compared with numerical prediction in Fig. 10. The numerical predictions are carried out using model presented in previous section. The predictions were also made considering crack closure. Numerical predictions are found to be in good agreement with experimental results. Higher deviations in numerical predictions are observed in the crack coalescence phase. In this phase, the model with closure predicts 24.9% higher aspect ratio whereas model without closure predicts 29.9% higher aspect ratio at a/t=0.216. This is principally due to uncertainty involved in transition from interaction and coalescence phases, and secondly due to effect of crack curvature at coalescence region.

The model including closure predicts smaller coalescence region relative to model neglecting closure phenomenon. The model including closure predicts the beginning of coalescence phase at a/t=0.196 whereas the model neglecting closure predicts the beginning of coalescence phase at a/t=0.189. The higher closure levels at the surface (Uc/Ua=0.9158) leads to slower crack growth at location 'C' relative to model excluding closure. Thus, the model considering crack closure predicts the beginning of coalescence phase at higher thickness ratio. The closure effects also leads to higher aspect ratio for same thickness ratio as observed in study of single crack. Since the interaction factor increases with aspect ratio the model with closure phenomenon predicts early completion of coalescence region.

Fig.11 shows the variation of interaction factor during crack progression. The crack growth starts with positive t_s/c i.e. from right side of the graph marked by an arrow. Interaction was not observed till t_s/c is approximately equal to 0.4 and then interaction factor increases and reaches to maximum value of 1.88. The interaction factor increases to 4.64 at the beginning of coalescence phase but drops very rapidly and subsequently it decreases and reaches to asymptotic value close to one.

The experimental life with crack lengths 'a' and 'c' is compared with predictions in Fig. 12. The numerical model predicts a higher growth rate in c-direction in coalescence region. The coalescence phase consumed more than 25% of total propagation life. The predicted life is found to be 3.4% higher relative to experimental life of this specimen.

7. Conclusion

Improved MVCCI method for three-dimensional cracks has been used to study interaction effects between two surface cracks in a finite body under uniform tension. Experiments were conducted to study the effects of multiple cracks on crack shape development and propagation life. It is observed that the centers of cracks shift towards the coalescence plane during the fatigue growth of the cracks. It is found that similar to a single crack the crack fronts in presence of multiple cracks can be approximated by part-ellipses if the shift in the centers of cracks is taken into account. A three-parameter crack growth model for estimation of crack shape development and life of multiple cracks has been proposed. The model was verified using the experimental results. The proposed three degrees of freedom model for interacting and coalescing crack found to predict the crack propagation life accurately.

Acknowledgment

The authors express their gratitude to the Director, Gas Turbine Research Establishment, Bangalore, for permitting to publish this work.

References

[1] W. D. Cowie, Structural Integrity of Aging Aircraft, 1995.

[2] Scott, P. M., and Thorpe, T. W., "A critical review of crack tip stress intensity factors for semi-elliptical cracks", Fat. Engng. Mat. Struct., Vol. 4, pp. 291-309, 1981.

[3] Newman Jr., J. C., and Raju, I. S., "An empirical stress intensity factor equation for the surface flaw", Engng. Fract. Mech., Vol. 15, pp. 185-192, 1981.

[4] Kim, J., and Song, J., "Crack growth and closure behaviour of surface cracks under axial loading", Fat. Fract. Engng. Mater. Struct., Vol. 15, pp. 477-489, 1992.

[5] M. Jolles and V. Tortoriello, "Effects of constraint variation on the fatigue growth of surface flaws", Fracture Mechanics, Fiftieth Symposium", ASTM STP 833, pp300-311, 1984.

[6] A. E. Grandt Jr., "An experimental and numerical investigation of the growth and coalescence of multiple fatigue cracks at notches", ASTM STP 905, 239-252 (1986).

[7] W. O. Soboyejo, J. F. Knott, M. J. Walsh and K. R. Cropper, "Fatigue crack-propagation of coplanar semi-elliptical cracks in pure bending", Engng. Fract. Mech., 37, 323-340 (1990).

[8] Y. Murkami and S. Nemat-Nasser, "Interacting dissimillar semi-elliptical surface flaws under tension and bending", Engng. Fract. Mech. 16, 373-386 (1982).

[9] M. Isida, T. Yoshida and H. Naguch, "A finite thickness plate with a pair of semi-elliptical surface cracks", Engng. Fract. Mech. 35, 961-965 (1990).

[10] K. Kishimoto, W. O. Soboyejo, R. A. Smith and J. F. Knott, "A numerical investigation of the interaction and coalescence of two coplanar semi-elliptical cracks", Int. J of Fat., 11, 91-96 (1989).

[11] K. Badari Narayana, B. Dattaguru, T. S. Ramamurthy and K. Vijaykumar, "A general procedure

for modified crack closure integral in 3-d problems with cracks", Engng. Fract. Mech.

[12] K. N. Sivakumar, P. W. Tan and J. C. Newman Jr, "A virtual crack closure technique for calculating stress intensity factors for cracked three-dimensional bodies", Int. J. of fracture 36, R43-R50 (1988). Mech. 48, 167-176 (1994).

[13] M. D. Raizenne, "Fatigue crack growth results for Ti-6Al-4V, IMI685, and Ti-17, AGARD Engine Disc Cooperative Test Program", AGARD R-766 (Addendum), (1993).

Description	ID	Crack 1 a_0	Crack 1 c_0 (mm)	Crack 2 a_0	Crack 2 c_0 (mm)
Deep equal#	Al-9	1.6623	0.8523	1.7045	0.8717

Table 1: Details of Multiple Surface crack test configurations.

Description	t_s/c	σ_{max} (MPa)	R for loading cycle	R for marker cycle
Deep equal#	0.942	674.54	0.05	0.8

Table 1 Continued

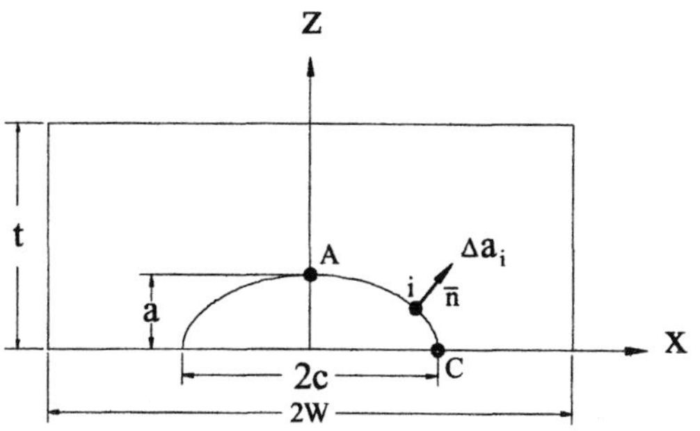

Fig. 1. Basic geometric parameters of semi-elliptical surface crack.

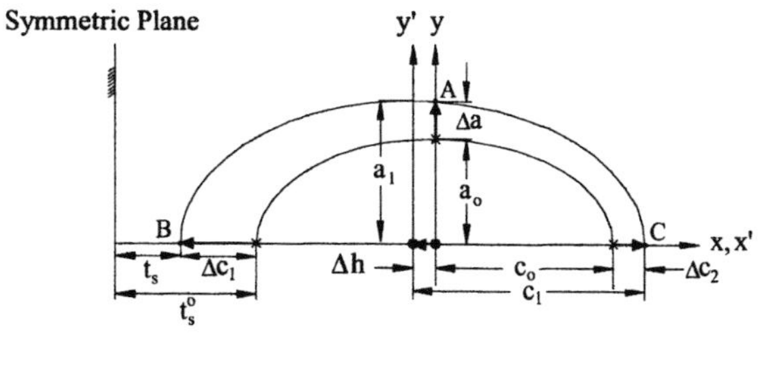

Fig. 2. Semi-elliptical interacting surface cracks before and after crack extension.

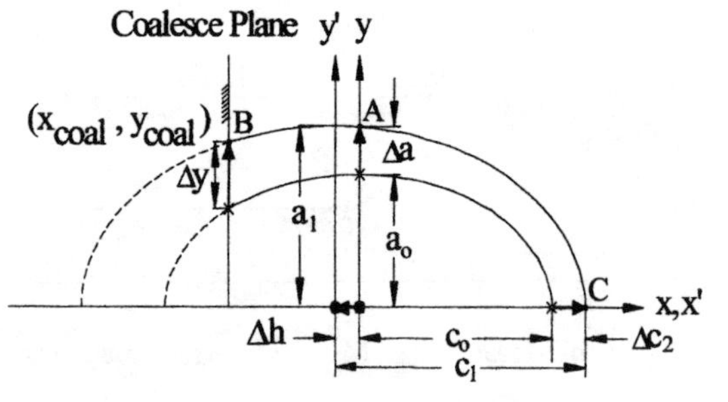

Fig. 3. Semi-elliptical coalescing surface cracks before and after crack extension (only symmetric crack is shown).

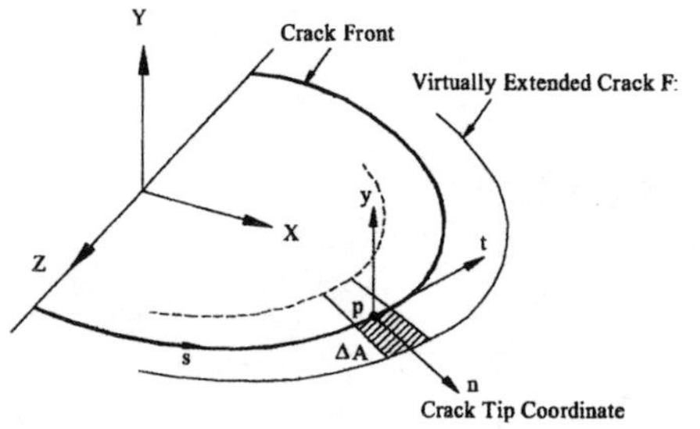

Fig. 4. Definition of coordinate system at crack front and virtual extension.

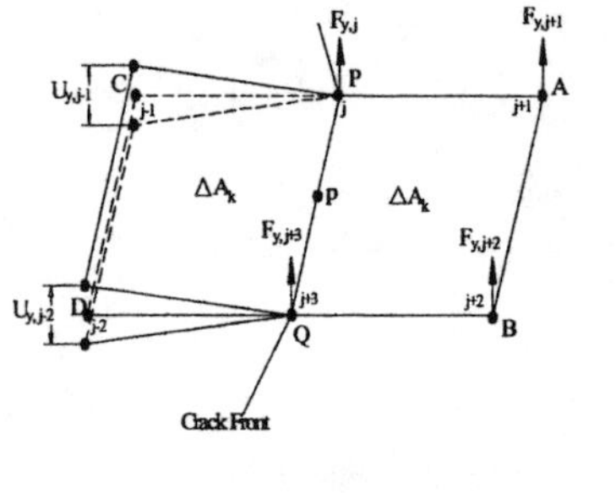

Fig. 5. Nodal displacements and forces on a face of kth eight noded brick element at crack front.

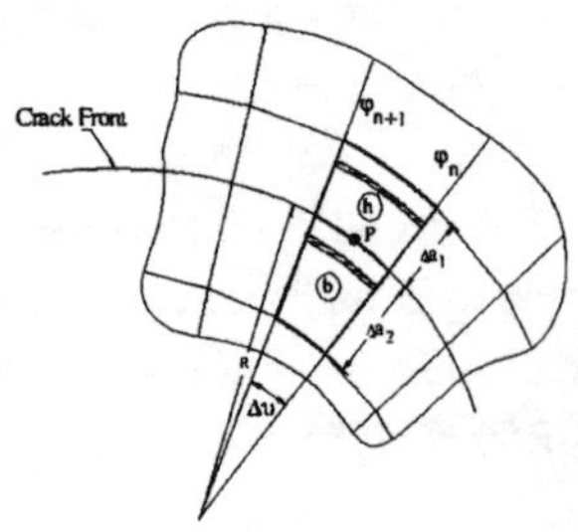

Fig. 6. Schematic of finite element mesh at crack front

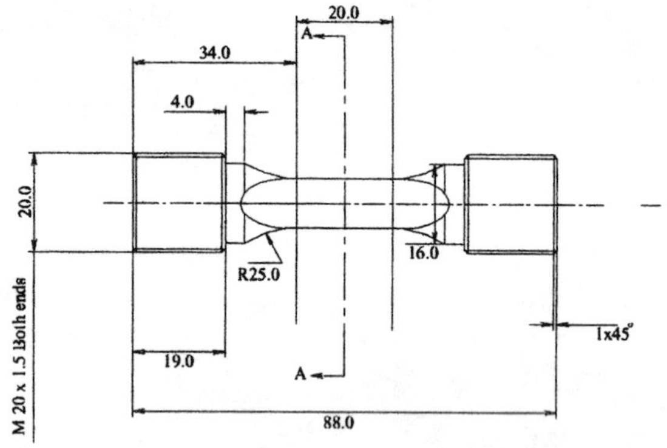

Fig. 7. Specimen geometry.

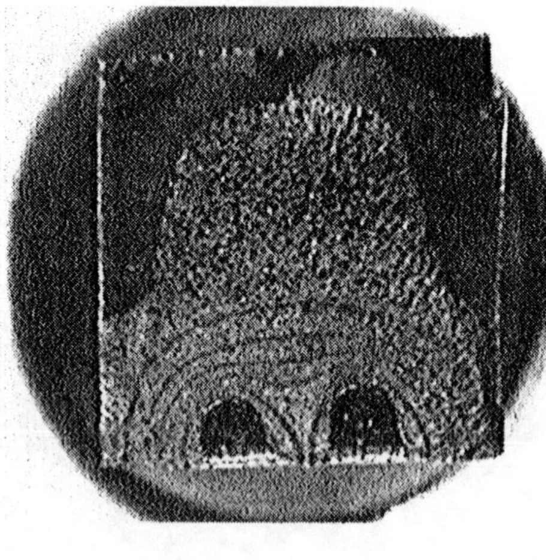

Fig. 8. A digital photograph of fracture surface of specimen AI-9.

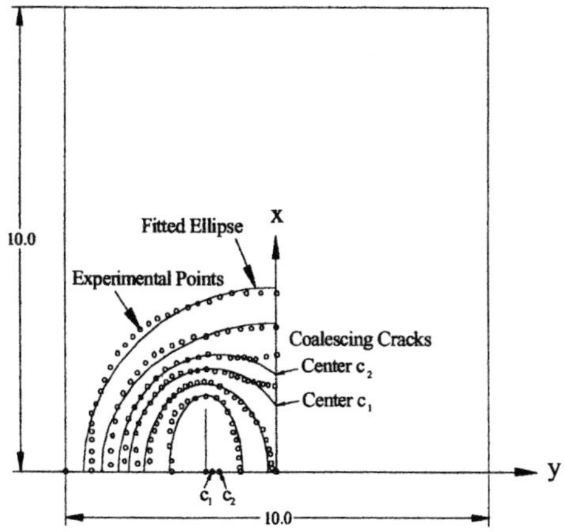

Fig. 9. Measured points on the crack front and fitted ellipse.

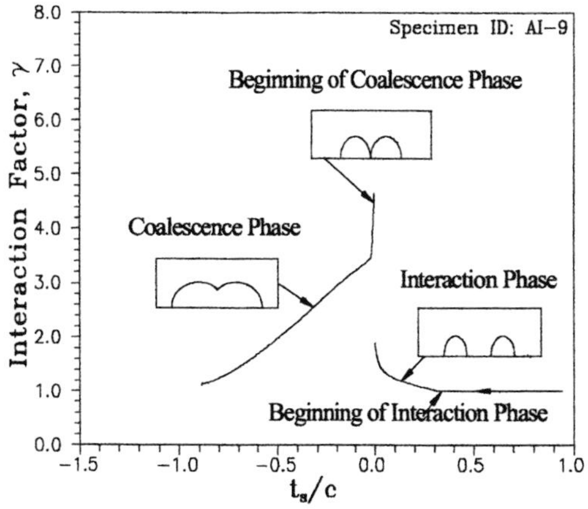

Fig. 11. Variation of interaction factor.

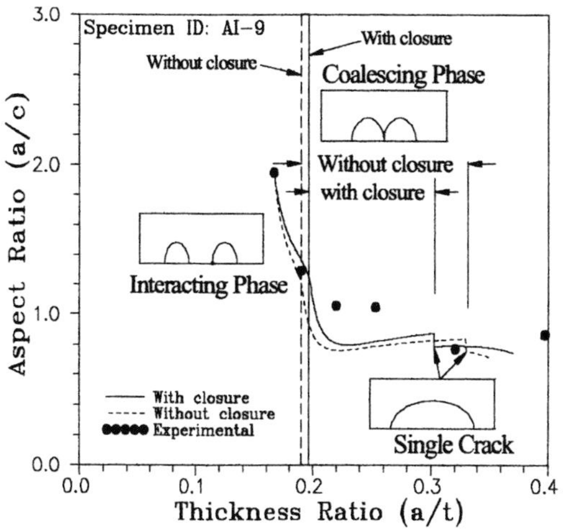

Fig. 10. Crack shape development for twin cracks.

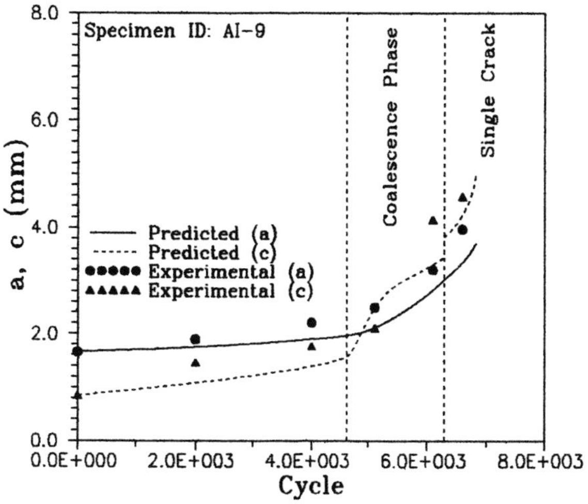

Fig. 12. Experimental and predicted life for specimen AI-9.

PVP-Vol. 458, Computer Technology and Applications

PVP2003-1910

DQEM ANALYSIS OF IN-PLANE DEFLECTION OF CURVED BEAM STRUCTURES

Chang-New Chen
Department of Naval Architecture and Marine Engineering
National Cheng Kung University, Tainan, Taiwan
Email: cchen@mail.ncku.edu.tw

ABSTRACT

The development of differential quadrature element method in-plane deflection analysis model of arbitrarily curved nonprismatic beam structures was carried out. The DQEM uses the extended differential quadrature to discretize the differential eigenvalue equation defined on each element, the transition conditions defined on the inter-element boundary of two adjacent elements and the boundary conditions of the beam. Numerical results solved by the developed numerical algorithm are presented. The convergence of the developed DQEM analysis model is efficient.

INTRODUCTION

The analysis of in-plane deflection of nonprismatic curved. Numerical methods can be used to solve this deflection problem. An efficient method that can be used to develop a solution algorithm for this structural problem is the differential quadrature element method (DQEM).

The DQEM adopts the differential quadrature (DQ) related discretization techniques. The method of DQ approximates a partial derivative of a variable function with respect to a coordinate at a node as a weighted linear sum of the function values at all nodes along that coordinate direction [1]. DQ is an effective method for approximating the partial derivatives and solving partial differential equations [2,3]. The application of DQ to the analysis of continuum mechanics problems is very limited because only problems with regular domain configurations and simple environmental conditions can be treated.

The author has proposed the extended differential quadrature (EDQ) [4,5]. In the EDQ, a certain order derivative or partial derivative of the variable function with respect to the coordinate variables at a node is expressed as the weighted linear sum of the values of function and/or its possible derivatives or partial derivatives at all nodes. The node and discrete point can be different. Consequently, more analytical functions can be used to define the EDQ discretization. By using certain analytical functions such as Hermite polynomials, not only derivatives can be considered as independent variables but also only simple algebraic operations are necessary for computing the weighting coefficients. The author has also generalized the DQ which results in obtaining the generic differential quadrature (GDQ) [6,7]. The weighting coefficients for a grid model defined by a coordinate system having arbitrary dimensions can also be generated. The configuration of a grid model can be arbitrary.

There is a discrete element analysis technique QEM which also adopts the DQ. The original QEM was proposed to solve truss and frame structures [8]. In this method, the truss element is limited to a three-node second-order approximation, while a δ-grid arrangement is used to define the DQ discretization of the flexural deformation. The δ-grid is designed to approximately define certain boundary conditions at a point close to the boundary, and certain inter-element transition conditions at a certain point close to the inter-element boundary. Consequently, the definition of boundary conditions and inter-element boundary conditions is inconsistent. When developing the plane stress and plate bending QEM models, Striz *et al* adopted a hybrid technique to incorporate the DQ discretization into a Galerkin finite element formulation and define a discrete element analysis procedure [9].

The author has also proposed a discrete element analysis method for solving a generic engineering or scientific problem having an arbitrary domain configuration [10]. Like the finite element method (FEM), in this method the analysis domain of a problem is first separated into a certain number of subdomains or elements. Then the DQ or GDQ discretization is carried out on an element basis. The governing differential or partial differential equations defined on the elements, the transition conditions on inter-element boundaries and the boundary conditions on the analysis domain boundary are in computable algebraic forms after the DQ or GDQ discretization. By assembling all discrete fundamental equations the overall algebraic system can be obtained which is used to solve the problem. The interior elements can be regular. However, in order to solve the problem having an arbitrary analysis domain configuration, elements connected to or near the analysis domain boundary might need to be irregular. The mapping technique can be used to develop irregular elements. It results in the DQEM [10]. The GDQ can also be used to develop the irregular elements. It results in the generalized differential quadrature element method (GDQEM) [6].

The theoretical basis of DQEM is rigorous since all fundamental relations are numerically satisfied. It has been proved that DQEM and GDQEM are efficient [10-13]. The convergence rate of these two methods is excellent. The DQEM and GDQEM also have the same advantage as the finite element method of general geometry and systematic boundary treatment. The two methods need less computer memory requirements than the FEM. The EDQ-basis DQEM model of the in-plane deflection analysis of arbitrarily curved nonprismatic beams was developed. The numerical procedures are summarized. Numerical results are also presented.

EXTENDED DIFFERENTIAL QUADRATURE

In using the EDQ to solve a problem, the number of total degrees of freedom attached to the nodes is the same as the number of total discrete fundamental relations required for solving the problem. A discrete fundamental relation can be defined at a point which is not a node. Then a certain order of derivative or partial derivative, of the variable function existing in a fundamental relation, at an arbitrary point with respect to the coordinate variables can be expressed as the weighted linear sum of the values of variable function and/or its possible derivatives at all nodes [4-5]. Thus in solving a problem, a discrete fundamental relation can be defined at a point which is not a node. If a point used for defining discrete fundamental relations is also a node, it is not necessary that the number of discrete fundamental relations at that node equals the number of degrees of freedom attached to it. This concept has been used to construct the discrete inter-element transition conditions and boundary conditions in the differential quadrature element analyses of beam bending problem, warping torsion bar problem.

Let $\pi(\xi)$ denote the variable function which is the displacement for the present vibration analysis associated with a one-dimensional problem. The EDQ discretization for a derivative of order m at discrete point α can be expressed by

$$\frac{d^m \pi_\alpha}{d\xi^m} = D^{\xi^m}_{\alpha i}\,\bar{\pi}_i, \quad i = 1, 2, ..., \bar{N} \tag{1}$$

where $\bar{N}$ is the number of degrees of freedom and $\bar{\pi}_{\bar{\alpha}}$ the values of variable function and/or its possible derivatives at the nodes. The variable function can be a set of appropriate analytical functions denoted by $\Upsilon_p(\xi)$. The substitution of $\Upsilon_p(\xi)$ in Eq. (1) leads to a linear algebraic system for determining the weighting coefficients $D^{\xi^m}_{\alpha i}$. If interpolation functions are used, explicit expression of weighting coefficients can be obtained. The variable function is approximated by

$$\pi(\xi) = \Psi_p(\xi)\bar{\pi}_p, \quad p = 1, 2, ..., \bar{N} \tag{2}$$

where $\Psi_p(\xi)$ are the corresponding interpolation functions of $\bar{\pi}_p$. Adopting $\Psi_p(\xi)$ as the variable function $\pi(\xi)$ and substituting it into Eq. (1), a linear algebraic system for determining $D^{\xi^m}_{\alpha\bar{\alpha}}$ can be obtained. And the mth order differentiation of Eq. (2) at discrete point α also leads to the extended GDQ discretization Eq. (1) in which $D^{\xi^m}_{\alpha i}$ is explicitly expressed by $D^{\xi^m}_{\alpha i} = \frac{d^m \Psi_i}{d\xi^m}|_\alpha$. Using this equation, the weighting coefficients can easily be obtained by simple algebraic calculations.

The variable function can also be approximated by

$$\pi(\xi) = \Upsilon_p(\xi)c_p, \quad p = 1, 2, ..., \bar{N} \tag{3}$$

where $\Upsilon_p(\xi)$ are appropriate analytical functions and c_p are unknown coefficients. The constraint conditions at all nodes can be expressed as $\bar{\pi}_p = \chi_{p\bar{p}}c_{\bar{p}}$, where $\chi_{p\bar{p}}$ are composed of the values of $\Upsilon_p(\xi)$ and/or their possible derivatives at all nodes. Then the variable function can be rewritten as $\pi(\xi) = \Upsilon_p(\xi)\chi^{-1}_{\bar{p}p}\bar{\pi}_{\bar{p}}$. Using this equation, the weighting coefficients can also be obtained

$$D^{\xi^m}_{\alpha i} = \frac{\partial^m \Upsilon_{\bar{p}}}{\partial \xi^m}|_\alpha\, \chi^{-1}_{i\bar{p}} \tag{4}$$

Various analytical functions such as the sinc functions, the Lagrange polynomials, the Chebyshev polynomials, the Bernoulli polynomials, the Euler polynomials, the rational functions, ..., etc. can be used to define the weighting coefficients. To solve problems having singularity properties, certain singular functions can be used for the EDQ discretization. The problems having infinite domains can also be treated.

FUNDAMENTAL RELATIONS

Consider that an isotropic and homogeneous curved beam, shown in Fig. 1, with the Young's modulus E, the cross-section area A and the moment of inertia of the cross-section I. Let r and ϕ denote the coordinate variables. Also let w and v denote the radial modal displacement and tangential modal displacement of cros-section, respectively. Refer to Fig. 2 and let p_r, p_t and t denote the distribution of radial force, tangential force and moment, respectively. Also let N, M and V denote the axial force, bending moment and shear force, respectively. These stress resultants are expressed as

$$N = \frac{EA}{r}\left[\left(\frac{dv}{d\phi} + w\right) + \frac{I}{r^2 A}\left(\frac{d^2 w}{d\phi^2} + w\right)\right],$$

$$M = -\frac{EI}{r^2}\left(\frac{d^2 w}{d\phi^2} + w\right),$$

$$\begin{aligned} V &= \tfrac{1}{r}\tfrac{dM}{d\phi} - t \\ &= -EI\left[\tfrac{1}{r^3}\left(\tfrac{d^3 w}{d\phi^3} + \tfrac{dw}{d\phi}\right) - \tfrac{2}{r^4}\tfrac{dr}{d\phi}\left(\tfrac{d^2 w}{d\phi^2} + w\right)\right] \\ &\quad -\tfrac{\rho I \omega^2}{rA}\left(\tfrac{dw}{d\phi} - v\right) \end{aligned} \tag{5}$$

The dynamic equilibrium equation in the tangential direction is expressed as

$$\frac{dN}{d\phi} + V = -rp_t$$

or

$$-\frac{EI}{r^4}\frac{dr}{d\phi}\frac{d^2w}{d\phi^2}+\frac{EA}{r}\frac{dw}{d\phi}-\left(\frac{EA}{r^2}+\frac{EI}{r^4}\right)\frac{dr}{d\phi}w$$
$$+\frac{EA}{r}\frac{d^2v}{d\phi^2}-\frac{EA}{r^2}\frac{dr}{d\phi}\frac{dv}{d\phi}=-rp_t \quad (6)$$

The dynamic equilibrium equation in the radial direction is expressed as

$$\frac{dv}{d\phi}-N=-rp_r$$

or

$$-\frac{EI}{r^3}\frac{d^4w}{d\phi^4}+\frac{5EI}{r^4}\frac{dr}{d\phi}\frac{d^3w}{d\phi^3}+EI\left[\frac{2}{r^4}\frac{d^2r}{d\phi^2}-\frac{8}{r^5}\left(\frac{dr}{d\phi}\right)^2\right.$$
$$\left.-\frac{1}{r^3}\right]\frac{d^2w}{d\phi^2}+\frac{5EI}{r^4}\frac{dr}{d\phi}\frac{dw}{d\phi}+\left\{EI\left[\frac{2}{r^4}\frac{d^2r}{d\phi^2}-\frac{8}{r^5}\left(\frac{dr}{d\phi}\right)^2\right.\right.$$
$$\left.\left.-\frac{1}{r^3}\right]-\frac{EA}{r}\right\}w-\frac{EA}{r}\frac{dv}{d\phi}=-rp_r \quad (7)$$

The kinematic boundary conditions are expressed as

$$w=\bar{w},\quad \frac{1}{r}\frac{dw}{d\phi}=\frac{1}{r}\frac{d\bar{w}}{d\phi},\quad v=\bar{v} \quad (8)$$

Assume that $\bar{N}$, $\bar{M}$ and $\bar{V}$ are tangential force, moment and radial force, respectively. The natural boundary conditions are expressed as

$$\frac{EA}{r}\left[\left(\frac{dv}{d\phi}+w\right)+\frac{I}{r^2A}\left(\frac{d^2w}{d\phi^2}+w\right)\right]=\bar{N},$$
$$-\frac{EI}{r^2}\left(\frac{d^2w}{d\phi^2}+w\right)=\bar{M},$$
$$-EI\left[\frac{1}{r^3}\left(\frac{d^3w}{d\phi^3}+\frac{dw}{d\phi}\right)-\frac{2}{r^4}\frac{dr}{d\phi}\left(\frac{d^2w}{d\phi^2}+w\right)\right]=\bar{V} \quad (9)$$

DQEM FORMULATION

The fundamental relations are referred to the physical coordinate system while the element basis EDQ discretization is carried out on the natural coordinate system. Therefore, in using the EDQ technique to discretize the fundamental relations, the transformation operations of coordinates and derivatives of displacements between two different coordinate systems, have to be carried out. Let ϕ_1^e and $\phi_{N^e}^e$ denote the global coordinates of node 1 and node N^e of element e, which are two end nodes, respectively. The range angle Φ^e of the element equals $\phi_{N^e}^e-\phi_1^e$. Let ϕ^e be the coordinate variable of the local coordinate system with the origin located at node 1 of the element. Also let the range of the natural coordinate ξ be $0. \leq \xi \leq 1.$. Then the coordinate transformation is expressed as

$$\phi^e=\Phi^e\xi \quad (10)$$

Using Eq. (10), the differential of ϕ^e can be expressed as:

$$d\phi^e=\Phi^e d\xi \quad (11)$$

Then the mth order derivative of the variable function π with respect to ϕ^e can be written as

$$\frac{d^m\pi}{d(\phi^e)^m}=\frac{1}{(\Phi^e)^m}\frac{d^m\pi}{d\xi^m} \quad (12)$$

In addition to the degrees of freedom for representing displacement components, the degrees of freedom for representing derivatives of local displacement component with respect to ϕ^e at an element boundary node can also be assigned to that element boundary node. The selection of derivatives can be flexible. In order to automatically set the kinematic transition conditions by only using the degree of freedom assigned to the element boundary nodes, the degrees of freedom representing the first order derivative of w must be assigned to the element boundary nodes. In the present DQEM curved beam analysis model, only the degrees of freedom which are necessary for automatically setting the kinematic transition conditions are assigned to the element boundary nodes.

Since the orders of w and v related differentiations existing in the fundamental relations are four and two, respectively, the orders of the corresponding approximate displacements must at least be four and two, respectively, and each of the two equilibrium equations needs at least one discrete point for defining one of its discrete equation. The discrete points for defining the discrete equilibrium equations can be either in the interior of the element or on the element boundary.

In the present analysis model, each element boundary node has three DOF representing w, v and the first derivative of w, respectively. The DOF representing the derivative of a displacement is not assigned to an interior node. Only interior discrete points are used to define the discrete equilibrium equations. Let N_w^e denote the number of nodes for defining the w-related discretization, $\bar{N}_w^e$ the number of the corresponding element degrees of freedom, $\bar{w}_i^e$ the w-related element basis displacements, and $D_{\alpha i}^{e\xi^m}$ the corresponding weighting coefficients. Also let N_v^e and v_j^e denote the number of nodes for defining the v-related discretization and the corresponding element basis displacements, respectively, with the related weighting coefficients expressed by $D_{\alpha j}^{e\xi^m}$. In defining the discrete fundamental relations, the number of discrete points at which the discrete equilibrium equation in the radial direction is $\bar{N}_w^e-4$. At the N_v^e-2 interior nodes, N_v^e discrete equilibrium equations in the tangential direction are defined. If a discrete point at which a discrete fundamental relation needs to be defined is not an element node, the interpolation is necessary for expressing w or v existing in the fundamental relation. Let $M_i(\xi)$ and $N_j(\xi)$ denote the interpolation functions of w and v, respectively. Then, $w^e(\xi)$ and $v^e(\xi)$ in an element can be expressed as $w^e(\xi)=M_i(\xi)\bar{w}_i^e$ and $v^e(\xi)=N_j(\xi)v_j^e$.

The introduction of Eq. (12) in Eq. (6) and the use of EDQ discretization at an interior node α in an element e lead to the following discrete equation

$$\left\{-\frac{(EI)_{(\alpha)}^e}{(r_{(\alpha)}^e)^4(\Phi^e)^2}\frac{dr_{(\alpha)}^e}{d\phi}\sum_{i=1}^{\bar{N}_w^e}D_{\alpha i}^{e\xi^2}+\frac{(EA)_{(\alpha)}^e}{r_{(\alpha)}^e\Phi^e}\sum_{i=1}^{\bar{N}_w^e}D_{\alpha i}^{e\xi}\right.$$

$$-\left[\frac{(EA)^e_{(\alpha)}}{(r^e_{(\alpha)})^2}+\frac{(EI)^e_{(\alpha)}}{(r^e_{(\alpha)})^4}\right]\frac{dr^e_{(\alpha)}}{d\phi}M_i(\xi_\alpha)\Bigg\}\bar{w}^e_i$$

$$+\left[\frac{(EA)^e_{(\alpha)}}{r^e_{(\alpha)}(\Phi^e)^2}\sum_{j=1}^{N^e_v}D^{e\xi^2}_{\alpha j}-\frac{(EA)^e_{(\alpha)}}{(r^e_{(\alpha)})^2\Phi^e}\frac{dr^e_{(\alpha)}}{d\phi}\sum_{j=1}^{N^e_v}D^{e\xi}_{\alpha j}\right]v^e_j$$

$$=-r^e_{(\alpha)}p^e_{t\alpha},\quad \alpha=1,2,...,N^e_v-2 \tag{13}$$

Similarly, the introduction of Eq. (12) in Eq. (7) and the use of EDQ discretization at a discrete point β lead to the following discrete equation

$$\Bigg\{-\frac{(EI)^e_{(\beta)}}{(r^e_{(\beta)})^3(\Phi^e)^4}\sum_{i=1}^{\bar{N}^e_w}D^{e\xi^4}_{\beta i}+\frac{5(EI)^e_{(\beta)}}{(r^e_{(\beta)})^4(\Phi^e)^3}\frac{dr^e_{(\beta)}}{d\phi}\sum_{i=1}^{\bar{N}^e_w}D^{e\xi^3}_{\beta i}$$

$$+\frac{(EI)^e_{(\beta)}}{(r^e_{(\beta)})^3(\Phi^e)^2}\left[\frac{2}{r^e_{(\beta)}}\frac{d^2r^e_{(\beta)}}{d\phi^2}-\frac{8}{(r^e_{(\beta)})^2}\left(\frac{dr^e_{(\beta)}}{d\phi}\right)^2-1\right]\sum_{i=1}^{\bar{N}^e_w}D^{e\xi^2}_{\beta i}$$

$$+\frac{5(EI)^e_{(\beta)}}{(r^e_{(\beta)})^4\Phi^e}\frac{dr^e_{(\beta)}}{d\phi}\sum_{i=1}^{\bar{N}^e_w}D^{e\xi}_{\beta i}+\left[\frac{(EI)^e_{(\beta)}}{(r^e_{(\beta)})^3}\left(\frac{2}{r^e_{(\beta)}}\frac{d^2r^e_{(\beta)}}{d\phi^2}\right.\right.$$

$$\left.\left.-\frac{8}{(r^e_{(\beta)})^2}\left(\frac{dr^e_{(\beta)}}{d\phi}\right)^2-1\right)-\frac{(EA)^e_{(\beta)}}{r^e_{(\beta)}}\right]M_i(\xi_\beta)\Bigg\}\bar{w}^e_i$$

$$-\frac{(EA)^e_{(\beta)}}{r^e_{(\beta)}\Phi^e}\sum_{j=1}^{N^e_v}D^{e\xi}_{\beta j}v^e_j=-r^e_{(\alpha)}p^e_{r\beta},\quad \beta=1,2,...,\bar{N}^e_w-4 \tag{14}$$

The stress resultants N^e_α, M^e_α and V^e_α at a discrete point α of an element e can thus be calculated by using the following equations

$$N^e_\alpha=\left[\frac{(EI)^e_{(\alpha)}}{(r^e_{(\alpha)})^3(\Phi^e)^2}\sum_{i=1}^{\bar{N}^e_w}D^{e\xi^2}_{\alpha i}+\left(\frac{(EA)^e_\alpha}{r^e_\alpha}+\frac{(EI)^e_\alpha}{(r^e_\alpha)^3}\right)\delta_{(\alpha)i}\right]\bar{w}^e_i$$

$$+\frac{(EA)^e_{(\alpha)}}{r^e_{(\alpha)}\Phi^e}\sum_{j=1}^{N^e_v}D^{e\xi}_{\alpha j}v^e_j,$$

$$M^e_\alpha=-\frac{(EI)^e_{(\alpha)}}{(r^e_{(\alpha)})^2}\left[\frac{1}{(\Phi^e)^2}\sum_{i=1}^{\bar{N}^e_w}D^{e\xi^2}_{\alpha i}+\delta_{\alpha(i)}\right]\bar{w}^e_i,$$

$$V^e_\alpha=\frac{(EI)^e_{(\alpha)}}{(r^e_{(\alpha)})^3}\left[-\frac{1}{(\Phi^e)^3}\sum_{i=1}^{\bar{N}^e_w}D^{e\xi^3}_{\alpha i}+\frac{2}{r^e_{(\alpha)}(\Phi^e)^2}\frac{dr^e_{(\alpha)}}{d\phi}\sum_{i=1}^{\bar{N}^e_w}D^{e\xi^2}_{\alpha i}\right.$$

$$\left.-\frac{1}{\Phi^e}\sum_{i=1}^{\bar{N}^e_w}D^{e\xi}_{\alpha i}+\frac{2}{r^e_\alpha}\frac{dr^e_{(\alpha)}}{d\phi}\delta_{\alpha i}\right]\bar{w}^e_i \tag{15}$$

The compatibility and conformability conditions at the inter-element boundary of two adjacent elements are autmatically satisfied. The equilibrium conditions of stress resultants and external forces at the inter-element boundary or natural boundary also have to be satisfied. Each equilibrium condition is either a natural transition condition or a natural boundary condition. Assume that $P^{i,i+1}_r$, $P^{i,i+1}_t$ and $T^{i,i+1}$ shown in Fig. 3 are external radial force, tangential force and moment, respectively, applied on the inter-element boundary $i, i+1$. The equilibrium condition of moments is expressed as

$$M^{i+1}_1-M^i_{N^i}=T^{i,i+1} \tag{16}$$

Introducing the second one of Eq. (2) in the above equation to express the moments as displacements, then by using Eq. (12) and the EDQ discretization the following explicit discrete equilibrium condition of moments can be obtained:

$$\frac{(EI)^i_{N^i}}{(r^i_{N^i})^2}\left[\frac{1}{(\Phi^i)^2}\sum_{i=1}^{\bar{N}^i_w}D^{i\xi^2}_{N^i i}+\delta_{N^i(i)}\right]\bar{w}^i_i$$

$$-\frac{(EI)^{i+1}_1}{(r^{i+1}_1)^2}\left[\frac{1}{(\Phi^{i+1})^2}\sum_{i=1}^{\bar{N}^{i+1}_w}D^{(i+1)\xi^2}_{1i}+\delta_{1(i)}\right]\bar{w}^{i+1}_i=T^{i,i+1} \tag{17}$$

The equilibrium condition of forces in the radial direction is expressed as

$$V^i_{N^i}-V^{i+1}_1=P^{i,i+1}_r \tag{18}$$

The explicit discrete equation is expressed as

$$\frac{(EI)^i_{N^i}}{(r^i_{N^i})^3}\left[-\frac{1}{(\Phi^i)^3}\sum_{i=1}^{\bar{N}^i_w}D^{i\xi^3}_{N^i i}+\frac{2}{r^i_{N^i}(\Phi^i)^2}\frac{dr^i_{N^i}}{d\phi}\sum_{i=1}^{\bar{N}^i_w}D^{i\xi^2}_{N^i i}\right.$$

$$\left.-\frac{1}{\Phi^i}\sum_{i=1}^{\bar{N}^i_w}D^{i\xi}_{N^i i}+\frac{2}{r^i_{N^i}}\frac{dr^i_{N^i}}{d\phi}\delta_{N^i i}\right]\bar{w}^i_i$$

$$+\frac{(EI)^{i+1}_1}{(r^{i+1}_1)^3}\left[\frac{1}{(\Phi^{i+1})^3}\sum_{i=1}^{\bar{N}^{i+1}_w}D^{(i+1)\xi^3}_{1i}-\frac{2}{r^{i+1}_1(\Phi^{i+1})^2}\frac{dr^{i+1}_1}{d\phi}\sum_{i=1}^{\bar{N}^{i+1}_w}D^{(i+1)\xi^2}_{1i}\right.$$

$$\left.+\frac{1}{\Phi^{i+1}}\sum_{i=1}^{\bar{N}^{i+1}_w}D^{(i+1)\xi}_{1i}-\frac{2}{r^{i+1}_1}\frac{dr^{i+1}_1}{d\phi}\delta_{1i}\right]\bar{w}^{i+1}_i=P^{i,i+1}_r \tag{19}$$

The equilibrium condition of forces in the tangential direction is expressed as

$$N^i_{N^i}-N^{i+1}_1=P^{i,i+1}_t \tag{20}$$

The explicit discrete equation is expressed as

$$\left[\frac{(EI)^i_{N^i}}{(r^i_{N^i})^3(\Phi^i)^2}\sum_{i=1}^{\bar{N}^i_w}D^{i\xi^2}_{N^i i}+\left(\frac{(EA)^i_{N^i}}{r^i_{N^i}}+\frac{(EI)^i_{N^i}}{(r^i_{N^i})^3}\right)\delta_{N^i i}\right]\bar{w}^i_i$$

$$+\frac{(EA)^i_{N^i}}{r^i_{N^i}\Phi^i}\sum_{j=1}^{N^i_v}D^{i\xi}_{N^i j}v^i_j$$

$$-\left[\frac{(EI)^{i+1}_1}{(r^{i+1}_1)^3(\Phi^{i+1})^2}\sum_{i=1}^{\bar{N}^{i+1}_w}D^{(i+1)\xi^2}_{1i}+\left(\frac{(EA)^{i+1}_1}{r^{i+1}_1}+\frac{(EI)^{i+1}_1}{(r^{i+1}_1)^3}\right)\right.$$

$$\left.\times\delta_{1i}\right]\bar{w}^{i+1}_i+\frac{(EA)^{i+1}_1}{r^{i+1}_1\Phi^{i+1}}\sum_{j=1}^{N^{i+1}_v}D^{(i+1)\xi}_{1j}v^{i+1}_j=P^{i,i+1}_t \tag{21}$$

Letting element m be an element consisting of the kinematic boundary, and I^m equal to 1 for the left boundary and N^m for the right boundary, the discrete kinematic boundary conditions are expressed by

$$w^m_{I^m} = \bar{w}^m_{I^m}, \quad \frac{1}{\Phi^m}\sum_{i=1}^{\bar{N}^m_w} D^{m\xi}_{I^m i}\bar{w}^m_i = \frac{d\bar{w}^m_{I^m}}{d\phi},$$

$$v^m_{I^m} = \bar{v}^m_{I^m} \tag{22}$$

The discrete natural boundary conditions can be obtained by discretizing Eq. (9)

$$\left[\frac{(EI)^n_{I^n}}{(r^n_{I^n})^3(\Phi^n)^2}\sum_{i=1}^{\bar{N}^n_w} D^{n\xi^2}_{I^n i} + \left(\frac{(EA)^n_{I^n}}{r^n_{I^n}} + \frac{(EI)^n_{I^n}}{(r^n_{I^n})^3}\right)\delta_{I^n i}\right]\bar{w}^n_i$$

$$+\frac{(EA)^n_{I^n}}{r^n_{I^n}\Phi^n}\sum_{j=1}^{N^n_v} D^{n\xi}_{I^n j}v^n_j = \bar{N},$$

$$-\frac{(EI)^n_{I^n}}{(r^n_{I^n})^2}\left[\frac{1}{(\Phi^n)^2}\sum_{i=1}^{\bar{N}^n_w} D^{n\xi^2}_{I^n i} + \delta_{I^n(i)}\right]\bar{w}^n_i = \bar{M},$$

$$\frac{(EI)^n_{I^n}}{(r^n_{I^n})^3}\left[-\frac{1}{(\Phi^n)^3}\sum_{i=1}^{\bar{N}^n_w} D^{n\xi^3}_{I^n i} + \frac{2}{r^n_{I^n}(\Phi^n)^2}\frac{dr^n_{I^n}}{d\phi}\sum_{i=1}^{\bar{N}^n_w} D^{n\xi^2}_{I^n i}\right.$$

$$\left.-\frac{1}{\Phi^n}\sum_{i=1}^{\bar{N}^n_w} D^{n\xi}_{I^n i} + \frac{2}{r^n_{I^n}}\frac{dr^n_{I^n}}{d\phi}\delta_{I^n i}\right]\bar{w}^n_i = \bar{V} \tag{23}$$

With the kinematic transition conditions in mind, then assemble the discrete governing equations (13) and (14) for elements having more than two nodes, discrete natural transition conditions, and discrete natural boundary conditions, an overall discrete governing/transition/boundary equation can be obtained. It is the overall stiffness equation. Consider the kinematic boundary conditions and solve the overall discrete governing/transition/boundary equation, transverse displacements and bending rotations at all nodes can be obtained. Like FEM, the assemblage is based on an element by element procedure. In assembling the discrete equations of element e, the discrete governing equations (13) and (14), and the six discrete element boundary forces of moments and shear forces, expressed by displacements, at the two element boundary nodes are directly assembled to the overall discrete equation system [10]. An element basis explicit matrix equation, containing the discrete governing equations and the discrete element boundary forces placed at the first and last three rows, is not necessary to be formed in the assemblage process. This element basis explicit matrix equation is an element stiffness equation which can be expressed by

$$[k^e]\{\delta^e\} = \{r^e\} \tag{24}$$

where $[k^e]$ is a $(\bar{N}^e_w + N^e_v) \times (\bar{N}^e_w + N^e_v)$ element stiffness matrix,

$$\{\delta^e\} = \lfloor w^e_1\ \theta^e_1\ v^e_1\ \ldots\ w^e_{\bar{N}^e_w}\ \theta^e_{\bar{N}^e_w}\ v^e_{N^e_v} \rfloor^T \tag{25}$$

is the element displacement vector, and

$$\{r^e\} = \lfloor -V^e_1\ -M^e_1\ -N^e_{z1}\ -r^e_1 p^e_{r1}\ -r^e_1 p^e_{t1}\ \ldots$$
$$-r^e_{\bar{N}^e_w-4}p^e_{r\bar{N}^e_w-4}\ -r^e_{N^e_v-2}p^e_{rN^e_v-2}\ V^e_{N^e}\ M^e_{N^e}\ N^e_{N^e} \rfloor^T \tag{26}$$

is the element load vector. In Eq. (25), θ^e_i represents the value of $\frac{1}{r}\frac{dw^e}{d\phi^e} = \frac{1}{r\Phi^e}\frac{dw^e}{d\xi}$ at the node i. As Eq. (24) contains discrete resultant forces at the two element boundary nodes, equilibriums of resultant forces and external forces at the inter-element boundary of two adjacent elements and the natural boundary are exactly satisfied in the assemblage process. Consequently, the DQEM is different from FEM which needs to form the element stiffness equation, and which neglects the exact equilibriums.

NUMERICAL EXAMPLES

In solving the problem, the elements are equally spaced. The DOF of flexural deformation, $\bar{N}^e_w$, and the DOF of tangential deformation, N^e_v, are the same. Defining $\Delta\xi = 1./(\bar{N}^e_w - 1)$, the interior discrete points for defining the element-based radial equilibrium equations are located at $\xi = (p-1)\Delta\xi$, $p = 3, \ldots, \bar{N}^e_w - 2$ while the interior discrete points for defining the element-based extensional equilibrium equations are located at $\xi = (p-1)\Delta\xi$, $p = 2, \ldots, \bar{N}^e_w - 1$. Only one DOF representing the radial displacement and one DOF representing the tangential displacement are assigned to an interior node. Polynomials are used to calculate the weighting coefficients.

The first problem solved involves a simply supported prismatic circular arch, subjected to a concentrated force P and shown in Fig. 4, with the values of radius r, area of cross-section A, moment of inertia of the cross-section I, Young's modulus E, and the concentrated force P equal to 1. The opening angle of the arch is 180^0. The analyses by both the p refinement procedure of increasing the number of DOF per element and the h refinement procedure of increasing the number of elements are carried out, separately. Numerical results are summarized and listed in Table 1. The results are compared with the analytical solutions. It shows that the results converge fast to the exact solutions by either increasing the DOF per element or the number of elements. It also shows that the solution procedure of increasing the DOF per element is more efficient than the solution procedure of increasing the number of elements. The distributions of converged displacements and stress resultants are shown in Figs. 5 to 9.

The problem is resolved by changing the boundary conditions and loading condition. The new model problem is shown in Fig. 10. The analyses by both the p refinement procedure of increasing the number of DOF per element and the h refinement procedure of increasing the number of elements are also carried out, separately. Numerical results are summarized and listed in Table 2. The results are compared with the analytical solutions. It also shows that the results converge fast to the exact solutions by either increasing the DOF per element or the number of elements. The distributions of converged displacements and stress resultants are shown in Figs. 11 to 15.

CONCLUSIONS

The DQEM in-plane deflection analysis model of nonprismatic curved beams was developed. The numerical model was summarized and presented. Numerical results proved that the DQEM has excellent convergence properties. Since the theoretical basis of this DQEM is rigorous, the performance of the developed DQEM analysis model is excellent. The DQEM is suitable for developing various solution algorithms for solving generic scientific and engineering problems.

REFERENCES

1. Bellman, R. E. and Casti, J., 1971, "Differential quadrature and long-term integration," *J. Math. Anal. Appls.*, Vol. 34, pp. 235-238.

2. Bellman, R. E. and Adomian, G., 1985, Partial Differential Equations, Dordrecht, The Netherland, D Reidel Publishing.

3. Bellman, R. E. and Roth, R. S., 1986, Methods in Approximation, Dordrecht, The Netherland, D Reidel Publishing.

4. Chen, C. N., 1998, "Extended Differential Quadrature," In: Proceedings of the Sixth Pan-American Congress of Applied Mechanics, Rio de Janeiro, Brazil, 6, 389-392.

5. Chen, C. N., 1999, "Differential Quadrature Element Analysis Using Extended Differential Quadrature," *Comput. Maths. Appls.*, Vol. 39, pp. 65-79.

6. Striz, A. G., Chen, W. L. and Bert, C. W., 1994, "Static Analysis of Structures by the Quadrature Element Method," *Intl. J. Sol. Struc.*, Vol. 31, pp. 2807-2818.

7. Striz, A. G., Chen, W. L. and Bert, C. W., 1995, "High Accuracy Plane Stress and Plate Elements in the Quadrature Element Method," Proceedings 36th AIAA/ASME/ASCE/AHS/ASC Structures, Structural Dynamics, and Materials Conference and AIAA/ASME Adaptive Structures Forum, New Orleans, USA, 2, 957-965.

8. Chen, C. N., 1998, "A Generalized Differential Quadrature Element Method," Proceedings of the International Conference on Advanced Computational Methods in Engineering, Gent, Belgium, 721-728.

9. Chen, C. N. 1999, "Generalization of Differential Quadrature Discretization," *Num. Alg.*, Vol. 22, pp. 167-182.

10. Chen, C. N., 1995, "A Differential Quadrature Element Method," Proceedings of the First International Conference on Engineering Computation and Computer Simulation, Changsha, CHINA, 25-34.

11. Chen, C. N., 1998, "The Warping Torsion Bar Model of the Differential Quadrature Element Method," *Computers & Structures*, Vol. 66, 249-257.

12. Chen, C. N., 1998, "Solution of Beam on Elastic Foundation by DQEM," *J. Engr. Mechs.*, Vol. 124, pp. 1381-1384.

13. Chen, C. N., 1999, "The Differential Quadrature Element Method Irregular Element Torsion Analysis Model," *Appl. Math. Model.*, Vol. 23, pp. 309-328.

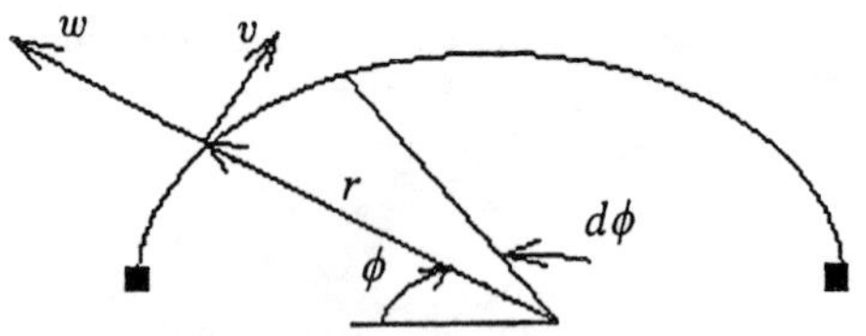

Fig. 1. Coordinates of an arbitrarily curved beam.

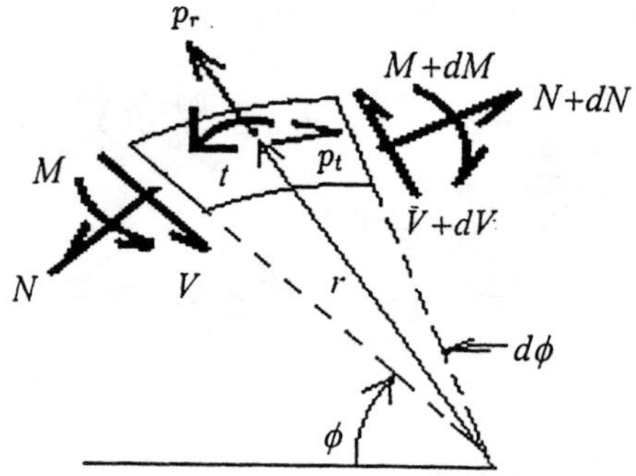

Fig. 2. Stress resultants and external loads.

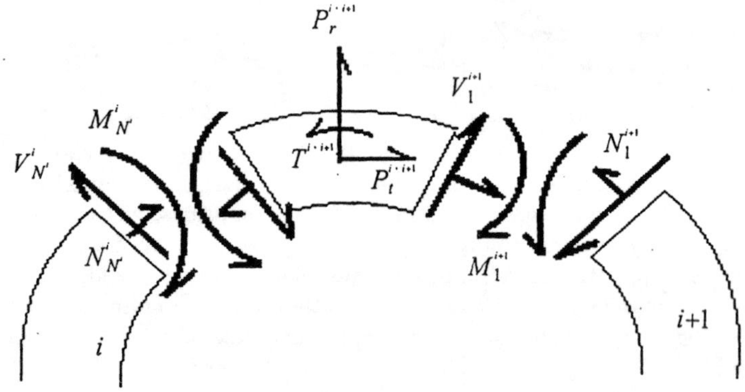

Fig. 3. Forces at the inter-element of two adjacent elements i and $i+1$.

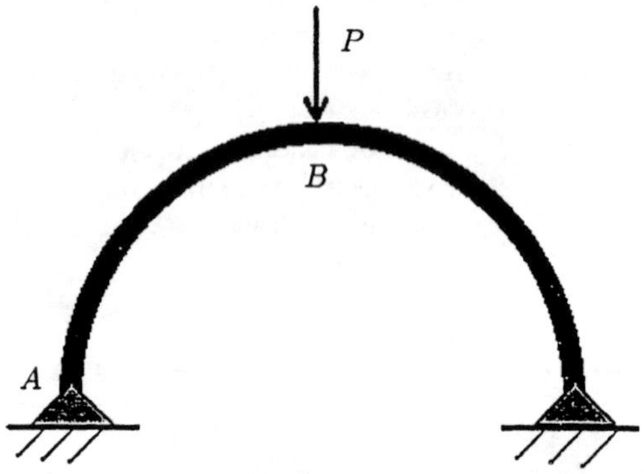

Fig. 4. A simply supported curved beam.

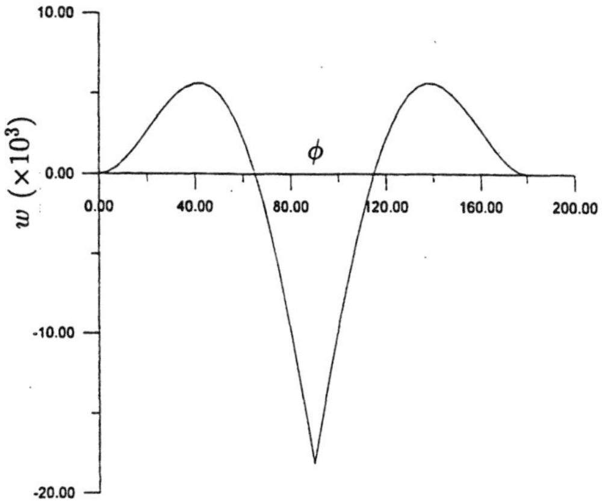

Fig. 5. Radial displacement of the simply supported curved beam.

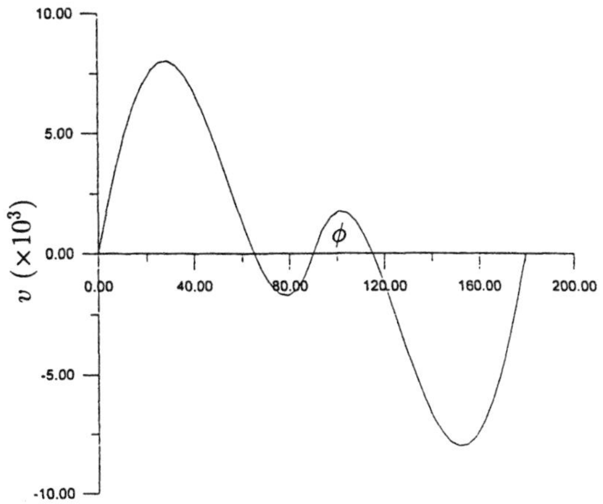

Fig. 6. Tangential displacement of the simply supported curved beam.

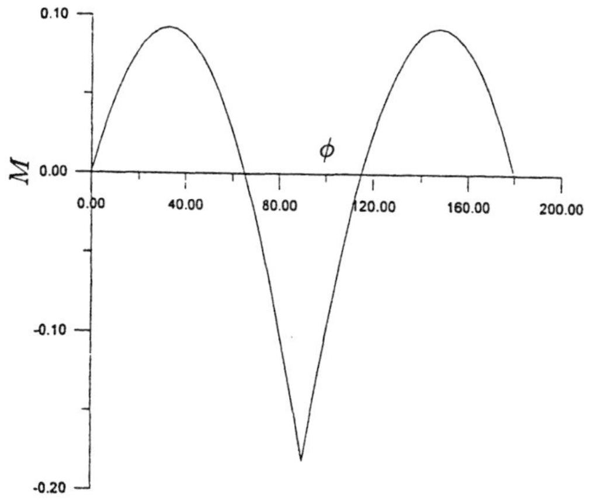

Fig. 7. Bending moment of the simply supported curved beam.

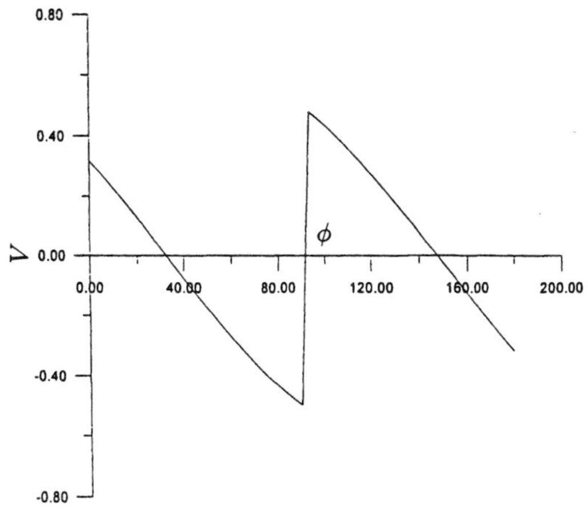

Fig. 8. Shear force of the simply supported curved beam.

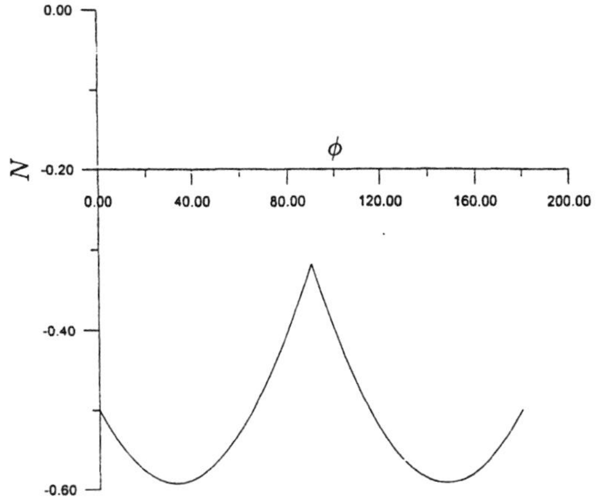

Fig. 9. Axial force of the simply supported curved beam.

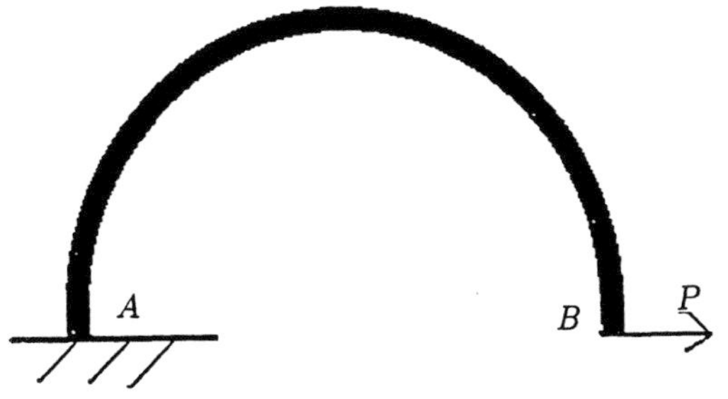

Fig. 10. A cantilever curved beam.

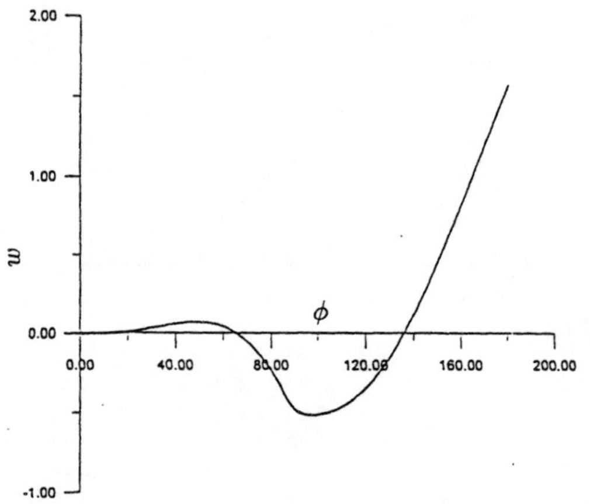

Fig. 11. Radial displacement of the cantilever curved beam.

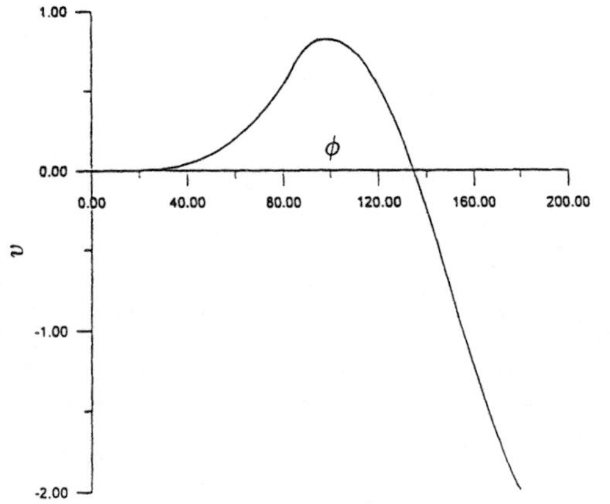

ig. 12. Tangential displacement of the cantilever curved beam.

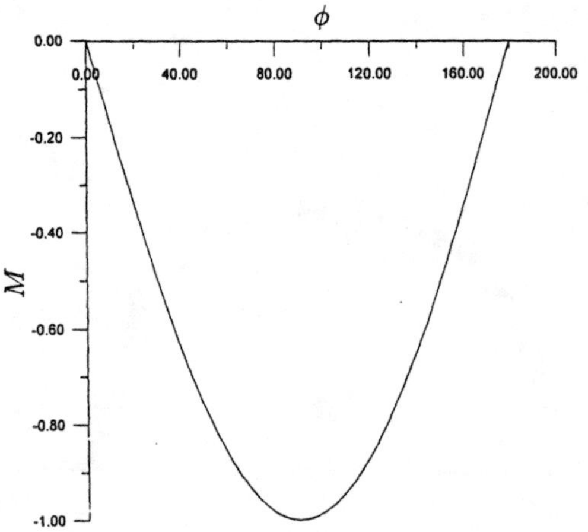

Fig. 13. Bending moment of the cantilever curved beam.

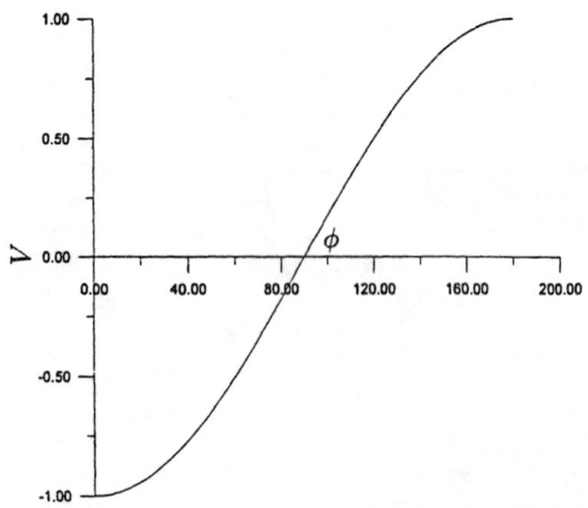

Fig. 14. Shear force of the cantilever curved beam.

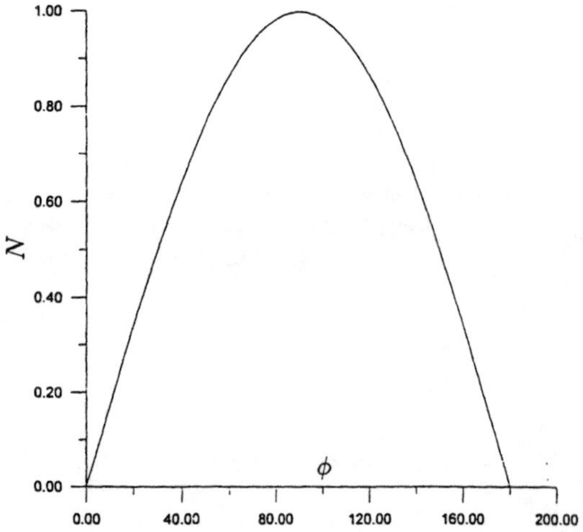

Fig. 15. Axial force of the cantilever curved beam.

Table 1. Results of a simply supported curved beam

DOF per element	Number of elements	w_B	N_A	M_B
10	2	$-.185194870\times10^{-1}$	$-.526387466\times10^{0}$	$-.185164252\times10^{0}$
	4	$-.184702325\times10^{-1}$	$-.508564260\times10^{0}$	$-.184500372\times10^{0}$
	6	$-.183169531\times10^{-1}$	$-.504656312\times10^{0}$	$-.184092580\times10^{0}$
14	2	$-.183099455\times10^{-1}$	$-.500194872\times10^{0}$	$-.183857117\times10^{0}$
	4	$-.183019268\times10^{-1}$	$-.500005198\times10^{0}$	$-.183198376\times10^{0}$
	6	$-.182851892\times10^{-1}$	$-.500001875\times10^{0}$	$-.182913655\times10^{0}$
18	2	$-.182751443\times10^{-1}$	$-.500002427\times10^{0}$	$-.182538425\times10^{0}$
	4	$-.182654927\times10^{-1}$	$-.500001253\times10^{0}$	$-.182377047\times10^{0}$
	6	$-.182550496\times10^{-1}$	$-.500000362\times10^{0}$	$-.182373975\times10^{0}$
Analytical solution		$-.182550423\times10^{-1}$	$-.500000000\times10^{0}$	$-.182373661\times10^{0}$

Table 2. Results of a cantilever curved beam

DOF per element	Number of elements	w_B	v_B
10	2	$.159003985\times10^{1}$	$-.200847195\times10^{1}$
	4	$.157855171\times10^{1}$	$-.200275481\times10^{1}$
	6	$.157337413\times10^{1}$	$-.200059755\times10^{1}$
14	2	$.157125648\times10^{1}$	$-.200008742\times10^{1}$
	4	$.157086784\times10^{1}$	$-.200004857\times10^{1}$
	6	$.157079756\times10^{1}$	$-.200002541\times10^{1}$
18	2	$.157076585\times10^{1}$	$-.200001985\times10^{1}$
	4	$.157077854\times10^{1}$	$-.200000647\times10^{1}$
	6	$.157077339\times10^{1}$	$-.200000134\times10^{1}$
Analytical solution		$.157077215\times10^{1}$	$-.200000000\times10^{1}$

PVP-Vol. 458, Computer Technology and Applications

PVP2003-1911

ASME PVP 2003 CONFERENCE
July 20-24, 2003

Failure Mode of Pipes Containing Circumferential Cracks Under Bending and its Consequence on the Application of NSCM Criterion

P.P. Milella*, N. Bonora**, D. Gentile**

*APAT, Rome, Italy, ** University of Cassino, Italy

Abstract

The results of some 60 tests performed in Italy on 2", 4", 6" and 8" pipes of A 106 B and 304 stainless steel, carrying circumferential through-wall cracks of various size under four point bending conditions (FPB), at room temperature and 300° C, have been analyzed using the Net Section Collapse Moment Criterion (NSCM) and the dimensionless plastic zone parameter (DPZP). Most of the test results have shown that the NSCM applies even though the DPZP is lower than unity. This apparent inconsistency is due to the fact that cracked pipes under bending fail by plastic hinge formation of the type occurring in FPB specimens carrying notches, like the Charpy VN ones, as predicted by the slip line theory. Under these conditions, two half circle plastic zones develop at both sides of the notch while the plastic zone straight ahead the notch tip is almost negligible. FE Calculations have confirmed this behavior: the plastic zone underneath the crack tip has not yet reached the neutral axis when the plastic hinge is formed on the sides of the piping, making the NSCM applicable. This, actually, implies that the DPZP as presently used in the screening criteria is not precisely the proper parameter to adopt in the assessment of the NSCM criterion applicability.

1. Introduction

According to the Net Section Collapse Load (NSCL) criterion *[1]*, a pipe containing a circumferential through-wall crack, under internal pressure and pure bending, fails when the section containing the crack reaches a limit load condition, as schematized in figure 1. This actually implies that the material is ductile enough that fully plastic conditions exist.

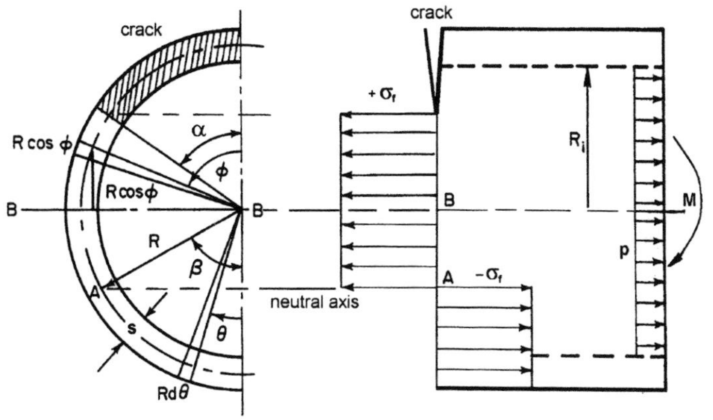

Figure 1 – Internal reactions on the net section of a pipe subjected to internal pressure *p* and external bending moment *M*.

Moreover, the criterion assumes that the material is elastic-perfect plastic, without strain hardening. If the only load acting on the system is the moment *M*, then failure occurs when the internal reactions on the cracked section balance the external moment:

$$M_{NSCM} = 2\sigma_f R^2 t \cdot (2 \sin\beta - \sin\alpha) \tag{1}$$

where M_{NSCM} is the collapse moment according to the NSCL criterion, σ_f is the flow stress of the material, *R* the mean radius of the pipe, *t* the thickness, 2α the overall crack aperture and β the angle of the neutral axis with the vertical axis. The flow stress is defined as half the sum of the yield strength σ_{ys} and the rupture strength σ_r:

$$\sigma_f = \frac{\sigma_{ys} + \sigma_r}{2} \tag{2}$$

One of the most surprising results obtained applying the NSCM criterion to real materials and piping is the fact that, in many cases, it works even when the crack tip plastic zone doesn't reach the neutral axis of the piping. In all these cases, the dimensionless plastic zone parameter (DPZP) is lower than one. This is apparently in contrast with the fundamental hypothesis of complete plasticity of the pipe section ahead of the crack, as shown in fig. 1. To solve this puzzle, a finite element analysis of the piping under bending has been done. The FE analysis has shown that the failure mode is the same as in a bar containing a notch under bending, i.e., plastic hinge formation that occurs before the crack tip plastic zone reaches the neutral axis of the pipes.

2. Experimental verification of NSCM

The extreme simplicity of NSCM criterion makes it attractive for design purpose. To this end, the results of about 60 tests on 2", 4", 6" and 8" pipes of A 106 B and 304 stainless steel, carrying circumferential through-wall cracks of various size under four point bending (FPB) conditions, at room temperature and 300° C, have been analyzed *[2]*. Figure 2 is a schematic of the test rig. An example of results obtained on 8" pipes, 11 mm thick, of A 106 B carbon steel at room temperature, carrying circumferential trough-wall cracks of 45°, 90°, 135° and 180°, respectively, is shown in figure 3. It can be seen how the theory is under predicting the

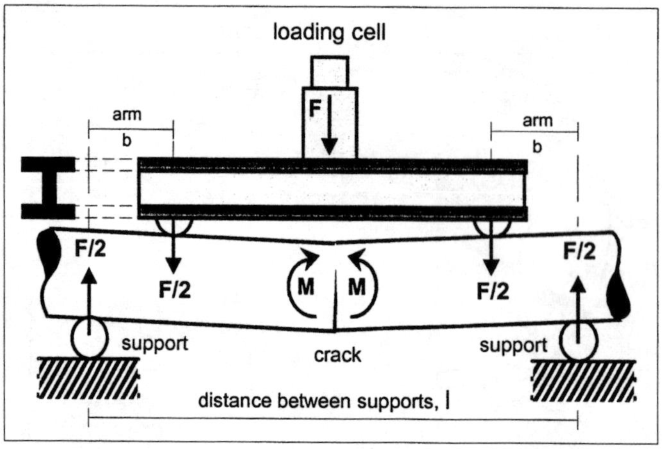

Figure 2 – Schematic of the four point bend test configuration adopted.

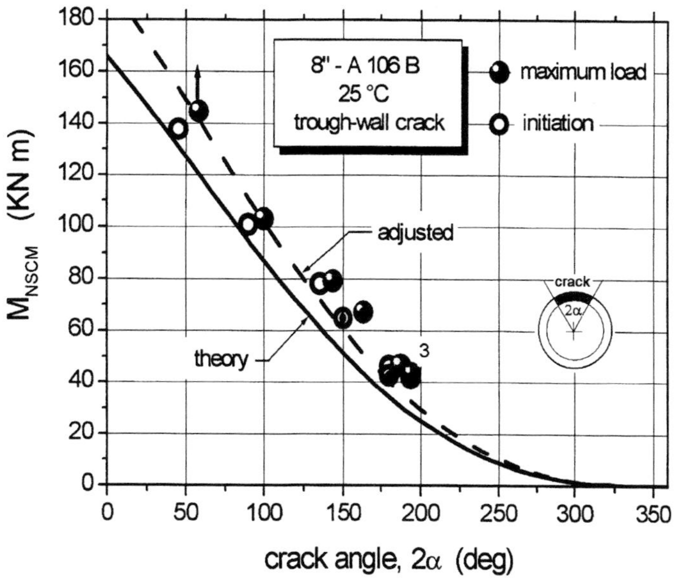

Figure 3 – Experimental results obtained on A 106 B, 8" pipes carrying circumferential trough-wall cracks of different sizes, at room temperature.

actual capability of the pipes to sustain the maximum moment. Probably, this is due to the strain hardening properties of the material that are not considered in eq. (1). The dotted line is drawn by increasing the flow stress σ_f from 383.5 MPa, the actual value, to 450 MPa, using 1.7 instead of 2 in eq. (2). Analogous results were obtained on all carbon steel pipes at room temperature, indicating how the NSCM approach, for ductile materials, can be conservative. Only at 288° C the A 106 B pipes were behaving as predicted by the NSCM and this is due to a dynamic strain aging that occurred at that temperature *[3]* that makes the steel more hard and less ductile, reducing the conservatives of the NSCM criterion. The same results were obtained on type 316 austenitic stainless steel at room temperature. This is due to the cold work hardening properties of the 316 stainless steel. All the experimental results obtained on A 106 B carbon steel are shown in figure 4 in terms of the ratio of the experimental maximum moment to the NSCM predicted by the theory, eq. (1), versus the dimensionless plastic zone parameter (DPZP). The DPZP is defined as:

$$DPZP \quad = \quad \left(\frac{EJ_{Ic}}{2\pi\sigma_f^2}\right) \Big/ \left(\frac{\pi-\alpha}{4}\right) D \tag{3}$$

where J_{Ic} is the critical toughness of the material, measured by the *J-integral* at maximum load. It is interesting to note that all larger pipes (6" and 8") fail when the DPZP is lower than one. Experiments run in the US on carbon steels and stainless steels containing through-wall cracks, have also shown that in some cases the maximum bending moment was reached when the DPZP was lower than one *[4]*.

3. FE calculations

The reason for that discrepancy (DPZP<1) has to be found in the failure mode of pipes containing a crack under bending. According to the slip line theory, a bar containing a notch loaded under four or three-point bending fails by plastic hinge formation, as schematized in

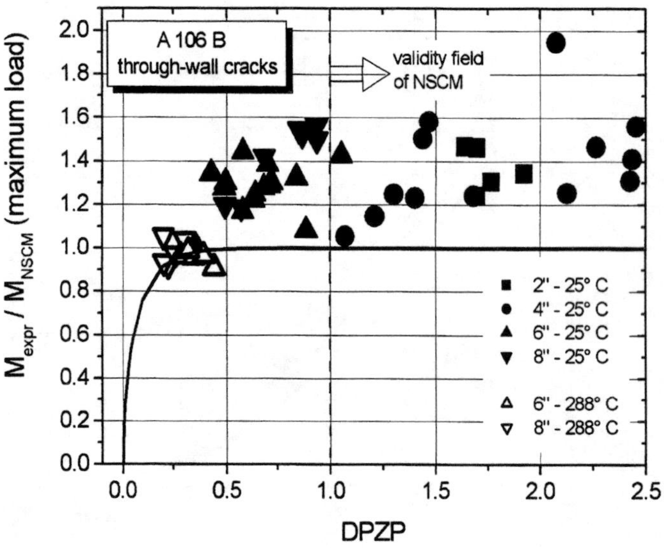

Figure 4 – The Battelle plastic-zone screening criteria for through-wall circumferentially cracked pipes of A 106 B at room temperature and maximum load.

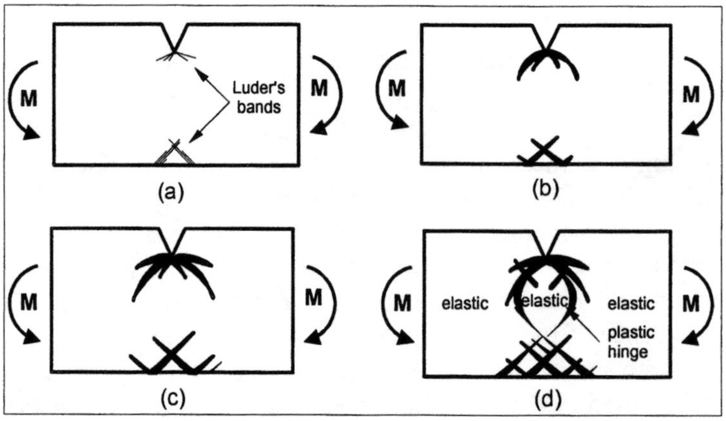

Figure 5 – Schematic of the plastic hinge formation in a bar containing a notch under four-point bending and growing moment *M*.

figure 5. What is important to note is the fact that at maximum moment, when the plastic hinge is completely formed, as shown in fig. 5(d), and the bar undergoes instability, the notch tip plastic zone is still negligible. It must be recalled that the slip-line theory considers plain strain conditions that do not exist in a pipe of small size and, in particular, small thickness. Therefore, the hypothesis of negligible crack tip plastic zone will not apply to small pipes and thickness. This is consistent with the results shown in fig. 4 where only larger pipes, 11mm and 18 mm thick, are closer the plain strain conditions indicating that the DPZP fails. The same data of figure 4, re-plotted in figure 6 in terms of DPZP versus crack angle 2α, show that for larger pipes (6" and 8") already at $2\alpha = 240°$ the DPZP is equal or smaller than one, while for smaller pipes (2" and 4") it is always larger than one. To better check the actual failure mode of the pipes, 3D-FE calculations have been performed. The results for a 6" pipe containing a $2\alpha = 140°$ crack are shown in figure 7. As it can be seen, the lateral plastic hinge develops along with the crack tip plastic zone. At low moments, the crack tip plastic zone is larger than the plastic hinge, but as the moment increases the plastic hinge develops more

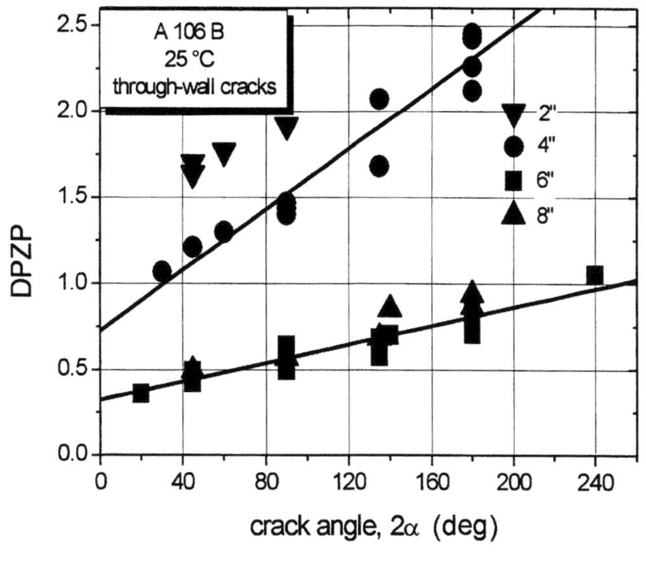

Figure 6 – Dependence of DPZP on crack angle 2α and pipe diameter.

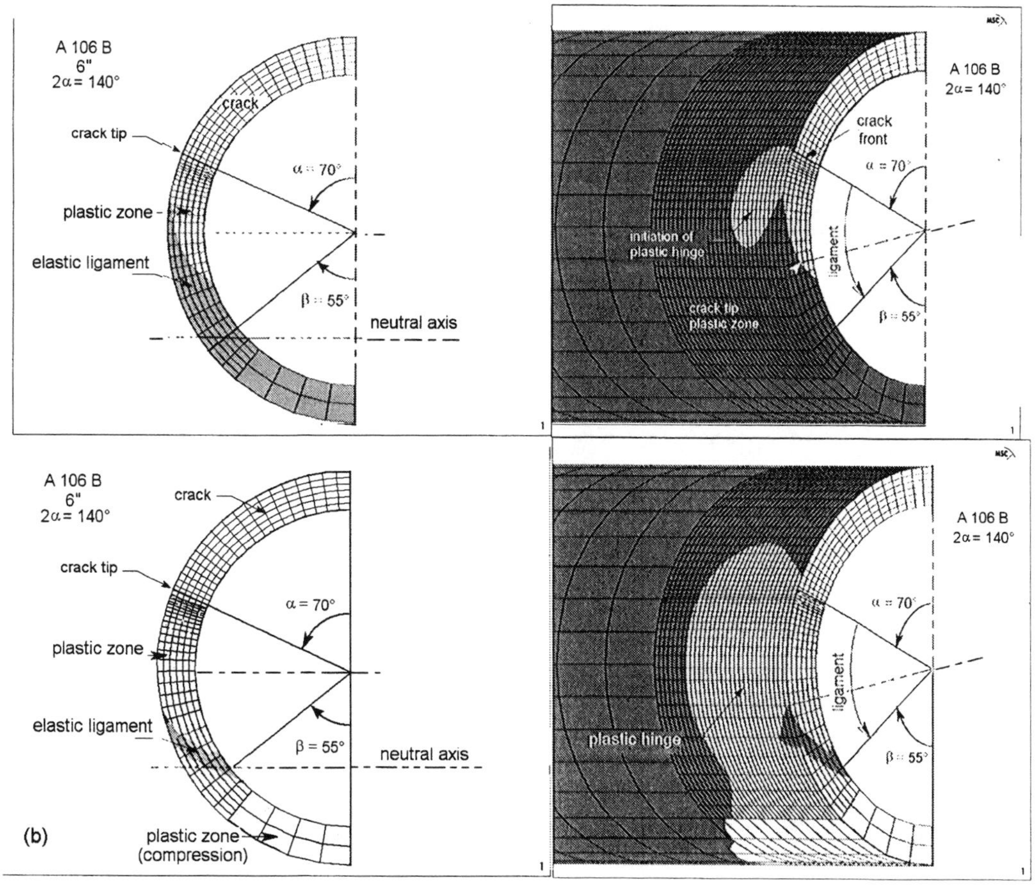

Figure 7 – FE cross section and lateral view of a 6" A 106 B pipe containing a trough-wall crack of 140°. It can be seen how the plastic hinge develops along with the crack tip plastic zone, but beyond the maximum moment (b) the plastic hinge is completely formed while the crack tip plastic zone has not reached yet the neutral axis.

rapidly than the former. Finally, when the maximum moment is passed the plastic hinge is completely formed and closed while the crack tip plastic zone has not reached yet the neutral axis, leaving an elastic ligament. At variance, at 288° C when the ratio of the maximum experimental moment M_{exp} to the M_{NSCM} is lower than one, indicating that the NSCM is not applicable, the FE calculations show that both the crack tip plastic zone and the plastic hinge are not completely formed.

4. Conclusions

The following conclusion can be drawn:

1. The failure mode of pipes containing a through-wall crack under bending is by plastic hinge formation. The plastic hinge forms as a ring emanating from the crack tip.
2. In larger and thicker pipes, where deformation conditions are closer to plain strain, the plastic hinge is formed before the crack tip plastic zone reaches the neutral axis, making DPZP be smaller than one even though conditions exist for the applicability of the NSCM criterion. Therefore, in larger pipes the NSCM criterion applies even when the DPZP is smaller than one.
3. Smaller pipes (2" and 4") are loaded under plain stress conditions making the crack tip plastic zone develop rapidly through the entire ligament. The DPZP is, by far, larger than one.
4. The DPZP parameter should be substituted by a criterion based on the size of the plastic lobe forming the plastic hinge.

5. References

[1] Kanninen, M.F., et al. "Towards an elastic plastic fracture mechanics predictive capability for reactor piping", Nuclear Engineering and Design, Vol. 48, pp. 117-134, 1978.

[2] Milella, P.P., "Outline of Nuclear Piping Research Conducted in Italy ", Nuclear Engineering and Design, 98, pp.219-229, 1987

[3] Milella, P.P., "Fracture Behaviour of Carbon Steel Pipes Containing Circumferential Cracks at Room Temperature and 300 °C ", International Workshop on LBB, Tokyo, Japan, May 18-19, 1987, Published on Nuclear Engineering and Design, 111, pp. 35-46, 1989

[4] Wilkowski, G.M., Olson, R.J., Scott, P.M., " State-of the-Art-Report on Piping Fracture Mechanics", US-NRC, NUREG/CR-6540, p. 2-106, 1998.

PVP-Vol. 458, Computer Technology and Applications

PVP2003-1912

Generalized Application of Periodic Symmetry in Molecular Simulations

By
Li Pan, Don Metzger
Department of Mechanical Engineering, McMaster University, 1280 Main Street West, Hamilton L8S 4L7, Canada

Marek Niewczas
Department of Materials Science and Engineering, McMaster University, 1280 Main Street West, Hamilton L8S 4L7, Canada

ABSTRACT

Periodic symmetry is widely used in molecular simulations to mimic the presence of an infinite bulk surrounding an N-atom model system. However, the traditional methods of applying periodic symmetry end up enforcing over-restrictive kinematic constraints between the periodic boundaries. After a brief overview of the periodic symmetry, the nature of the constraint is discussed briefly in this paper. Thereafter, the objective is to provide a means to ensure that periodicity is upheld while avoiding unnecessary constraint of the repeating cell boundaries.

This paper demonstrates the usual application of periodic symmetry into a molecular simulation algorithm through a typical example. Meanwhile, a novel method is introduced, which uses equivalent external forces applied to physical boundary atoms. Comparisons between the classic treatment and the new method using one-dimensional and two-dimensional models are made. Moreover, the potential application of the new method in regular Finite Element Analysis is discussed.

INTRODUCTION

Periodic symmetry (or periodic boundary condition) is a key method in molecular simulations to mimic the presence of an infinite bulk in an N-atom model system. The model containing N atoms is treated as the primitive cell of an infinite periodic lattice of identical cells. In figure 1, any atom *i* interacts with all other atoms including the atoms in the primitive cell and in all its image cells within the cut-off radius.[1] Therefore, certain methods are needed to generate forces in order to deal with the symmetry condition in the molecular model. Fortunately, methods to account for symmetry in Finite Element Analysis (FEA) have already been developed, and some of these methods may be applied to the molecular simulations.

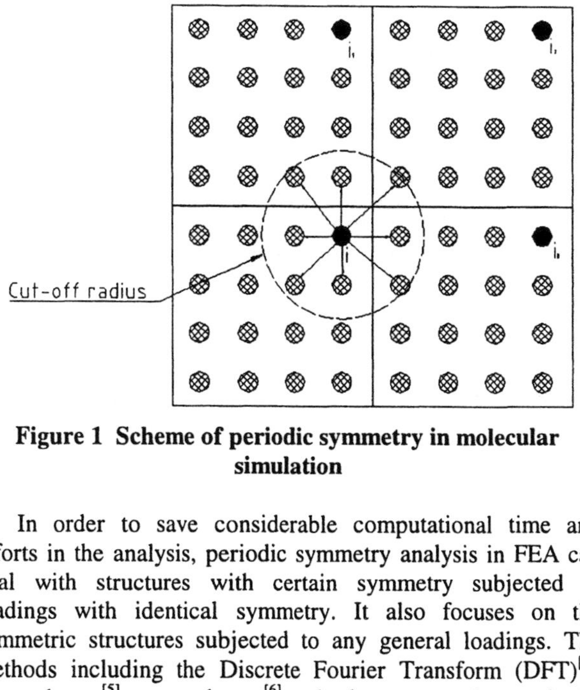

Figure 1 Scheme of periodic symmetry in molecular simulation

In order to save considerable computational time and efforts in the analysis, periodic symmetry analysis in FEA can deal with structures with certain symmetry subjected to loadings with identical symmetry. It also focuses on the symmetric structures subjected to any general loadings. The methods including the Discrete Fourier Transform (DFT)[4], group theory[5] , wave theory[6] and others are used to study the static and dynamic properties of structures. However, these methods cannot directly be implemented for the N-atom system because (i) the molecular model is meshless[2][3][7] and (ii) the periodic treatments in molecular simulation are needed to extend all the atoms not only limited to physical boundary atoms in the primitive cell due to the effect of interaction between atoms.

Another technique in FEA called 'tied nodes' could be used in the molecular case. This technique forces two nodes (or atoms, in this case) to follow identical motions during the simulation. However, this technique is not suitable for all circumstances and has certain limitations. The details will be discussed in a later section.

Thus, this paper first briefly overviews the traditional method for applying periodic symmetry in the molecular model with a typical application, and then introduces a novel method to overcome the pitfall of traditional method and satisfy various conditions in the atomistic simulation.

OVERVIEW OF TRADITIONAL METHOD

Method

The N-atom example studied currently is presumed to have its non-equilibrium state caused only by its internal defects. This implies that no external forces act on the boundaries, and only internal forces push atoms into new equilibrium places. The internal force is first derived from the gradient of potential energy, which is the function of distances between atoms.[2]

When applying periodic boundary conditions to the model, the resultant internal force on a certain atom and on all of its corresponding images should be identical. Therefore, the atom and all its images should have the same acceleration (motion). A method, herein called the traditional method, has been developed[1] to form the appropriate resultant forces required to assume that the motion is compatible with the symmetry.

Followings are some key points of the traditional method for the calculation of periodic symmetry:

(1) For 3-D models, applying a periodic boundary condition to one coordinate direction is widely used. For a given atom *i*, there are two images in one direction, which is shown in figure 2.

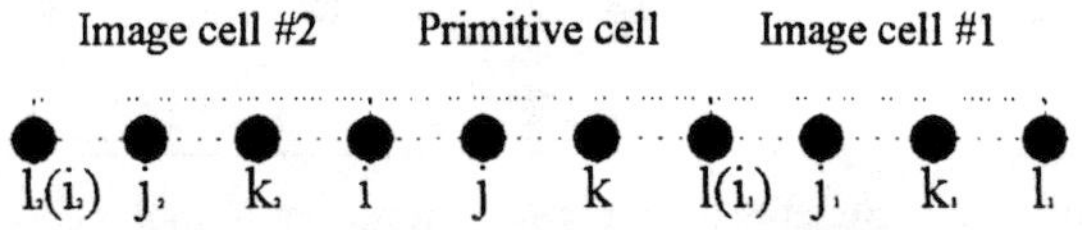

Figure 2 Periodic symmetry in one coordinate direction

(2) Generally, the cut-off distance r_c must be less than half the length of the primitive cell r_L (periodic length) in any coordinate direction. Otherwise, some image atoms could be missed or double counted.

(3) The distance r_{ij} between any atom *i* and *j* now is determined by atom $i(x_i, y_i, z_i)$, atom j (x_j, y_j, z_j) and all *j*'s images if fixing the position of atom *i*. For one direction (*x* direction) periodic symmetry, first evaluate the distance r_{ij}. If $r_{ij} > r_c$, then evaluate the distance between atom *i* and two images of atom *j*: j_1 $(x_j + r_L, y_j, z_j)$ and j_2 $(x_j - r_L, y_j, z_j)$. Three circumstances are listed as follows:

   (a) If neither r_{ij1} nor r_{ij2} is within the cut-off radius r_c, atom *j* is skipped;

   (b) If either r_{ij1} or r_{ij2} is within the cut-off radius r_c, this value is the distance between atom *i* and *j*;

   (c) if both r_{ij1} and r_{ij2} are within the cut-off radius r_c, choose minimum value in r_{ij1} and r_{ij2} to determine the distance r_{ij}.

(4) If periodic symmetry is applied to the model in more than one coordinate direction of the model, the procedure is the same as above except that there are more image atoms involved in the calculation.

Defining the primitive cell

For atomic models using periodic symmetry, there is a problem with coincident atoms, which happens to the pair of physical boundary atoms. For example, in figure 3, there are 6×6 atoms in the model (primitive cell 1) and the cut-off distance r_c is to cover two nearest neighbors. Figure 3 shows how to calculate the properties of the atom *i* (black circle) in a given time step. The white circles indicate atoms located in the primitive cell and image cells, which interact with atom *i* within r_c. After calculating the distance r_{ij} between atom *i* and *j*, it is found that $r_{ij} > r_c$. Then all the atom *j*'s images are searched. Atom *k*, as one of the images of atom *j*, has minimum distance with atom *i*. Thus, atom *k* is counted. Meanwhile, atom *k* itself is within the cut-off distance with atom *i*, so it is counted again. Therefore, the position of atom *k* is calculated twice, which is not correct in the simulation.

The solution for the coincident atoms problem could be to build a neighbor list for atoms while calculating the distance with their neighbor atoms. If the position of one neighbor atom is on the list, that means that this position was calculated before and this atom is skipped. Second alternative solution is to increase one more unit cell length (half in the each side) to the original primitive cell (primitive cell 1, solid line in figure 3), which makes new primitive cell (primitive cell 2, dash line in figure 3). While the first solution is straightforward, the second one is perhaps more easily implemented but restricted to the situation that the spacing between atoms in periodic direction is evenly distributed.

Example

Here is an example of typical atomic model to present the determination of the length of primitive cell with consideration of the cut-off distance in the traditional method. The boundary condition of a 3D model is that:

(i) All the atoms at the physical boundary of the model are fixed in X and Y direction

(ii) In Z direction, periodic boundary condition is applied to all the atoms of the model.

The atom configuration in Z direction is shown in figure 4, in which D1 is the unit length between adjacent atoms and cut-off radius is a little larger than D1 ($D1 < r_c < D2$). Atoms A and B form a pair of boundary atoms.

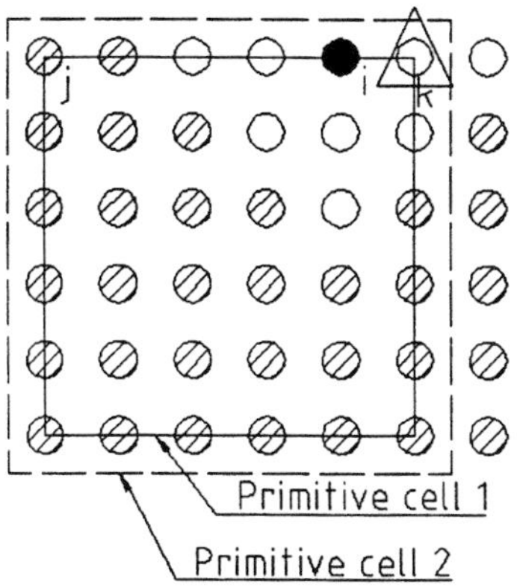

Figure 3 Scheme of coincident atoms

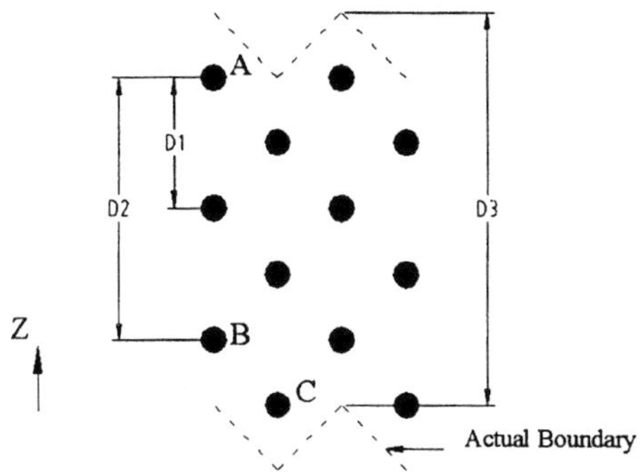

Figure 4 Scheme of atom configuration in Z direction

Given of the orientation of the model, the periodic length in Z direction cannot be taken to be the distance between the top atom A and the bottom atom C, which is 2.5 times of unit length D1. This is because atom A and its image atom B lose their uniform motion under the periodic length of $2.5 \times \mathrm{D1}$. The reason for this is that the periodic length cannot be a non-integer multiple of the unit length D1.

For the configuration in this model, the periodic length in Z direction also could not be adjusted to D2, which is two times of unit length D1. At this stage, the cut-off distance r_c is larger than half the length of periodic length D2, which violates the rule (2) above of periodic boundary condition so that some atoms are double counted.

In the particular case, when the cut-off distance is just more than half of the current periodic length but less than next integer multiples of unit length, in order to save computational effort, the alternative solution is to increase one more unit cell length (half in the each side) to the original primitive cell. Therefore, in this model, since cut-off distance r_c is more than half of periodic length D1, but less than next integer multiples of unit length D2 $(2 \times \mathrm{D1})$, periodic length is increased to three times of unit length D1, as D3 shown in figure 4. Using D3 as periodic length in Z direction ensures the atoms and their images have uniform motion.

Discussion

In the traditional method of applying periodic symmetry to the model, in order to preserve the periodicity, e.g. in figure 4, the periodic length must be kept constant during the simulation. Otherwise, atoms and their images lose uniform motion. Therefore, in the usual treatment, the periodic length could not be stretched or compressed but the corresponding periodic boundaries are allowed to move and deform, but at a constant spacing. Consequently, periodic symmetry generates a kind of restrictive 'kinematic constraint' between the periodic boundaries. Figure 5 shows an example that atoms with even spacing cannot relax their high potential energy after applying two-dimensional periodic symmetry.[2] The relaxation under periodic symmetry in traditional method takes place only within these boundaries until even spacing is achieved.

Figure 5 Configuration of two-dimensional model

The technique of 'tied nodes' provides another way to obtain the motion of corresponding pair of atom and its image. However, it can only apply to the pair of physical boundary atoms while inner atoms in the model still need use traditional method to calculate their movements. In the 'tied node' method, forces on the boundary atoms, $\mathbf{F}_i$, are first obtained without any regard for symmetry (no images considered). Then, the resultant force $\mathbf{F}_i^R$ on each boundary node i is calculated by adding (assembling) the force from its image node i',

$$\mathbf{F}_i^R = \mathbf{F}_i + \mathbf{F}_{i'} = \mathbf{F}_{i'}^R \tag{1}$$

This ensures that the resultant force on a boundary node and its image are the same. Therefore, this method is equivalent to the traditional method, but since searching for the pairs of boundary atoms is required, the 'tied node' method is seldom used for the molecular simulation.

INTRODUCTION OF NEW METHOD

The usual treatments including 'tied nodes' method are acceptable when the change of atomic structure is only dependent on the internal forces. However, if external forces are involved into the simulation, e.g. tension test, the lattice parameter is changed. The periodic length can follow the same trend and be the identical parameter throughout the model. But the periodic boundary under the traditional method is inflexible. On the one hand, the traditional method itself enforces over-restrictive kinematic constraints between periodic boundaries since it requires any pair of boundary atoms be in the same motion. Actually, they don't have to as long as all pairs of boundary atoms in the model could maintain unique periodic length at any time during the simulation. On the other hand, if randomly distributing external forces directly on the atoms without any special rules, at most of the cases, the model easily loses unique periodic length so that the periodicity of the model is subverted. This is because every pair of boundary atoms maintains their own motion subjected to the integration of external and internal forces according to their positions in the model. Hence, a new method is needed to ensure that the periodicity is upheld, while avoiding unnecessary constraints of the repeating cell boundaries. The following one-dimensional and two-dimensional models are used to demonstrate the new method and its applications.

One-dimensional model

The approach in the new method is to substitute equivalent external forces for the internal forces generated by the image atoms at the physical boundary atoms. One-dimensional model shown in figure 6 is used to explain the scheme of the new method.

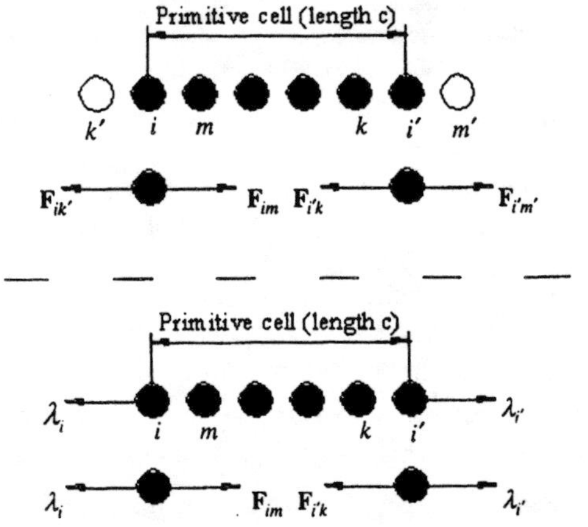

Figure 6 Scheme of the new method

In figure 6, the primitive cell is from boundary atom i to boundary atom i' and its length is c. The cut-off distance is supposed to extend to the one nearest neighbor. As the traditional method, atom i (or atom i') should interact with two atoms, one of which is the image atom (white circle in the figure 6). In the new method, forces on the boundary node i and its image i' are found without regard for any symmetry (just like the 'tied method'). However, the boundary force λ_i is used in place of the interaction between boundary atoms and image atoms. This gives resultant forces on i and i',

$$\mathbf{F}_i^R = \mathbf{F}_i + \lambda_i \tag{2}$$

$$\mathbf{F}_{i'}^R = \mathbf{F}_{i'} + \lambda_{i'} \tag{3}$$

Since the new method does not require that the periodic length to be constant during the simulation, and the physical boundary atom i and its image i' have identical motion, the acceleration of the boundary nodes are related by

$$\ddot{\mathbf{x}}_i - \ddot{\mathbf{x}}_{i'} = \ddot{c} \tag{4}$$

Therefore,

$$\frac{\mathbf{F}_i + \lambda_i}{m_i} - \frac{\mathbf{F}_{i'} + \lambda_{i'}}{m_{i'}} = \ddot{c} \tag{5}$$

Internal forces are obtained from the atom in the primitive cell within cut-off distance. The boundary forces have in the same magnitude and opposite direction in order to be compatible between the periodic boundaries, i.e.

$$\lambda_i = -\lambda_{i'} \tag{6}$$

By combining equation (5) and (6), the external forces could be obtained as follows:

$$\lambda_i = -\lambda_{i'} = \left(\frac{m_i m_{i'}}{m_i + m_{i'}}\right)\left[\ddot{c} - \left(\frac{\mathbf{F}_i m_{i'} - \mathbf{F}_{i'} m_i}{m_i m_{i'}}\right)\right] \tag{7}$$

When $\ddot{c} = 0$, equation (7) becomes

$$\lambda_i = -\lambda_{i'} = -\frac{\mathbf{F}_i m_{i'} - \mathbf{F}_{i'} m_i}{m_i + m_{i'}} \tag{8}$$

Equation (8) reveals how much external force needs to be applied on the boundary atoms to keep the periodic length constant throughout the simulation. Therefore, the relaxation result based on equation (8) should be in line with the results from the traditional method and 'tied node' method.

In order to examine the suitability of the new method, a one-dimensional model with 7 atoms in a row with randomly different spacing is tested so that relaxation could happen under the periodic symmetry. The initial spacing between adjacent atoms is shown in the figure 7(a). Then, two methods, the traditional method and the new method with $\ddot{c} = 0$, are both applied to the model above. The results are demonstrated in the figure 7 (b) and (c). (Note: in the new method, external forces are only applied to the boundary atoms. Positions of other atoms in the primitive cell are calculated in the same way as in the traditional method.)

As one can see from the figure 7, the new method is proved to be effective for applying periodic symmetry to the molecular model because both methods converge to the state that the distances between adjacent atoms are equal. The system momentum is balanced using both methods since the external (reactions) forces are in the equilibrium state. The apparent rigid motion is due to the relative motion of the boundaries not the movement of the Center of Gravity in the model. The periodic length is also not changed as is the case with the traditional method. Therefore, the new method not only includes the features of the traditional method, but also provides a way of applying external force to the model when $\ddot{c}$ is not equal to zero.

Two-dimensional model

A two-dimensional model is used to demonstrate the details of applying external forces to the molecular model while keeping the periodicity. Figure 8 shows a pair of tension forces applied to the boundary of two-dimensional model, in which the shape of boundary is random but periodic length is unique. Periodic symmetry is applied to X direction of the model while free boundary condition is used in the Y direction. Consequently in the simulation, the tension forces are distributed on the boundary atoms.

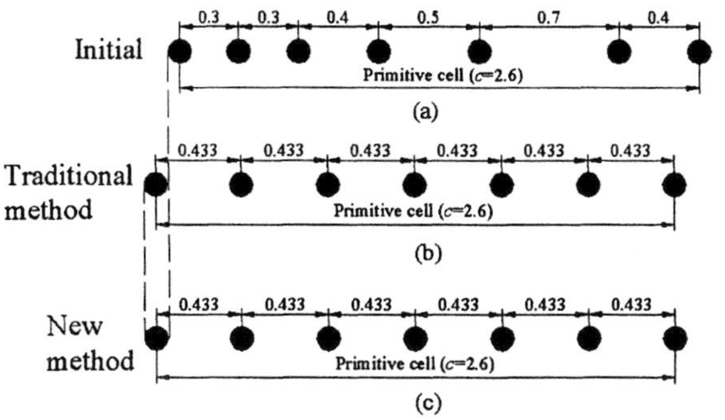

Figure 7 1-D model (a) Initial configuration (b) New equilibrium configuration after using traditional method (c) New equilibrium configuration after using new method

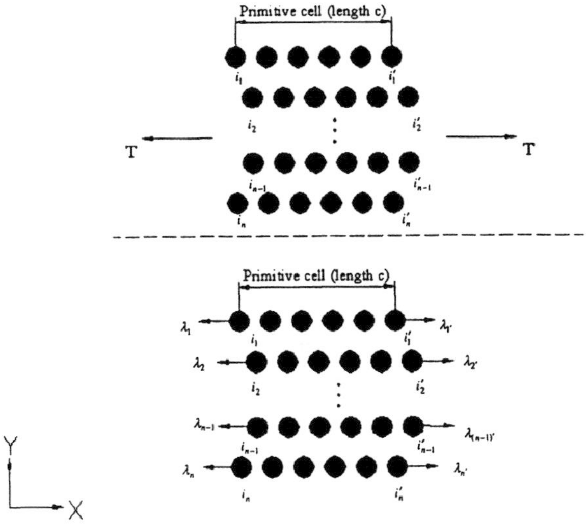

Figure 8 Two-dimensional model with external tension force

Therefore,

$$(\pm)\sum_{i=1}^{n}\lambda_i = (\pm)\mathbf{T} \quad (9)$$

Based on the equation (4) and (5), after substituting equation (6) and (7) to (9), it is obtained

$$\sum_{\substack{i=1\\i=i'}}^{n}\left\{\left(\frac{m_i m_{i'}}{m_i + m_{i'}}\right)\left[\ddot{c} - \frac{\mathbf{F}_i m_{i'} - \mathbf{F}_{i'} m_i}{m_i m_{i'}}\right]\right\} = \mathbf{T} \quad (10)$$

There are two unknowns, **T** and $\ddot{c}$, in equation (10). One of these must be specified so that the other unknown could be calculated. It could change every time step to adapt to the change of internal forces until convergence.

Two examples using small block of atoms are used to illustrate the application of the new method for periodic symmetry. The first example applies periodic symmetry to the rectangular atomic model with even spacing in the columns as shown in figure 9(a). The periodic length is 4.0-arbitrary-units. A pair of 5.0-arbitrary-unit external tension forces is applied to the boundary. Periodic symmetry is applied to the X direction of the model while free boundary condition is used in the Y direction. The results of new equilibrium configuration using the new method are shown in figure 9(b). The periodic length in all rows becomes 4.36-arbitrary-units, which is in the line with the expectations.

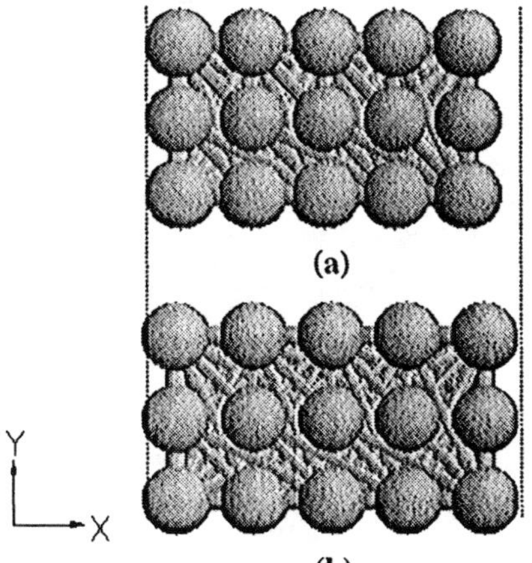

Figure 9 Rectangular model (a) Initial configuration (b) New equilibrium configuration after applying external force

As discussed in the previous work (reference [2] and [3]), rectangular or cubic configuration of atoms in 2D model is kind of the local minimum potential energy position with certain spacing. Therefore, after applying periodic symmetry in the X direction, the spacing between atoms is equally increased and the unique periodic length is maintained during the simulation. In the Y direction, the distances between atoms are free to be adjusted to the equilibrium state. Due to the new periodic symmetry treatment, the initial model is stretched and relaxed under the external forces applied to the boundaries but lower minimum potential energy positions (closed-packed configuration) is not achieved.

Another example is obtained by applying periodic symmetry to the parallelogram atomic model with even distance between adjacent atoms in the X direction shown in figure 10(a). The periodic length, external force and boundary conditions are the same as in the previous model. The results of new equilibrium configuration using the new method are shown in figure 10(c). The periodic length in all rows becomes 4.40-arbitrary-unit.

The Parallelogram configuration is another kind of minimum potential energy position since it could generate closed-packed arrangements of the atoms. The minimum potential energy value of the model is dependent on the angle θ and spacing r between the atoms.[2][3] In the figure 10(c), it is found that the angle θ of the parallelogram is changed due to the relaxation of the model to lower potential energy positions while the periodic length is stretched. This means that the new method allows for applying external forces to the model without preventing the model from relaxing to lower potential energy. However, if the traditional method is used, the relaxation in the X direction will not happen since even spacing is already achieved between atoms, which is shown in figure 10(b). Moreover, just like other boundary conditions, the new method causes the relaxation to be in the restrictive way, i.e. the relaxation must be compatible to boundary conditions.

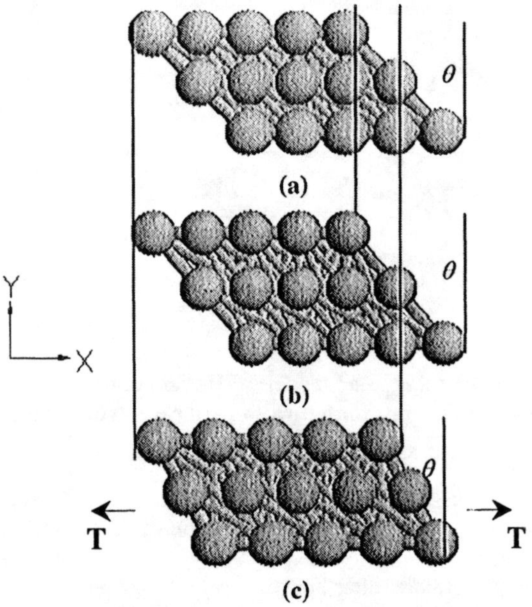

Figure 10 Parallelogram model (a) Initial configuration (b) New equilibrium configuration after using traditional method (c) New equilibrium configuration after using new method

CONCLUSION

The new method for periodic symmetry in atomistic simulation is proposed and successfully implemented to the molecular simulations. Compared with the traditional method, the new method overcomes the over-restrictive constraint within the repeating cell boundaries and meantime keeps the periodicity of the model so that it supplies a flexible treatment for periodic symmetry and is suitable for broad cases in molecular simulation especially when external forces are applied.

The new method could be used fully in place of the traditional method when $\ddot{c} = 0$. However, the traditional method, with advantages of being straightforward and easily implemented, could still be the choice for the cases that the change of the atomic structures is dominated by internal forces. The new method needs computational effort to search for the pairs of boundary atoms, but this could be offset by the need to calculate fewer pairs of internal forces for the boundary atoms. Therefore, the computational burden of the new method is about the same level as the traditional method in the simulation.

The new method has already shown its potential capability for various periodic symmetry cases in molecular simulations. The future work is to add this new approach to the meshless Dynamic Relaxation algorithm to simulate practical material models with more complicated situations.

The new method could be also used in the regular FEA. The scheme of the new method, which utilizes equivalent external forces, allows this new approach to be easily incorporated into the current FEA packages.

Reference

[1] Frenkel, D. and Smit, B., 1996, "Understanding molecular simulation: from algorithms to application," Academic Press, Inc

[2] Pan, L., Metzger, D. R. and Niewczas, M., 2002, "The meshless Dynamic Relaxation techniques for simulating atomic structures of materials," *ASME-PVP, v 441, Computational Mechanics: Developments and Applications (2002 ASME Pressure Vessels and Piping Conference)*, pp. 15-26

[3] Pan, L., 2002, "Meshless Dynamic Relaxation techniques for simulating atomic structures of materials" Masters thesis, Mechanical Engineering Department, McMaster University, Hamilton, Ontario, Canada

[4] Moses, E., Ryvkin, M. and Fuchs, M. B., 2001, "A FE methodology for the static analysis of infinite periodic structures under general loading," *Computational Mechanics*, Vol. 27, pp. 369-377

[5] Kangwai, R. D., Guest, S. D. and Pelegrino, S., 1999, "An introduction to the analysis of symmetric structures," *Computers and Structures*, Vol 71, pp. 671-688

[6] Li, D. and Benaroya, H., 1992, " Dynamic of periodic and near-periodic structures," *Appl. Mech. Rev.*, Vol. 45, No. 11, pp. 447-459

[7] Sauvé, R. G. and Metzger, D. R., "Advances in dynamic relaxation techniques for nonlinear finite element analysis", *Journal of Pressure Vessels and Piping, Transactions of ASME*, Vol. 117, pp. 170-176, 1995

PVP-Vol. 458, Computer Technology and Applications

PVP2003-1913

Periodic Symmetry for Explicit Methods for Large Deformation Twisting and Stretching

Don R. Metzger
McMaster University, Department of Mechanical Engineering
1280 Main St. W., Hamilton, Ontario, Canada, L8S 4L7

ABSTRACT

The presence of repetitive symmetry in continua and structures allows for the modeling of a single symmetric cell. Typically, the symmetry conditions are enforced with homogeneous kinematic constraints, but these may preclude some otherwise admissible response. Also, in large deformation situations, the original symmetric boundaries can distort significantly. The objective of this work is to provide a means for explicit finite element methods to ensure periodic symmetry while resultant loads may be prescribed on the symmetric boundaries. A method is devised to determine external forces that concurrently maintain dynamic equilibrium of the system with compatible deformations at the boundaries. The method requires only internal forces and nodal positions, which are accessible in an explicit finite element formulation.

INTRODUCTION

The use of symmetry in finite element modeling is paramount to efficient and accurate analysis. Most commonly, planes of symmetry or antisymmetry are incorporated with global constraints. More generally, symmetry in the form of periodic, or repeating, structures can be accommodated by modifying the stiffness matrix to "connect" the symmetric boundaries. This approach was first applied to the static analysis of rotationally periodic (cyclic symmetry) geometry of impellers [1]. Since then, many others have applied cyclic symmetry in both static and dynamic analysis. For dynamic simulations, more sophisticated treatment of the connectivity between the boundaries has been considered to capture the general harmonic response. References [2-7] represent some of the previous work for linear response, with reference [7] giving a quite thorough list of related literature.

The basic procedure of applying cyclic symmetry is given a clear description in [8]. In particular, cyclic symmetry is encountered frequently enough that it has been facilitated directly into general purpose finite element software. However, the resulting structural equations (including the symmetry conditions) are typically based on the initial undeformed geometry, and large deformation problems are less accessible. More general symmetry may be accommodated with the use of multi-point constraints (MPC), although the user is most likely restricted to linear homogeneous functions of displacement degrees of freedom. Such constraints are usually applied with the Lagrange multiplier method, which is best suited for so called implicit formulations.

Aside from cyclic symmetry, applications involving structures comprised of an array of holes in a two dimensional continuum have been addressed in detail with direct kinematic constraints (planes of periodic symmetry move but remain plane) [9], and with MPC's [10,11]. In all of these cases, the analysis is static but nonlinear due to elastic-plastic material response. Ultimately, all of these works address rather special cases of symmetry in that the constraints are treated as linear for small deformations.

Recent popularity of explicit methods for large deformation problems apparently has not yet encountered the need for general symmetric capabilities. Prominent explicit software packages still only accommodate simple constraint conditions such as tied nodes. This situation is likely due to the tendency of such programs to be applied mainly to full models of complex structures where symmetry cannot be exploited. There are some cases where rotational cyclic symmetry was considered in large deformation simulations. In analyses of roller expanded joining of tubes [12-13], an explicit finite element method was used, and the symmetric boundaries were seen to deform considerably. Figure 1 shows how boundaries may deform while maintaining a periodic structure.

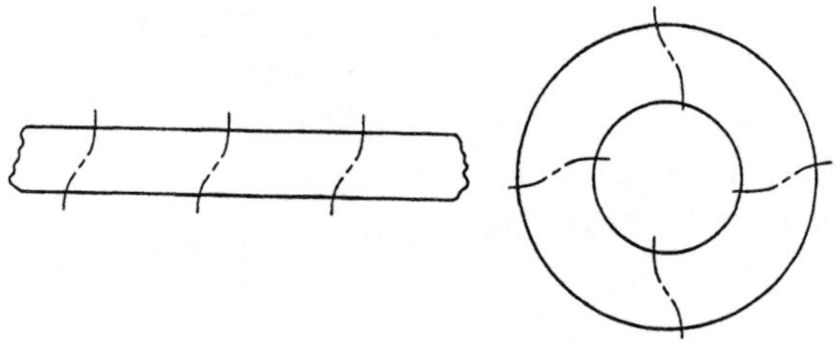

a) translational b) cyclic

Figure 1 – Periodic symmetry with large deformation

In [12-13], the method used to maintain the symmetry conditions is essentially an explicit implementation of the method in [8]. Forces acting on corresponding boundary nodes are made to be consistent by assembling forces and masses shown in figure 2 so that the corresponding nodes will have the same (but for a rotation transformation) equation of motion. Thus, the force and mass for the boundary nodes is obtained as

$$F_i^* = F_i + R^T F_{i'}, \qquad F_{i'}^* = F_{i'} + RF_i \tag{1a}$$

$$m_i^* = m_{i'}^* = m_i + m_{i'} \tag{1b}$$

Where the * indicates the quantities that are actually used in the equations of motion and R is a rotation tensor corresponding to the angle between the boundaries. Although the boundaries are free to distort, the included angle between cell boundaries is constrained by this approach to remain constant. This is necessary for a closed circular geometry, but in the most general terms, the angular period could be allowed to change.

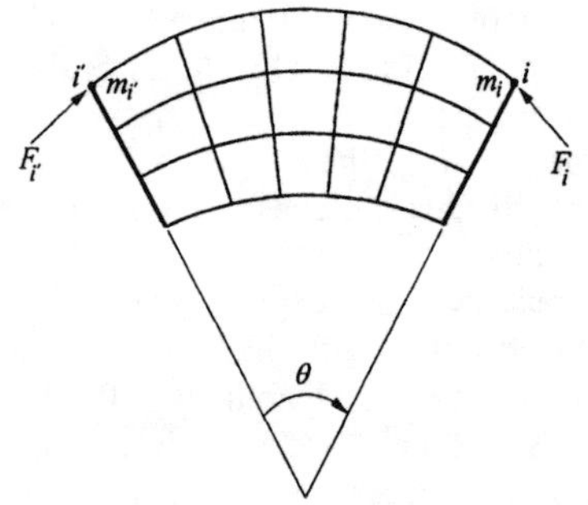

Figure 2 – Cyclic symmetry with tied node approach

The main problem with the approach in figure 2 is that the symmetric boundaries are treated as internal boundaries. This precludes the application of tractions on the boundaries, and the tractions realized will merely be the reactions corresponding to the kinematic constraint. In the case of tied nodes, the reaction forces make the tied nodes move in unison, with no change in separation. In effect, the method of figure 2 has overconstrained the general problem in the sense that it has enforced compatibility of the boundaries (always necessary) and enforced the angular motion of the boundaries to be the same (not generally necessary, only true for no moments). In order to avoid this pitfall, a review of the kinematic relationship between the boundaries is required. This is the main focus of this paper.

PERIODICITY BY TRANSLATION

The most simple case of periodic symmetry is the case of a linear array of symmetric cells subjected to resultant tractions on the free surfaces as well as a net traction applied at a far distance. On the free surfaces, the distribution of the tractions must be known, whereas on the symmetric boundaries only the resultant traction can be specified. It is imperative that the loads are in equilibrium. Figure 3 shows a free body diagram of a finite element model of a single cell. There must exist a distribution of "external" forces on the symmetric boundaries such that equilibrium is satisfied along with the compatible motion of the boundaries.

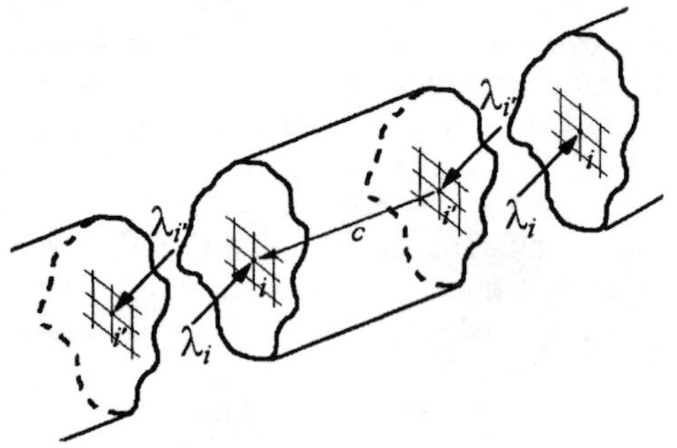

Figure 3 – Periodic symmetry by translation

Kinematically, the boundary nodes must move such that they fit together. The method of figure 2 (equation (1) effectively enforces the relative position between corresponding nodes on each boundary to remain fixed. This proves to be overly restrictive, and the condition can be relaxed by including an unspecified relative displacement vector, c, such that

$$x_i - x_{i'} = c \tag{2a}$$

or equivalently in terms of velocities and accelerations,

$$\ddot{x}_i - \ddot{x}_{i'} = \ddot{c} \tag{2b}$$

Equation (2) will be used to represent the constraints of symmetry in a regularized form. This approach is a type of constraint regularization and provides a nonsingular inertial matrix that can be used with standard time integration procedures.

The general approach of applying constraints to the accelerations was largely advocated by Baumgarte [14], and upon consideration of potential numerical difficulties for highly nonlinear systems, considerable attention has been given to the stable integration of the constrained equations of motion. While this stability issue bears directly on the behaviour of this work, it is not the main thrust of this investigation. At this time it is assumed to be sufficient that several options are available, the most relevant of which are in [15-18]. Also, it is anticipated that linear and weakly nonlinear constraints should pose no problems according to the success [12,13] in which an alternate (but mathematically equivalent) form of equation (2b) was used directly without any stabilization terms. Some theoretical consideration for the well behaved nature of linear constraints is given in [19].

From symmetry conditions, the constraint forces, λ, due to symmetric boundary tractions acting on corresponding nodes must be equal and opposite. Thus,

$$\lambda_{i'} = -\lambda_i \tag{3}$$

Note that at the corners where the symmetric boundaries meet a free surface, force due to traction on the free surface must be included in the equations of motion.

The acceleration of each degree of freedom is given by

$$\ddot{x}_i = \frac{F_i + \lambda_i}{m_i}, \qquad \ddot{x}_{i'} = \frac{F_{i'} + \lambda_{i'}}{m_{i'}} \tag{4}$$

where $F_i = F_i^{ext} + F_i^{int}$ with F_i^{int} being the internal forces known from the usual finite element calculations, and F_i^{ext} is the part of external force from tractions on the external surfaces and from body forces as determined without consideration of the symmetry. Combining equations (2b), (3) and (4) gives

$$\frac{F_i + \lambda_i}{m_i} - \frac{F_{i'} - \lambda_i}{m_{i'}} = \ddot{c} \tag{5}$$

from which the symmetry boundary external force is determined to be

$$\lambda_i = \frac{m_i m_{i'}}{m_i + m_{i'}}\left(\ddot{c} - \frac{F_i}{m_i} + \frac{F_{i'}}{m_{i'}}\right) \tag{6}$$

From equilibrium, the external force on the boundary must correspond to the applied traction, T, such that $\Sigma\lambda_i = T$. Therefore,

$$T = \ddot{c}\sum_i \frac{m_i m_{i'}}{m_i + m_{i'}} - \sum_i \frac{m_i m_{i'}}{m_i + m_{i'}}\left(\frac{F_i}{m_i} - \frac{F_{i'}}{m_{i'}}\right) \tag{7}$$

Upon finding $\ddot{c}$ from equation (7), the individual external forces may be determined from equation (6) and the accelerations can be used in the central difference equations to advance the motion of the degrees of freedom. The conditions of this symmetry enforcement are not restricted to small deformations as long as geometry is updated accordingly.

In general, the net effect of forces applied to the system must satisfy force and moment equilibrium, assuming no rigid motion due to unbalanced forces. More specifically, it can be shown that these conditions require

$$\sum_n F_n^{ext} = 0 \tag{8}$$

This is not surprising since the periodic boundary forces must balance each other due to equation (3).

PERIODICITY BY TRANSLATION AND ROTATION

Kinematics

In general, the periodic cells may be related by a translation and rotation rather than a simple translation. Seemingly, the most general shape that could be obtained with a single direction of repetition is a helix around an axis, which will be assumed (without loss of generality) to be parallel to the global z axis. Figure 4 shows a periodic cell which generates the complete geometry by translation along and rotation about the z-axis.

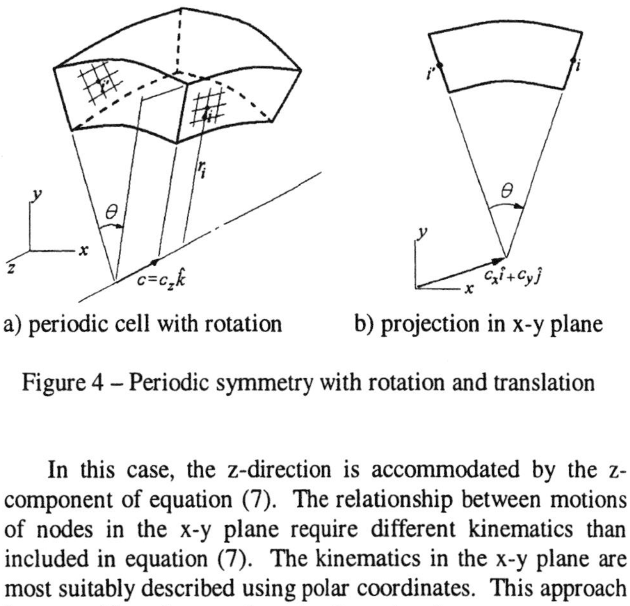

a) periodic cell with rotation b) projection in x-y plane

Figure 4 – Periodic symmetry with rotation and translation

In this case, the z-direction is accommodated by the z-component of equation (7). The relationship between motions of nodes in the x-y plane require different kinematics than included in equation (7). The kinematics in the x-y plane are most suitably described using polar coordinates. This approach is acceptable as long as the axis of rotation does not move.

In order for the periodic boundaries to remain compatible, the radial component of corresponding node pairs must remain equal. In polar coordinates, this is stated as

$$r_i - r_{i'} = 0 \qquad (9)$$

However, the angular component of corresponding node pairs must differ by a finite rotation. In the usual cyclic symmetry method, this condition is

$$\theta_i - \theta_{i'} = \theta \qquad (10)$$

where the unsubscripted quantity represents the relative angle between periodic surfaces. Differentiation of equations (9) and (10) gives

$$\dot{r}_i - \dot{r}_{i'} = 0 \qquad (11a)$$

$$\dot{\theta}_i - \dot{\theta}_{i'} = \dot{\theta} \qquad (11b)$$

and

$$\ddot{r}_i - \ddot{r}_{i'} = 0 \qquad (12a)$$

$$\ddot{\theta}_i - \ddot{\theta}_{i'} = \ddot{\theta} \qquad (12b)$$

Note that setting the right hand side of equation (12b) to zero reduces the symmetry condition to the usual cyclic symmetry as applied to small deformations. Here, a more general condition allows for change in the relative orientations of the periodic boundaries. Although this is still a special case, it is substantially more general than cyclic symmetry.

Kinetics

The next step is to use the kinetic relationships

$$\frac{F_{ir} + \lambda_{ir}}{m_i} = \ddot{r}_i - r_i\dot{\theta}_i^2 \qquad (13a)$$

$$\frac{F_{i'r} + \lambda_{i'r}}{m_{i'}} = \ddot{r}_{i'} - r_{i'}\dot{\theta}_{i'}^2 \qquad (13b)$$

and

$$\frac{F_{i\theta} + \lambda_{i\theta}}{m_i} = r_i\ddot{\theta}_i + 2\dot{r}_i\dot{\theta}_i \qquad (14a)$$

$$\frac{F_{i'\theta} + \lambda_{i'\theta}}{m_{i'}} = r_{i'}\ddot{\theta}_{i'} + 2\dot{r}_{i'}\dot{\theta}_{i'} \qquad (14b)$$

which are the polar coordinate counterparts of equation (4). Combining equations (9) to (14) gives

$$\frac{(F_{ir} + \lambda_{ir})}{m_i} - \left(\frac{(F_{i'r} + \lambda_{i'r})}{m_{i'}}\right) = r_i\left(\dot{\theta}_i^2 - \dot{\theta}_{i'}^2\right) = r_i\dot{\theta}\left(2\dot{\theta}_i - \dot{\theta}\right) \qquad (15)$$

$$\frac{(F_{i\theta} + \lambda_{i\theta})}{m_i} - \left(\frac{(F_{i'\theta} + \lambda_{i'\theta})}{m_{i'}}\right) = -(r_i\ddot{\theta} + 2\dot{r}_i\dot{\theta}) \qquad (16)$$

But, from the symmetry conditions,

$$\lambda_{i'r} = -\lambda_{ir}, \quad \lambda_{i'\theta} = -\lambda_{i\theta} \qquad (17)$$

so that equation(15) simplifies to

$$\lambda_{ir} = \frac{m_i m_{i'}}{m_i + m_{i'}}\left(-r_i\dot{\theta}\left(2\dot{\theta}_i - \dot{\theta}\right) - \frac{F_{ir}}{m_i} + \frac{F_{i'r}}{m_{i'}}\right) \qquad (18a)$$

$$\lambda_{i\theta} = \frac{m_i m_{i'}}{m_i + m_{i'}}\left(-(r_i\ddot{\theta} + 2\dot{r}_i\dot{\theta}) - \frac{F_{i\theta}}{m_i} + \frac{F_{i'\theta}}{m_{i'}}\right) \qquad (18b)$$

Equation (18) defines the boundary forces in terms of known internal forces and inertial forces corresponding to the present motion. Notice that there is no need to determine a kinematic quantity corresponding to the relative radial motion of boundary pairs, but the angular quantity represents an additional unknown in the system.

Momemt Equilibrium

The angular motion between the periodic boundaries must be related to applied moments due to the λ_i . It is convenient to take moments about the z axis (origin of cylindrical coordinate system) since the radial components will pass through z axis and the axial components are parallel to the z axis. Then the z component of moment applied to a boundary node *i* depends only on the $\lambda_{i\theta}$ component according to

$$\left(M_z\right)_i = r_i\lambda_{i\theta} \qquad (19)$$

The total moment applied to the boundary follows from summing over all *i* on the boundary so that

$$M_z = \sum_i r_i\lambda_{i\theta} \qquad (20)$$

Combining with equation (18b) gives the z component of moment

$$M_z = \sum_i \frac{m_i m_{i'} r_i}{m_i + m_{i'}}\left(-\frac{F_{i\theta}}{m_i} + \frac{F_{i'\theta}}{m_{i'}}\right) - \ddot{\theta}\sum_i \frac{m_i m_{i'} r_i^2}{m_i + m_{i'}} - 2\dot{\theta}\sum_i \frac{m_i m_{i'} r_i \dot{r}_i}{m_i + m_{i'}} \qquad (21)$$

This provides an equation relating the kinematic quantity $\ddot{\theta}$ to the applied moment M_z at the boundary.

Equations for Boundary Constraints

In satisfying force equilibrium, only the z direction is relevant. From equation (7), the z component of force is

$$T_z = \ddot{c}_z \sum_i \frac{m_i m_{i'}}{m_i + m_{i'}} - \sum_i \frac{m_i m_{i'}}{m_i + m_{i'}} \left(\frac{F_{iz}}{m_i} - \frac{F_{i'z}}{m_{i'}} \right) \qquad (22)$$

Note that equations (21) and (22) are decoupled and provide a simple means to compute the necessary quantities. Only a relatively small set of inner product like operations are required that only involve boundary nodes. Since the number of boundary nodes will typically be much smaller than the total number of nodes (and elements), these calculations will be more or less overwhelmed by the internal force calculations.

Starting from known initial conditions where θ and $\dot{\theta}$ are given, $\ddot{\theta}$ can be found from equation (21). Subsequently, $\dot{\theta}$ and θ must be updated so that present values are known. Note that the relationship between the constraint and net reactions is clear. If c_z is specified, T_z will follow from the solution, while if T_z is specified, c_z will respond accordingly. Similarly, either one of M_z or θ may be specified.

Additional restrictions on the basis of global equilibrium must be considered. For force equilibrium of the system,

$$\sum_{n=1}^{nodes} F_n^{ext} + (I - R)T = 0 \qquad (23)$$

where F_n^{ext} are the external forces applied to the system, R is a rotation tensor depicting the relative orientation between periodic boundaries, T is the resultant force on one symmetric boundary and $-RT$ is the corresponding resultant force on the opposite boundary. This shows that the net external forces can only be zero if $R=I$ or if T is orthogonal to R (i.e., in the z direction only). In the limiting case where $R=I$, the net external force must be zero. If T and M are specified, then the kinematic variables are determined. Otherwise, if the kinematic variables are given, T and M must follow as the resultant reactions complementary to the kinematic constraints. It is possible to mix these quantities (i.e., if M and d are specified, then θ and T will be determined during the solution).

Regardless of how the constraints and tractions are applied, the boundary conditions are ultimately applied as natural boundary conditions. That is, each degree of freedom is unconstrained, but forces are applied to be consistent with the inhomogeneous constraints. The fact that there are no essential boundary conditions specified does not pose any problem for explicit methods, since rigid motion naturally follows from the equations of motion. Given stationary initial conditions, the arbitrary rigid motion that can accompany an unconstrained system corresponds to a stationary centre of mass (since from equilibrium, the external forces sum to zero). Rather than posing a problem, the entirely natural boundary conditions actually improve the rate of propagation of a solution through the model. This condition is studied in detail in [20] and shows that loading opposite ends of a model leads to faster convergence. This is contrary to the usual "rule" of constraining rigid modes.

IMPLEMENTATION

Input Data

Implementation into a finite element program is relatively straightforward with equations (21) and (22). Mainly, the issue is how the data structure is accessed through the user interface. The necessary data is described below.

1. *NP*=number of nodes on periodic boundary

2. i and i' defining *NP* pairs of nodes on the periodic boundary and its image.

3. A flag to specify either translational symmetry or the more general rotational case.
   NTYP=1 for translation only (x, y and z)
   NTYP=2 for rotation and z translation

4. Mask for kinematic or natural degree of freedom.
   NKIN(I) = 0 for specifying T or M
   1 for specifying $\ddot{c}$ or $\ddot{\theta}$

5. Initial values for c, $\dot{c}$ and $\theta, \dot{\theta}$.

6. Components of $\ddot{c}$ or T (or $\ddot{\theta}$ or M) according to the corresponding values of NKIN(I). Note that the T components must be compatible with global equilibrium.

Computational Procedure

At each time step, after natural external and internal forces have been calculated for each element, additional steps that must be included are listed in Table I

Table I – Procedure for applying symmetry

1	Compute the sums of equations (21) and (22) (or equation (7) for translation only).
2	Depending on NKIN(I), specified resultants or kinematic quantities are obtained from a user defined function.
3	The equations of step 1 are solved for the unknown parameters. This provides all boundary kinematics and resultant reactions.
4	Equation (6) or (18) is used to find the λ_i which are assembled into the global force vector.
5	Kinematic velocities and positions are updated with central difference formulas.

After the steps of Table I, the usual explicit solution procedure determines accelerations from the global force vector and updates the velocity and position of all degrees of freedom.

Upon programming and testing the procedure in a general purpose finite element progam [21], a few important numerical issues were noticed. In the central difference method used to solve the equations of motion, velocity quantities are known only up to one half step behind the present time. While this is of no consequence to translational symmetry (velocity terms are not present in the constraint equations or definition of boundary forces, equation (6,7)), this naturally leads to some error in calculations using equations (19-22). Velocity terms required in the rotational method can be updated to allow for interpolation to the present time. This requires use of a predictor-corrector procedure, but since no updating of internal force is involved, the computational burden is minimal. However, the method seems to work effectively with the use of asynchronous velocity quantities, and no predictor-corrector process was used.

While errors due to velocity lag were tolerable, care was still required in computing the boundary forces. The Cartesian coordinates quantities were transformed to polar coordinates to facilitate the rotational method. The transformation is based on the actual position of a node, regardless of which boundary it is on. Although corresponding node pairs should have the same radial component, small differences may appear due to the velocity terms. The use of a single radius to normalize the direction vectors for both nodes in a boundary pair results in numerical errors that ultimately destabilize the system.

While the method has been tested successfully, there could be ill-conditioned problems for which difficulties could be encountered. For such cases, the predictor-corrector approach can be used. Furthermore, the stabilization method of Baumgarte [14] may be required, and its implementation is straightforward in an explicit method. These items are considered for future work if the need is recognized.

EXAMPLES

Translational Symmetry

An example of translational periodic symmetry is a strip of elastic material with a repeating pattern of holes. Figure 5 shows the undeformed geometry of the unit cell along with portions of adjacent cells. In this case, the external boundaries are free of tractions and a net resultant tension is applied across the periodic boundaries. The deformed geometry resulting from this loading is shown unmagnified in figure 5. Note that the load is very high in relation to the modulus of elasticity so that large deformation occurs. Due to the pattern of holes, the boundaries experience warping, as well as an increase in separation. It is evident that the opposite boundaries have warped in a compatible manner.

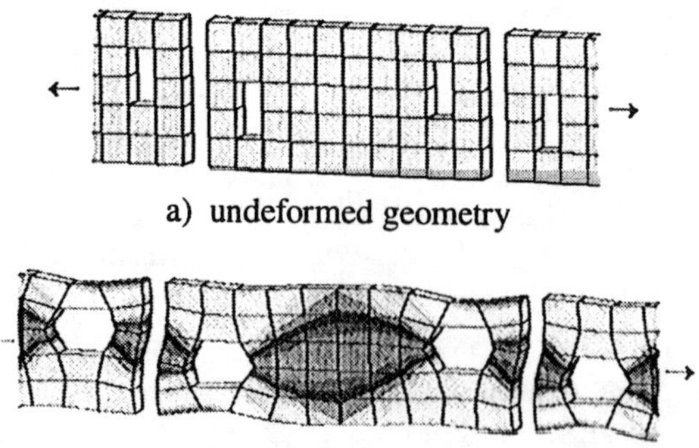

a) undeformed geometry

b) deformed geometry

Figure 5 – Periodic strip under applied tension

The entire model has no essential boundary condition applied, and is free to move rigidly under the action of the applied forces. Since the net force is zero (equilibrium), the unit cell does not experience any rigid translation with respect to its center of mass. Since equilibrium is met numerically to within roundoff error, any drift accumulates very slowly.

Translation and Rotation

Another example is considered to test the rotational symmetry method. This case corresponds to the situation of twisting a long rectangular bar while maintaining a net tension of zero in the direction along the axis of twist. According to the derivation above, the axis of twist is taken as the z direction. Aside from a rotation, the deformation field does not change with axial position. Therefore, only a short representative length of the bar needs to be modeled.

The finite element model considers a bar with a width to thickness ratio of 10. While only a single layer of elements in the z direction is necessary, here two layers have been used to aid in the visualization. The angular motion is specified by imposing an initial relative angular velocity, $\dot{\theta}$, between the periodic boundaries and by setting $\ddot{\theta}$ to zero. Also, the net force in the z direction is specified to be zero. This represents a case where both kinematic and resultant force quantities are used concurrently in a single boundary.

The deformed geometry along with contours of stress components shown in figures 6a-c is based on elastic properties. The unit cell model is the strip shown with contours, while a representation of a length of the bar has been created by stacking the unit cell model with the appropriate rotation applied. The relative rotation between symmetric boundaries is just 5°, but the warping accompanying twist has caused considerably more rotation of the edge of the bar. Although the actual model is processed in Cartesian coordinates, the orientation of the unit cell allows for the stress distributions shown in figure 6 to approximately correspond to the cylindrical coordinate components indicated. Shear stress is as expected, but tension developed along the edge of the bar reverts to compression near the center. Also, radial stress is predominantly compressive and can ultimately lead to buckling instability like the one depicted in figure 7. Buckling modes would not generally have the same spatial periodicity as the arbitrarily selected unit cell, so the use of periodic symmetry is likely to preclude detecting such behavior. However, in such a case as shown in figure 7, end effects are present due to the application of the boundary conditions, and a long slender model (very disadvantageous aspect ratios) must be used. Periodic symmetry applied to a relatively long unit cell would avoid end effects and allow for a somewhat reduced model that would allow for local instabilities.

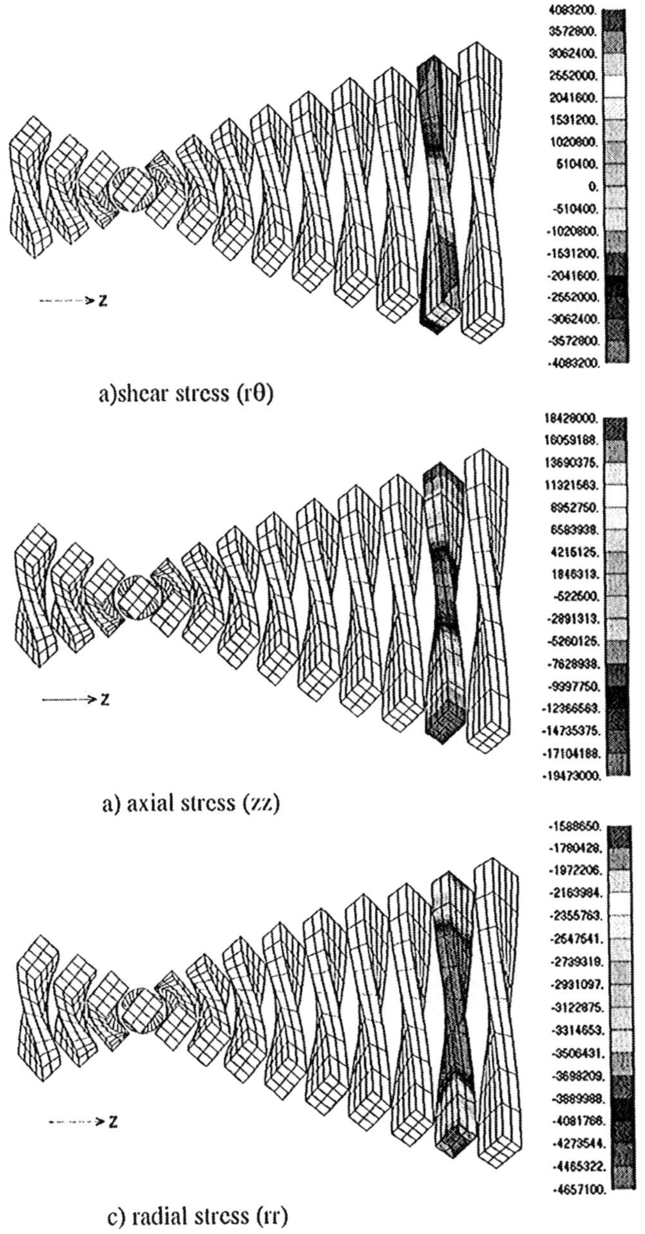

Figure 6 – Deformed geometry of strip under torsion

CONCLUSIONS

A generalized approach for applying periodic symmetry in explicit finite element methods has been developed. The new method allows for application of net tractions and moments on periodic boundaries while maintaining compatible deformations of the boundaries. Provision is made to model twisting and stretching through the action of imposed relative motion of the boundaries or through applied forces or moments. The method has been integrated into a finite element program, and addition computations are minimal in comparison with the rest of the finite element calculations. Examples considered have demonstrated that the method is effective and stable.

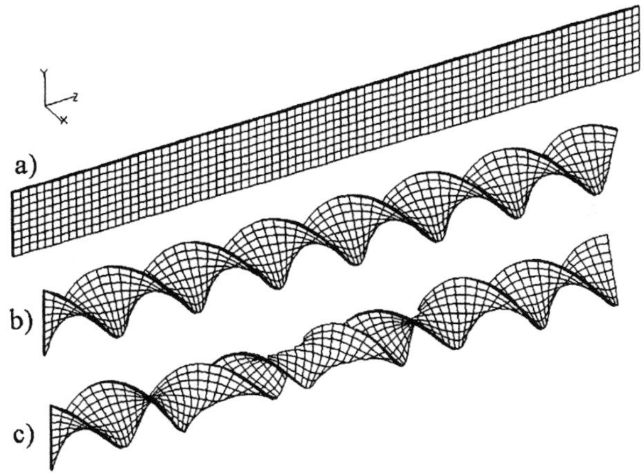

Figure 7 – Full model for twist of flat bar, a) undeformed geometry, b) uniform twist, c) localized buckling

REFERENCES

[1] Zienkiewicz, OC, Scott, FC, "On the Principle of Repeatability and its Application in Analysis of Turbine and Pump Impellers", International Journal for Numerical Methods in Engineering, Vol. 4, No. 3, pp.445-448, 1972.

[2] Thomas, DL, "Dynamics of Rotationally Periodic Structures", International Journal for Numerical Methods in Engineering, Vol. 14, pp.81-102, 1979.

[3] Fricker, DL, Potter, S, "Transient Forced Vibration of Rotationally Periodic Structures", International Journal for Numerical Methods in Engineering, Vol. 17, pp.957-974, 1981.

[4] Elchuri, V, Smith, GCC, Gallo, AM, "NASTRAN Forced Vibration Analysis of Rotating Cyclic Structures", Journal of Vibration, Acoustics, Stress, and Reliability in Design, ASME, Vol. 26, pp. 224-234, 1984.

[5] Ramamurti, V, Rao, MA, "Dynamic Analysis of Spur Gear Teeth", Computers and Structures, Vol. 29, No. 5, pp.831-843, 1988.

[6] Dickens, JM, Pool, KV, "Modal Truncation vectors and Periodic Time Domain Analysis Applied to a Cyclic Symmetry Structure", Computers and Structures, Vol. 45, No. 4, pp.685-696, 1992.

[7] Jacquet-Richardet G, Ferraris G, Rieutord P, "Frequencies and Modes of Rotating Flexible Bladed Disc-shaft Assemblies: A Global Cyclic Symmetry Approach, Journal of Sound and vibration, Vol. 191, No. 5, pp.901-915, 1996.

[8] Cook, RD, Malkus DS, Plesha, ME, "Concepts and Applications of Finite Element Analysis", John Wiley & Sons, 1989.

[9] Jones DP, Gordon JL, Hutula DN, Banas D, Newman JB, "An Elastic-Perfectly Plastic Flow Model for Finite Element Analysis of Perforated Materials", Journal of Pressure Vessel Technology, Vol. 123, pp. 265-270, 2001.

[10] Reinhardt WD, "A Fourth-Order Equivalent Solid Model for Tubesheet Plasticity", ASME PVP-Vol. 385, Computer Technology, G.M. Hulbert, Ed. , pp.151-157, 1999.

[11] Reinhardt WD, Mangalaramanan SP, "Efficient Tubesheet Design Using Repeated Elastic Limit Analysis Technique", Journal of Pressure Vessel Technology, Vol. 123, pp. 197-202, 2001.

[12] Metzger, DR, Sauve, RG, "Computation of Residual Stress Distribution in Tubes Due to a Rolled Joint Forming Process for Steam Generators", ASME PVP-Vol. 235, Design and Analysis of Pressure Vessels, pp.209-214, 1992.

[13] Metzger, DR, Sauve, RG, Nadeau, E, "Prediction of Residual Stress by Simulation of the Rolled Joint Manufacturing Process for Steam Generators", ASME PVP-Vol. 305, Current Topics in Computational Mechanics, J.F. Cory and J.L Gordon, Eds., pp.67-74, 1995.

[14] Baumgarte J, "Stabilization of Constraints and Integrals of Motion in Dynamical Systems", Computer Methods in Applied Mechanics and Engineering Vol. 1, pp.1-16, 1972.

[15] Yoon S, Howe RM, Greenwood DT, "Geometric Elimination of Constraint Violations in Numerical Simulation of Lagrangian Equations", Journal of Mechanical Design, Vol. 116, pp.1058-1064, 1994.

[16] Petzold L, Ren Y, Maly T, "Regularization of Higher-index Differential-Algebraic Equations with Rank Deficient Constraints", SIAM Journal of Scientific Computing, Vol. 18, No. 3., pp. 753-774, 1997.

[17] Rosen A, Edelstein E, "Investigation of a New Formulation of the Lagrange Method for Constrained Dynamic Systems", Journal of Applied Mechanics, Vol. 64, pp.116-122, 1997.

[18] Lin ST, Hong MC, "Stabilization Method for Numerical Integration of Multibody Mechnical Systems", Journal of Mechanical Design, Vol. 120, pp.565-572, 1998.

[19] Gear CW, Petzold LR, "ODE Methods for the Solution of Differential/Algebraic Systems", SIAM Journal of Numerical Analysis, Vol. 21, No. 4, pp. 716-728, 1984.

[20] Metzger DR, Sauve, RG, "Effect of Discretization and Boundary Conditions on the Convergence of the Dynamic Relaxation Method", 354, Current Topic in the Design and Analysis of Pressure Vessels and Piping, D.K. Williams, Ed., pp105-110, 1997.

[21] Metzger DR, Sauve, RG, "H3DMAP Version 6 - A General Three-Dimensional Finite Element Computer Code for Linear and Nonlinear Analysis of Structures", Ontario Power Technologies Report No. A-NSG-96-120, Rev. 1., 1999.

PVP-Vol. 458, Computer Technology and Applications

PVP2003-1914

ON A PROPOSED NEW KINEMATIC ASSUMPTION FOR ANALYSIS OF LAMINATED, FIBER-REINFORCED COMPOSITE SHELLS - PART 1

Scott E. Steinbrink
Department of Mechanical Engineering
Gannon University
Erie, Pennsylvania, 16541
steinbrink@gannon.edu

ABSTRACT

A proposal is made for a new form of kinematic assumption for inclusion into a theory of laminated shells. The proposed form is a variant of the first-order transverse shear deformation theory, incorporating a stiffness-based discontinuity term at lamina interfaces, in order to provide complete, *a priori* satisfaction of interlaminar continuity of transverse shear and transverse normal stresses, and correspondent interlaminar discontinuity of thickness-direction gradients of displacement. Continuity of displacement is maintained, as is the possibility of interlaminar discontinuity of in-plane stresses. It is hoped that this kinematic form will enhance the utility of shell theories for analysis of laminated structures. The paper deals only with the development of the kinematic assumption, and not its implementation into analysis code.

INTRODUCTION

In the analysis of structural shells, it is common to utilize the finite element method. FEM presents a convenient, robust tool, and provides accurate results when performed competently. Yet FEM is, to a significant extent, a "black box" approach: the day-to-day user of a code often has little or no fundamental understanding of the mathematics used to derive element equations, nor the limitations imposed on the element as a result of assumptions made. Thus, it is easy for an end user to misuse a finite element code, and there is little for that user to look at, in terms of equations, in order to verify results.

On the other hand, the theory of elasticity can be used to provide exact solutions to problems of shells, but such solutions are usually quite numerically complicated. In many cases, indeed most cases, elasticity solutions are simply too difficult to be practical.

In the middle ground between the robust but opaque finite element approach, and the exact but cumbersome elasticity approach, lies the shell theory approach. Shell theories are approximations, hence inferior to elasticity approaches, and are more challenging to program and solve than FEM. But they provide equations which can be used to draw physical understanding - equations which are often lacking in FEM - and are much more readily solvable than the equations of 3-D elasticity. It is within that middle ground that the current paper resides.

The fundamental goal of the current work is to create a new statement of shell theory, applicable to laminated, fiber-reinforced polymer (FRP) shells of revolution. The new theory will differ from previously published results by incorporation of interlaminar continuity of transverse stress components. Failure to ensure interlaminar continuity of transverse stresses, particularly the transverse shear stresses, results in poor estimation of these stresses, with consequent misestimation of gross laminate load-bearing capacity. The present work will ensure continuity of all transverse stresses by asserting discontinuous thickness-direction gradients of displacements. The completed theory will ideally be written in a convenient first-order state vector form. The work will assume linear elastic material properties, and will initially be limited to geometrically linear analysis. It is hoped that the newly derived equation will be useful in its own right for solution of problems, as well as a theoretical tool for the interpretation and/or verification of finite element results.

The current paper deals with the kinematic assumption of shell theory, only. The paper is a description of first-phase work. Future works will deal with incorporation of the kinematic assumption into shell theory and with finite element development.

NOMENCLATURE

In what follows, we will assume a laminate of N_L plies of material, as depicted in Fig. 1. The thickness of the generic "k^{th}" ply will be

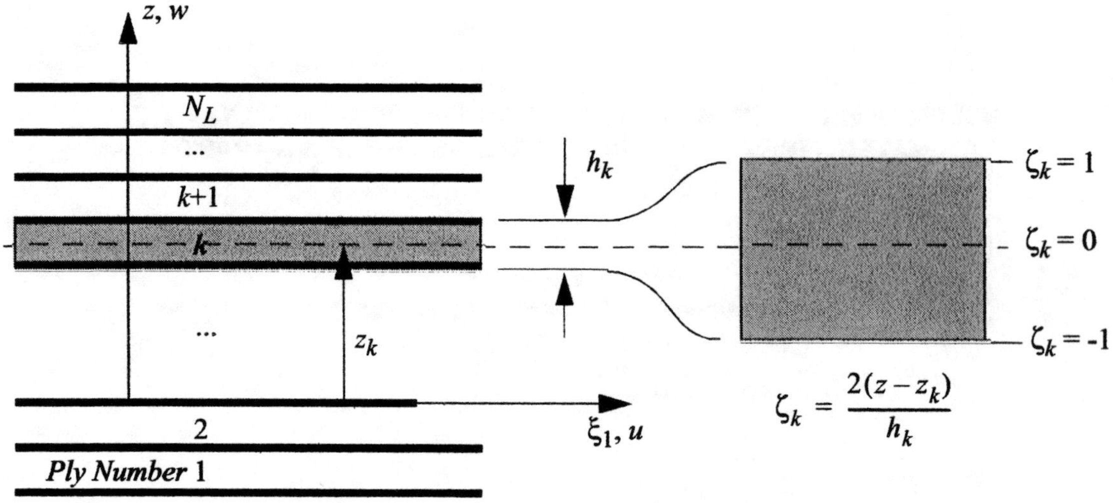

Figure 1: Lamina Numbering and Local Coordinate

denoted by h_k. The coordinate "z" will be used to define the thickness direction within the laminate, and gaussian coordinates (ξ_1, ξ_2) will describe position within the reference surface. Position of any material point within the laminate will be thus denoted by (ξ_1, ξ_2, z). The reference surface may be anywhere within the laminate or on its lateral faces. (Note that in Fig. 1, the reference surface has been arbitrarily placed at the interface of plies 2 and 3.) Plies will be numbered in the usual ordering of lamination theory: from 1 to N_L, with ply number 1 being the innermost ply. Within each ply we define position z_k, the z-directed center.

KINEMATIC ASSUMPTIONS

Previous work has been done by the present author and others, in derivation of the state vector form of linear shell theory, for laminated composite shells, for classical and first-order transverse shear deformation (FSDT) theories (Steele and Kim, 1992 [1], Steinbrink and Johnson, 1997 [2]). The equations presented therein were also solved numerically using various techniques, and shown to compare well to theory for several test cases. These results, however, were based on assumed through-the-thickness variations of displacement which do not allow for discontinuous displacement gradients. That is, it is assumed that a straight line initially drawn normal to the undeformed shell reference surface will remain straight as the reference surface deforms. The associated kinematic assumption is

$$u = u_0 + z\varphi_u \tag{1}$$

In equation (1), "u" is a generic displacement, and may thus represent any of the three-dimensional translation components (u, v or w), though FSDT also assumes that z-directed displacement is constant through the thickness, thus $\varphi_w = 0$ .The subscript "0" is representative of a reference surface value.

For laminated shells, the assumption of equation (1) is not necessarily valid. It is true that each point in the shell must have a single-valued displacement (assuming no fracture or delamination,) but gradients may be discontinuous - the straight line may zig and zag under deformation, while always connecting the same set of material particles. In order to incorporate this effect, Murakami [3] modified the standard form of the assumed displacement fields of first-order transverse shear deformation theory, to add a zig-zag term. That is, he used

$$u = u_0 + (-1)^k \zeta_k u_z + z^r \varphi_r \qquad r = 1, 2, \ldots, N \tag{2}$$

in which k indicates the kth ply of the laminate, and r is a summation index, ranging from 1 to N, a free parameter of the model. ζ_k is a ply-level thickness coordinate, ranging from -1 to 1 over the thickness of a ply:

$$\zeta_k = \frac{2(z - z_k)}{h_k} \tag{3}$$

The alternating term of equation (2) provides for the desired discontinuous gradient of displacement with respect to coordinate z. Note that Murakami's formulation collapses to the form for FSDT, equation (1), if we let u_z be zero and take $N = 1$. For a shell theory, the assumption (2) poses a difficulty. Specifically, u_0, u_z and φ_r must be constants with respect to z in order to maintain continuity of displacement at lamina interfaces. The alternating term then enforces a discontinuity in displacement gradient at all lamina interfaces, regardless of whether or not such discontinuity is appropriate. Murakami's kinematic assumption is depicted in Fig. 2, along with FSDT and the currently proposed new kinematic assumption.

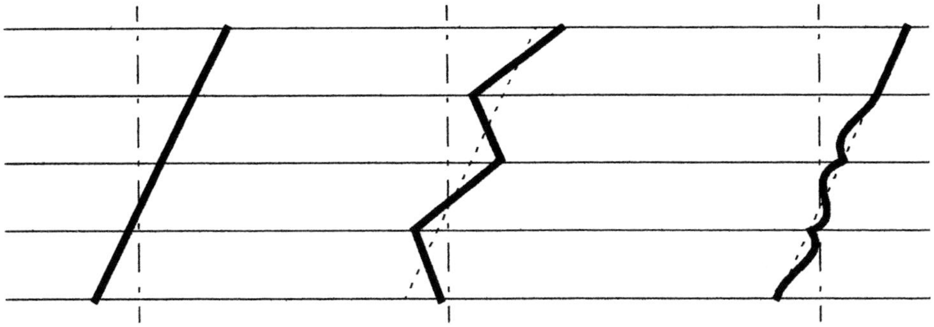

Figure 2: Comparison of Displacement Assumptions (Dotted line represents FSDT result)

An alternative to the method presented by Murakami is described by Carrera [4]. In this alternative method, displacement is described on a layer-by-layer basis, and thus there is no need for any artificial zig-zag term. The equation for displacement is written as

$$u^k = F_t u_t^k + F_b u_b^k + F_r u_r^k \tag{4}$$

where, again, k is the ply number, and u is a generic displacement term. Here, subscript "t" refers to the top of the ply, subscript "b" refers to the bottom of the ply, and subscript "r" is a summation variable, $r = 1, 2, \ldots, N$. The F_i, ($i = t, b, r$) terms are functions of the thickness-direction coordinate. This form is well suited to finite element formulation, but has the distinct disadvantage of requiring introduction of new displacement quantities, the number of which is dependent upon the number of layers in the laminate. This variable number of introduced displacement quantities is problematic for development of a shell theory.

Each theory for displacement distribution has its drawbacks. The question to be addressed in this work is, then, "*Is there a displacement assumption which provides continuity of displacement at lamina interfaces, and allows for appropriate discontinuous gradients at those locations, without introduction of a variable number of unknown displacement terms?*" The current author respectfully suggests that the answer to this question is "yes", and that this displacement assumption can ensure the desired continuity of transverse stresses. Specifically, the following kinematic assumption is proposed:

$$u = u_0 + z\varphi_u + \left(1 - \zeta_k^2\right) u_z^k \tag{5}$$

In equation (5), u_0, φ_u are reference surface values, whereas u_z^k is ply-specific. Derivation follows in the next section of this paper. Note, however, that this expression does provide continuity of displacement as required, and has the potential to provide discontinuous displacement gradients with respect to the normal coordinate, z. At lamina interfaces, $\zeta_k = \pm 1$, displacements from equation (5) match the form of FSDT, equation (1), although u_0, φ_u will likely have different values than those of FSDT.

DERIVATION OF GRADIENT TERM u_z^k

To begin, assume that each ply of a laminate may be treated as a homogeneous, orthotropic, linear elastic material, with known material constants. If this is so, then Hooke's law reads as:

$$\{\sigma\} = [Q]\{\varepsilon\} \tag{6}$$

for any ply, with

$$\{\sigma\} = [\sigma_{11}, \sigma_{22}, \sigma_{33}, \sigma_{23}, \sigma_{13}, \sigma_{12}]^T,$$

$$\{\varepsilon\} = [\varepsilon_{11}, \varepsilon_{22}, \varepsilon_{33}, \gamma_{23}, \gamma_{13}, \gamma_{12}]^T,$$

and

$$[Q] = \begin{bmatrix} Q_{11} & Q_{12} & Q_{13} & 0 & 0 & Q_{16} \\ Q_{12} & Q_{22} & Q_{23} & 0 & 0 & Q_{26} \\ Q_{13} & Q_{23} & Q_{33} & 0 & 0 & Q_{36} \\ 0 & 0 & 0 & Q_{44} & Q_{45} & 0 \\ 0 & 0 & 0 & Q_{45} & Q_{55} & 0 \\ Q_{16} & Q_{26} & Q_{36} & 0 & 0 & Q_{66} \end{bmatrix}$$

In the above stress and strain notations, subscripts "1" and "2" refer to the global coordinates ξ_1, ξ_2, not to the fiber direction and in-plane transverse directions. Subscript "3" refers to the transverse, out-of-plane direction. The material law is assumed to be valid for any ply, though the constants Q_{ij}, which represent the transformed stiffnesses, will in general vary from ply to ply. In order for transverse stresses to be continuous, then, we must make the transverse strains discontinuous. Gradient discontinuity may be provided by the equation of Murakami, equation (2), but, as previously noted, this approach enforces that there will ALWAYS be a discontinuity in displacement gradient at the interface. Hence, it is desired to find a new assumption, which provides the discontinuity only when appropriate, and in an appropriate amount.

The following function is proposed for the ply-dependent term u_z^k:

$$u_z^k = \frac{h_k}{8}\left(\frac{1-\zeta_k}{2}\Delta u_{,z}^k + \frac{1+\zeta_k}{2}\Delta u_{,z}^{k+1}\right) \tag{7}$$

The $\Delta u_{,z}^j$ terms represent discontinuities in displacement gradients with respect to normal coordinate z, at the bottom of the jth ply. These are assumed to be functions only of the gaussian coordinates ξ_1, ξ_2. The form of equation (7) is chosen to provide the following features:

- when used in conjunction with equation (5), it provides continuity of displacement at lamina interfaces,
- it allows for discontinuous thickness-direction gradients,
- the presence of the linear ζ_k terms inside the parentheses provides that each interface is affected only by the difference in stiffness at that location, while in the ply interior, stiffness changes at interfaces both above and below the current ply are considered.

The leading term $h_k/8$ is placed so that the discontinuity of $u_{,z}$ at lamina interfaces is shared equally by both interfacing plies.

In the form of the displacements as defined by equations (5) and (7), we have introduced a new complexity: u_0 and φ_u are reference surface values, but u_z^k is ply-dependent - there is one value of u_z^k for each ply-ply interface. Thus, we have introduced $(N_L - 1)$ displacement variables to be found. This is contrary to the spirit of the shell theory, in which only reference surface values should be required. So a necessary goal for further exploration of this idea is the introduction of a new set of equations, which will allow for expression of u_z^k in terms of reference surface values.

Borrowing a bit of nomenclature from Carrera [4], we here denote the set of transverse stresses by the term σ_n: $\sigma_n = [\sigma_{33}, \sigma_{23}, \sigma_{13}]^T$. This will be made to be continuous at the interface of the kth and $(k+1)^{th}$ laminae, by use of the equation

$$\sigma_{nt}^k = \sigma_{nb}^{k+1} \tag{8}$$

in which "t" represents the top of the ply, "b" represents the bottom of the ply. This will be done by utilizing Hooke's law, in the form shown following equation (6). It is also a feature of the current theory that in-plane strains are continuous at lamina interfaces (See the Appendix for definition of strain measures). This then allows us to write the following equation:

$$\begin{bmatrix} 1 & 0 & 0 & 0 & 0 & 0 \\ 0 & 1 & 0 & 0 & 0 & 0 \\ Q_{13} & Q_{23} & Q_{33} & 0 & 0 & Q_{36} \\ 0 & 0 & 0 & Q_{44} & Q_{45} & 0 \\ 0 & 0 & 0 & Q_{45} & Q_{55} & 0 \\ 0 & 0 & 0 & 0 & 0 & 1 \end{bmatrix}^k \begin{Bmatrix} \varepsilon_{11} \\ \varepsilon_{22} \\ \varepsilon_{33} \\ \gamma_{23} \\ \gamma_{13} \\ \gamma_{12} \end{Bmatrix}_t^k = \begin{bmatrix} 1 & 0 & 0 & 0 & 0 & 0 \\ 0 & 1 & 0 & 0 & 0 & 0 \\ Q_{13} & Q_{23} & Q_{33} & 0 & 0 & Q_{36} \\ 0 & 0 & 0 & Q_{44} & Q_{45} & 0 \\ 0 & 0 & 0 & Q_{45} & Q_{55} & 0 \\ 0 & 0 & 0 & 0 & 0 & 1 \end{bmatrix}^{k+1} \begin{Bmatrix} \varepsilon_{11} \\ \varepsilon_{22} \\ \gamma_{33} \\ \gamma_{23} \\ \gamma_{13} \\ \gamma_{12} \end{Bmatrix}_b^{k+1} \tag{9}$$

or, in shortened form,

$$Q_n^k \varepsilon_t^k = Q_n^{k+1} \varepsilon_b^{k+1} \tag{10}$$

where Q_n is the ply-specific (6-by-6) matrix shown in equation (9), and ε is the strain vector. This latest expression defines the discontinuity in transverse strain that is necessary to have proper continuity of transverse stress. This amount of discontinuity of strain can then be used to determine the values of the ply-specific "$\Delta u_{,z}$" terms of equation (7).

The strain-displacement relations are detailed in the appendix to this work. Note, however, that the strain-displacement relations and the kinematic assumptions of the current model combine to allow strains to be written as a combination of two parts: ε_C which is continuous at ply interfaces, and ε_N which provides the necessary discontinuity in gradient. Then, equation (10) can be written as

$$Q_n^k\left\{\varepsilon_{Ct}^k + \varepsilon_{Nt}^k\right\} = Q_n^{k+1}\left\{\varepsilon_{Cb}^{k+1} + \varepsilon_{Nb}^{k+1}\right\} \tag{11}$$

Continuity of ε_C,

$$\varepsilon_{Ct}^k = \varepsilon_{Cb}^{k+1} \tag{12}$$

can be used in equation (11) to get

$$\left(Q_n^k - Q_n^{k+1}\right)\varepsilon_{Ct}^k = Q_n^{k+1}\varepsilon_{Nb}^{k+1} - Q_n^k\varepsilon_{Nt}^k \qquad (13)$$

Into equation (13) are asserted appendix equations (A5), (A8), (A9). Recognizing that "φ" terms are reference surface values and thus clearly continuous at lamina interfaces, we can rewrite equation (13) as

$$\left(Q_n^k - Q_n^{k+1}\right)\left[\varepsilon_C + \begin{Bmatrix} 0 \\ 0 \\ \varphi_w \\ \varphi_v \\ \varphi_u \\ 0 \end{Bmatrix}\right]_t^k = \frac{1}{2}\left(Q_n^{k+1} + Q_n^k\right)\begin{Bmatrix} 0 \\ 0 \\ \Delta w_{,z} \\ \Delta v_{,z} \\ \Delta u_{,z} \\ 0 \end{Bmatrix}_b^{k+1} \qquad (14)$$

Finally, we recognize that, when evaluated at lamina interfaces, all strain terms contained in ε_C are describable in terms only of reference surface displacement quantities and their derivatives. Thus, the "$\Delta u_{,z}$" terms are describable in terms of reference surface values, only, and *no new displacement terms have been introduced by the current kinematic assumption.* Furthermore, it is clear from equation (14) that the "$\Delta u_{,z}$" terms are identically zero wherever two similarly oriented plies meet, and non-zero otherwise. *This provides the required appropriate discontinuity.* Figure 2 depicts this continuity, where the top two plies are assumed to have similar orientation. For all other interfaces, stiffness discontinuity leads to discontinuity of displacement gradients.

REMAINING ISSUES AND NEXT STEPS

It has been shown in the work above that the proposed kinematic assumption does not result in introduction of a large number of displacement terms. It remains to be seen, though, what the effect of the kinematic assumption will be on the number and type of stress resultants contained within the completed shell theory. Thus, the next step will entail investigation of stress resultants. Clearly, strain measures will be cubic in ζ_k, seemingly leading to the introduction of more stress resultants than those of most shell theories. This will be limited, though, by the symmetric limits on ζ_k on each ply: integrals of odd powers of ζ_k will vanish. Other tools may be necessary to deal with remaining "additional" resultants.

The previously noted works by Steele and Kim [1], Steinbrink and Johnson [2] and Carrera [4], all utilize Reissner's [5-7] mixed variational principle to derive their basic equations. References [1] and [2] utilized the theorem first presented in 1950 [5], and achieved a statement of shell theory in the desired first-order state vector form.

Carrera [4] uses the "new" theorem, presented in Reissner (1984) (reference [6]) and Reissner (1986) (Reference [7]). The original intent of Reference [5] was to introduce a method which resulted in equations of equilibrium and equations of stress-strain relation which were of equal quality. To achieve this goal, Reissner suggested a functional form which possessed both stress and displacement variables which could be varied simultaneously, but independently. The modification made to that principle, and which was described in [6] allows for independent variation of displacement and *transverse* stresses. Distribution of transverse stresses is assumed in a manner similar to the assumption of the distribution of displacements, equation (4) (see Carrera, [4]):

$$\sigma_n^k = F_t\sigma_{nt}^k + F_b\sigma_{nb}^k + F_r\sigma_{nr}^k \qquad (15)$$

In-plane stresses are not subject to independent assumption, and are not varied when the variational expression is formed. This modification was introduced in order to accommodate the analysis of laminated structures.

The current work will also utilize Reissner's mixed variational theorem, but will not require assumption of the form of the transverse stresses, having instead accomplished the same goal through the kinematic assumption. Boundary conditions on transverse stresses will be defined by the equation

$$\sigma_{nb}^1 = \overline{\sigma_{nb}} \qquad \sigma_{nt}^{N_L} = \overline{\sigma_{nt}} \qquad (16)$$

for the laminate with N_L layers. Additional boundary conditions should result from application of the variational principle.

CONCLUSION

A new kinematic assumption has been proposed for incorporation into shell theory, for analysis of laminated structures. The assumption is based on discontinuous stiffnesses from ply to ply of the laminate, and allows for complete, *a priori* satisfaction of interlaminar continuity of transverse stresses, not previously achieved in shell theories. The assumption is thought to be appropriate for shell theories, as all "new" displacement terms are describable in terms of reference surface displacement values.

REFERENCES

1. Steele, C. R. and Kim, Y. Y., 1992, "*Modified Mixed Variational Principle and the State Vector Equation for Elastic Bodies and Shells of Revolution,*" Journal of Applied Mechanics, **59**, September 1992, pp 587-595

2. Steinbrink, S. E. and Johnson, E. R., 1997, "*Analysis of Axisymmetric Composite Pressure Domes by the Multiple Shooting Method,*" 38th AIAA/ASME/ASCE/AHS/ASC Structures, Structural Dynamics and Materials Conference, AIAA Paper 97-1366

3. Murakami, H., 1986, "*Laminated Composite Plate Theory with Improved In-Plane Responses*", Journal of Applied Mechanics, September 1986, pp 661-666

4. Carrera, E., 2001, *"Developments, ideas and evaluations based on Reissner's Mixed Variational Theorem in the modeling of multilayered plates and shells"*, Applied Mechanics Reviews, **54**, Number 4, pp 301-328

5. Reissner, E., 1950, *"On a Variational Theorem in Elasticity"*, Journal of Mathematics and Physics, **29**, pp 90-95

6. Reissner, E., 1984, *"On a Certain Mixed Variational Theorem and a Proposed Application"*, International Journal for Numerical Methods in Engineering, **20**, pp 1366-1368

7. Reissner, E., 1986, *"On a Mixed Variational Theorem and on Shear Deformable Plate Theory"*, International Journal for Numerical Methods in Engineering, **23**, pp 193-198

APPENDIX: STRAIN MEASURES

Strain-displacement relations of the linear theory of shells used for this work are as included below. The following notation is used:

A_1, A_2: Reference surface metrics in the ξ_1, ξ_2 directions.
H_1, H_2: Surface metrics for surfaces parallel to the reference surface, in ξ_1, ξ_2 directions.
R_1, R_2: Radii of curvature of the surface, with respect to ξ_1, ξ_2 directions
$(\,)_{,1}$, $(\,)_{,2}$, $(\,)_{,z}$: Partial derivative of the item in the parentheses, with respect to the coordinate following the comma
u, v, w: Displacement in the direction of ξ_1, ξ_2, *z*, respectively.
Ψ_1, Ψ_2: Geometric parameters of the shell shape, as defined below:

$$\Psi_1 = A_{1,2}/A_2$$
$$\Psi_2 = A_{2,1}/A_1$$

Then:

$$\varepsilon_{11} = \frac{1}{H_1}\left[u_{,1} + \Psi_1 v + \frac{A_1}{R_1}w\right]$$
$$\varepsilon_{22} = \frac{1}{H_2}\left[\Psi_2 u + v_{,2} + \frac{A_2}{R_2}w\right]$$
$$\gamma_{12} = \frac{1}{H_1}[v_{,1} - \Psi_1 u] + \frac{1}{H_2}[u_{,2} - \Psi_2 v]$$
$$\varepsilon_{33} = w_{,z} \tag{A1}$$
$$\gamma_{23} = \frac{1}{H_2}\left[w_{,2} - \frac{A_2}{R_2}v\right] + v_{,z} = \varepsilon_{23} + v_{,z}$$
$$\gamma_{13} = \frac{1}{H_1}\left[w_{,1} - \frac{A_1}{R_1}u\right] + u_{,z} = \varepsilon_{13} + u_{,z}$$

Note that the inclusion of ε_{33} is unusual for a shell theory. It is included here in order to maintain proper continuity of σ_{33} at lamina interfaces. Also, discontinuity of ε_{33} at interfaces requires displacement *w* not to be constant through the thickness. Thus, the current theory differs from FSDT in that the laminate must be assumed to have through-the-thickness thinning under load.

The displacements used in the strain-displacement relations are as in equation (5), here expanded to include *u*, *v* and *w*:

$$u = u_0 + z\varphi_u + \left(1 - \zeta_k^2\right)u_z^k$$
$$v = v_0 + z\varphi_v + \left(1 - \zeta_k^2\right)v_z^k \tag{A2}$$
$$w = w_0 + z\varphi_w + \left(1 - \zeta_k^2\right)w_z^k$$

The "0" subscripted terms are reference surface values. The "φ" terms are also reference surface values, thus providing linear variation of displacement through the thickness when multiplied by coordinate *z*. The form of u_z^k, etc., is given in equation (7), repeated and expanded here for completeness:

$$u_z^k = \frac{h_k}{8}\left[\left(\frac{1-\zeta_k}{2}\right)\Delta u_{,z}^k + \left(\frac{1+\zeta_k}{2}\right)\Delta u_{,z}^{k+1}\right]$$
$$v_z^k = \frac{h_k}{8}\left[\left(\frac{1-\zeta_k}{2}\right)\Delta v_{,z}^k + \left(\frac{1+\zeta_k}{2}\right)\Delta v_{,z}^{k+1}\right] \tag{A3}$$
$$w_z^k = \frac{h_k}{8}\left[\left(\frac{1-\zeta_k}{2}\right)\Delta w_{,z}^k + \left(\frac{1+\zeta_k}{2}\right)\Delta w_{,z}^{k+1}\right]$$

The "Δ" terms are assumed *z*-independent. Thus, *z*-dependence comes only as a result of the explicit inclusion of ζ_k. This implies that, except for gradients with respect to *z*, all of the gradients within equation (A1) are *z*-independent when evaluated at a ply interface, where $\zeta_k = \pm 1$.

It is convenient to rewrite equation (A1) as

$$\{\varepsilon\} = \{\varepsilon_C\} + \{\varepsilon_N\} \tag{A4}$$

in which

$$\{\varepsilon_N\} = \{0, 0, w_{,z}, v_{,z}, u_{,z}, 0\}^T \tag{A5}$$

and $\{\varepsilon_C\}$ contains all of the other terms of equation (A1), which are continuous at lamina interfaces. That is,

$$\{\varepsilon_C\}_t^k = \{\varepsilon_C\}_b^{k+1} \tag{A6}$$

Using equations (A2) and (A3), we can write

$$u_{,z} = \varphi_u - \frac{\zeta_k}{2}\left[\frac{1-\zeta_k}{2}\Delta u_{,z}^{k} + \frac{1+\zeta_k}{2}\Delta u_{,z}^{k+1}\right] + \frac{1}{8}\left(1-\zeta_k^2\right)\left(-\Delta u_{,z}^{k} + \Delta u_{,z}^{k+1}\right)$$

$$v_{,z} = \varphi_v - \frac{\zeta_k}{2}\left[\frac{1-\zeta_k}{2}\Delta v_{,z}^{k} + \frac{1+\zeta_k}{2}\Delta v_{,z}^{k+1}\right] + \frac{1}{8}\left(1-\zeta_k^2\right)\left(-\Delta v_{,z}^{k} + \Delta v_{,z}^{k+1}\right) \quad \text{(A7)}$$

$$w_{,z} = \varphi_w - \frac{\zeta_k}{2}\left[\frac{1-\zeta_k}{2}\Delta w_{,z}^{k} + \frac{1+\zeta_k}{2}\Delta w_{,z}^{k+1}\right] + \frac{1}{8}\left(1-\zeta_k^2\right)\left(-\Delta w_{,z}^{k} + \Delta w_{,z}^{k+1}\right)$$

which, evaluated at the top of ply k, $\zeta_k = 1$ yields

$$u_{,z} = \varphi_u - \frac{1}{2}\Delta u_{,z}^{k+1}$$

$$v_{,z} = \varphi_v - \frac{1}{2}\Delta v_{,z}^{k+1} \quad \text{(A8)}$$

$$w_{,z} = \varphi_w - \frac{1}{2}\Delta w_{,z}^{k+1}$$

and at the bottom of ply (k+1), $\zeta_k = -1$ yields

$$u_{,z} = \varphi_u + \frac{1}{2}\Delta u_{,z}^{k+1}$$

$$v_{,z} = \varphi_v + \frac{1}{2}\Delta v_{,z}^{k+1} \quad \text{(A9)}$$

$$w_{,z} = \varphi_w + \frac{1}{2}\Delta w_{,z}^{k+1}$$

Finally, using equations (A2) in equations (A1), and recognizing that $\left(1-\zeta_k^2\right) = 0$ at lamina interfaces, it can be seen that strains ε_{11}, ε_{22}, ε_{23}, ε_{13} and γ_{12} are all describable in terms of reference surface values only, when evaluated at lamina interfaces.

RISK AND RELIABILITY TOPICS

Introduction

John A. Farquharson

ABSG Consulting Inc. (ABS Consulting)
Knoxville, Tennessee

This section presents five technical papers presented in the sessions for "Risk and Reliability Topics" during the 2003 ASME PVP Conference. These papers address a variety of subjects in the growing area of risk and reliability. In addition to the papers, there is an additional presentation and a panel discussion on how the risk-based approach fits in the traditionally deterministic world of ASME.

Dr. Bill Cho's paper, "Enhancing Probabilistic Risk Assessment Methodology for Solving Reactor Safety Issues" reveals a paradigm of analyzing the consequential effects of severe nuclear reactor accident, radionuclides fraction and source terms release in performing Level 3 probabilistic risk assessments (PRAs) for commercial nuclear facilities. His insight in combining the different consequence states associated with a nuclear plant PRA is an important perspective for a holistic view of risk.

The paper, "A New Graphical Tool for Building Logic-Gate Trees," by Bott, Eisenhawer, Kingson, and Key shows how traditional risk assessment techniques, such as the fault tree, can be used to model possible sabotage scenarios. The growing concern of security risk has created a need for risk and reliability analysts to use their problem-solving mentality and tools to address concerns that were traditionally "out of scope" for previous efforts.

Professor Timashev's paper, "Human Factor in the Life Cycle and Safety of Machines and Pipelines," highlights how the human element is often the most important contributor to risk and reliability studies. The paper discusses the details of how to (1) select human error probabilities and (2) account for psychological reasons for errors and how the errors affect the overall system.

The other two remaining papers combine the traditional deterministic world of stress and strain with the risk-based approach of estimating the stochastic value of when components will fail. The paper by Li, Metzger, and Nye, "Reliability Analysis of the Tube Hydroforming Process Based on Forming Limit Diagram," presents a method to assess the probability of failure of the process based on reliability theory and the forming limit diagram. The paper by Kirkpatrick and Klopp, "Risk Assessment for Damaged Pressure Tank Cars," evaluates the validity of existing programs for assessment of damaged pressure tank cars.

I humbly thank all authors, reviewers, presenters, and panel members for the risk and reliability sessions. I especially acknowledge Dr. Bill Cho for his many contributions over the years in the field of risk and reliability analysis.

PVP-Vol. 458, Computer Technology and Applications

PVP2003-1915

A NEW GRAPHICAL TOOL FOR BUILDING LOGIC-GATE TREES

Terry F. Bott and Stephen W. Eisenhawer
Los Alamos National Laboratory

Jonathan Kingson and Brian P. Key
Innovative Technical Solutions, Inc.

ABSTRACT

Tree structures that use logic gates to model system behavior have proven very useful in safety and reliability studies. In particular process trees are the basic structure used in a decision analysis methodology developed at Los Alamos called Logic Evolved Decision modeling (LED). *LED TOOLS* is the initial attempt to provide LED-based decision analysis tools in a state of the art software package. The initial release of the software, Version 2.0, addresses the first step in LED – determination of the possibilities. *LED TOOLS* is an object-oriented application written in Visual Basic for Windows NT based operating systems. It provides an innovative graphical user interface that was designed to emphasize the visual characteristics of logic trees and to make their development efficient and accessible to the subject matter experts who possess the detailed knowledge incorporated in the process trees. This eliminates the need for the current interface between subject matter experts and logic modeling experts. This paper provides an introduction to *LED TOOLS*. We begin with a description of the programming environment. The construction of a process tree is described and the simplicity and efficiency of the approach incorporated in the software is discussed. We consider the nature of the logical equations that the tree represents and show how solution of the equations yield natural language "paths." Finally we discuss the planned improvements to the software.

INTRODUCTION

Tree structures that use logic gates to model system behavior have proven very useful in safety and reliability studies. The fault tree, in particular, has seen wide application as an analysis tool for risk assessment. In our research we have extended the basic ideas used in the fault tree to create a related class of logic-gate trees that we call process trees [1]. Process trees provide a framework for deductively generating possible causes of a final state or possible outcomes of an initial state.

Process tree construction has a proven record in successfully addressing a wide range of complicated and difficult problems. Process trees have been used to describe complex physical processes, such as cook-off of a heated high explosive [2]. They have been used in a forensic mode to determine what the possible causes of an observed outcome could have been [3]. They have been used to identify what features a system needs to perform its mission successfully [4] and the order in which upgrades should be performed to assure a process' continued success [5]. Process trees have been used in analysis of spare parts stocking priorities [6] and in setting research priorities in a safety program. They have been used to enumerate the attack modes of a saboteur or terrorist [7] and the ways a spy could obtain protected information. Process trees have also been used to describe abstract processes. For example, we used a process tree to determine the analytical procedure for evaluating the risk of different information compromise scenarios during a visit by a foreign delegation to a secure facility (the information compromise scenarios were also identified by a different process tree) [8].

Process trees are the basic structure used in a decision analysis methodology developed at Los Alamos called Logic Evolved Decision modeling (LED) [6,9]. The basic elements of a decision model are

- Determine the possibilities or alternatives
- Select the metric to rank the possibilities
- Design an inferential model for the metric
- Rank the possibilities according to the metric
- Express the degree of uncertainty in the results
- Express the results in a form useful to the decision maker

LED uses coupled process trees to develop a comprehensive set of possibilities and to evaluate them in a consistent and traceable manner. In the past LED analyses have been performed using a combination of general-purpose software as well as a fault tree drawing application, *SEATREE* [10]. This approach was acceptable when the number of analyses being performed was small and much of the emphasis was on methodology development. However it has become clear that in order to perform analyses more efficiently and to make the LED approach to decision analysis available to a wider audience, application specific software would be needed.

LED TOOLS is the initial attempt to provide LED-based decision analysis tools in a state of the art software package. The initial release of the software, Version 2.0, addresses the first step in LED – determination of the possibilities. *LED TOOLS* is an object-oriented application written in Visual Basic for Windows NT based operating systems. It provides an innovative graphical user interface that was designed to emphasize the visual characteristics of logic trees and to make their development efficient and accessible to the subject matter experts who possess the detailed knowledge incorporated in the process trees. This eliminates the need for the current interface between subject matter experts and logic modeling experts.

This paper provides an introduction to *LED TOOLS*. We begin with a description of the programming environment. The construction of a process tree is described and the simplicity and efficiency of the approach incorporated in the software is discussed. We consider the nature of the logical equations that the tree represents and show how solution of the equations yield natural language "paths." Finally we discuss the planned improvements to the software.

OVERVIEW OF THE *LED TOOLS* INTERFACE

LED TOOLS uses a hierarchical format familiar to users of Windows to display tree structures. A Multiple Document Interface allows multiple files to be opened and viewed at the same time. Two windows are displayed for each project as shown in Fig. 1. One is the main tree window and the other is a replicants window, to be discussed in more detail below. The associated project name appears at the top of each window. In the main tree window, the tree starts with a top node and branches down. This is called the tree stem. The tree stem is populated by gates, terminal nodes, and replicant nodes. Each node has an icon representing its logical functionality and a text line that is the node display name. The output of each node is shown by means of light dashed connecting lines. Levels of development within the tree are shown by the extent of indentation. This format allows very large logic-gate trees with many levels of hierarchy to be displayed in a compact format.* Sub-sections of the tree are expanded or collapsed by double clicking on the top node for that sub-tree. Double clicking a top node in the main window will collapse the tree structure below the selected node and replace the node with a diamond symbol. A collapsed sub-tree can be expanded by double clicking on the diamond symbol. Double clicking on a replicated node will collapse the sub-tree below that node to a triangle.

Just below the standard menu is the gate palette. A gate can be dragged from the palette to the tree in a manner similar to a graphics program. Control-dragging a gate onto an existing node changes the gate type. The preferred method of tree construction is visual, using drag and drop. However these operations can also be performed using commands on the EDIT menu or by right clicking. All of the gates used in the process tree are either of the OR-family or the AND-family. However as we shall see below, process trees are not restricted to the logic gates used in fault trees. The *LED TOOLS* architecture is designed to allow for the addition of new logic gates as needed.

Because the software will be used by subject matter experts who are unfamiliar with many of the details of mathematical logic and graph theory, an extensive Help system is available via the HELP menu. The Help system contains a detailed tutorial on tree construction as well as descriptions of the various logic gates available.

CONSTRUCTING A PROCESS TREE USING *LED TOOLS*

As noted earlier, process trees have been used in many different applications. In this paper we discuss the construction of a process tree for a problem of current interest [8]. In this case the possibilities to be deduced are associated with the question – What are possible attack scenarios using weapons of mass destruction (WMD) against a set of targets? Figure 2 shows an *LED TOOLS* process tree for the WMD problem. This tree is logically hierarchical in form; most of the hierarchy has been collapsed in this view. Development of the process tree requires the deduction of a series of logical steps from cause to effect or from effect to basic causal events and one could consider this to be a very preliminary version of the final tree.

A WMD attack requires a weapon and an attacker to use it. These requirements are indicated by the use of an AND gate at the top of the tree with the two inputs *utilizing* and *The attack is perpetrated by*. Each of these is in turn developed in more detail using deduction: successive deductive steps produce the characteristic hierarchical logic structure. The gates describe how each deduction is related to the previous one. We can quickly deduce that various attackers are possible and these appear under *The attack is perpetrated by*. This set of possibilities forms a taxonomy. Taxonomies are often encountered in process trees and a TAXONOMY gate is available in *LED TOOLS* for this purpose. It can be considered to be a specific type of an EXCLUSIVE OR (EOR) gate. As noted earlier many other gate types are available to express logical relationships. The taxonomy here has three members; each appears as a solid circle. This indicates that they are terminal nodes. That is, no further development occurs below it at present. Note that two of them are colored blue and one is green. Blue indicates that a node has been designated as excluded. The effect of an exclusion will be discussed below. Returning to the input, *utilizing,* we see that it is also a taxonomy with three terminal nodes as members. Two of these have been excluded as well. The third member, *radionuclides* is shown as a solid green diamond. A diamond indicates that this node has been contracted – there are additional nodes below this one. The fact that the diamond is solid is the visual indication that this node has been terminated. That is, the nodes that appear below it in the logical hierarchy are all considered to be neglected. Terminal nodes are also, by definition, terminated. They mark the end of a logical development, and this is why they appear as solid.

* This outline format is also used in *SEATREE.*

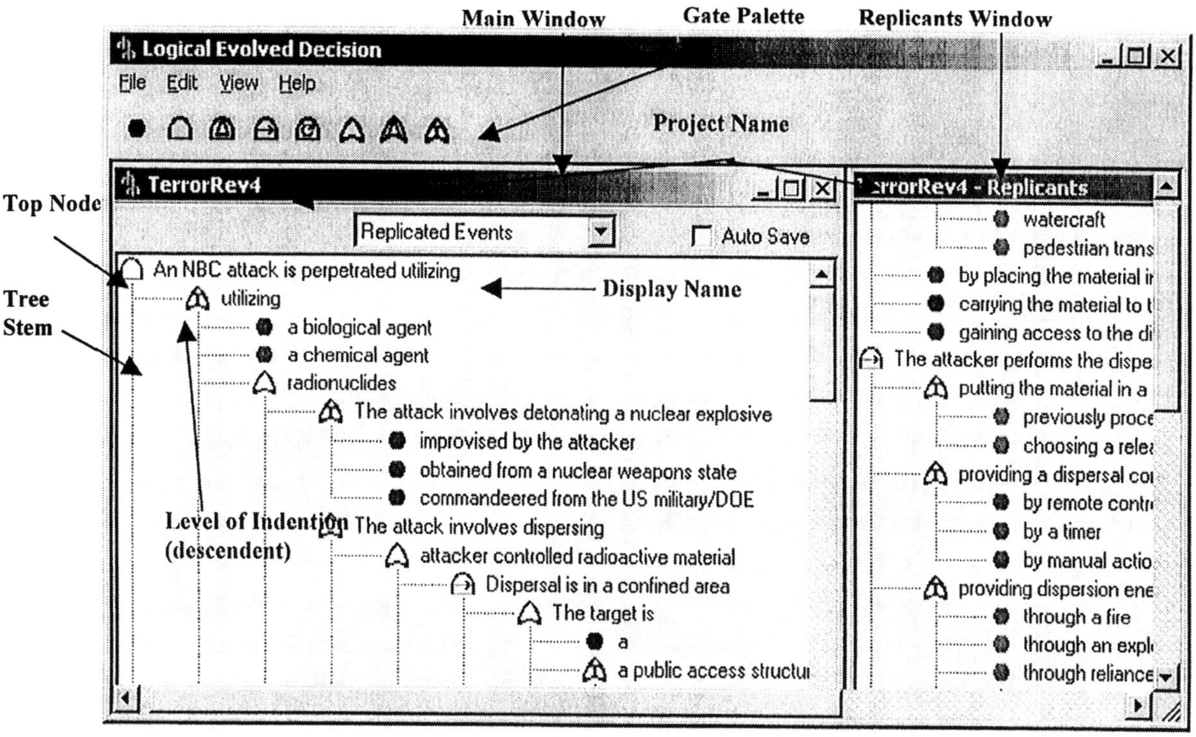

Figure 1. The LED TOOLS programming environment

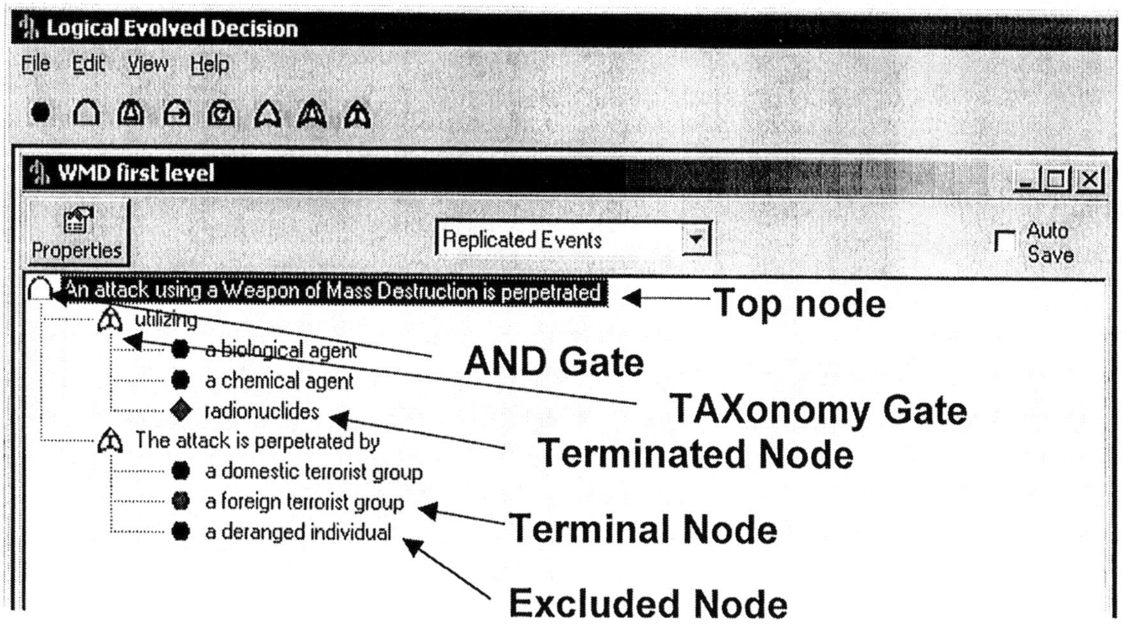

Figure 2. Top level of process tree for WMD attack possibilities.

Of course the process tree that we've build so far isn't particularly interesting. Recall that the node *radionuclide* is collapsed. Figure 3 shows what happens when this node is changed from terminated to included and is then expanded (by double clicking) – new levels of logical relationships are revealed. At the first level under *radionuclides* appear three nodes connected by EXCLUSIVE OR logic that describe what the material source of the radioactive material is, e.g. a nuclear explosive. Note that this node is blue so it is excluded and is an open diamond. An open diamond indicates that the node is collapsed. In Fig. 2, the node *The attack is perpetrated by* is collapsed so the details about the attacker groups that we saw in Fig. 1 are now hidden. The use of the expand and collapse functions allows the tree builder (or viewer) to concentrate on particular aspects of the tree as desired.

In Fig. 3 the expand/contract feature has been used to emphasize the logic associated with the sub-path

...The attack involves dispersing -- attacker controlled radioactive material—Dispersal occurs in an unconfined location....

The last node in this sub-tree, *Dispersal occurs in an unconfined location*, is connected to its five inputs by a CAUSAL gate. A CAUSAL gate is a variant of an ordered AND gate where it is understood that the order of the siblings, from top to bottom represents a causal chain that taken together produces the parent [11].

The first input of this CAUSAL gate is *The released material contaminates a region that includes*, and we see that the only region contained in this taxonomy that is included with the tree in its current state is a downtown area. Note that the TAX gate is enclosed by a box outline. This indicates that *The released material contaminates a region that includes* is a replicant.

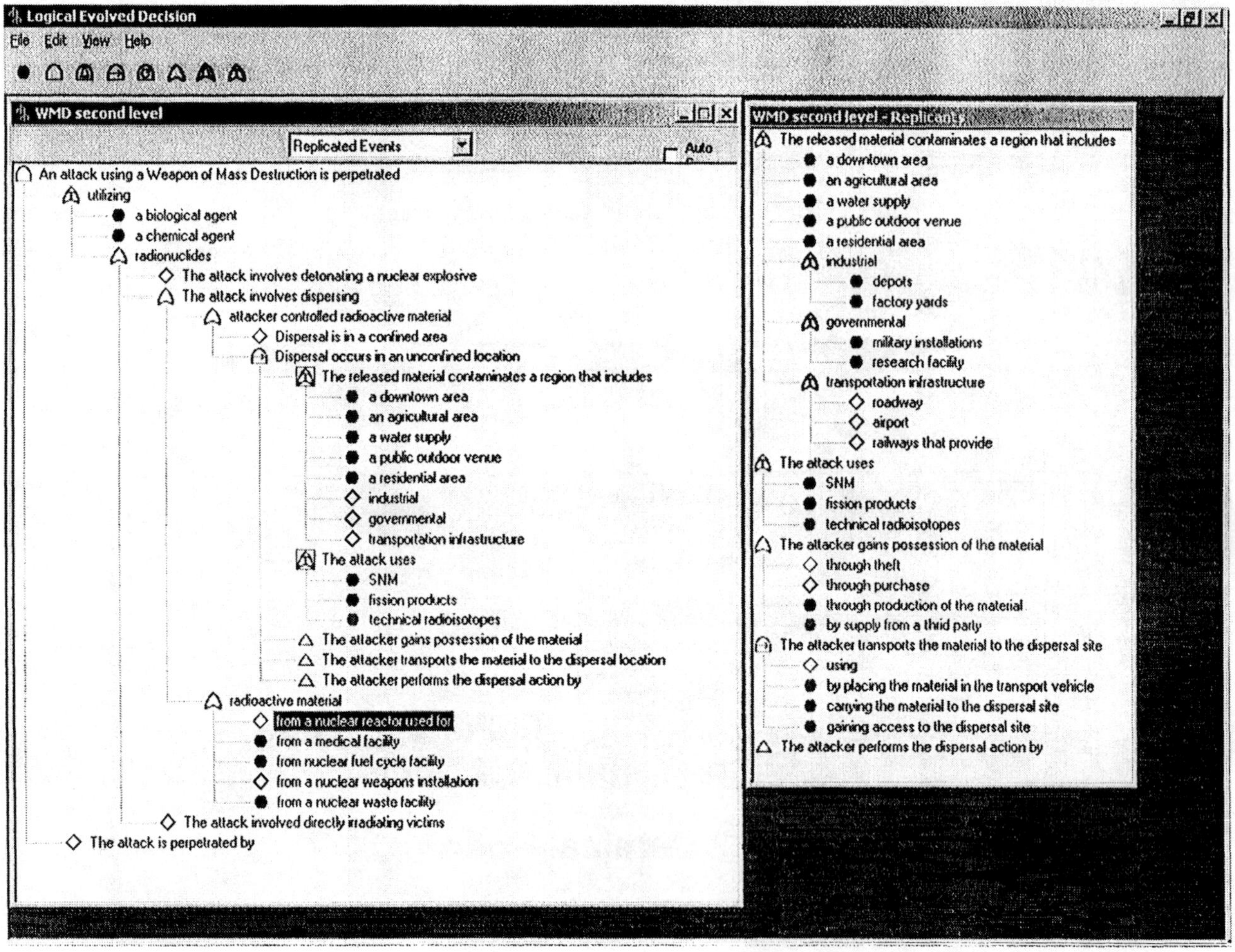

Figure 3. Second level development for WMD process tree. File: WMD second level.

USING REPLICANTS

A replicant is a sub-tree that has the property that it can be used more than once in the main tree.* The ability to create and use replicants is a very powerful feature of *LED TOOLS*. The individual replicant sub-trees are contained in a separate window with "- Replicants" appended to the main window file name. *The released material contaminates a region that includes* appears at the top of the replicant window in Fig. 3. The image of a replicant in the main window is referred to as an instance. All editing of a replicant occurs in the replicant window. Any edit operation thus performed is immediately reflected in all associated instances in the main window. A replicant can only be added to the main tree by dragging it from the replicant window to the desired location in the main window. This is to avoid possible confusion as discussed below.

A collapsed replicant is denoted by a triangle in either window. Note that the degree of expansion may differ between the replicant and an instance; the logic remains the same. Terminations and exclusions performed in the replicant window are also applied to all the associated instances as well. Individual instances of a replicant may be terminated as well. This option is available so that repetitions of a sub-tree in a path can be suppressed to improve readability.

SOLUTIONS AND PATHS

The properties of inclusion, exclusion and termination are associated with two very important properties of a process tree, the paths and the implicants. A process tree is a set of simultaneous logical equations. Nodes that have been excluded or terminated in a tree will be ignored when the equations are solved. With the tree as shown in Fig. 2, there are three logical equations:

1. *An attack using a Weapon of Mass Destruction is perpetrated* AND *utilizing, The attack is perpetrated by*
2. *utilizing* TAX *a biological agent, a chemical agent, radionuclides*
3. *The attack is perpetrated by* TAX *a domestic terrorist group, a foreign terrorist group, a deranged individual*

Here blue indicates the node that is the logical result (output) of the inputs in black operated upon by the logic gate in red. The set of path solutions of these equations with the given exclusions and termination has only one element, the path

> *An attack using a Weapon of Mass Destruction is perpetrated utilizing radionuclides. The attack is perpetrated by a foreign terrorist group.*

Conceptually, the paths are found by successively substituting into the logical equations and preserving the partial results at each step. The number of paths depends upon the gate logic. For example, if all of the attacker types were included (green indication) there would be three paths differing after "*perpetrated by...*" If all of the terminal nodes were green there would be nine paths, accounting for all possible combinations of agent and weapon type. If one takes a path and removes all of the elements except terminals, then the result is the implicant set that is the analog of the cut set for a fault tree.* For the path given above the implicant set is {radionuclides, a foreign terrorist group}.

LED TOOLS was designed to facilitate the construction of logical equations using natural language expressions. In addition to making it easier to perform the basic deductive steps, this feature also results in paths that can be read as paragraphs; each path is a unique and complete story describing how the top node in the tree occurs. One of the paths that results from the solution of the tree in Fig. 3 is.

> *An NBC attack is perpetrated utilizing radionuclides. The attack involves dispersing attacker controlled radioactive material. Dispersal is in a confined area. <u>The target is a public access structure housing a mass audience sports venue</u>. The attack uses <u>technical radioisotopes</u>. The attacker gains possession of the material <u>through theft from a source of stored material by means of insider diversion</u>. The attacker transports the material to the dispersal site <u>using a road vehicle by placing the material in the transport vehicle carrying the material to the dispersal site gaining access to the dispersal site</u>. The attacker performs the dispersal action by putting the material in a dispersible form <u>by previously processing the material into a dispersible form</u> <u>providing a dispersal command signal by manual action of the</u> attacker providing dispersion energy through reliance on natural dispersive forces <u>providing a dispersion path through the ventilation system</u>. The attack is perpetrated by <u>a foreign terrorist group</u>.*

Each path solution is a fairly detailed "story" about an individual WMD attack. The underlined elements comprise the implicant set for this path and can be regarded as a "summary" of the story. The collection of paths can be further elaborated by adding additional structure in the tree. The use of the exclusion and termination features reduces the total number of paths that are obtained by solving the tree and allows the user to control the number and level of detail in the paths. Removing all of the solution restrictions in the complete WMD tree would yield a set of paths that number in the thousands.

Solving a process tree for the paths rather than the cut-sets may appear strange to readers familiar with fault trees. However it is clear that we can address many different questions using process trees beyond 'How does a system fail?' In doing so the details and interrelationships amongst the nodes are important. For example, in considering the risk of the scenario above, one would have to consider the capabilities of the attacker relative to the intricacy of the attack. Experts capable of performing such evaluations will need to understand scenario details and how scenarios are related. The combination of the process tree and the solution paths provides the necessary information.

* "Replicant" is taken from the novel <u>Do Androids Dream of Electric Sheep</u> by Philip K. Dick,, the basis for the movie Bladerunner.

* Note however that a process tree path is not the analog of the path set for a fault tree.

WORKING WITH NATURAL LANGUAGE TREES

The combination of natural language nodes and the visual paradigm implemented in *LED TOOLS* has a number of implications that are not obvious from the discussion thus far. In a complicated path it might be useful to use the same phrase several times but where the phrase is a different logic node. For example, the word "utilizing" could appear as the parent in sub-trees associated with biological, chemical and/or nuclear threats. Of course all of these sub-trees are different logically and therefore "utilizing" appears in different contexts. Each *utilizing* is a unique node. *LED TOOLS* allows for context-dependent node names by using the rule that all nodes are born unique. Consider the situation in Fig. 4. In the first OR gate at the second level, each node is a unique logical object, although the words are the same. The second two inputs were created by dropping and dragging the first. All drag and drop, or equivalently copy and paste operations within the main window actually reproduce the structure of the node copied. All drag and drop operations from the replicant window to the main window copy an exact instance of the replicant. All instances of a replicant are logically identical. On occasion it is useful to take an instance of a preexisting replicant and "make it unique." In this case the instance loses its replicant status. That is, it can be edited in the main window and can be dragged, copied, etc. like any other non-replicant node.

FUTURE WORK

Development of *LED TOOLS* continues. These developments include:

Fast Path Solver: The normal need to solve for all of the paths associated with a particular combination of terminations and exclusions requires a fast solution algorithm. A *LED TOOLS* file contains the complete ordering of the tree. Knowledge of this structure makes it possible to determine the number of paths quickly and to calculate paths much faster than by using a classic successive substitution approach [12].

Digraph Drawing Interface: A process tree is actually a directed graph and the software provides a tool to develop digraphs deductively. The capability to convert an *LED TOOLS* file into the equivalent digraph has been demonstrated. Future versions of the software will have the capability to do this in real time. Together with the fast path solver this will provide the analyst with multiple ways to study the logical structure of a problem.

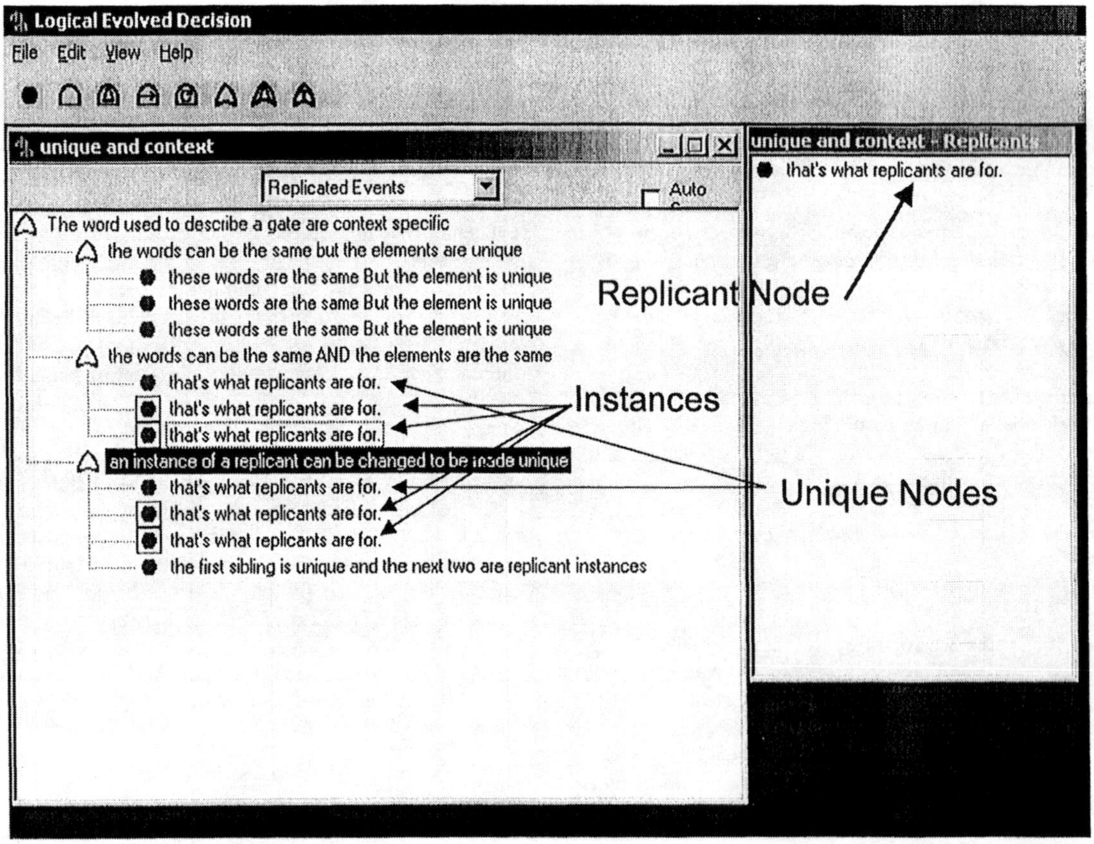

Figure 4. Context specific gate text.

Hyperlink Utility: A process tree is a very powerful database structure. Subject matter experts can use it to organize and archive information associated with a problem. We are currently implementing a hyperlinking feature in the software to provide this capability.

Inferential Process Tree Designer: Inferential models – used to rank order a set of possibility paths, are efficiently represented as process trees. Such models have associated with them additional variables that are not easily represented in tree form. We are developing utilities to make it possible to create and edit these variables while building the inferential process tree.

Gate Designer: As noted earlier, the basic software architecture allows for the addition of new gates to the gate palette. A new gate requires the definition of an icon, the basic logical equation and an equivalent digraph form. Preliminary design of a utility to provide these functions is planned for the near future.

CONCLUSIONS

Process trees have proven to be valuable tools in describing complex systems. Applications have included a broad range of technical, physical and even administrative systems. To date, however, all of these applications have required the involvement of the developers of the process tree. This is because the concepts used in process tree development have been under development and the available tree drawing tools required the translation from process tree to fault tree terminology and concepts. The use of the process tree has thus been dependent on the availability of a few people who interact with experts and translate their expertise into process tree logic. This is an inefficient use of the technique. In many cases it would be preferable for the subject matter experts to construct their own process trees with guidance from the process tree specialists only when required. The development of *LED TOOLS* makes this a viable option. the process tree methodology available to a much wider audience of subject matter experts. The software has been designed using a modern, visual approach so that a new user can quickly "learn by doing" without an excessive amount of manual reading. The availability of the software has in turn also spurred new development of the process tree methodology that we think will continue as the tool is more widely used.

ACKNOWLEDGMENT

The authors thank Drs. Philip Howe, Deanne Idar, and Larry Luck for their support in the development of LED Tools.

REFERENCES

[1] T. F. Bott and S. W. Eisenhawer, "Programme Planning with Logic Trees," **International Journal of Quality and Reliability Management**, Vol. 6, 1989, pp. 14-24.

[2] S. W. Eisenhawer, T. F. Bott, L. B Luck, J. Kingson and B. P. Key, "A Logic Model for Cook-off Phenomenology in High Explosives," to appear *21st International System Safety Conference*, Ottawa, Canada, August 4–8, 2003.

[3] 4. S. W. Eisenhawer and T. F. Bott, "Accident Reconstruction using Process Trees," *Risk and Safety Assessment: Building Viable Solutions*, PVP-Vol. 320/SERA-Vol. 5, SME 1995.

[4] T. F. Bott, S. W. Eisenhawer, "An Approach to Assessing the Need for High Explosives Replacement in Aging Nuclear Weapons," *27th International Pyrotechnics Seminar*, Grand Junction, July 2000

[5] T. F. Bott and S. W. Eisenhawer, "A Logic Model Approach to the Conceptual Design of a Scientific/Industrial Complex," *ASME-PVP Annual Meeting*, Vancouver, 2002, PVP-444, pp 119-127.

[6] S. W. Eisenhawer, T. F. Bott, and J. W. Jackson, "Prioritizing the Purchase of Spare Parts Using an Approximate Reasoning Model," *Proc. International Symposium on Product Quality and Integrity*, January 2002.

[7] S. W. Eisenhawer, T. F. Bott, and D. V. Rao, "Assessing the Risk of Nuclear Terrorism Using Logic Evolved Decision Analysis," to appear ANS Topical Meeting on Risk Management, San Diego, June 1-4, 2003.

[8] K. B. Christiansen, T. F. Bott, J. L. Darby, and S. W. Eisenhawer, "The Approximate Reasoning (AR) Based Method for Information Loss Path Analysis," *Institute for Nuclear Materials Management Annual Meeting*, Orlando, June 2002.

[9] T. F. Bott, S. W. Eisenhawer, "Evaluating Complex Systems When Numerical Information Is Sparse," *Words Automation Conference*, Orlando, June 2002.

[10] B. Bingham, J. Hutchinson, and B. Lopez, "SEATREE Version 2.62 User Manual," Science and Engineering Associates, Albuquerque, New Mexico, 1994.

[11] S. W. Eisenhawer, T. F. Bott, "Application of Approximate Reasoning to Safety Analysis," *Proc. International System Safety Conference*, Orlando, Aug 1999, pp 374-382.

[12] Richard B. Worrell and Desmond W. Stack, A SETS User's Manual for the Fault Tree Analyst, NUREG/CR-0465, Sandia Laboratories, November 1978.

PVP-Vol. 458, Computer Technology and Applications

PVP2003-1916

RELIABILITY ANALYSIS OF THE TUBE HYDROFORMING PROCESS BASED ON FORMING LIMIT DIAGRAM

Bing Li, Don R. Metzger, T. J. Nye

Department of Mechanical Engineering
McMaster University
1280 Main Street West
Hamilton, Ontario, Canada L8S 4L7

ABSTRACT

Tube hydroforming has become an increasingly attractive manufacturing process in automotive industry due to it having several advantages over alternative methods. The forming limit diagram has been extensively used in metal forming as the criteria of formability. A method to assess the probability of failure of the process based on reliability theory and the forming limit diagram is proposed in this paper. The tube hydroforming process is affected by many parameters such as geometry, material properties, and process conditions. Finite element simulation was used to predict the relationship between the strain and these parameters, and a numerical method was applied to get the statistical distribution of the strain. Based on the forming limit band in the forming limit diagram, the reliability of the forming process can be evaluated. A tube hydroforming process of free bulging is then introduced as an example to illustrate the approach. The results show this reliability evaluation technique to be an innovative approach for product designers and process engineers to avoid failure during tube hydroforming.

INTRODUCTION

Hydroforming technology has found a great acceptance and wide application in many industrial fields, especially in the automobile industry, during recent years. Compared with conventional metal forming technology, hydroforming technology has the potential advantages of reducing workpiece cost, tool cost and product weight [1,2]. In tube hydroforming, the goal is to form a tubular blank of usually uniform cross-section into a die cavity of a complex shape through the use of internal hydrostatic pressure without causing any kind of forming instability like fracture, necking, wrinkling or buckling.

The forming limit diagram (FLD) has proven to be a useful tool in the formability analysis for sheet metal forming [3]. The forming limit curve (FLC) separates the strain for the onset of failure from that which will not fail. As for the tube hydroforming limiting strains, due to a shortage of experimental data, the conventional sheet forming limit data is usually applied [4-6].

The purposes of this study are to (1) establish a reliability analysis model by combining the traditional reliability theory with FLD, and (2) predict the failure probability of a given tube hydroforming process at the initial design stage. A simple tube hydroforming process of free bulging will be analyzed to illustrate our method.

THEORETICAL BACKGROUND

Forming Limit Diagram

The forming limit in FLD is conventionally described as a curve in a plot of major strain vs. minor strain. If the maximum principal strain is above the forming limit curve, it indicates that the necking or fracture failure will happen; otherwise failure will not occur and the process is safe [7]. As shown in Fig. 1, for two postulated forming processes for the same material, the maximum strain obtained in process one crosses the forming limit curve, where necking failure will happen, while the maximum strain obtained in process two is below the forming limit cure. Thus, process two is assumed to be safe.

The forming limit curve of a material is determined by experimental measurement. Errors will exist due to the uncertainty in experimental measurements, the material properties and the process conditions, etc., so the forming limit curve is more aptly described as a forming limit band [8]. With the forming limit band (e.g., as shown in Fig. 2) if a strain point is located under the lower forming limit band boundary, the forming process is safe. If the strain point is located above the upper forming limit boundary, then necking failure is predicted to occur. If the strain point is located between the upper and the lower boundaries, then the forming process is marginal and there is a probability that failure will happen. In the following Section, we use reliability theory to calculate the necking failure probability of the process when the maximum strain obtained is located among the forming limit band. In the particular example shown in Figs. 1 and 2, while the conventional FLD predicts the process is safe, use of the forming limit band allows a probability of failure to be calculated.

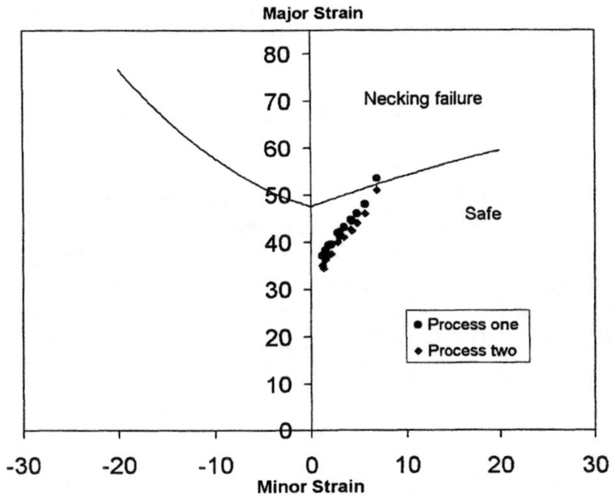

Fig. 1 Example of forming limit diagram

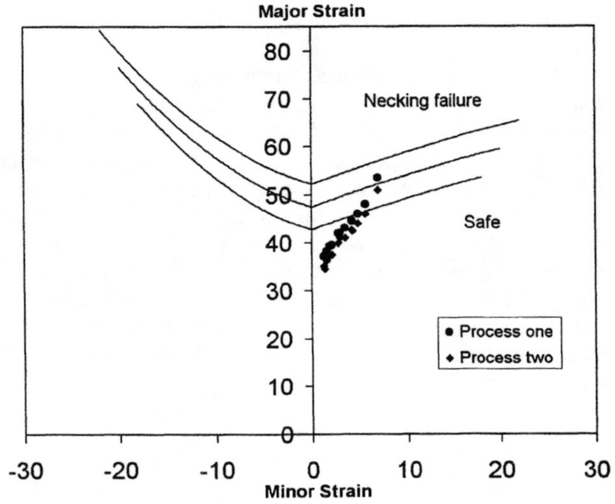

Fig. 2 Forming limit band in FLD

Nominal Strain and Nominal Limit Strain

Before introducing the reliability analysis method based on FLD, we present two terms (i.e., nominal strain and nominal limit strain). We assume there is a forming process and a particular strain point is plotted on the FLD (point "1" in Fig. 3), and the forming limit band for the material is as shown in Fig. 3. A proportional strain path is assumed through this strain point. Because of inherent process variability, the strain at a given location of a formed part will vary from part to part during production, thus the nominal strain of that point describes the median of this strain distribution.

The proportional strain path (also shown on the right of Fig. 3) has three intersection points with the forming limit band, i.e., points "2", "3" and "4". We define the distance from the origin in the FLD to the point "1" as the nominal strain of the strain point "1". We also define the distance from the origin to any point between the point "3" and "4" in the strain path as the nominal limit strain of the material, the distance from origin point to point "2" will be the mean value of the nominal limit strain, and the distance from original point to point "3" and point "4" will be the maximum and minimum values of the nominal limit strain along the given strain path.

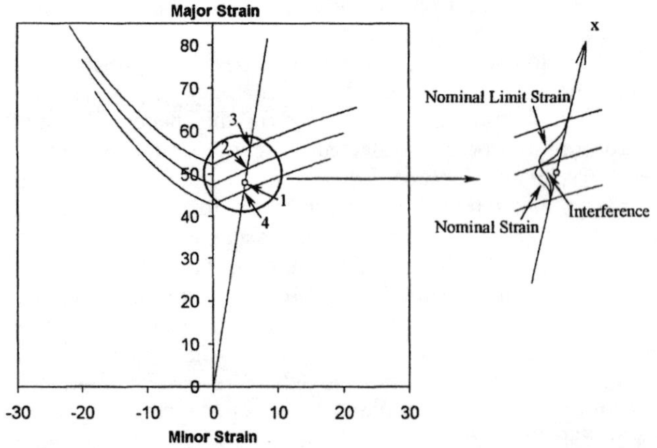

Fig. 3 Concept of nominal strain and nominal limit strain

Reliability Analysis Based on FLD

In our research, we assume that both the nominal strain of the tubular blank and the nominal limit strain of the material obey the normal distribution, then the reliability of the hydroformed tube based on the FLD will be [9]:

$$R(Z_R) = \frac{1}{\sqrt{2\pi}} \int_{-\infty}^{Z_R} \exp(-\frac{t^2}{2})dt \tag{1}$$

where Z_R is the reliability coefficient:

$$Z_R = \frac{\mu_{le} - \mu_{ne}}{\sqrt{\sigma_{le}^2 + \sigma_{ne}^2}} \tag{2}$$

and μ_{ne}, μ_{le}, and σ_{ne}, σ_{le} are the mean values and the standard deviations of nominal strain and nominal limit strain, respectively.

Traditional reliability theory [9] uses the stress-strength interference model to predict reliability. Here we present the nominal strain-nominal limit strain interference model as shown in Fig. 4. Nominal strain is a function of process variability while nominal limit strain represents the distribution of material strength. The overlap area in Fig. 4 represents the necking failure probability of the forming process. In this case the axis 'x' of the interference model represents the proportional strain path shown in Fig. 3.

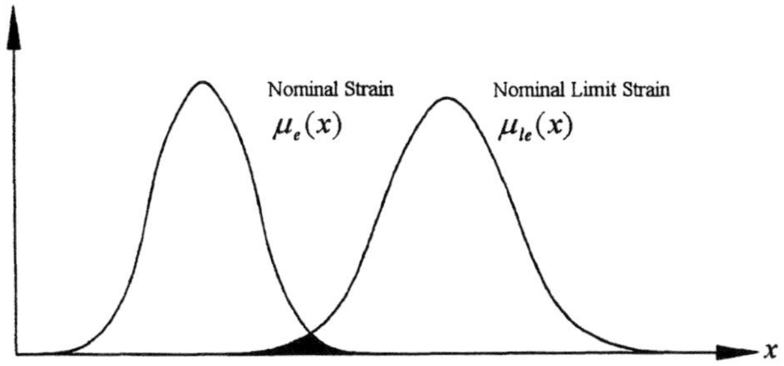

Fig. 4 Nominal strain – nominal limit strain interference model

Distribution of Nominal Limit Strain

We have assumed the nominal limit strain as a random variable following the normal distribution, according to the definition of nominal limit strain in Section 2.2. The width of the forming limit band is typically 10% to 20% true strain, depending on the experimental results for a particular material. Assuming this band represents plus or minus three standard deviations of the distribution, the standard derivation of the nominal limit strain can be obtained approximately as:

$$\sigma_{le} = p\mu_{le}/3 \tag{3}$$

where p is the percent of the true strain of the half width of forming limit band.

Distribution of Nominal Strain

The strain distribution of one forming process is dependent on many parameters; we can express the nominal strain ε as

$$\varepsilon = f(x_1, x_2, \cdots, x_n) \tag{4}$$

where x_i (i=1,...,n) are the process parameters that affect the nominal strain. Assuming these parameters also obey the normal distribution, then the mean value and the variance of the nominal strain can be obtained approximately as [10]:

$$\mu_{ne} = f_{ne}(\mu_{x_1}, \mu_{x_2}, \cdots, \mu_{x_n}) \tag{5}$$

$$\sigma_{ne}^2 = \sum_{i=1}^{n} \left[\left. \frac{\partial f_{ne}}{\partial x_i} \right|_{x_i = \bar{x}_i} \right]^2 \sigma_{x_i}^2 \tag{6}$$

where $\mu_{x_i}, \sigma_{x_i}^2 \ (i = 1, \cdots, n)$ are the mean value and the variance of x_i.

APPLICATION

FEM Simulation

Simulations have been performed using the explicit FEM code H3DMAP [11]. Figure 5 shows the hydroforming process of free bulging of a straight tube with simultaneously applied internal pressure and axial force. Due to the axial symmetry of the tube, only one-eight of the tubular blank and the tooling were modeled. The finite element model consisted of 1800 elements for the tooling and 3200 for the tube. The tooling was modeled as a rigid body. The tube material was assumed to be isotropic elastic-plastic obeying the Ludwik-Hollomon hardening relationship, $\sigma = k\varepsilon^n$. Table 1 shows the material properties of the tubular blank.

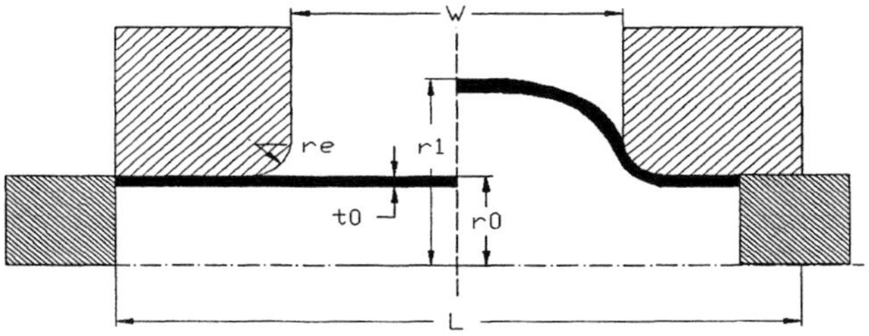

Fig. 5 Schematic view of free bulging hydroforming

Table 1 Mechanical properties used in the simulation

ρ, kg/m^3	E, GPa	K, MPA	n	σ_y, MPa	σ_u, MPa	ν
7850	205	537	0.227	240	350	0.3

During simulation of the forming, the loading path strategy described by Manabe and Amino [12] was used. In this strategy, the internal pressure is applied independently and the axial load is applied according to the flowing equation of nominal stress ratio m [13]:

$$m = \frac{\sigma_\phi}{\sigma_\theta} = \frac{(F - P_f A_0)/A}{P_f R_i / t_0} \tag{7}$$

where σ_θ is the circumferential stress, σ_ϕ is the axial stress, P_f is the forming internal pressure, F is the axial load, R_i is the inner radius of the tube, A is the initial cross-sectional area of the tube, A_0 the inner cross-sectional area of the tube, and t_0 is the thickness of the tube.

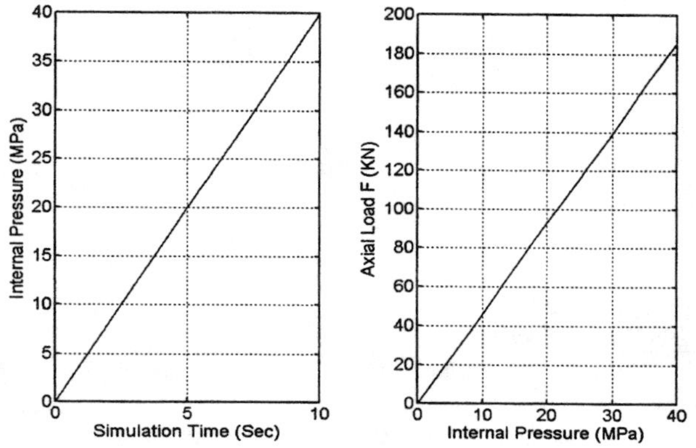

Fig. 6 Loading path in simulation

Figure 6 shows the loading path used in the simulation. Internal pressure is applied linearly with the simulation time and axial load is applied in proportion to the internal pressure.

Effects of Different Parameters on Hydroformability

Generally, there are three groups of parameters that affect the formability of the tube hydroforming process: (a) geometrical parameters: length, radius, thickness, etc.; (b) material parameters: various material, strain hardening exponent, etc.; and (c) process parameters: internal pressure, axial load, friction coefficient, etc. [14].

The parameters of interest in our study are the geometrical parameters including length of the tube (L_0), thickness (t_0), die entry radius (r_e) and bulge width (W) (as shown in Fig. 5), the material parameter including the hardening exponent (n) and the process parameters including internal pressure (P_f), nominal stress ratio (m) and friction coefficient (μ).

In the simulation, each parameter was varied over a range of +/-10% from the nominal value, while keeping the rest of the parameters constant, in order to investigate the effect of each parameter on the process. Table 2 presents the different parameter values used in the FEM simulation. The nominal values are those in the center column.

Table 2 The Forming conditions used in the simulation

Geometrical parameters					
Length of tube L_0 (mm)	180	190	200	210	220
Outer radius of tube r_0 (mm)			30		
Thickness of tube t_0 (mm)	1.35	1.425	1.5	1.575	1.65
Die entry radius r_e (mm)	9	9.5	10	10.5	11
Bulge width W (mm)	90	95	100	105	110
Material parameters					
Hardening coefficient K (MPa)			537		
Hardening exponent n	0.2043	0.2157	0.227	0.2384	0.2497
Process parameters					
Internal pressure P_f (MPa)	36	38	40	42	44
Nominal stress ratio m	0.36	0.38	0.4	0.42	0.44
Friction coefficient μ (Coulomb)	0.054	0.057	0.06	0.063	0.066

In order to verify the FEM results, the bulge height, i.e., r_l in Fig. 5, was compared to and found to be in good agreement with experimental results reported in the literature [12,14-16]. Figure 7 illustrates the effect of the different parameters on the bulge height. From Fig. 7, we can see that the length of the tube, the bulge width and the internal pressure have the greatest effect on the bulge height, with the other parameters having relatively smaller effects.

Reliability of Tube Hydroforming Process

The principle strains of all the elements of the hydroformed tube from the FEM simulation using the nominal parameter values are shown in Fig. 8. Because of rotational symmetry of this process, each strain point in Fig. 8 actually represents a series of similar strain points of the elements in the same hoop. What appears to be two strain 'paths' in the FLD is the result of using two elements in the thickness direction in our FEM model. The critical (maximum) strain was found to be in the center of the expansion region, element 3660 in Fig. 9. As the parameter values in Table 2 were used in the simulation, the strain distribution of the critical element (in this case, always element 3660) was found (Fig. 10).

The strain path corresponding to the distribution axis in Fig. 4 was taken through the strain point of element 3660 for the nominal parameter value case. Using a polynomial representation of the forming limit curve, the mean value of nominal limit strain along this strain path is μ_{le}= 49.3337. Assuming the width of the forming limit band as 20% of the mean, then p in Eq. (3) is 0.1 so the standard derivation of nominal limit strain is σ_{le}= 1.6445.

All the strain points of element 3660 corresponding to the various values of the process parameters are shown in Fig. 10. (In every simulation case, the critical strain occurred in element 3660.) Taking the strain path through the nominal case strain point as a local coordinate axis, the strain points for all cases were projected to this axis. The projections were made parallel to the FLC, using the dummy curve "Temp FLC". Points along this curve have the same probability of necking failure. For instance, the point "2" in Fig. 10 is the projected point of the strain point "1". Applying this method results in all strain points being projected onto the strain path, and consequently a normal distribution may be fitted to the strain data.

The mean value of nominal strain of element 3660 is obtained directly. Equation (6) is used to evaluate the standard derivation of nominal strain as a function of each parameter. Figure 11 shows the relationship of the strain of element 3660 to each process parameter. This data is used to fit a second-order polynomial, to represent the function f_{ne} in Eq. (6). After calculating, we get the mean value of nominal strain of element 3660 μ_{ne}= 47.9065 and the standard deviation σ_{ne} = 4.4530.

From Eq. (1), we can obtain the reliability of the element 3660 R=61.82%. Since the strain of element 3660 is the critical strain of the hydroformed tube, then the reliability of this hydroforming process is 61.82%, i.e., the necking failure probability of the process is 38.18%.

If a specific reliability is required for this process, the designer can alter the process and product design at the initial design stage and use this approach to predict the resulting reliability. Indeed, data generated during the analysis can be plotted as in Fig. 11 to present sensitivity information that would aid in selecting the parameters to change, and the directions of change.

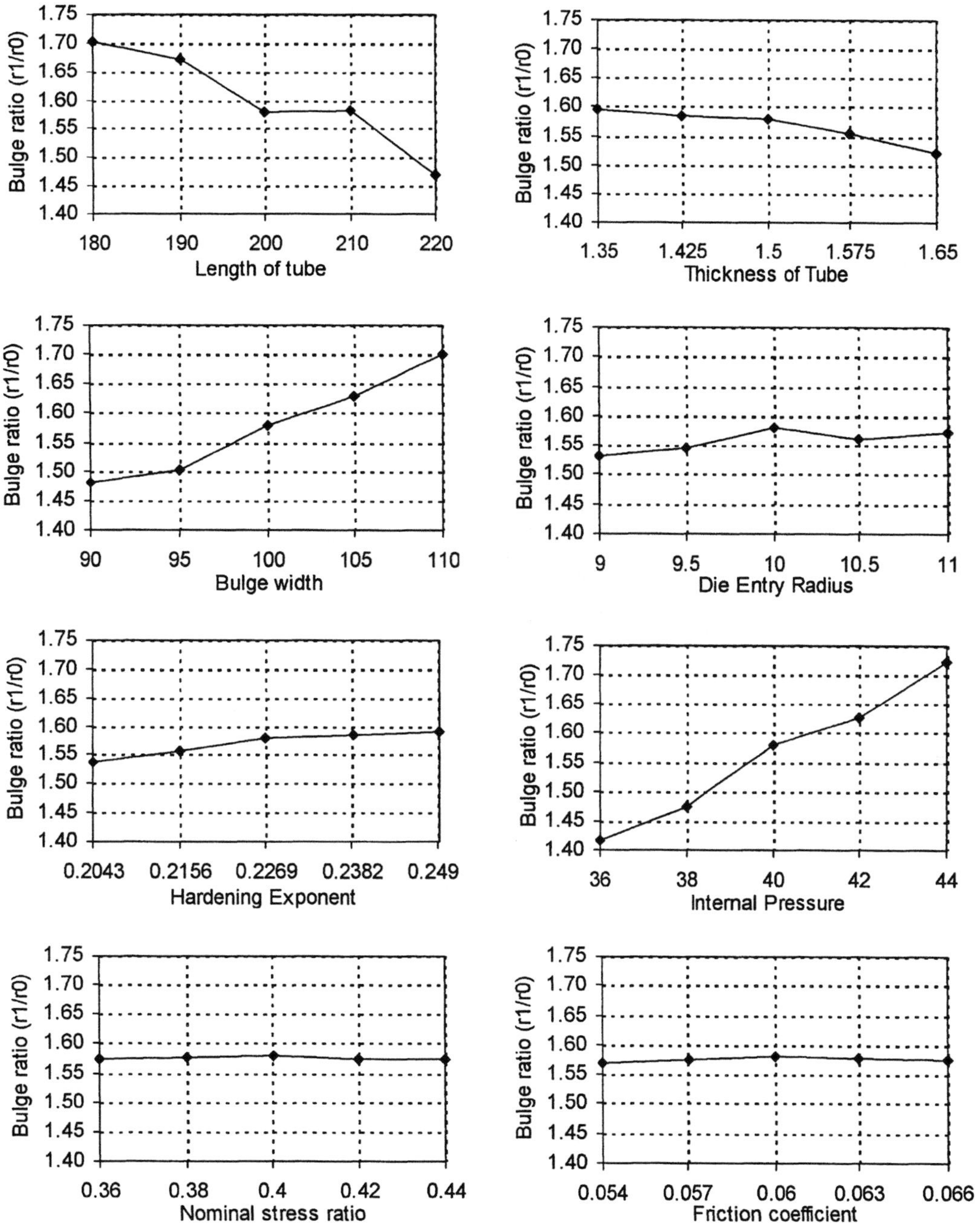

Fig. 7 Effect of the different parameters on the bulge height

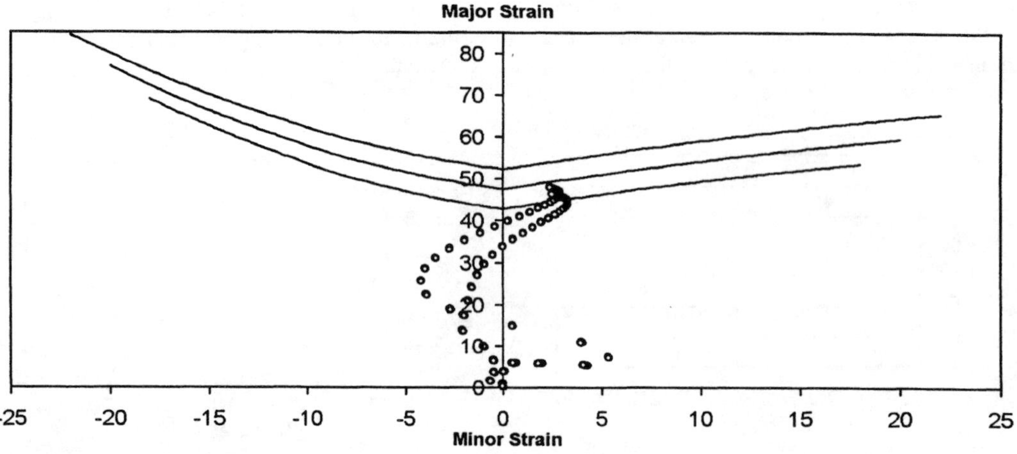

Fig. 8 Strain distribution of hydroformed tube

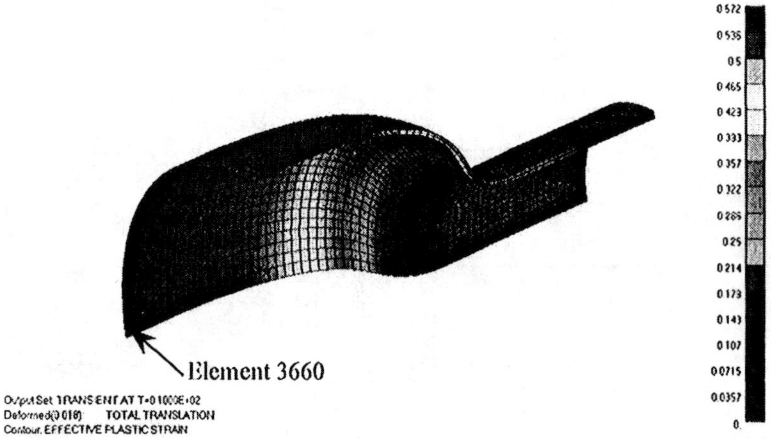

Fig. 9 Effective strain contour of hydroformed tube

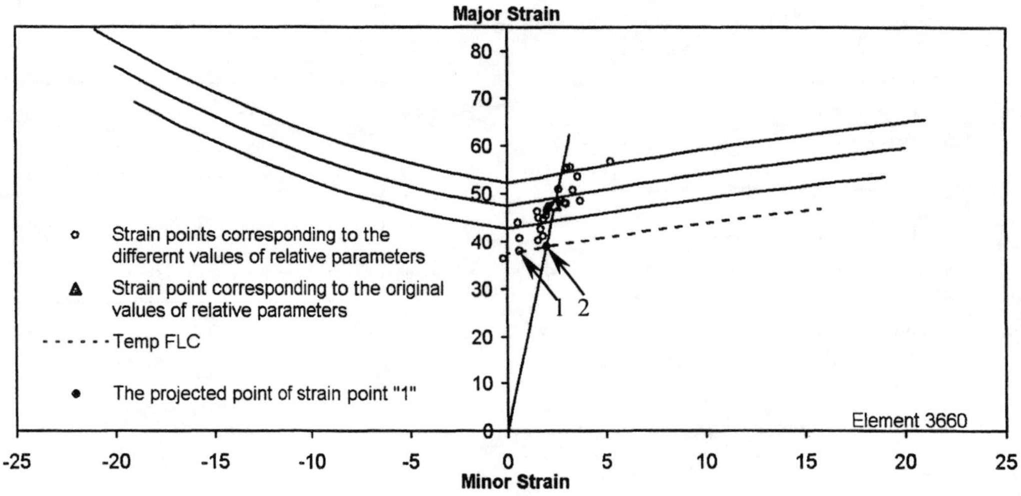

Fig. 10 Strain distribution of element 3660

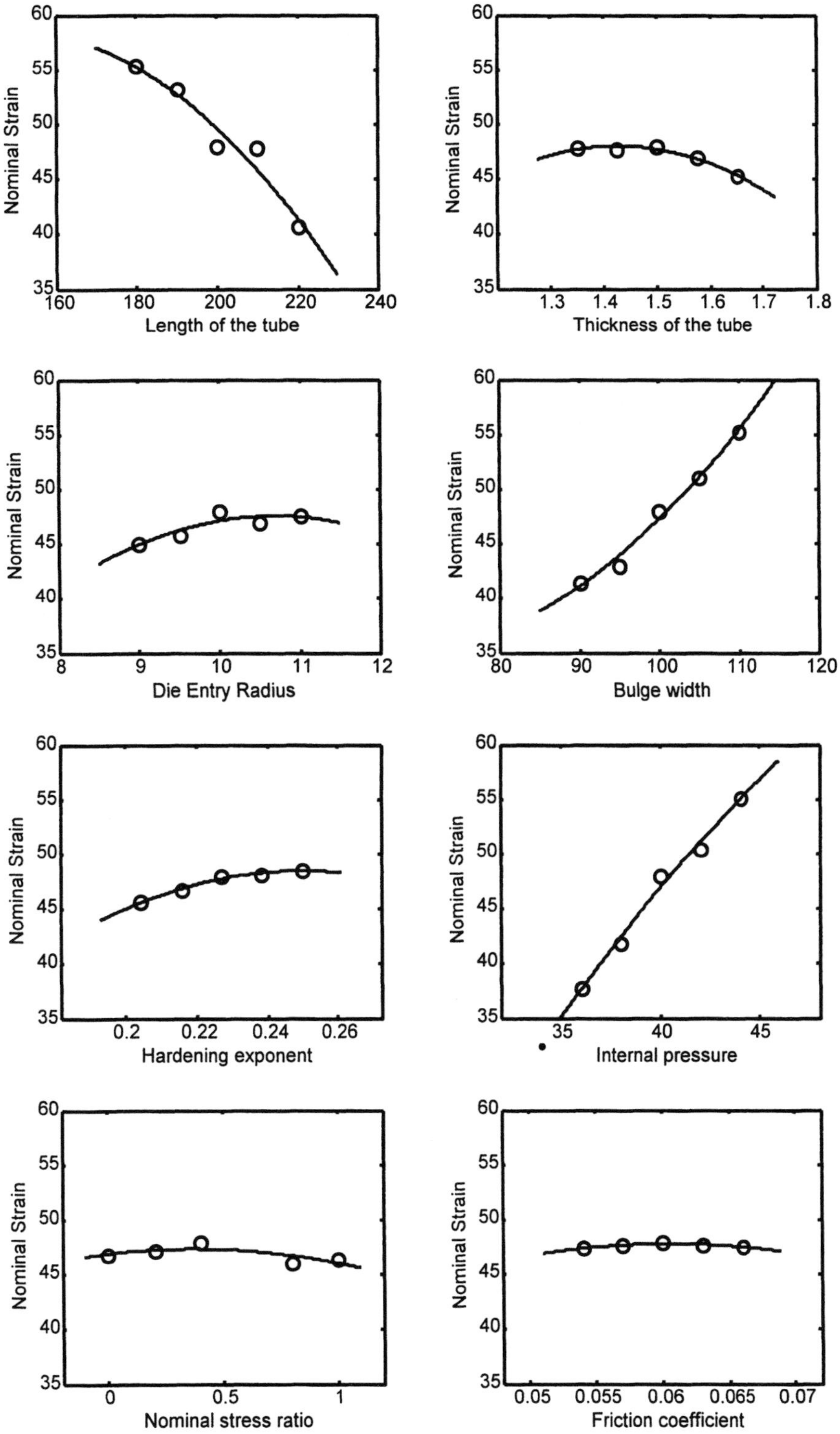

Fig. 11 Relationship between nominal strain and the parameters

DISCUSSION AND CONCLUSION

Because of the variability encountered when creating FLDs for a given material and the process variability found during production, both the forming limits and strains found within a process are random variables. By taking advantage of this idea, traditional reliability theory can be applied to the task of evaluating process quality and used to predict the probability of part failure during forming.

An advantage of using probability distributions such as the normal distribution rather than hard limits such as the upper and lower forming limit curves is that the distribution tails can evaluate cases where small, but non-zero, probabilities of failure occur.

The approach described here is meant to be applied at the initial product design stage. As such, it allows process validation to be performed before any time or money has been spent building tooling. In addition to predicting process reliability, this approach also provides sensitivity information that can be used to modify process designs to improve the reliability.

The particular failure criteria used in this work has been a strain-space FLD. Work is also being undertaken now to study how reliability methods can be applied to analyses based on other failure criteria.

REFERENCES

1. F. Dohmann, C. Hartl, 1996, "Hydroforming - A method to manufacture light-weight parts," Journal Of Materials Processing Technology, 60(1-4), pp. 669-676
2. M. Ahmetoglu, T. Altan, 2000, "Tube hydroforming: state-of-the-art and future trends," Journal Of Materials Processing Technology, 98(1), pp. 25-33
3. W.M. Sing, K.P. Rao, 1993, "Predication of Sheet-metal Formability Using tensile-test Results," Journal of Materials Processing Technology, 37, pp. 37-51
4. H.L.Xing, A.Makinouchi, 2001, "Numerical Analysis and Design for Tubular Hydroforming," International Journal of Mechanical Sciences, 43, pp.1009-1026
5. Koc, M. and Altan, T., 2002, "Prediction of Forming Limits and Parameters in the Tube Hydroforming Process," Journal of Machine Tools & Manufacture, 42, pp.123-138
6. G.Nefussi, A.Combescure, 2002, "Coupled Buckling and Plastic Instability for tube Hydroforming," International Journal of Mechanical Sciences, 44, p899-914
7. Thoms B. Stoughton, 2000, "A General Forming Limit Criterion for Sheet Metal Forming," International Journal of Mechanical Sciences, 42, pp.1-27
8. Koen Janssens, Frouke Lambert, 2001, "Statistical Evaluation of the Uncertainty of Experimentally Characterised Forming Limits of Sheet Steel," Journal of Materials Processing Technology, 112, pp.174-184
9. K. C. Kapur and L. R. Lamberson, 1977, "Reliability in Engineering Design," John Wiley, New York.
10. Li Bing, Zhang Yong, Zhu Meilin, 1999, "A Study on the Structure Design and Reliability of Cylinder Head for Vehicle Engines," SAE No.1999-01-0828
11. Sauve, R.G., 1999, "H3DMAP Version 6 — A General Three-Dimensional Linear and Nonlinear Analysis of Structures," Ontario Hydro Technologies Report No. A-NSG-96-120, Revision 1, Toronto, Canada.
12. Ken-ichi Manabe, Masaaki Amino, 2002, "Effects of process parameters and material properties on deformation process in tube hydroforming," Journal of Materials Processing Technology, 123, pp.285-291
13. Y. Yoshitomi, 1987, "Estimation method for application of bulge forming to circular tubes," Journal Of The Japan Society For Technology Of Plasticity, 28(316), pp.432-437
14. Koc, M. and Altan, T., 2002, "Application of Two Dimensional (2D) FEA for the Tube Hydroforming Process," Journal of Machine Tools & Manufacture, 42, pp.1285-1295
15. B.Carleer, G. van der Kevie, L. de Winter, B. van Veldhuizen, 2000, "Analysis of the Effect of Material Properties on the Hydroforming Process of Tubes," Journal of Materials Processing Technology, 104, pp.158-166
16. Yeong-Maw Hwang, Yi-Kai Lin, 2002, " Analysis and Finite Element Simulation of the Tube Bulge Hydroforming Process," Journal of Materials Processing Technology, 125-126, pp.821-825

PVP-Vol. 458, Computer Technology and Applications

PVP2003-1917

COMPUTER SIMULATION OF PIPELINE DEFORMATIONS ON THE BASIS OF DATA FROM AN INTELLIGENT CALIPER INSPECTION TOOL

Boris Blyukher
Health and Safety Department
Indiana State University
Terre Haute, IN 47809

Tadeusz Niezgoda, Jerzy Malachowski, Wieslaw Szymczyk
Department of Applied Mechanics
Faculty of Mechanics
Military University of Technology
Kaliskiego Street 2
00-908 Warsaw, Poland

ABSTRACT

The method of pipeline inspection data usage for needs of numerical analysis of technical condition of pipeline is considered. A real crude oil pipeline was taken into considerations to make numerical assessment of stress state in case of large deformations which were measured by an intelligent caliper inspection tool. The pipeline was rested on concrete supporting blocks in a boggy terrain. The tool detected very large deformations of the pipe in the areas of these supports which were caused by washouts. Data from the tool were processed into the format readable for MSC/PATRAN - graphical pre-processor of the computational system MSC/NASTRAN based on the Finite Element Method - FEM. Then a mesh of discrete model was generated by means of MSC/PATRAN.

INTRODUCTION

Pipelines for crude oil, natural gas and fuel transportation purposes are exposed on various factors and impacts, which may cause their damage, malfunction and failure. Factors that increase risk in exploitation of pipelines are: corrosion, dents, cracks in welds, ground shift changing location and shape of the centre line (including huge and sudden displacements caused by seismic phenomena or mining activities, explosions, changes in water logging of surrounding ground enabling washouts) and many others. Implementation of appropriate risk management strategies into pipelines exploitation is necessary for their integrity protection, keeping them in effective operational condition and avoiding ravages originated by severe defects and flaws.

Programs of regular inline inspections are made with the use of intelligent tools (so called "pigs"- pipeline inspection gauges) which are specially designed for these purposes. They detect flaws and damages when moving inside the pipe like a pistons pushed by pressure of transported medium without interruption of pipelines operation.

Providers of such services manufacture their own inspection tools which are equipped with specialized software for data collecting, transforming and analyzing. Their software packages are strictly dedicated for their own fleet of pigs. A typical fleet of pigs consists of cleaning and intelligent tools, such as caliper (geometry measuring – geo-pigs) and damage detecting tools. Intelligent tools are equipped with on board computer systems which are able to collect, autonomically transform, compress and store a huge amount of data possessed from gauges working with higher and higher resolutions. Damage detection may engage different methods: ultrasonic, magnetic flux leakage, magnetic remanence or eddy currents. There are examples of tools engaging a combination of few methods. The usable signals produced by gauges may also be affected and distorted in a specific way by pure mechanical contractions between deformed wall of the pipe and the supporting and sensor carrying system. These contractions may produce additional effects in visualization of the data that are unique for the tools made by particular manufacturers.

After inspection, an off-line analysis of data possessed from intelligent tools is carried out to identify damages and they are classified accordingly to their severity. Damages may be identified automatically with the use of software based on neuron networks. But the final interpretation and verification is always done by experienced analysts prepared for pipeline operators by inspection service provider, describe results of inspection and give estimation of maximum allowed operating pressure (MAOP) which is calculated in agreement with appropriate ASME standard. Such normalized and obligatory procedures may be completed by numerical analysis based on FEM (finite element method).

Commercial software packages based on FEM are used widely for developing of advanced constructions and for simulations of their response for complex states of exploitational loads. Despite they are not obligatory they can be used for computer aided analysis of operational condition of pipelines. Effective usage of such software in this field may be improved when the modeling process is automated and supported by available libraries of schemes or parametric models that were built and experimentally verified for typical cases of damages.

It is recommended to elaborate methodologies of model building and results analysis for typical pipeline engineering constructions and terms of their exploitation. They would be especially desired when a fast analysis is needed to undertake proper prevention or repair activities immediately after a severe damage detection.

The method of FEM model building should be verified experimentally. Examinations done with the use of pigs may be completed by the field, test stand, and laboratory research of stress state (e.g., with the method of Bärkchausen noise) and rheological properties of materials used for the pipeline building.

The use of associated numerical and experimental methods (with taking into consideration rheological phenomena) should provide more detailed assessment of technical condition of pipelines and ensure assistance for making decisions concerning the way of damage repairs (urgent or subsequent) which weigh strongly on expenses of pipelines exploitation.

Thus we have the situation where at the field of pipelines technical condition analysis two types of software differently originated may be used: the specialized one, which is dedicated for particular fleets of intelligent inspection tools and the commercially worldwide available one, based on FEM and destined for strain and stress analysis of wide range of other types of constructions. The problem is to use them together despite the FEM analysis is usually not demanded by standard regulations.

A method of numerical FEM model building and boundary conditions definition may be considered which advantage is the direct use of data collected by pigs. This should make it attractive to providers of pipeline inspection services because it should enhance capabilities of their own inspection tools and software in the most natural way.

EXAMPLE OF ANALYSIS

For the needs of the method elaboration the real incident was considered.

The pipeline of 526mm in diameter, transporting crude oil, was buried in a boggy terrain. It rested on concrete supporting blocks placed in distances of 12÷15m from each other. The caliper inspection tool (geopig) detected huge deformations of the pipe in areas of these supporting blocks. The change in diameter of the pipe reached ~15%. Deformations were expected to be plastic. An excavation showed out, that the indications from the inspection tool were proper. Damages had shape of dents and transverse folds. The center line of the pipe was strongly deformed too. The displacement measured on the top surface of the pipe between the supporting blocks was ~0.7m. It was caused by the ground overlay in conjunction with the washout under the pipe.

Design of the caliper pig was simple. It's task was to measure only diameters along the pipe that was needed to determine ability of safe pass of other, more tools such as ultrasonic or magnetic ones. The geopig measured six internal radii – so, it's circumferential resolution was only 60°. The longitudinal resolution was 0.05m. The tool was not able to detect deformation of the center line of the pipe. It had no pendulum to detect self rotation. In such circumstances, the capability of geometry inspection data usage for the pipe condition analysis with the use of FEM based software package was investigated.

FEM MODELING

The aim of the ongoing investigations is to recognize capabilities of direct usage of data collected by geopigs for FEM model mesh generation and proper loads establishment for highly deformed pipes. The MSC PATRAN and MSC/NASTRAN software was used for FEM models generation, calculations and post processing of results.

Preliminary calculations were made in 3 stages with the models using bar and 2D finite elements.
In the first stage the center line deformation was estimated with the use of the bar FEM model for the reason that the pig did not provide such information. The bar model was loaded by displacements occurring between the supporting blocks that were measured in the field. In the second stage the model engaging 2D finite elements was used to make more exact assessment of strain and stress state in the area of the supporting block selected on the base of calculations with the use of the bar model.

The collaboration of the pipe and the supporting block was simulated. For this purpose the FEM model with the use of 2D finite elements was built. It was the model non deformed pipe undergoing simulated loads of estimated ground pressure, weight of transported medium (crude oil), and deadweight of the pipe. The supporting blocks displacements were not taken into consideration because of data lack. Preliminary calculations showed out that the stresses exceed yield stress of the material. It suggested that rheological path should be recognized in future, more advanced calculations.

The results obtained from the second stage were used for the comparison with results from the third stage where the final FEM model geometry was built with the direct use of data taken from the geo-pig. The data possessed from the geo-pig were processed and recorded in the format readable by the graphical preprocessor MSC/PATRAN.

The estimated center line deformations were taken from the bar model from the first stage and combined with the indications of pipe surface damages collected by the geo-pig. They may be introduced in the way of direct measures in the field when the pipe is excavated. Of course, problems with the center line deformations are not present if the more advanced geo-pig is used and they are inherent part of the data collected by the intelligent tool itself.

Since the data from the geo-pig were read by the MSC/PATRAN, the six generating lines in longitudinal direction (that reflects the pig sensors trajectories) were established with the use of b-spline technique that is implemented in PATRAN options of curves generation. It was assumed that the pig did not rotate in the pipe when passed along the investigated area. Then, on the base of these lines, the surface of the pipe was interpolated and generated in circumferential direction with the use of the same technique of b-spline. In spite of the fact that the resolution of the pig in circumferential direction was very low, the final FEM mesh of strongly deformed pipe was still reasonable. For the mesh building there were 8 node 2D elements used to provide more smooth surface.

The essential problem for making calculations with the use of the FEM model is to establish proper boundary conditions of loads. The initial configuration of model nodes is not known. The history of deformations is not known too (the pipeline is 30 years old and the previous inspection was made 3 years earlier).

To illustrate the problem we can consider two different ways of kinematical load generation when the nodes of the model relocate from the initial (non deformed) configuration to the final one (represented by the data from the geo-pig).

Let the initial configuration be the circle of nominal pipe diameter. In the case A, nodes relocate straightly along radii of the nominal cross section. In the case B nodes relocate still remaining in the same proportional location on the circumference. These methods of kinematical load simulation produces results quite different in qualitative sense and not complying with real state of strain and stress. Especially, the circumferential component is strongly distorted in both of the cases. In areas of transverse folds the axial component is distorted too.

The more convenient method of kinematical load generation is developed in ongoing works. It is based on simulation of contact between two surfaces. The former one (master) is a dummy structure and behaves as described in the case A. The nodes of the master surface reach positions accordingly to the data from the geo-pig providing appropriate constrain of the latter surface (slave), which is proper surface of the modeled pipe. It is assumed that there is no friction between these surfaces and the nodes of the slave surface can freely move in the circumferential direction. It is expected that results obtained with such kinematical load simulation should be more real. Then the rheology path in calculations may be activated. For such a way of calculations the more convenient software package is needed (i.e., MSC MARC), which is directly oriented on huge plastic deformations and complex contact problems analysis.

CONCLUSIONS

The method of building of FEM model of pipeline fragment is proposed for a case of huge geometric deformations of a pipe. The main idea of the method is the direct use of data collected by the intelligent inspection tool (so called “geo-pig”). The geo-pig is destined for ovality of pipes measuring to ensure safe passage more advanced intelligent tools passages. There were taken into considerations deformations of the real pipe which was buried in a boggy terrain. The areas of huge geometrical distortions that were detected by the geo-pig were caused by washouts between supporting concrete blocks. The indicated damages had shape of dents and transverse folds. For the reason that the pig did no measures of center line deformations it is proposed to make additional calculations with the use of simplified bar FEM model loaded by displacements between supporting blocks measured at the field when an excavation was made.

The centerline deformations interpolated with the use of bar model were combined with those measured by the pig. Then the FEM mesh was generated with the use of MSC/PATRAN on the basis of data from the geo-pig. Then the strain/stress state calculations were possible with the use of MSC/NASTRAN. The base problem in ongoing works is to establish proper boundary conditions when the history of damages origination is not known. The two extreme ways of kinematic load simulations showed out the need of elaboration of more convenient method of load definition which would be able to provide obtaining reliable results. The final suggestion is to engage in a specific way an option of contact problems simulation. It is also recommended to use another FEM software package which is specially oriented on huge plastic deformations analysis and complex contact problems simulations (MSC MARC).

BIBLIOGRAPHY

T.Niezgoda, J.Malachowski, A.Derewonko, W.Szymczyk, 1999. The role of residual stress in exploitation of pipelines. In polish, conf. proc., II Polish Technical Conference on Risk Management in Pipelines Exploitation, Plock, Poland,

T.Niezgoda, J.Malachowski, A.Derewonko, W.Szymczyk, 2000. Computer simulation of pipeline deformations on the basis of data from an inspection tool. In polish, conf. proc., III Polish Technical Conference on Risk Management in Pipelines Exploitation, Plock, Poland,

T.Niezgoda, J.Malachowski, W.Szymczyk, 2001. Analysis of operational condition of barrier-tapping set of WC DN700 gas pipeline. In polish, conf. proc., IV Polish Technical Conference on Risk Management in Pipelines Exploitation, Plock, Poland.

Mackerle J., 1996. Finite elements in the analysis of pressure vessels and piping – a bibliography (1976-1996), Int. J. Press. Vess. Piping; 69:279-339

Mackerle J., 1999. Finite elements in the analysis of pressure vessels and piping – a bibliography (1996-1998), Int. J. Press. Vess. Piping; 76:461-485

Mackerle J., 2002. Finite elements in the analysis of pressure vessels and piping – a bibliography (1999-2001), Int. J. Press. Vess. Piping; 79:1-26

PVP-Vol. 458, Computer Technology and Applications

PVP2003-1918

Enhancing Probabilistic Risk Assessment Methodology for Solving Reactor Safety Issues

by

F. L. (Bill) Cho
RiskSolver Communications - Consultants
Irvine, CA U.S.A.
Phone/Fax: (949) 654-1734
E-mail: fuliong@yahoo.com
http://groups.yahoo.com/group/Peers-of-RiskSolver/

INTRODUCTION

This paper reveals a paradigm of analyzing the consequential effects of severe nuclear reactor accident, radionuclides fraction and source terms release, that will influence the MACCS2 codification [1], by coupling with the results of SAPHIA-PSA Levels 1& 2 quantification process [2], MELCORE [3], STCP [4], PST [5], and XSOR [6]. Those codes are mutually exclusive and useful. However, it lacks of the closed interface and linkage for addressing Plant Damage States (PDS), Severe Accident Sequences, and Risk Consequence. Thus, it is imperative to formulate the consistent baseline information for MACCS2, PSA Levels 1, 2 and 3, and then linking to a new algorithm of NCM.

The principal task is to generate the input of radionuclides fraction and source terms release for MACCS2 by using of MELCOR or other thermal hydraulic codes, which are capable of computing the transport of fission products release for specific plant damage states and/or sever accident event sequences. Through above working processes, we can identify or ascertain the root causes of all probable risk outcomes, which could likely be initiated and/or caused by single or multiple fatal failures and damages of some safety-related Structure, System and Component (SSC), Equipment, and/or Man-made Error/Fault. Subsequently, those affected PDS or Accident Event Sequences could induce the degradation of RPV and Containment failures, if the whole process not been controlled and managed.

A newly developed algorithm, NCM, as intended, will facilitate the necessary links for risk quantification process for the above inter-related codes. Hence, the working procedure of NCM will associate with those degraded PDS and Accident Event Sequences by computing the λ-Indexes (Cho's Weighted Factors), represented by λ_a, λ_b, $\lambda_c \ldots \lambda_g$, for the seven blocks of Normalized and Combined State Variable Terms (NCSVT), which designate the entire nuclear reactor accident phenomenological process. That definition of those NCSVT blocks and the λ-Indexes computation will be illustrated at a later section of this paper.

UNDERSTANDING THE NATURES OF SEVERE NUCLEAR ACCIDENTS

The performance of a Probabilistic Risk Assessment (PRA), Level 2 Analysis for Large Early Release Frequency (LERF), as stated in the ASME Standard for PRA for Nuclear Power Plant Application [7] and NUREG/CR-6595 [8], is a formidable task, if we fail to understand the characteristics of severe accident phenomena. The principal concerns are that we have to deal with multi-disciplinary tasks and to understand various kinds of complex phenomenological scenarios involved.

Starting with the PRA Level 1 risk assessment, we must consider that the threshold of Core Damage Frequency (CDF) from the analysis must be less than the value of 1.0E-06/year to assure safe reactor operation. By using this CDF value as the standard, the likelihood of an accident leading to a core melt damage sequence (CMD) will be minimized. It is universally accepted that each nuclear plant must meet minimum plant operability standards and maintain procedures for systems, equipment and components. Additionally, each nuclear plant must maintain human performance requirements and operate under the constraints of the Reliability Centered Maintenance (RCM) Program. These operational considerations also act to minimize the probability of a CMD event.

Historically the consideration of core melt effects and its consequences has prompted the development of Severe Accident Technology by the various NRC research groups, National Laboratories, working with the Industry Core Rulemaking (IDCOR) committees, have addressed many issues resulting from severe core damage scenarios, core melt progression phenomenology, and in-vessel and ex-vessel accident sequences.

Thermal-hydraulic analyses for studying the effects of the reactor core melt phenomena and its severe accident consequences (source term and offsite release consequences), have been extensively undertaken by the various analysis codes such as MELPROG/TRAC, CORMLT, MAAP(IDCOR), SCDAP/RELAP5, MELCOR and MARCH (BMI).

At present, there still exist many technical uncertainties in these studies. A discussion of those uncertainty and sensitivity issues can be found in references [8, 9, 10, and 11]: some of the areas of uncertainty remaining are to be studied.

E ROLE AND FUNCTION OF PSA LEVELS 1, 2 AND 3 EVALUATION, ASSOCIATED WITH RADIONUCLIDES FRACTION AND SOURCE TERMS RELEASE, WHICH SHALL BE DERIEVED FROM A SEVERE NUCLEAR REACTOR ACCIDENT EVALUATION

The nature of severe accident phenomena studies contains deterministic and knowledge uncertainties in analyzing damage mechanisms, and other physical phenomenon. It has been a continuing United States national effort to resolve such issues. However, it remains difficult to reach any comprehensive and conclusive results. With the development of the ASME PRA methodology [7], we have begun to learn a new procedure for PRA Levels 1 and 2 analyses and quantification. However, we still must understand the mechanics of linking Plant Damage States (PDS) and its associated severe accident sequences.

Our current knowledge of containment response can be credited to the U.S. NRC leadership and management, resulting in the characterization of the phenomena of containment failure modes for BWR and PWR nuclear power plants. We are now in a position to adopt the PRA/PSA methodology to link the radionuclides fraction and source terms release from thermal-hydraulic analyses for PRA/PSA Levels 1, 2 and 3, and then linking to MACCS code. The overall strategy of implementing risk assessment process is listed below:

PSA Level-1 Results: Core Damage Frequency or Plant Damage State Frequencies; (SAPHIRE)

PSA Level-2 Results: Containment Failure Frequency, Conditional Containment Failure Probability, Large Early Release Frequency (LERF-L2 Quantification); (SAPHIRE), Parametric Source Term (PST), (Licensee IPE)

PSA Level-3 Results: Identifying and Determining the Safety of Radionuclides Release Level for Assessing Its Health Effects, Environmental Impact, and Financial Liability for the Public Safety and Concerns; (MACCS2)

In order to understand how the results of PSA Levels 1, 2 and 3 evaluations could be affected by the radionuclides fraction and source terms release, we must resolve the following reliability and safety issues, which are encountered with each nuclear power plant, as stated below:

1. What are PRA/PSA Level 1 attributes that contribute or impact Levels 2 and 3 analyses?

2. What are the interactions between frontline systems and its support systems for supporting nuclear reactor safety operations?

3. What are the risk mitigation strategies and measures of ECCS safety functions, so that it can prevent a core melt damage?

THE DETERMINATIONS OF THERMAL HYDRAULIC ANALYSES AND THE UNDERSTANDING OF PHYSICALLY CHAOTIC PHENOMENA, PROMPTING RADIONUCLIDES FRACTION AND SOURCE TERMS RELEASE TO THE ENVIRONMENT

The Sandia National Laboratories studies on Source Terms Analysis [9] had identified the significant State Variables, which can define the characteristics of radionuclides release process and transport phenomena during the Reactor Pressure Vessel (RPV) and Containment failures.

With an understanding of the physical conditions in a severe accident phenomena, we can select the most important representative elements of the State Variables and their constitutive relations to reflect the whole severe accident phenomena, expressed by five explicit equations (Equations 1 through Equation 5 below). These equations are primarily from Sandia's source term studies published in the NRC topical reports. We believe that beginning with these five governing equations a methodology can be developed to simplify the assessment of the consequences of radionuclides release to the environment, resulted from RPV and Containment failures. The goal is to derive a meaningful synthesis of the five fundamental equations into one integral expression (Equation 6) to measure the total likelihood of release consequence to the environment.

These fundamental equations are as follows:

Equation #1: for early per RCS iodine release:

(ST1) = [FCOR * FVES * FCONV/DFE] + DST

For the late release or CCI iodine release:

(ST2) = [(1 – FCOR) * FPART * FCCI * FCONC/DFL] + FLATE + LATE1

Where:

STi = fraction of the core iodine in the Reactor Coolant System (RCS) release to the environment; for accident regime i; where, i = 1 for early release state, i = 2 for late release state;

FCOR = fraction of the iodine in the core released to the vessel before Vessel Breach (VB);

FCONV = fraction of the iodine in the containment from the RCS release that is released by the containment in the absence of any mitigating effects;

FVES = fraction of the iodine released to the vessel that is subsequently released to the containment;

DFE = decontamination factor for RCS release (sprays, etc.);

DST = fraction of core iodine released to the environment due to direct containment heating at vessel breach;

FPART = fraction of the core that participates in the Core Concrete Interaction (CCI);

FCCI = fraction of the iodine from CCI released to the containment;

FCONC = fraction of the iodine in the containment from the CCI release that is released from the containment in the absence of any mitigation effects;

DFL = decontamination factor for late release (spray, etc.);

FLATE = fraction of core iodine in the RCS that is revolatilized and released late in the accident;

LATE1 = fraction of core iodine remaining in the containment that is converted to volatile forms and released late in the accident.

Some of the above State Variables will be expressed later by Equation 6 by a ratio of the specific accident scenario value to its baseline value.

Equation #2 - The fraction of radionuclides released:

FCONV = mVout/mVin

Where:

mVin = mass of radionuclide (or class) released from the vessel to containment atmosphere at or before VB;

mVout = mass of a radionuclide (or class) released from the vessel to containment atmosphere at or before VB that is subsequently released from containment.

Equation #3 - The calculation of the late release of iodine in volatile form:

LATE1 = XLATE* [{FCOR*FVES + (1-FCOR)*FPART*FCCI} - ST - STL + FLATE]

Where:

XLATE = fraction of iodine in the containment late in the accident that assumes a volatile form and is released to the environment;

FCOR = fraction of iodine in the core released to the vessel before VB;

FVES = fraction of the iodine released to the vessel that is subsequently released to the containment;

FPART = fraction of the core that participates in the CCI;

FCCI = fraction of the iodine from CCI released to the containment;

ST = fraction of the core iodine in the RCS released to the environment;

STL = fraction of core iodine released to environment.

Equation #4 - The conversion of each isotope into an equivalent amount of I-131:

$EQN_k = CF_k * I_k * STN_k * \exp[-\lambda_k \ (TN + DTN/2)]\ EQ = \Sigma_k EQ1_k + \Sigma_k EQ2_k$

Where:

N = 1 for early release and N = 2 for the late release;

EQN_k = the equivalent amount of I-131 for isotope $_k$ for release N;

CF_k = the isotope conversion factor for isotope $_k$;

I_k = the inventory of isotope $_k$;

STN_k = the release fraction of isotope $_k$ for release N;

λ_k the decay constant for isotope $_k$;

TN = the time of the start of release N, and;

DTN = the duration of release N.

Equation #5 - the chronic fatality weight for each isotope:

$CFW_k = I_k (ST1_k + ST2_k) * (ELCF_k + CLCF_k)/R_k$

Where:

I_k - the inventory of isotope $_k$;

N = early release (1) or late release (2);

STN_k - the release fraction of isotope $_k$ for release N;

CFW_k = the chronic fatality weight, in latent cancer fatalities, for a release of an amount $I_k (ST1_k + ST2_k)$ of isotope $_k$;

$ELCF_k$ = the number of latent cancer fatalities due to the early exposure from a release of an amount R_k of isotope $_k$;

$CLCF_k$ = the number of latent cancer fatalities due to the late exposure from a release of an amount R_k of isotope $_k$;

R_k = the amount of isotope $_k$ released in the MACCS calculation per MACCS2 Code [1], used to determine $ELCF_k$ and $CLCF_k$.

AN EMERGING PARADIGM, EXPRESSED BY THE EQUATION #6 FOR ASSESSING "NUCLEAR CONSEQUENCE MEASURE (NCM)"

An attempt is made to establish a quantifiable methodology to assess the overall risk consequence of radionuclides released to the environment for PSA Levels 2 and 3 considerations. By considering key components of the above five fundamental equations, a simplified equation is proposed to formulate a Nuclear Consequence Measure (NCM) for the consequence determination for any probable accident sequence. This NCM will reflect the severity of any in-site and/or offsite consequences following a radionuclides release from the existing nuclear power plant. A larger NCM indicates a more severe offsite consequence than a lower number.

The proposed Equation #6 is expressed as follows:

$$NCM = \{(st1/ST1)\lambda_a + (st2/ST2)\lambda_b + (fconv/FCONV)\lambda_c + (late1/LATE1)\lambda_d + [\Sigma_k(eq1_k/EQ1_k)]\lambda_e + [\Sigma_k(eq2_k/EQ2_k)]\lambda_f + [\Sigma_k(cfw_k/CFW_k)\lambda_g]\}$$

Where:

NCM = for assessing the in-site and offsite consequences measure by seven Normalized and Combined State Variables Terms (NCSVT) blocks.

Lower Case Terms = measured, calculated or estimated values;

Upper case Terms = baseline value.

λ_i = Seven λ-Indexes (Cho's Weighted Factors) for quantifying the Normalized and Combined State Variable Terms (NCSVT) blocks in Equation #6. The characteristics of those $\lambda_{i=a,b,c...g}$ shall be discussed at the subsequent section for those seven NCSVT blocks, activated by any accident phenomena process at the times of: "Early or Late Release before Vessel Breach [8]" states.

Each block of NCSVT in Equation #6 consists of a scenario-specified calculated, measured or observed value (lower case variable names), divided by its respective baseline value (capitalized variable names). In addition, each term is further multiplied by the "$\lambda_{i=a,b,c,d..g}$ Indexes" respectively for calculating a mixed representation and consideration of reliability and safety for the entire nuclear plant facility.

Each the λ-Index can express the plant degradation process in all Plant Damage State (PDS) or Accident Event Sequences, associated with the physical damage states of Structure, System, Component (SSC), Equipment, and/or Man-made Fault/Error, which shall be expressed by an algorithm of Successive Failure Probability Ratio Multiplication (SFPRM) between the current and previous values.

THE CHARACTERIZATION OF λ-INDEXES (Cho's Weighted Factors) FOR THE NORMALIZED AND COMBINED STATE VARIABLES TERM (NCSVT) BLOCKS IN EQUATION #6

The characteristic of λ-Indexes (Cho's Weighted Factors) has an implicit time-dependent representation, which can express a continued series of the "Successive Failure Probability Ratio Multiplication (SFPRM)" computation, by tracing any change of the reliability value change between the new (current) and old (previous) from those affected PDS and failed Accident Event Sequences in SSC, Equipment, and Man-made Errors/Faults. Therefore, it is applicable to monitor the change of the plant time history management by tracking any successive reliability changes for the daily, weekly, monthly, semi-annual or annually operation, and concurrently for following up the surveillance or inspection activities to determine any degree of the degraded conditions of malfunction , damage, or loss of system and component safety function, etc.

The general form is expressed as follows:

$$\lambda_{PDS=1,2...N} \text{ or } \lambda_{AES=1,2..m} = \int_{\theta=1...\theta} [FP_\theta(new)/FP_\theta(old)\bullet\bullet\bullet]\bullet(RF_\theta)\bullet(CDF_\theta)d\theta$$

Where:

" $_{\theta=1,....\theta}$": It denotes the certain elements # ($_\theta$) within its failed PDS, or Sequence Cut Sets, which contain the base event failure probability being changed for the participated Accident Event Sequence, resulting from the degraded or damaged cause in Structure, System, Component (SSC), Equipment and Man-made Fault/Error, etc;

RF_θ : Reduction Factor for the individual λ-Index will be required by multiplying a decimal percentage factor of its element CDF contribution, whenever the failed element (PDS) using a value of "1.0" as a new Failure Probability for replacing the previous (old) value.

CDF_θ: a CDF for PDS, contained the "$_\theta$" element, or a CCDP for its Sequence Cut Set Events group, contained the failed basic event for the Accident Event Sequence.

The necessary computational features for RF_θ and CDF_θ will be demonstrated in the subsequent section of An Illustrated Example..........

Since the objective of risk assessment is to evaluate whole phenomenological process of a severe nuclear accident, we have to compute a total summation of the $\lambda_{NCSVT=a,b,c,d....g}$ (Cho's Weighted Factors), that can be expressed as shown below, by treating each block of Equation #6 respectively, per the extent of each PSA Levels 1,2 and 3 evaluation:

For PDSs:

$$\lambda_{NCSVT=a,b,c...g} = \Sigma_{NCSVT=a,b..g} \{\Sigma_{pdsn=1..n}\{\int_{\theta=1...\theta} [FP_\theta(new)/FP_\theta(old)\bullet\bullet\bullet](RF_\theta)\bullet(CDF_\theta)d\theta\}_{NCSVT=a,b...g}\}$$
$$\bullet \text{ (1.0/The Limit Value, Based on the Plant CDF);}$$

For Accident Event Sequences:

$$\lambda_{NCSVT=a,b,c...g} = \Sigma_{NCSVT=a,b..g} \{\Sigma_{AES=1....m}\{\int_{\theta=1.....\theta} [FP_\theta(new)/FP_\theta(old)\bullet\bullet\bullet]\bullet(RF_\theta)\bullet(CDF_\theta)d\theta\}_{NCSVT=a,b.....g}\} \bullet \text{ (1.0/The Limit Value, Based on the Plant CDF;}$$

Where:

Each λ-Index (λ_a, $\lambda_{b...}$ and λ_g) represents the individual sum of all Successive Failure Probability Ratio Multiplication (SFPMR) for the respective block of Normalized and Combined State Variables Terms (NCSVT), so that we can assess the PSA Levels 1, 2 and 3 evaluation. Then, the Limit Value of Plant Base CDF shall be referred to the latest Plant Base Modified CDF, being reported in the Licensee's IPE Reports, which had submitted to the NRC.

The algorithm of Successive Failure Probability Ratio Multiplication (SFPRM) can fully express the process of physical changing state of the failed PDS or Accident Sequence through the means of tracking down the path of RPV and Containment failure progression, occurred at the times of "Before, At and Late" Vessel Breach, and then progressed to a Core Melt Damage (CMD) and probably might reach the total damage of RPV/Containment.

Since the determination of λ_a (the first block of NCSVT in Equation #6) basically covers the scope of PSA Levels 1 and 2 quantification, thus we must first comprehend about the underlying root causes of CMD phenomenological occurrence, and then solve the following open tasks, as been discussed in the preceding sections:

1. What are PRA/PSA Level 1 attributes that contribute or impact Level 2 analyses?
2. What are the interactions between frontline systems and its support systems for supporting nuclear reactor safety operations?

3. What are the risk mitigation strategies and measures of ECCS safety functions to prevent a core melt damage?

Now, it becomes apparent that by having well thought processes for above tasks, it will help us to formulate a strategy of linking PSA Levels 1 and 2 results into MACCS2 code. Thus, the most important concern is that whether or not both PRA/PSA and non-PRA/PSA practitioners be able to trace any latest information of radionuclides fraction and source terms release from MELCOR or other thermal-hydraulic codes for performing PSA Level 2 Analysis from the Licensee's latest Individual Plant Examination (IPE) Reports, which had submitted for the USNRC?

AN ILLUSTRATED EXAMPLE FOR QUANTIFYING AND QUALIFYING THE λ_a INDEX AND NCM (NUCLEAR CONSEQUENCE MEASURE), BASED ON THE LICENCEE'S IPE-RELATED PSA LEVELS 1, 2 AND 3 DATA

Part 1: For PSA Level 1Consideration:

Quantification Process: λ_a (Cho's Weighted Factor) for the block #1 of NCSVT in Equation #6:

Step #1 - Computing the Changes of Failure Probability for All Affected PDS: for the following Cases 1 and 2, which had been reported and occurred at the same time instant:

Case 1: Assumed that a "PDS #23 (B14S)" had a 0.54% (1.67E-07) contribution to the plant CDF:

PDS	A Series of Events
#23- (B14S):	LOOP; AC141; AC142; SX; SBO; XCC

A failure of XCC , indicated "core uncover prior to 24 hrs w/RCS cool down being designated as "SEALLOCA"; which was excerpted from a plant IPE;

A conditional failure frequency of "SEALLOCA" had been reported as "3.22E-02", resulting from XCC failure;

Thus, computing the failure probability for the "new" and "old" damage states as follows:

Old failure probability of XCC: $FP_{\#23}$ (old)= 3.22E-002;
New failure Probability of XCC: $FP_{\#23}$ (new)= 1.00E-000; assumed as a worst case

Case 2: Assumed that there were another set of the affected PDS, which were: PDS #55 (SE6K), PDS #74 (SL6E), and PDS #85 (L14E), that being reported, and had contributed a sum of 0.3.% (1.24E-07) to the plant CDF jointly by three PDS from the previous state, and their present state been affected by the damages of FC and CI as shown below:

PDS	A Series of Events
#55-(SE6K):	SLOCA; ESF-A; ESF-B; CCP; SIP; AFW; ORF; FC
#74-(SL6E):	SLOCA; ESF-A; ESF-B; CCP; SIP; AFW; LPI; FC; CI
#85-(L14E):	LOOP; AC141; AFW; CCP; CI

A series of the multiple damage states were reported as follows:

(1) - PDS #55: FC (All FAN COOLERS - TRAIN A & B OR SX FAILED) with a Failure Probability (FP) equals "1.0"

(2) - PDS #74: CI (CTMT ISO FAILS - ALL POWER AVIL - SIS FAILED, CS ON) with a Failure Probability (FP) value of "2.31E-01";

(3) - PDS #85: CI (CTMY ISO FAILS - ALL POWER AVAIL, SIS SUCCESS, CS ACTIVATED) with a Failure Probability (FP) value of "8.01E-03.

Step #2 - Computing the Increase of Failure Probability Ratio from those affected PDS by its respective Reduction Factor, since a worse Failure Probability being assumed to be the value of "1.0":

- $\lambda_{\#23}$ for PDS #23 = (1.0/3.22E-02)•0.0054 = 0.168
- $\lambda_{\#55}$ for PDS #55 = (1.0/1.0) = 1.0
- $\lambda_{\#74}$ for PDS #74 = (1.0/2.31E-01•0.003 = 0.0129
- $\lambda_{\#85}$ for PDS #85 = (1.0/8.01E-03)•0.003 = 0.372

Step #3 - Computing the Increase of Plant CDF by the individually damaged PDS:

	Old CDF_θ	Per Step #2	Step #3
• - by	($\lambda_{\#23}$ * 1.67E-07) =	(0.1680 * 1.67E-07) =	2.67E-08
• - by	($\lambda_{\#55}$ * 5.76E-08) =	(1.0000 * 5.76E-08) =	5.76E-08
• - by	($\lambda_{\#74}$ * 3.84E-08) =	(0.0129 * 3.84E-08) =	4.95E-10
• - by	($\lambda_{\#85}$ * 2.75E-08) =	(0.3720 * 2.75E-08) =	1.01E-08

A Sum of the Plant CDF Increased = 9.49E-08,

Step #4 - Calculating the value of λ_a, affected by all above degraded PDS for NCM:

: Quantifying the Index of λ_a based on the Plant's System/Component Reliability, not yet accounting the possible effects by the releases of Radionuclide Fraction and Source Terms Release, since there was no report of "Vessel Breach (VB)"being occurred. However, we like to see how the Licensee's IPE results had assessed their concerns on the possible release issue: this can be assessed by λ_a for the block #1 of NCSVT in Equation #6:

λ_a = 9.49E-08 /3.09E-05 = 0.00307 < 1.0 ; an insignificant case , which had no radionuclides released

Where: "3.09E-05" which was available data for the Plant Base CDF from an existing nuclear power plant, considered this as the basis of the Limit Value.

Since there will likely be no CMD Phenomena, so that the In-vessel and Ex-vessel Analyses are not needed. Thus, there will be no any further off-site consequence by Radionuclids Fraction and Source Terms Release to the environment. But we still check the Licensee's IPE assessment.

Step #5 - Tracing other available source of Radionuclides Fraction and Source Terms Release, which had been determined by the Licensee in their IPE Report to the NRC.

Based the above evaluation of λa, which is one of seven NCSVT blocks in Equation #6, we may look into any possibility of Radionuclides Fraction Release in the Reactor Coolant System. As being studied, the Licensee's IPE report had verified its Reactor Coolant System. At present, there were no baseline information available for NCSVT block #1 for comparing with the Licensee's findings as listed below:

PDS	I	CS	TE	SR
#55-(SE6K)	9.98E-01	9.6E-04	9.6E-04	8.0E-010
#74-(SL6E):	7.43E-03	3.0E-06	3.0E-06	6.0E-08
#85-(L14E):	1.4E-03	5.0E-06	5.0E-06	9.0E-09

; no fraction of radionuclides release for RU, LA, CE,and BA.

ADDITIONAL CONSIDERATIONS FOR ENHANCING THE PART 1 OF ABOVE PSA LEVEL 1 QUANTIFICATION PROCESS

Important Note #1:

As intended, the λ-Indexes shall be used for weighing some indefinite failure probability input data and their baseline information from the existing nuclear power plant, thus requiring to use the algorithm of Successive Failure Probability Ratio Multiplication (SFPRM) for tracking down between the present (new) and previous (old) values of failure probability of the elements for the failed or damaged PDS or Accident Event Sequence, and then to be divided by the baseline data. It is worthy to note that the end result of λ_a has an inverse relation between the nuclear power plant reliability (failure probability) and the safety level of Radionuclides Fractions and Source Terms Release.

Important Note #2

The essence of above proposed λ_a-Index computation is cited on the observation or surveillance of the factual degradation phenomena and the failure-damage occurrences in SSC, Equipment and Man-made Fault/Error during the operation of nuclear power plant facilities.

Part 2: For PSA Levels 2 and 3 Considerations

We will still adopt the same quantification and qualification processes, shown in Part 1 above. However, we must focus on the failures of specific SSC, equipment, human performance reliability (errors and faults) from those affected PDS and Accident Sequence by associating with the failure scenarios and damage information, being considered and used in ASP/PST and XSOR [2, 3, 4, 5 and 6] in order to establish the necessary links between NCM and PSA Levels 1, 2 and 3 Results consistently per the following concerned areas::

1. Containment Failure Mode
2. Containment Heat Removal System Available
3. Core Concrete Interaction
4. Mode of Vessel Breach
5. SGTR
6. Amount of Oxidation
7. Containment Failure Size
8. Time of Core Damage
9. Others not yet being specified

Basically, PST [5] had used above damage domains to establish the characteristics of Source Term Vector (STV), Source Terms Group (STG), Source Term Event Tree (STET), and Risk Event Tree (RET) in conjunction with the NRC Accident Sequence Precursor (ASP). All of those source information can be integrated and interfaced with NCM, while adopting the same methods of qualification and quantification, as shown in Part 1. Thus, all risk-based information are accountable and consistent between the baseline and quantified values, regardless of the analyst's preference of using their adopted computational code and method.

CURRENTLY APPLICABLE BASELINE INFORMATION FOR DETERMINING THE RADIONUCLIDE RELEASE AND SOURCE TERMS, AVAILABLE FOR MACCS2 , PSA LEVELS 2 AND NCM CODIFICATIONS

Recently, the USNRC had issued a draft regulatory Guide DG-113, "METHODS AND ASSUMPTIONS FOR EVALUATING RADIOLOGICAL; CONSEQUENCES OF DESIGN BASES ACCIDENTD AT LIGHT-WATER NUCLEAR POWER REACTORS", in which defined the transport phenomena and the governing equations for radionuclides fraction release and source term determination, and also stated the Release Fractions for DBA LOCA per Table 1 and for Non-LOCA Fraction of Fission Product Inventory in Gap per Table 2.

Furthermore, the USNRC Regulatory Guides [11, 12, 13, 14] had addressed nine (9) DBA for evaluating radiological consequences for PWR and PWR nuclear plants for:

- - LWR Loss of Coolant Accident
- - Fuel Handling Accident
- - BWR Rod Drop Accident
- - BWR Main Steam Line Break Accident
- - PWR Steam Generator Tube Rupture Accident
- - PWR Main Steam Line Break Accident

- - PWR Locked Rotor Accident
- - PWR Rod Ejection Accident

As can be seen from above list, those DBA were not discussed about whether or nor it had been occurred on or before or within the Core-Melt-Damage (CMD) progressions of Reactor Vessel and containment failures. Besides, the NRC had not discussed about whether or not those accidents could associate with any failed PDS Accident Event Sequences, any of ECCS safety functions, that could affect the increase of CDF beyond the Plant Limit, thus prompting a CMD, and then releasing the fractions of Radionuclides and Source Terms to the environment.

As I see now, the next level of the PSA Level 3 consideration, we have to rely on MASS2 in conjunction with the codes of Parametric Source Term, PST [5] and XSOR [6], which are capable of modeling the transport of radionuclides release and source terms from the core, through the reactor vessel, through the containment, and to the environment. At present, PST does provide a proper computational method for source term determination, but requiring extensive In-Vessel and Ex-Vessel modeling with the time-dependent progression. The merit and value of PST[5] can be coupled with SAPHIRE [2] /ASP Source Term Vector (STV) information by linking with the PSA Levels 1 and 2 Accident Event Sequences, so that it enables us to determine the fraction of radionuclides release and source term determination. The whole linking process for making a viable "Bridge Event Tree (BET)" for the PRA Levels 2 and 3 quantification is still an extremely rigorous, complex, and tedious effort.

Therefore the creation of Nuclear Consequence Measure (NCM) algorithm is necessary, so that it can integrate with all viable information of the joint Reliability and Safety considerations. As discussed in the preceding section, the fundamental of NCM paradigm is represented by an integral synthesis expression of the Equation #6, which associates with seven (7) Normalized and Combined State Variables Terms (NCSVT) blocks by multiplying λ-Indexes (Cho's Weighted Factors) by the means of Successive Failure Probability Ratio Multiplication algorithm. Thus, this newly developed paradigm can compile and evaluate all risk-based information from the sources of the existing nuclear plant operations and management activities, particularly enabling us to measure the degradation process, occurred in any kind of the time frame through a continued and concerted examination for those affected PDS or degraded Accident Event Sequences, if required.

CONCLUDING REMARKS

So far, this paper presents a viable means for risk quantification and qualification by using the above discussed paradigm and algorithm of computing the λ-Indexes and NCM, interfaced with the PSA Levels 1, 2 and 3 [2 &3] results, and associated with all applicable thermal-hydraulic codes [3, 4, 5, & 6], whatever being used.

Thus I have demonstrated that the importance of MACC2 Input data have to be accountable and to have a complete transparent information throughout all quantification and qualification processes, so that we will be enabling to treat all pertinent risk-based information consistently, but not to be separated or de-coupled. It is important that we ought to link and interface with all applicable data, so that we are qualified to perform a truly integrated risk evaluation. By virtue of the NCM algorithm, we will be able to link all available data sources, while we are dealing with both reliability and safety issues concurrently.

The proposed methodology could reduce a great deal of times and efforts needed to quantify the results of source terms and radionuclides fraction release. It is my hope that the "NCM, Coupled with λ-Indexes Computation" will soon become a most practical process for risk quantification and qualification for assessing the reliability and safety issues for the nuclear energy generation communities to serve as a worldwide application technique.

REFERENCES

[1] - U.S. Nuclear Regulatory Commission, Code Manual for MACCS2, NUREG/CR-6613, Volumes 1,and 2, April 1998.

[2] - Idaho National Engineering and Environmental Laboratory, System Analysis Programs for Hands-on Integrated Reliability Evaluation (SAPHIRE), Version 6.7, July 2000

[3] - S. E. Dingman, C.J. Shaffer, A.X. Payne, and M.K. Carmel of Sandia National Laboratories, MELCOR ANALYSIS FOR ACCIDENT PROGRESSION ISSUES/NUREG/CR-5331, January 1991

[4] - J. A. Gieseke, P. Cybulskis, H. Jordan, K.W. Lee, etc. of Battelle's Columbus Laboratory, SOURCE TERM CODE PACKAGE- USER'S GUIDE-STCP, MODE 1/ NUREG/CR-4587, July 1986

[5] - Idaho National Engineering Laboratory (INEL), The Parametric Source Term (PST), INEL-96/0308, October 1996.

[6] - H. N. Jow, W. B. Murfin, and J. D. Johnson, XSOR Code Users Manual, NUREG/CR-5360, November 1993

[7] - American Society of Mechanical Engineering, "American National Standard for Probabilistic Risk Assessment for Nuclear Power Plant" Application, Rev 14A, May 11 2001.

[8] - U.S. Nuclear Regulatory Commission, An Approach of Estimating of Various Containment Failure Modes and Bypass Events, NUREG/CR-6595, January 1999.

[9] - U.S. Nuclear Regulatory Commission, Reactor Risk Reference Documents, , NUREG-1150, Volumes 1, Volume 2, and Volume 3 February 1987.

[10] - U.S. Nuclear Regulatory Commission, Alternative Radiological Source Terms for Evaluating Design Basis Accidents at Nuclear Power Reactors, Regulatory Guide 1.183, July 2000.

[11] - R.L. Iman, J.C. Helton, and J.D. Johnson, PARTITION: A Program for Defining the Source Term/Consequence Analysis Interface in the NUREG-1150 Probabilistic Risk Assessments, NUREG/CR-5253, SAND86-2940, November 1989.

[12] - USNRC, Assumption Used for Evaluating the Potential Radiological Consequences of a Loss of Coolant Accident for Boiling Water Reactors, Regulatory Guide 1.3, Revision 2, June 1974.

[13] - USNRC, Assumption Used for Evaluating the Potential Radiological Consequences of a Loss of Coolant Accident for Pressurized Water Reactors, Regulatory Guide 1.4, Revision 2, June 1974.

[14] - USNRC, Assumption Used for Evaluating the Potential Radiological Consequences of a Steam Line Break Accident for Boiling Water Reactors, Regulatory Guide 1.5, March 1971.

PVP-Vol. 458, Computer Technology and Applications

PVP2003-1919

HUMAN FACTOR IN THE LIFE CYCLE AND SAFETY OF MACHINES AND PIPELINES

Sviatoslav A. Timashev

Science & Engineering Center "Reliability & Resource of Large Machine Systems,"
Ural Branch, Russian Academy of Sciences
54 A, Studencheskaya St., Ekaterinburg, 620049, Russia
Ph./Fax: +7 (343) 274-1682. E-mail: wekt@r66.ru, timashevs@aol.com

ABSTRACT

The paper is an overview (using references listed below) and describes the main components, means and methods of a holistic and quantitative human reliability analysis (QHRA) using quantitative values of human error when performing Fault Tree Analysis (FTA) and Event Tree Analysis (ETA). It also deals with qualitative assessment of the influence of the human factor (HF) reliability on safety and risk analysis of potentially dangerous man-machine-structures-environment systems (PDMMSES). Qualitative risk analysis of such man-machine-structures-environment (MMSE) systems is based on using the event-decision technique in combination with a generalized socio-psychological model of the decision making person (DMP). Three types of DMP's are considered: members of maintenance/repair crews, diagnosticians and different rank DMP's that operate or own the PDMMSES.

INTRODUCTION. MAIN NOTIONS AND DEFINITIONS

Human factor as an object for research is studied by fundamental and applied sciences. A human error (HE) is an action that fails to meet some limit (criteria) of acceptability as defined for a concrete system. The action could be physical (e.g., opening a valve) or cognitive (e.g., wrong diagnosis or decision making). Human error can increase risk during different operation or maintenance procedures if they lead to increased demands on control and/or protective systems.

Human mistakes can also be defined as an unfulfilled posed problem (or execution of a forbidden action) which may be the cause of equipment or property damage or disruption of the standard flow of planned operations.

Fundamental research on human factor is being conducted in social sciences, psychology and natural science. The main goal here is to create a model of human behavior that would allow quantitative accounting for the occurrence of human error.

Before 9/11/01 HRA dealt only with causes of human error and with the probability assessment of that error. Now HRA is starting to consider malicious behavior of the "operator," taking explicitly into account acts of vandalism, sabotage and terrorism (Timashev, 2001, 2002 a, 2002 b).

INTRINSIC ENERGY-ENTROPY SOURCE OF RISK IN MMSES

According to the second principle of thermodynamics, creation of synthetic substances, chemically clean elements, production and accumulation of energy, processing refining and transportation of natural materials and the like are all "illegal", because they lower (not raise) the level of entropy. Moreover, they are dangerous, because they overcome natural energy barriers and by this virtue, bring the systems that perform the above processes into an unstable and, therefore, dangerous state.

Being unstable, these systems tend to stability with a corresponding release of energy and growth of its entropy.

When the problems imposed on people are on the brink of human capacity, and considered as life threatening, the human body, following the instinct of self-preservation, accelerates those biochemical processes that create adaptive body reactions. Such protective reaction of the organism is called stress response.

HOLISTIC APPROACH TO ASSESSING THE ROLE OF HUMAN FACTOR

Industrial activity is a process where specifics of a human organism and environmental factors tightly intermingle.

Technical systems become interdependent exactly because of the human connections between them. According to statistics, around 70-80% of all dangerous situations and failures directly or indirectly are due to human mistakes. Therefore in order to correctly assess the role of human factor a holistic approach should be used. In the case under consideration this means that the system should be a generalizedMMSES.

Modern MMSE systems are complex automated systems with control contours that are being locked (closed) by human beings.

MMSE systems during their life cycle are undergoing stages of design, manufacture installation, diagnostics and control, maintenance, repair, rehabilitation and upgrade, decommissioning and disposal. A correct and justified account of the human factor on each of the above stages is needed to achieve maximal effectivity and safety.

THE HOMEOSTASIS CONCEPT

Biological action of chemical substances on a human organism changes her/his homeostasis. By homeostasis one defines the relative stationarity of content and properties of the body's internal environment and stability of the main psychological functions of the organism; e.g., the ability of the organism to autoregulate itself with the change of the external environment (STAP, 2001). Homeostasis can also be violated by extreme psychological stress, caused by different types of external non-chemical causes – loss of self-control, fear, panic, anger, death threat, etc. Autoregulation of a biological system should be considered as a regulation of the dynamic state of an open system, influenced by a biological rhythm. At this, homeostasis includes not only the dynamic stationarity but also the stability of its main biological functions. Influence of a detrimental substance may cause not only changes of certain parameters of a biological object, but also damage of the homeostasis regulating systems; e.g., violation of the latter. In order to preserve the homeostasis under the influence of various chemical substances, during evolution a special system of chemical detoxication evolved. Relatively small amounts of detrimental substance or reasonable stress levels do not violate the homeostasis (Fig. 1) (STAP, 2001). In this figure $\{x_1,x_2\}$ is the homeostasis region. Usually, it is convex, with somewhat fuzzy boundaries, because the optimal parameters Y of the biological object themselves change in time. Outside the $\{x_1,x_2\}$region the homeostasis is violated, which means abrupt change of the Y values. The x_1 and x_2 are called the threshold values of X. The region of homeostasis is the region of the negative feedback, because here the organism works to bring itself back into the initial stationary condition/state. When homeostasis is severely violated, the object enters the region of positive feedback. In this case the changes caused by detrimental substances/stresses become irreversible and the object more and more deviates from the stationary state.

The outlined above permits one to take a step further, namely, to address the psychological problems of homeostasis in general, and human behavior and mistakes, specifically.

PSYCHOLOGY IN THE SAFETY PROBLEM

Psychology of work safety is an important part in the overall strategy of providing for human security. Statistics show that the root cause in 60-90% of casualties are of logistic-psychological nature (Kondratyev, and Zaitsev, 1986).

The psychology of safety discipline studies psychological processes, properties and states of human beings in the workplace. Among specific psychological states one should distinguish states of being alarmed, concerned, afraid, frightened, or even in a state of panic. Extraordinary forms of psychological stresses cause disintegration of psychological activity, which leads to decreasing levels of psychological performance capacity.

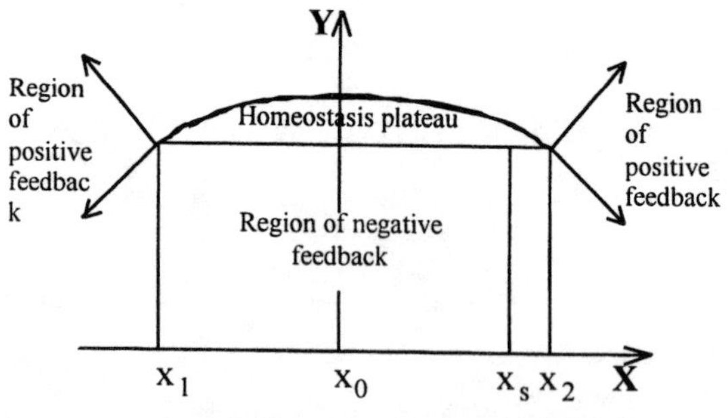

Fig. 1 The homeostasis scheme: Y is some property of a biological object; X is the concentration/dose of a harmful detrimental substance or stress intensity level; x_s is the ultimate safe level of influence by this substance stress; x_1 and x_2 are the threshold values of x

PSYCHOLOGICAL CAUSES OF MISTAKES

Performance of an operator in general consists of the following stages: perception of information, its assessment, analysis and generalization on the basis of criteria given beforehand and formulated criteria; decision making on actions; execution of the decision. In all these stages a human being can make mistakes.

By erroneous action one understands an action that deviates from normal; e.g., anticipated, expected and, by virtue of this, leads to grave consequences: casualties, fatalities, loss of property. Generally speaking, mistakes are unconscious dangerous actions. In some cases such dangerous or inadequate actions are committed consciously, deliberate and are classified as violations; in other cases when the operator does not realize he's executing a dangerous action, it is classified as a mistake. Going one step further, one can say that a malicious, conscious act aimed at putting in danger innocent human beings can be classified as a terrorist act.

The causes of mistakes can be divided into direct, main, and contributing. Direct causes depend on their placing in the psychological structure of operators' action (perception; decision making, return action, etc.) and the type of this action; e.g., on psychological pattern that defines optimal action: disparity of psychological ability of data processing (volume or speed, relation to the distinction threshold, insufficient duration of the signal etc.); insufficient skills (standard actions in a non-standard situation, etc) and attention structure.

Main causes are connected with the work place, work organization, operator preparedness, state of mind and health. Contributing causes depend on particular failures of the personality, state of her/his health, external environment, selection and training. In the MMSE system the human being is the most variable component. In any given problem his/her behavior is influenced by some million individual factors.

The causes of mistakes can be also classified as orientation mistakes (insufficient information); decision mistakes (wrong decision); action mistakes (wrong action). Using other principles mistakes can be classified as irreversible and reversible, systematic and random. In most cases mistakes are the result of change in operator's physical and mental state.

ERRORS CLASSIFICATION

The role of human error in structural failures is being extensively studied (Grigoriu, 1984, Melchers, 1987, Nowak, and Collins, 2000). Errors are categorized according to causes and consequences. Control of error can be achieved by reducing error frequency and/or minimizing its consequences.

The technology of error finding includes systemic checking and inspections of workplaces to assess the quantity of errors. Then the severity of their consequences in estimated through sensitivity analyses. The sensitivity functions are often used as a key element in error-control technique. A considerable discrepancy is being observed between the theoretical and actual failure rates This discrepancy is due to incompleteness of existing theoretical models. Most of the failures occur as a result of human errors which are not included in the analysis.

Classification of human errors made at different stages of their interaction in the MMSE system is important in identification of the error occurrence mechanisms and selection of efficient control measures. Errors can be classified by who made the error, at what phase of the system life (time), where the error was made, root cause (why was it made), frequency of occurrence, mechanism of occurrence (see Fig. 2). Additional types of classification are given below.

Surveys of structural failures indicate that human error is the major cause (Nowak, and Collins, 2000). Moreover, error in design or construction often greatly exacerbate the damaging effect of other hazards and may lead to cascade failures and domino effects. Even more so this is true for machines and facilities.

NOTION OF RISK INDUCED BY HUMAN ACTIONS

Risk is a quantitative characteristic of acting dangers that originate via concrete, human activity or nature in the form of number of fatalities, casualties (temporary or permanent disability), illness cases caused by the influence on humans of a concrete danger (moving objects, electrical current, detrimental substances, criminal and terrorist actions) referred to a certain amount of workers or people during concrete periods of time (shift, day, week, quarter, year...).

Dangers can materialize in the form of fatalities, casualties and illnesses only if the zone where the dangers are formed (the danger zone – sometimes it is called noxosphere) intersects with the human activity zone - the homosphere. In the case of industrial plants it is the work zone and the concrete human source of danger.

It is necessary to discern different types of risk. Individual risk characterizes realization of a certain type of dangerous activity for a concrete person.

Collective risk is defined as fatality/casualty of two and more individuals due to the influence of dangerous or detrimental industrial/natural factors.

In this paper only that part of risk that is induced by human activity is discussed.

MOST OFTEN USED HUMAN ERROR QUANTIFICATION METHODS

In this section major techniques of human error quantification are briefly described, following (CPQRA, 1985).

All the HRA basically come down to

- identification of relevant tasks performed by operators
- decomposition of the task to identify opportunities for error and points of interaction with the plant
- use of actuarial data
- identification of performance - shaping factors (stress, training, quality of operator's displays and controls).

Results of an HRA are usually expressed in the form of HE probabilities/rates:

$$HEP = \text{Number of errors} / \text{Number of opportunities for error} \quad (1)$$

$$HER = \text{Number of errors} / \text{Total task duration} \quad (2)$$

- Technique for Human Error Rate Prediction (THERP). This method requires breaking down a task into unit tasks and creating an event tree (ET) using this information. A flow chart for use of THERP is given in Fig. 3. The ET calculations are performed by estimating conditional probabilities of success/failure (S/F) for each branch of the tree (Fig. 4).
- Accidence Sequence Evaluation Program (ASEP) is used mainly for initial screening of the importance of human error. This short and conservative method estimates the HEP for two sequential stages: pre-accident and post accident.

The pre-accident screening analysis is performed to identify those components of a MMSES that are vulnerable to human errors. If the probability of system failure is acceptable, human error is not important. The steps in the pre-accident HEP analysis are shown in Table 1.

The post-accident HRA is performed to evaluate the probability of the operators to have detected and corrected their error. Once the accident sequence started, the most important variable is time the operators/DMPs have to detect and correct their errors before the accident turns into a catastrophe. Before corrective action can be taken, the operators/ diagnosticians/DMPs must diagnose that there is a problem. Figure 5 gives the idea how the probability of a correct diagnosis is related to the time that the DPMs of all kinds have to properly diagnose abnormal event.

The time available for diagnosis can be computed as

$$T_d = T_m - T_a \quad (3)$$

where T_d is the time available for the machine system operator to diagnose that an abnormal event has occurred; T_m is the maximal time available to correctly diagnose that the abnormal event happened and to have completed all corrective actions necessary to prevent the resulting accident; T_a is the time required to complete all post-diagnosis required actions to bring the system under control.

The maximum time T_m to diagnose and correct a problem is determined by a detailed analysis of each accident sequence, including possible time delays (due to rate of heat transfer, chemical reaction kinetics, flow rates etc.).

Next determined is the time T_a needed to correct the problem. This time is estimated using a list of the operator tasks that must be completed to correct the problem created in each accidence sequence, according to Table 2.

Once the times T_m and T_a have been estimated, the time available for the operators to correctly diagnose an abnormal event (T1) is determined using Eqn. (3). The probability that operators will fail to diagnose the problem can be assessed using Fig. 6. If the abnormal event is not annunciated in the control room by one or more signals, the probability of failing to properly diagnose the problem can be assessed using Table 3.

THREE TYPES OF HUMAN FACTOR

When analyzing reliability and safety of MMSES three categories of human factor (HF) should be accounted for: HF1, HF2 and HF3:

- HF1 describes repair/maintenance crew members behavior
- HF2 describes monitoring, NDT specialists and diagnosticians performance
- HF3 describes reliability of DMP's of different level – from plant line operators and up to DMP's that own or run the MMSES (influence the total plant modus operandi).

These three groups of human factors are discussed below.

Out of the three categories of human factor, the HF1 is the relatively easiest factor to account for.

The main parameters that are needed in this case are (Timashev, 2000)

- the quality of maintenance/repair, which either brings the system to the "as new" state or to a point somewhere in the safe space.
- time needed to perform the prescribed quality maintenance/repair. This time is random and is characterized by its mean m and variance (Fig. 7). The proper thing to do is to assess the actual times of performance of a maintenance team/repair crew of a given size and skill to do a specific maintenance/repair job and create a statistical database. It was shown that the probability density function (PDF) of the maintenance/repair time depends on the time of the day, health and stress level the maintenance/repair crew is exposed to during the operation (Fig. 8). Therefore, in general

RV(maintenance/repairtime) = function (maintenance/repair budget, team/crew size, maintenance technology, stress level, skill level, health conditions, time of day, etc.)

Table 1 Pre-accident HR Screening Analysis Procedure

Step	Description
1.	Identify critical human actions that can cause an accident
2.	Assume that the following basic conditions apply relative to each critical human action - no indication of a human error will be annunciated in the work place (control room, work shop, construction site, etc.) - the activity subject to human error is not checked - there is no possibility for the person to detect the error she/he had made - shift or daily checks of the activity subject to human error are not made/effective
3.	Assign a human error probability of 0.03 to each critical activity
4.	If two or more critical activities are required for an accident to occur, assign a HEP of $9 \cdot 10^{-4}$ for the entire sequence of activities. If these two or more critical activities involve two or more redundant safety systems (relief valves, etc.) assign a HEP of $3 \cdot 10^{-2}$ for the entire sequence of activities (a conservative assumption).

Table 2 Time required for Post-accident Activities

Description of Activity	Time required, min
Find and initiate written procedure (if not connected to memory)	5
Travel plus manipulation time on main control panel	1
Travel plus manipulation time on secondary control panel (if any)	2
Travel plus manipulation time of a manually operated field system	(*)
Process stabilization	(**)

(*) Use actual times, determined by walk-through simulations. Use twice the operator's estimate if no other information is available. Use 5 min if no information is available.

(**) The system may require some time to stabilize after the necessary actions have been taken. If no information is available, assume the system instantly returns to safe condition. This assumption must be flagged for further study.

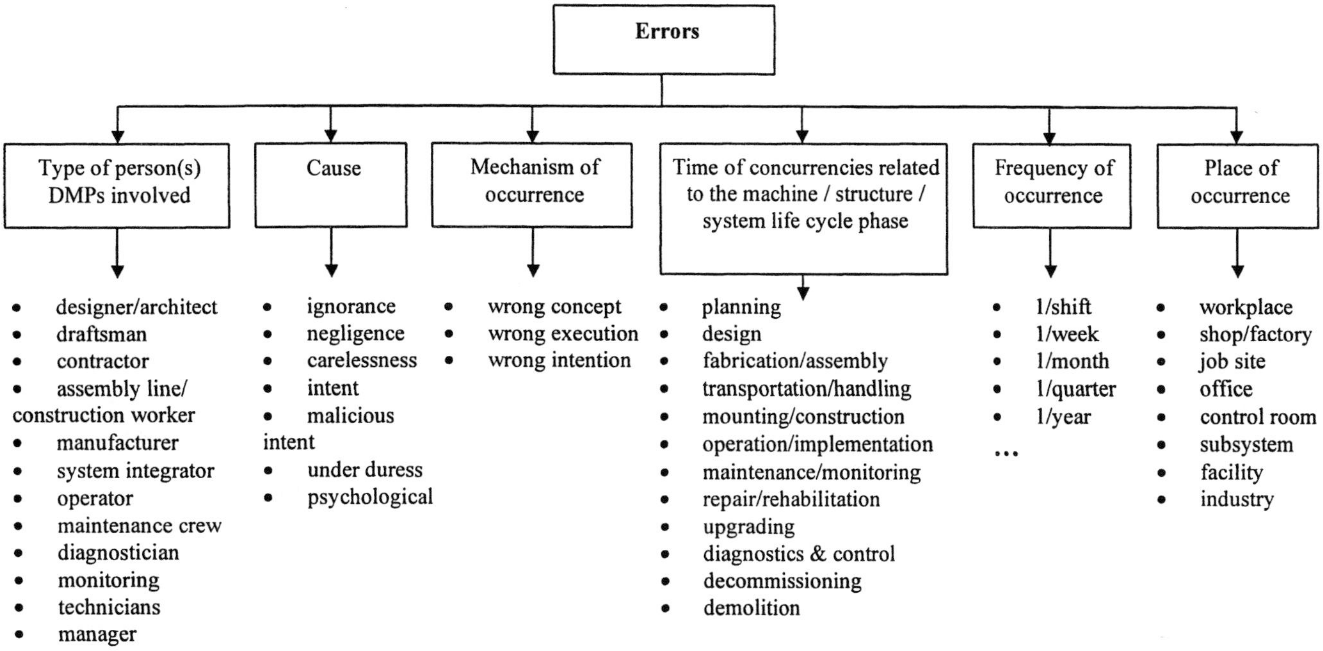

Fig. 2 Human error classifications

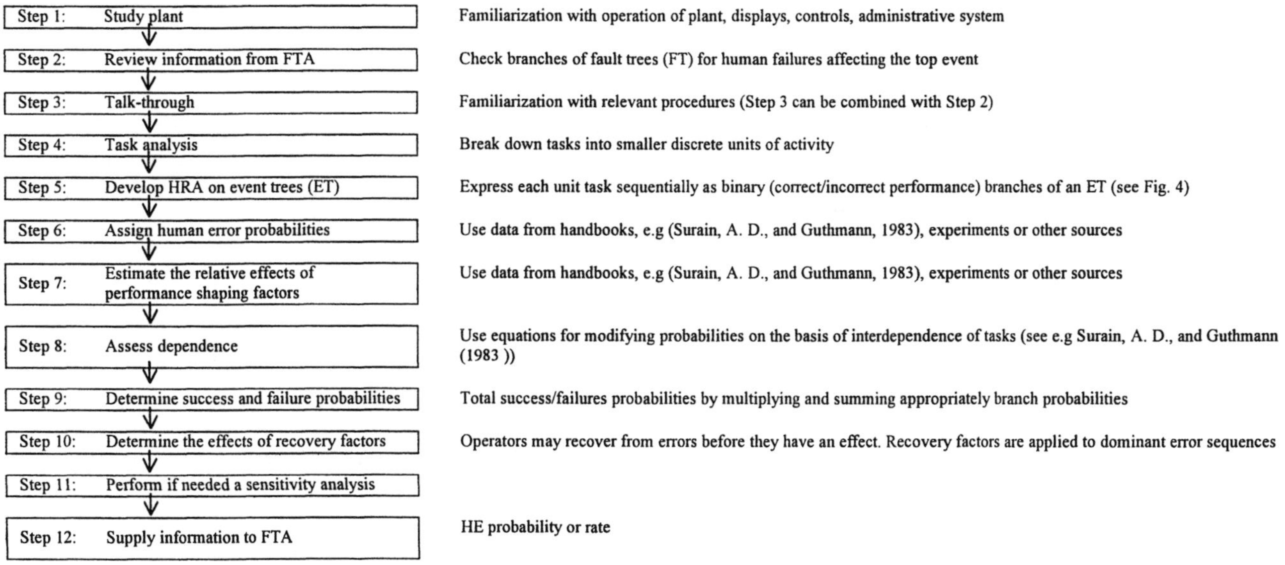

Fig. 3 Human reliability analysis flow chart

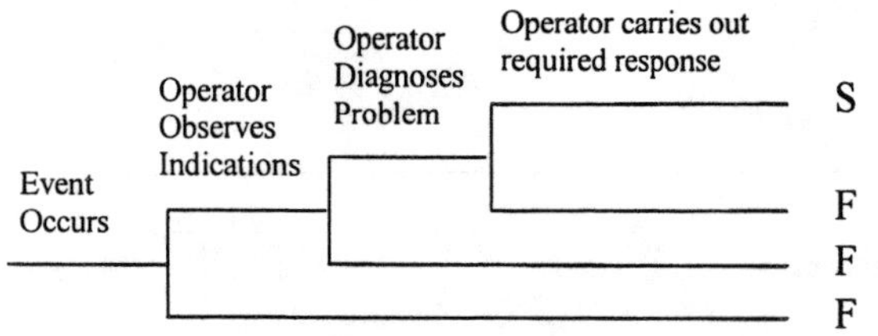

Fig. 4 Basic operator action tree

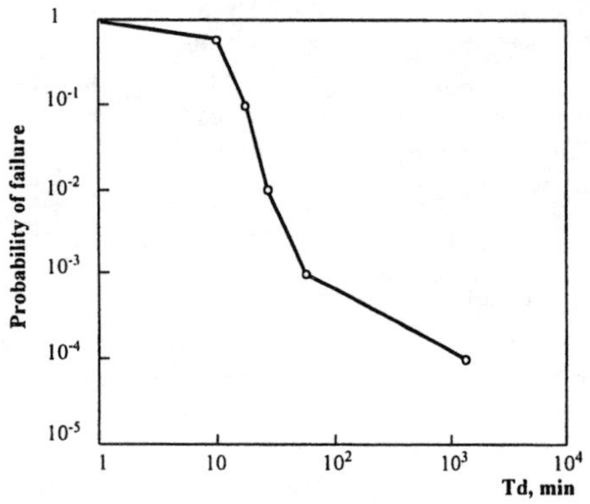

Fig. 5 Probability of operator's failure to correctly diagnose an abnormal event

Table 3 HEP in Recovery from an Abnormal Event

Description	HEP
Perform a required action outside of the control room	1,0
The same, while in radio contact with the control room operator [1)]	0,5
Perform a critical skill/rule based action when no written procedures are available [2)]	1,0
Perform a critical action under conditions of moderately high stress	0,05
The same under conditions of extremely high stress [3)]	0,25

[1)] HEP of 0,5 includes failure of the operator to either properly complete an assigned task or failure to receive instructions due to a disrupture in communications (noise, stress, radio failure etc.).

[2)] Conditions of moderately high stress would occur when dealing with an abnormal event that could result in a major loss of product, shutdown of the process unit, operator employment action such as a reprimand, or other adverse outcomes that are not life endangering to the operator.

[3)] Conditions of extremely high stress would occur when dealing with an abnormal event that could result in a major fire, runaway reaction, or toxical chemical release that could kill or seriously injure the operator or his friends.

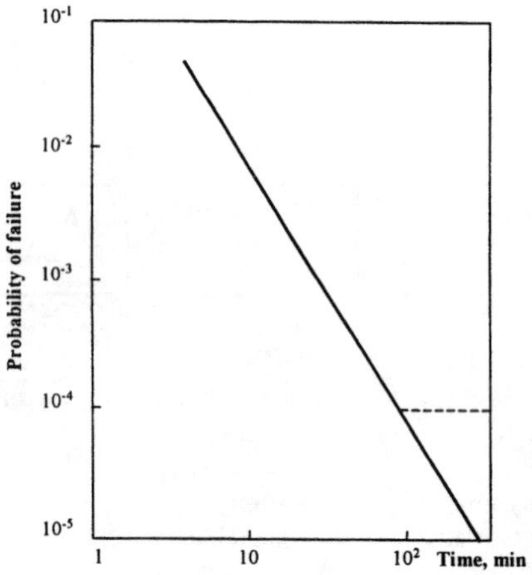

Fig. 6 Operator action tree reliability curve

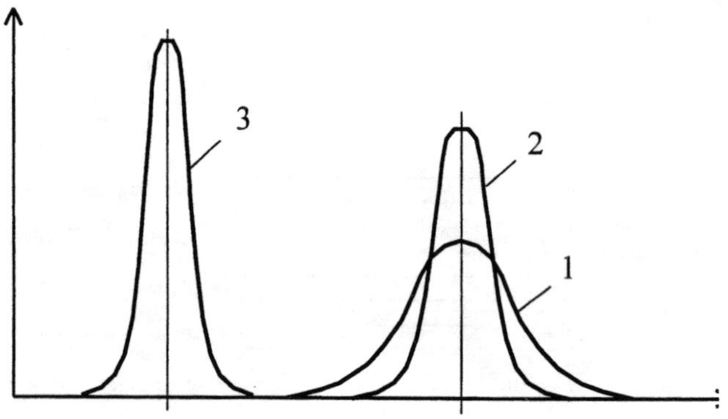

Fig. 7 Change of m as a random variable

Team/crew performance can be greatly enhanced, if special premium is awarded, and a set of incentives established for commitment and a fast, high quality job.

Concrete values of the time of performance PDF parameters should be assessed for every company and plant, but now, they practically do not exist. Establishing them is a special task that should be performed in every company. Knowledge of these parameters opens the door to MMSE systems reliability and maintenance optimization.

The same lines of argument are used in assessing the quality of maintenance/repair as a human factor.

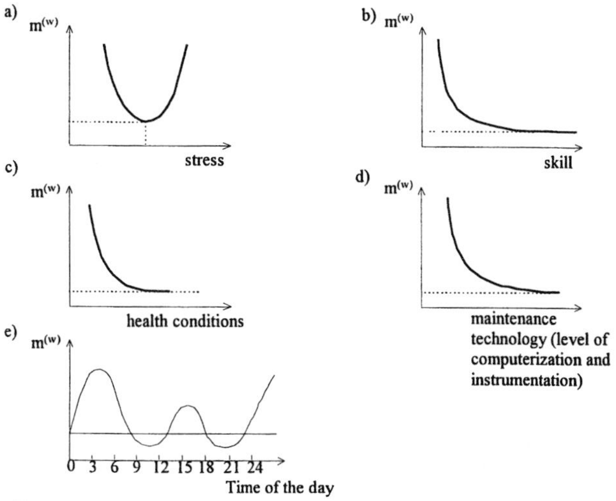

Fig. 8 Dependencies of m on levels of team stress (a), skill (b), health (c), maintenance technology (d) and the time of the day (e)

QUANTITATIVE ASSESSMENT OF DIAGNOSTICIAN'S ERRORS

The quantitative assessment errors of a NDE technician diagnostician are rather complicated because it depends on many interrelated factors which are not easily quantified. Usually the following are important characteristics for "good" NDT diagnosticians/technicians (NUREG/CR-4908, 1989):

- ability to concentrate
- understanding of NDT theory
- personal patience
- tolerance of environment conditions
- manual dextertity
- mathematical ability

The qualitative assessments of the above human characteristics are obtained through direct observations and critical incident interviews.

Quantitative assessment often is made by plotting the so called Relative Operating Characteristic (ROC) curves which plot Probability of Detection and the False Call Probability and are used in analyzing diagnosticians performance. Some results of measurements conducted in the nuclear energy industry by NDT diagnosticians is shown in Fig. 9. It can be seen that individual diagnosticians vary substantially in precision.

The ROC quantifies the relationship between detection performance and decision criteria. ROC curves usually are used to examine a team's (or procedure's) inherent detection probability (IDP). IDP is the performance that would be achieved if the optimum decision criteria were being employed. In actual field inspections a procedure with excellent IDP may still yield poor results because a poor decision criterion is being employed or the decision criterion (DC) cannot be controlled. The DC may also vary from inspection to inspection and/or team (person) to team (person). Therefore, the actual laboratory DC that is used for extrapolation to field experiments is very important.

HUMAN RELIABILITY AS A COMPONENT OF A COMPLEX TECHNICAL SYSTEM

The main causes of incident are human errors/mistakes, equipment failure and non-design technological and natural (external) influences. In a disaster or a catastrophe they form a causal chain. It is expedient, therefore, to simulate such chains and observe where they stop – at the disaster level or before it. During simulation most substantial properties of the HF3 are taken into account: perception, decoding and structuring of the data about operation as it develops; detection of discrepancies from the technological requirements; assessment of the necessity to interfere, comparison of possible behavior alternatives; execution of the control correction with respect to the technological process.

The basic simulation model (Belov, 1998, Grazhdankin, 2001) is based on a macro-level logical and linguistical model of an incident spring-up in a MMSE system.

The outlined simulation method can be enhanced by using more precise models of human behavior.

CONCLUSION

1. Current knowledge of human error in industry and construction constitutes a major part of knowledge of machinery and structural failure. This knowledge is mostly intuitive, based on experimental best practices, often proprietal and therefore largely inaccessible and perishable, disorganized and incomplete. There still are no means to organize this knowledge. Much research remains to be done to improve and create new frameworks for practical control of the hazard of human error.
2. The role of human error in reliability and safety of machines and structures and other man-made facilities and systems is starting to be an increasingly effective object for research, development and implementation. Exploration of the input of human component into the reliability and risk of performance of MMSES is a new challenge to researchers.
3. People are the major contributors to risk in any plant or facility. Measuring the human operational risk component and reducing risk contributions associated with humans is a new research field and new technologies.
4. Today there doesn't exist a generally accepted methodology that is able to identify and subsequently eliminate management practices and procedures that contribute to operational risk.
5. For effective safety control of any man-made system, a understanding perception of human error is crucial. The steady increasing emphasis on manufacturers' liability pushes up demands of flawless performance of engineered systems. Therefore, error control is a critical part of the overall strategy when enhancing the reliability and safety of any man-machine-system.

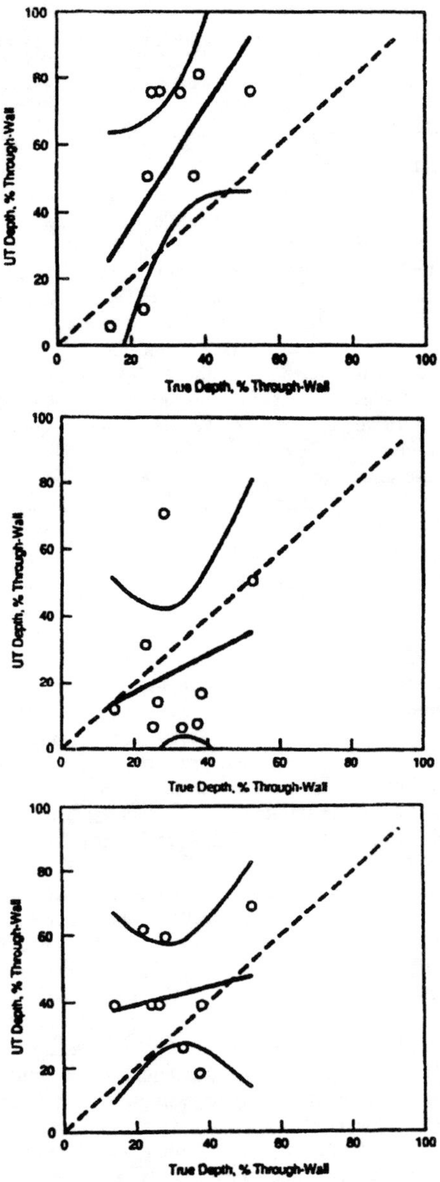

Fig. 9 Relative Sizing Results for different UT inspectors with 90% Confidence Bounds: a) - inspector #5; b) - inspector #6; c) - inspector #7

REFERENCES

1985, "Chemical Process Quantitative Risk Analysis (CPQRA)", ASChE, N.Y.

Belov, P. G., 1998, "Basics of Theory of System Safety Engineering," Kiev International University of Civil Aviation, Kiev, pp. 456, in Russian.

Grazhdankin, A. I., 2001, "Development of an Expert System for Assessment of Technogenic Risk and Optimization of Safety Measures on Dangerous Industrial Objects," PhD Thesis, Moscow, in Russian.

Grigoriu, 1984, M., Risk, "Structural Engineering and Human Factor," University of Waterloo Press, Waterloo, Ontario.

Kondratyev, S. V., and Zaitsev, K. S., 1986, "Engineering and Psychological Assessment of Man-Machine Systems," MIFI, Moscow, pp. 81, in Russian.

Melchers, R. E., 1987, "Structural Reliability Analysis and Prediction," Ellis Horwood, Chichester, England.

NUREG/CR-4908, 1989 Washington, D.C.

Nowak, A., and Collins, K., 2000, "Reliability of Structures," MC Graw Hill, Singapore, pp. 338.

Surain, A. D., and Guthmann, H. E., 1983, "Handbook of Human Reliability Analysis with Emphasis on Nuclear Power Plant Applications," DCNUREG/CR-1278, U.S. Nucleear Regulatory Commission, Washington, D.C.

Timashev, S. A., 2000, "Optimal Machinery Integrity and Maintenance Control," COMADEM.

Timashev, S. A., 2001, "To get to know Lucifer's Science," Nauka Urala, in Russian

Timashev, S. A., 2002 a, "It is possible to control risks," "Vecherny Ekaterinburg", in Russian.

Timashev, S. A., 2002 b, "The-Smart-Map-For-First-Responders Concept: The Russian Version," Paper presented at the Russian-American CRDF Special Competition Workshop #12112, Washington, D.C., July.

2001, "Safety of Technical Processes and Production (STAP)," Moscow, Visshaya Shkola, pp. 368, in Russian.

PVP-Vol. 458, Computer Technology and Applications

PVP2003-1920

RISK ASSESSMENT FOR DAMAGED PRESSURE TANK CARS

Steven W. Kirkpatrick
Applied Research Associates, Inc.
2672 Bayshore Parkway, Suite 1035
Mountain View, CA 94043

Richard W. Klopp
Exponent Failure Analysis Associates
149 Commonwealth Drive
Menlo Park, CA 94025 USA

ABSTRACT

Following an accident or derailment involving a freight train, emergency response, salvage, and repair personnel need to clear the right-of-way, repair the tracks, and remove the damaged rail cars. If the train contains pressure tank cars the potential hazard and safe handling procedures need to be determined prior to salvage and repair activities. The severity of the damage determines the appropriate course of action such as rerailing or unloading the damaged cars. The operations must avoid the risk to response personnel of a delayed rupture.

A research program was performed to establish the validity of the existing industry guidelines for assessment of damaged pressure tank cars. The research program first focused on evaluating the technical foundation for the existing guidelines and the degree to which they have been validated. We then designed a program of experiments and analyses to validate the guidelines and estimate their margins of safety. The experimental effort used laboratory specimens to provide material property data as well as validation data for the analyses. The analyses used three different approaches: nonlinear elasto-plastic finite element simulations for modeling denting behavior, elasto-plastic fracture mechanics for analysis of cracks, and nonlinear finite element simulations combined with local fracture theories to quantify the severity of scores, gouges, and rail burns (longitudinal damage features caused by sliding a tank along a rail). This paper emphasizes the latter aspect of the work.

INTRODUCTION

A research program was performed with the objective of validating industry guidelines for the proven and reliable assessment of damage to tank cars. In the rest of this paper, the damaged tank car assessment guidelines are referred to as "the Guidelines". Reference 1 contains a relatively recent version, and Giovanola and Shockey [2] present a full history of the Guidelines' development. The ultimate objective was to generate an updated Association of American Railroads (AAR) handbook, *Field Removal Methods for Tank Cars* , [3] based on the Guidelines. The handbook helps emergency response personnel at an accident scene evaluate the criticality of damage to tank cars and, based on this evaluation, decide on appropriate procedures for handling the damaged cars and their contents (e.g., unload or rerail the cars). The guidelines in this handbook have been used since 1985 by emergency response personnel to make judgments in the field regarding the severity of damage to tank cars involved in accidents.

TANK FRACTURE ANALYSIS

The objective of the fracture analyses is to develop and apply computational tools to assess the risk of a catastrophic failure of the tank for a given set of damage and loads.

The effect of dents on the structural integrity of pressure tank cars was assessed with dynamic nonlinear finite-element analysis. Using this approach the relationship between the dent deformation history and the final observable dent shape can be investigated. We used linear elastic fracture mechanics and elasto-plastic fracture mechanics to evaluate the safety threat caused by the presence of macroscopic cracks and determine whether leak will occur before break. In contrast, burns, scores, and gouges present a more complex problem to which more complex fracture analysis methods must be applied. This class of problem can be examined using the local fracture approach because it can, (1) predict fracture initiation in the absence of a preexisting macroscopic crack, (2) treat cases of multiaxial loading such as exist in a tank car, and (3) account for possible inhomogeneity in microstructure and properties. Therefore, we chose local fracture methodology as the core of our approach to burns scores and gouges. As structural finite element modeling and elasto-plastic fracture mechanics are well established the local fracture methodology will be the focus of this section.

Of the potential fracture modes in pressure tank cars, cleavage fracture presents the greatest risk. Cleavage cracks propagate with lower energy input, so results obtained are more conservative than for tearing. On page 5 of his report on rail burns [4], Pellini makes the case that as-rolled steels then in service (1983) (and still in service today) are susceptible to cleavage fracture. He then presents several actual incidents that support his case. None of these incidents occurred at unusually low temperatures. In fact, some were in the 60°F range yet cleavage fractures occurred. Thus, we conclude that, although ductile tearing is a likely and preferred response, cleavage failure is a significant risk in any accident, and the Guidelines must allow that any fracture could result in cleavage.

The cleavage model used in our analysis is the local cleavage criterion developed by Beremin and coworkers [5]. The term "local criterion" describes a modeling approach in which damage is calculated locally within the material based on the stress and strain states using a micromechanical model for the fracture processes. The Beremin approach yields what is essentially a zero-stress-threshold Weibull model of predicted fracture stresses.

Predicting cleavage fracture is different from predicting ductile fracture in that there is typically a large scatter in the measured cleavage fracture stress for a sample of identical tests on a single batch of material. Thus, a cleavage failure criterion will predict a probability of fracture for a given material, geometry, and stress level, rather than specify a deterministic failure stress.

The microstructural processes that produce a cleavage fracture are similar to many other classical fracture problems. The material is assumed to have a distribution of preexisting microcracks, typically initiated in the material from inhomogeneities. For example, in mild steels, the microcracks are produced by fracture of sulfide inclusions or grain boundary carbides. The catastrophic propagation of these cracks results in a cleavage fracture, which occurs when the stress normal to the microcrack planes reaches a critical value. Following Beremin [5] this critical stress, σ_c, can be approximated as

$$\sigma_c = \left[\frac{2E\gamma}{\pi(1-\nu^2)l_o} \right]^{1/2} \tag{1}$$

where E is Young's modulus, γ is the fracture surface energy, ν is Poisson's ratio, and l_0 is the microcrack length. See also Broberg [6].

The statistical nature of the cleavage criterion is introduced by the distribution of microcrack sizes within the material. Within a given microstructural characteristic volume (V_0), the probability of finding a crack with a characteristic length between l_0 and $l_o + dl_0$ is taken as

$$P(l_o)dl_o = \frac{\alpha}{l_o^{\beta}} dl_o \tag{2}$$

By integrating the microcrack distribution function over the range of crack lengths greater than or equal the critical crack length at a given stress level, we obtain the probability of failure as

$$P(\sigma) = \left(\frac{\sigma}{\sigma_u} \right)^m \tag{3}$$

where $m = 2\beta - 2$ and σ_u is a material constant.

The remaining step in determining the cumulative failure probability of the structure is to combine the probabilities in each of the small representative volumes. Because the probability in any single representative volume is small, the cumulative rupture probability, P_R, can be approximated in the context of finite elements as

$$P_R = 1 - \exp\left[\sum_j -\frac{V_j}{V_0} \left(\frac{\sigma_j}{\sigma_u} \right)^m \right] \tag{4}$$

where the summation, j, includes all finite elements with nonzero plastic strain, V_j is the element volume, and V_0 is a characteristic microstructural volume. The cumulative rupture probability is used to estimate cleavage failure probability. This probability function is a Weibull distribution with no threshold stress. A necessary condition for the validity of the approach is that the element volume V_j is larger than the microstructural volume, V_0, but small enough that element stress gradients are negligible. Additional information about this model can be found in Reference 4.

The local fracture model for cleavage rupture was incorporated into the DYNA3D finite element model. DYNA3D, developed at Lawrence Livermore National Laboratory [7], is an explicit nonlinear three-dimensional finite element code for analyzing the large deformation dynamic response of solids and structures. Although many of the simulations in this project were quasi-static, some were dynamic. In addition, the authors find that modifications are easier to implement in an explicit code like DYNA. In the quasi-static cases, DYNA3D was run at a sufficiently low loading rate that nominal equilibrium was achieved. To describe the modifications we made to DYNA3D, we first outline the solution procedure used in this code. DYNA3D is organized in a modular fashion. This procedure is common to most explicit finite element codes and is similar to those described by other authors (e.g., Owen and Hinton, [8]). The following simplified flow chart shows the tasks performed by DYNA3D.

Task 1: Input—Data are input to describe the geometry, materials, boundary conditions, loading, and solution control parameters.

Task 2: Initialize—Initial values read into the structural and material data arrays.

Task 3: Calculate nodal forces—Nodal forces are calculated as the difference between internal and external forces. External loading would include pressure, concentrated loads, and gravity. Internal forces are calculated from stresses.

Task 4: Solve equations of motion for accelerations—At each node, the acceleration is calculated from the force divided by the nodal mass. For explicit codes such as DYNA3D, no global stiffness matrix is ever assembled or inverted.

Task 5: Update velocities and displacements —Nodal velocities and displacements are calculated and updated by using accelerations and the current time step.

Task 6: Calculate strain increment —From the nodal velocities, strain rates are calculated from which strain increments can be calculated.

Task 7: Calculate element stresses —Element stresses are calculated in the material constitutive models from the strain increments. Compare with critical stresses for selected probability of failure.

Task 8: Calculate internal forces —Nodal internal forces are calculated by integrating the element stresses.

Task 9: Output and check for problem termination — Output results if specified. If the calculation time is less than the specified termination time, return to Task 3.

The code changes performed here to implement the Beremin cleavage model primarily take place in Task 7, in the routines where element stresses are calculated; however, other subroutines are also affected (such as material model input, summation of fracture probability, damage parameter output). The resulting model calculates a probability of fracture at each time step.

TESTING AND MODEL CALIBRATION

Two series of laboratory tests were performed on the tank car materials of interest. The two test series are referred to as calibration tests and engineering tests. The calibration tests were performed using both smooth and notched round bar tensile specimens to provide fracture data required to calibrate the Beremin local fracture model. The engineering tests were performed to measure the fracture behavior on laboratory scale specimens that simulated directly the types of damage indicated in the Guidelines [1]. The results of the engineering tests were used to validate the Guidelines and to validate the local fracture model. These included both gouge specimens and bend specimens. The bend specimens were used to investigate damage produced by denting of the tank.

Two different tank car materials were tested, A515 Grade 70 and TC-128B in the as-rolled condition. The results presented in this paper are primarily for the tests on the A515 Grade 70 material. The A515 Grade 70 steel is a replacement for A212 Grade B steel, an obsolete specification that was withdrawn by ASTM in 1966. Roughly, 15% of the current DOT 112 tank car fleet was built before 1966, when the A212 specification was in use. Most of these cars will be in service at least another decade and appear to represent the greatest risk of catastrophic rupture in accidents. A comparison of the various tank car steels and the rationale for selecting the steels that were tested in this program are given in Reference 8.

In choosing the temperature for the calibration and engineering laboratory tests, we needed to satisfy three competing constraints. First, the tests must be performed at a low enough temperature that cleavage fracture, rather than ductile tearing, is ensured. Second, we should test at a temperature that fairly represents the most brittle conditions likely to be found in the field yet does not lead to overly conservative results. Third, the tests must be performed at a temperature that is practical to reach in the laboratory.

In the AAR literature on A212B/A515-70 as-rolled steel, we find that the lowest expected NDT is around -7°C. Unfortunately, the NDT is measured under dynamic conditions and in the presence of a weld. Our tests were performed under quasistatic conditions mostly without welds, and steels are generally less prone to cleavage under such conditions. Thus, to ensure that cleavage occurred, we had to test at a temperature well below the lowest expected NDT. We found it necessary to test smooth, macroscopically unflawed specimens at around -150°C to ensure that cleavage fracture occurred. Thus, we chose -150°C as the temperature for most of our testing. Two methods were used to perform the low temperature tests. Larger specimens were immersed in liquid nitrogen and cooled to -196°C, placed in the testing machine and allowed to warm to the desired temperature for testing. Smaller specimens were placed in an insulated box surrounding the grip area of the testing machine and liquid nitrogen was piped into the box until the desired test temperature was reached.

A potential concern is that the results will be too conservative since temperatures of -150°C are never expected in the field. This concern can be eliminated, except perhaps for dents, because once typical tank car steels are cooled into the cleavage regime, the variation of fracture resistance with temperature can be predicted from knowledge of the variation of cleavage stress and flow strength with temperature. For any given steel, cleavage stress is relatively constant with changes in temperature [5]. Thus, once we are in the cleavage regime, results are expected to be similar for all cases in which cleavage occurs. That is, the Beremin model [5] parameters are assumed invariant with temperature. However, the flow stress rises with decreasing temperature (which is why lower shelf energy is not quite constant), so results must be adjusted to account for this effect. For example, if a gouge specimen cleaves under a certain stress at -150°C, we expect that, if cleavage were somehow initiated at a higher temperature, the cleavage stress would be nearly the same, although the strain necessary to reach that stress would be higher because the flow stress is reduced and more strain hardening is required to compensate. We needed to verify, however, that the specimen would not first fail at a lower stress by ductile fracture instead of cleavage due to the decreased flow stress at the higher temperature.

In the case of dents, if the tank steel is in the cleavage regime during dent formation, the tank will fail immediately. We showed this by performing cold bend tests at -150°C. TC-128B as-rolled, A515-70, and welded A515-70 plates all failed in cleavage at bend radii around 30 inches, far in excess of the 2-inch and 4-inch limits specified in the Guidelines. Apparently, if a dent is observed in a tank that has not ruptured, the dent was formed when the steel was in the ductile regime. Thus, to evaluate the validity of the dent guidelines, one must determine the subsequent behavior of the tank, particularly whether a delayed brittle rupture could occur.

Model Calibration and Validation

The parameters of the Beremin [5] model are developed in a two-step process. First, the stress-strain-temperature behavior is obtained from calibration to smooth round bar tensile tests. Then, the probabilistic cleavage parameters are obtained from calibration to notched round bar tensile tests.

Smooth round bar tensile tests and calculations were performed at various temperatures to establish the stress-strain behavior of the steels and the increase in flow strength with decreasing temperature. The finite element model of the smooth round bar test specimen is shown in Figure 1. During the tests, a 25-mm gauge length extensometer was installed to record engineering strain. Posttest, failed neck diameters were measured to provide the final true strain for input to the finite element model. All tests were performed at a strain rate of approximately 10^{-3} s^{-1}.

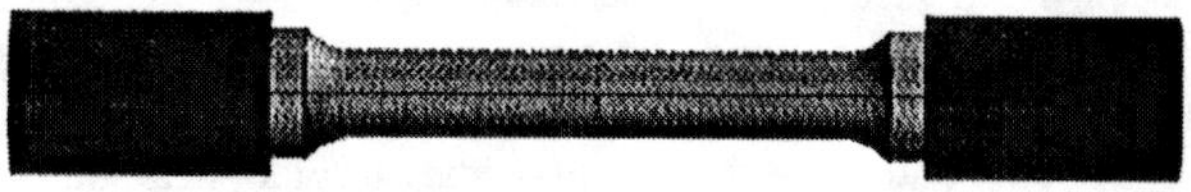

Figure 1. Finite Element Model of the smooth round bar tensile test specimen.

The smooth round bar tensile tests were performed on the A515-70 tank car steel at various temperatures ranging from 20°C to -160°C. The results of the coupon tests were used to calibrate the elastic-plastic stress-strain behavior in the constitutive model. An estimate of the true stress-strain curve is first obtained from the measurement of the engineering stress-strain behavior and periodic photographs of the necking behavior. This true stress-strain behavior is input into the finite element code's constitutive model in tabular form. Performing iterative simulations of the tensile test was used to further refine the true stress-strain curve by making small adjustments until the necking behavior was accurately reproduced. Using this approach, we believe we obtain a more accurate representation of the true tress-strain curve than obtained from the approximate correction of the engineering data. The resulting comparison of the measured and calculated engineering stress-strain curves for the A515-70 tank car steel at -150°C are shown in Figure 2.

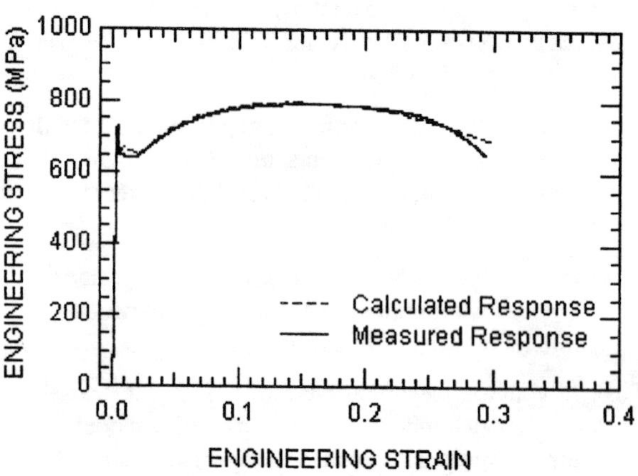

Figure 2. Measured and calculated behavior for the smooth round bar tensile test.

The next step is to use the calibrated stress-strain behavior from the smooth round bar to analyze the notched round bar specimens. The notched round bar tensile tests establish the stress-strain and fracture behavior of the steels under varying levels of triaxial constraint. Specimens were machined with three notch root radii: 6.35 mm, 2.54 mm, and 1.27 mm. Finite element models of the notched round bar tensile specimens were created as shown in Figure 3. Test data and analyses of these coupon tests were used to calibrate the Beremin [5] cleavage parameters for the two tank car materials. The finite element models of the tensile specimens had an element size of approximately 400 µm in the gauge section. With these simulations, the cleavage stress parameter (σ_u) was varied until a good match between predicted and measured rupture stresses was obtained. The microcrack exponent (m) was assigned a value of 22 based on results of Beremin's previous work [5].

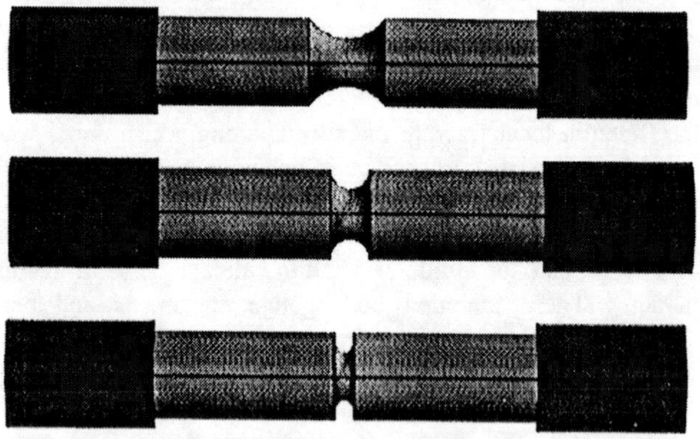

Figure 3. Finite Element Models of the notched round bar tensile test specimens.

During the notched round bar tests, a 25-mm gauge length extensometer was installed such that the gauge length straddled the specimen notch. The extensometer output was used to verify the displacement computed in the finite element model.

In a few room temperature tests, a radial extensometer was also applied to measure the change in specimen diameter as necking proceeded. For the tests in the cleavage regime (-150°C), the radial extensometer was unnecessary because the amount of necking was negligible. All tests were performed with a displacement rate of 25 μm/s.

For the given values of the cleavage model parameters, a probability of rupture can be determined at any load level using Eq. 4. For the A515-70 steel a value of 1.65 GPa for σ_u was used to fit the notched round bar data. The characteristic volume (V_0) was obtained as the cube of the characteristic microstructural length (customarily 200 μm). Figure 4 compares the resulting calculated notched round bar behaviors for the A515-70 steel with representative curves from tests. The stress-strain behavior is from the fitted tensile tests. The points on model the stress-strain curves at which load levels for 5%, 50%, and 95% rupture probability as obtained from the cleavage are indicated in the figure. These would represent the failure points of the specimens in a given test with 5, 50, or 95% probability.

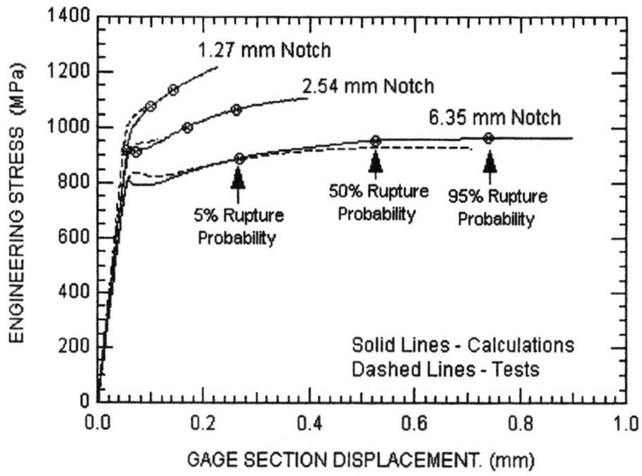

Figure 4. Measured and calculated behavior for the notched round bar tensile tests.

The statistical nature of cleavage fracture can be seen by normalizing all the notched round bar failure stresses by the predicted load level for a 50% probability of failure. These normalized failure stresses can then be plotted against a normalized rupture probability curve from Eq. 4 as shown in Figure 5 for all of the A515-70 steel tensile tests. This fit is good over the full range of tests performed considering the relatively small statistical sample of tests, and is comparable to fits shown in Beremin for A508 pressure vessel steel [5] and in Broberg for glass plates [6].

Engineering Test Models

As part of the engineering approach, we performed coupon tests with simulated gouges. We also performed bend tests to investigate dent damage. The results of the gouge specimen tests are presented here. These tests provided an opportunity to directly validate the gouge guidelines and to further validate the calibrated finite element models.

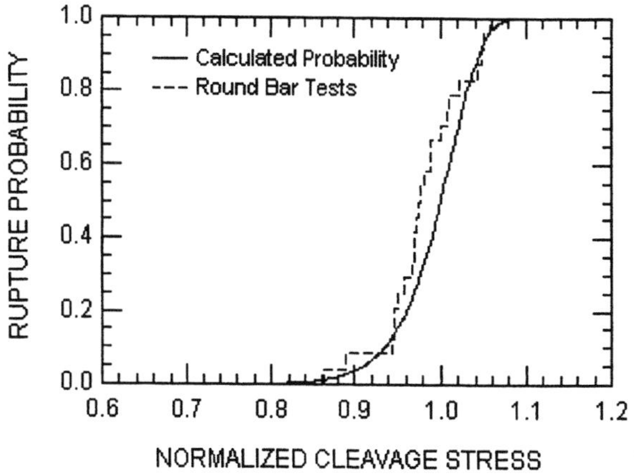

Figure 5. Measured and calculated behavior for the smooth round bar tensile test.

A finite element model of an engineering gouge specimen is shown in Figure 6. The specimens all had the same width (44.5 mm), the same gouge depth (6.35 mm) and were of the as-rolled thickness, but had two different gouge root radii, 12.7 mm and 3.2 mm. The typical element size around the gouge region is approximately 1-2 mm. All gouge tests were performed in the cleavage regime at a loading rate around 2.5 mm/s crosshead speed. Extensometers were installed on the front and back faces of the specimens, bridging the notches, to verify the displacements computed in the finite element models and to measure the amount of bending induced by the asymmetry of the specimen and loading. The simulations of the tests included models for the clamps, pins, and ball joints of the testing fixtures to accurately reproduce the loading conditions.

Figure 6. Finite Element Model of an engineering test gouge specimen.

Test data and analyses results of these engineering tests were used to validate the Beremin cleavage fracture model. The calculated and measured gouge test results for the A515-70 steel are summarized in Figure 7. The predicted load-displacement behavior for the tests agrees reasonably well with the measured values. The discrepancies between measured and predicted load-displacement behaviors can be attributed to slight temperature variations between specimens and the small increase in temperature that occurs during the duration of experiment. The cleavage model does a good job of predicting the cleavage rupture load of the gouge tests for the A515-70 steel. The greatest discrepancy is for the 12.7-mm-radius gouge, which appears to be more ductile than the calculated result. However, the measured cleavage stress of 486 MPa is below the calculated 95% rupture probability stress of 489 MPa. The additional ductility and lower stresses are a result of an increasing temperature of the specimen, a condition that was not included in the simulation.

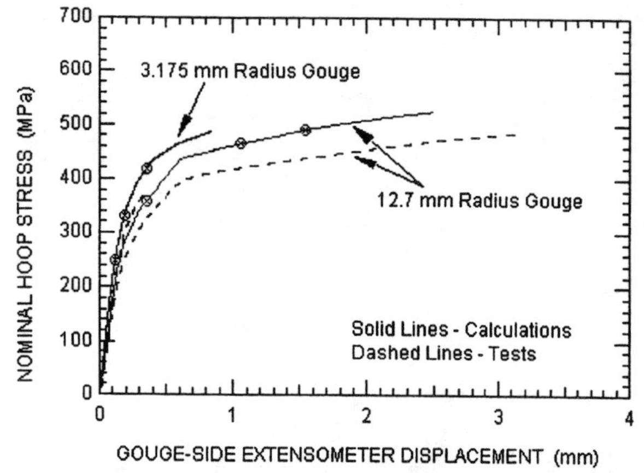

Figure 7. Finite Element Models of the engineering test gouge specimens.

Once the Beremin cleavage model was validated, it was applied to validate the Guidelines section on assessment of scores, gouges, and rail burns. The Guidelines [1] contain tables of allowable tank pressures for a given depth of damage. These pressures of course generate far-field hoop and axial shell stresses that are easily calculated. We gradually applied these stresses to finite element models of scores, gouges, and rail burns similar to Figure 6, and noted the corresponding tank pressures at which there was a 5% probability of failure. We found that, in general, predicted stresses for cleavage failure exceeded stresses for gouge root net section yielding at all realistic temperatures, and that, regardless of whether failure was by cleavage or yielding, predicted failure pressures were about a factor of two times allowable pressures in the tables. Since the safety factor originally used in constructing the tables was stated to be 2.0, the tables appear valid.

TANK CAR STRUCTURAL ANALYSES

A finite element model of a DOT 112A340W pressure tank car was generated to analyze the loads and stresses in the tank car as a result of service and salvage operating conditions. The DYNA3D finite element code was used for the tank car analyses described here. In addition to quasi-static evaluation of service and salvage loads, the tank car model is well suited for analysis of impact and dent damage behavior.

Two external views of the tank car model are shown in Figure 8. The structure includes the tank with ellipsoidal heads, manway, stub sill, bolster, and bogie structures. Figure 8b gives a detailed view of the model car end structures and the associated model mesh resolution. A pressure distribution on the inside of the tank and gravitational accelerations were used as boundary conditions to include the effects of the lading and pressure loads. The resulting simulations were used to investigate the effects of service and salvage loading on the stresses in the tank wall that could lead to an unstable tank rupture. In addition, the tank car model was used to calculate the behavior of dents in a tank car caused by impact in combination with the internal pressure. The results of these analyses are presented in the following sections.

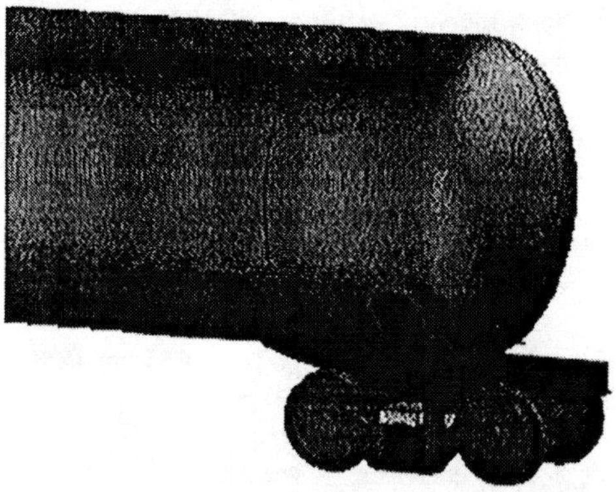

(a) Tank car model

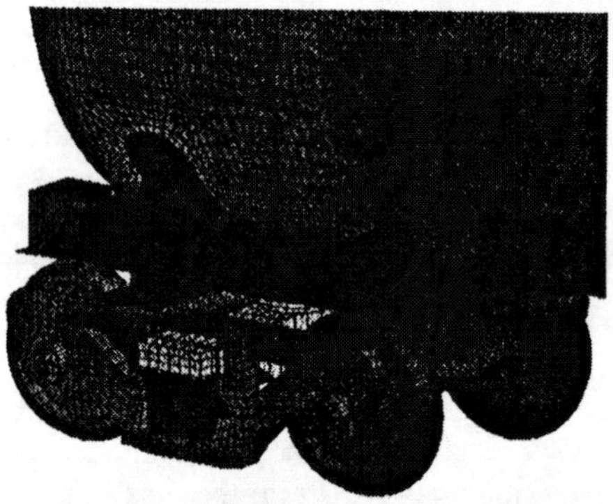

(b) Detail of the tank car mesh

Figure 8. Model for the 112A340W pressure tank car.

Effects of Lifting Methods and Pressurization

The initial calculations with the tank car model were performed to determine the tank wall stress magnitudes from service and salvage loading conditions. For these simulations, the tank car is in an undamaged condition. A pressure distribution was specified on the inside of the tank wall to simulate the combined effects of an internal pressure and the hydrostatic pressure from the lading. In addition, a gravitational acceleration was specified to include the dead weight loading of the tank car structure. The resulting simulations were used to determine the magnitude of the stresses under various conditions.

The first simulation corresponds to the loading conditions of an unpressurized tank with a lading load from a full tank of fluid of density equal to water. The model predicts stress concentrations around the bolster and stub sill doubler plate, but the majority of the tank wall has stresses below approximately 40 MPa. These results are consistent with simple estimates that can be made approximating the tank car as a simply supported beam.

Additional simulations were performed combining the gravity loading with internal pressures of 0.90 and 1.75 MPa. The addition of the internal pressure increases the stress in the majority of the tank wall to approximately 85 MPa and 170 MPa respectively. These results are consistent with strength-of-materials estimates of the shell stresses. The model again predicts stress concentrations around the bolster and stub sill doubler plate and near the tank head corner. The high stresses in the tank head are probably not characteristic because of the increased head thickness that is typical in practice.

The above calculations show that the stresses induced by lifting are small relative to stresses induced by tank pressure for the undamaged tank. We would expect this conclusion to hold for damaged tank cars as long as the damage is localized such that the tank mostly maintains its undamaged shape. We caution, however, that the lifting stresses, no matter how small, are additive to the pressure stresses, and could put critical stresses over the rupture limit.

Effects of Tank Pressure on Dent Behavior

Additional calculations were performed with the tank car model to investigate dent damage and the interaction of the internal pressure with a dent. The concern is that the pressure in a tank car can push out the dent such that the final dent shape does not indicate the full extent of deformation in the tank wall. The root of the dent undergoes a cycle of reversed plastic deformation. In addition, a tank car with dent damage could be at risk of further plastic deformation if the internal pressure within the tank were to increase during the salvage operation.

Only the stress-strain portion of the calibrated damage model was used in the denting simulations. The cleavage fracture predictions were ignored because the cleavage model was not calibrated for the case of reversed plastic deformation. Beremin [5] indicates that prior monotonic straining increases the cleavage resistance, but the reasons are not fully understood. The effects of reversed straining are an interesting topic for further investigation. To first order, the effects of strain reversal appear negligible, and the probability of dent rupture by cleavage would be driven by the surface residual stresses after strain reversal. On the other hand, dent failure by cleavage seems unlikely unless the metal temperature drops substantially after denting. This is because the dent would fail immediately during formation if cleavage were operative.

A series of simulations were performed using both vertical and longitudinal indentors. From the simulations, the longitudinal dent was found to be the more critical dent orientation in the presence of internal tank pressure. An example simulation is presented here that illustrates the effect of internal pressure on the dent response. The simulation is for a longitudinal indentor with an impact velocity of 18 m/s. For this simulation, the indentor has a rigid impact nose with a radius of 8 cm and a length of 3.0 m. The ends of the indentor were rounded away from the tank to avoid localized damage from the impact of sharp corners. The model used in this simulation is shown in Figure 9. The simulation performed calculates the dent formation for an unpressurized tank then smoothly increases the internal pressure to a maximum level of 1.75 MPa.

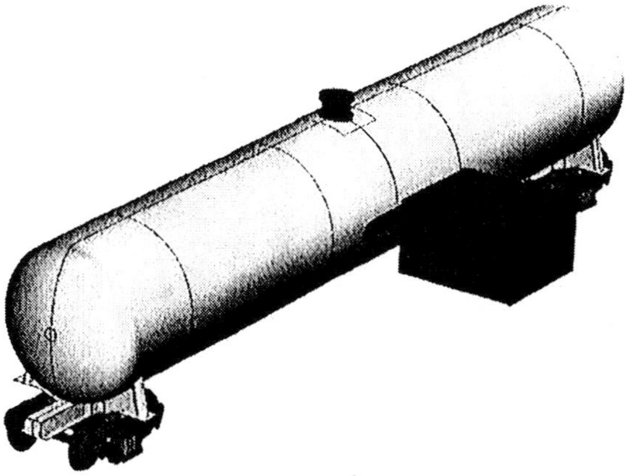

Figure 9. Model geometry for the dent formation and rebound simulations.

The calculated response is shown in Figure 10. The formation of the dent is shown in Figure 10(a). For visualization, only one half of the tank car model is shown in the figure. The maximum dent depth is approximately 15 cm at the center in the unpressurized tank. However, as the tank car pressure is increased to 1.76 MPa, the dent is pushed outward to the point where the original dent is no longer visible at the tank centerline. This final tank geometry is seen in Figure 10(b). Additional simulations indicate that a significant reduction in a longitudinal dent depth occurs for pressures greater than 0.69 MPa. As a result, the final dent geometry in a pressurized tank car is not a good indicator of the extent of damage in the tank wall. Note that similar observations may hold for denting of pipelines [10]. The Guidelines [1] contain limits on dent root radius. The current work shows that this approach is invalid.

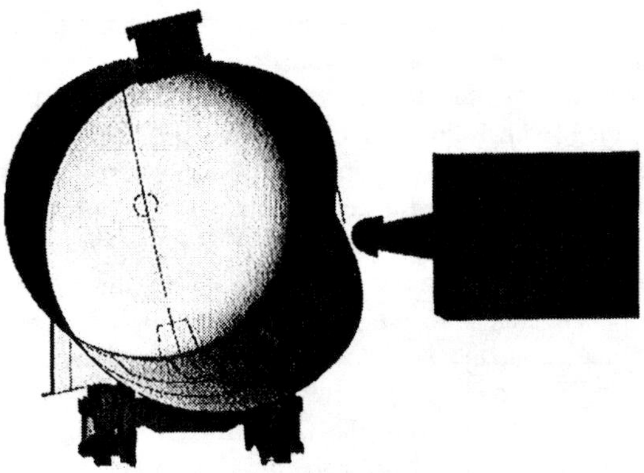

(a) Longitudinal dent formation

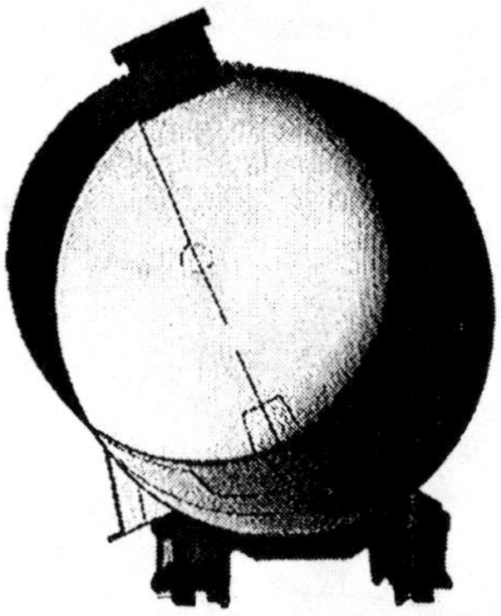

(b) Dent rebound with pressurization

Figure 10. Simulation of dent formation and rebound for a pressurized tank car.

The root of a dent represents site of potential high residual bending moment in the tank wall. The result of this residual moment is a significantly increased potential for a crack to extend the length of the dent. Therefore, during a rapid, visual damage assessment, it is the length of the vestigial dent that should be of interest, rather than the dent root radius, depth, or angle (the involvement of welds notwithstanding). The length of the vestigial dent should be compared with the known crack length for instability at a given pressure.

CONCLUSIONS

The existing Accident Damage Assessment Guidelines [1] lacked formal validation, although the Guidelines' 15-year history of reliable use provides some degree of confidence. Additionally, the level of conservatism in the Guidelines was unknown. Fortunately, very few catastrophic delayed ruptures have occurred, and those ruptures occurred only in cases where the tank damage would not have passed the Guidelines' rules.

To validate the Guidelines, a research program was performed combining laboratory tests and computer simulations. The approach was to focus on DOT 112A340W tanks and use laboratory tests on A515 Grade 70 and TC-128B as-rolled plate to calibrate a finite element and local cleavage fracture model. The resulting computational tool was helpful in analyses of tank damage such as scores, gouges, and rail burns, to which conventional fracture mechanics is inapplicable. This tool can be used to further analyze the Guidelines for other tank cars, damage scenarios, and salvage operations. The laboratory tests both provided data to validate the cleavage model and directly simulated damage in the form of gouges and dents.

The effect of rerailing and service loads on undamaged tank cars was investigated. The analyses found that rerailing stresses are small compared to pressure stresses for pressures of 0.69 MPa and above. Additional simulations were performed to investigate the interaction of internal pressure with dent formation. The calculations show that when internal pressures of 0.69 MPa or greater are present, significant rebound of the dent is likely. Therefore, the final dent geometry as observed in the field during damage assessment is not necessarily a good indicator of the total dent damage. It is the length of the vestigial dent, which potentially could become a crack, which should be compared with known lengths of unstable cracks.

REFERENCES

1. Association of American Railroads, Transportation Technology Center, Hazardous Material Training School, *Tank Car Safety Course Manual,* Pueblo, CO (1991).
2. Giovanola, J.H., and Shockey, D.A., "Literature Search and Evaluation Pertaining to Damage Assessment of Tank Cars Involved in Accidents," Association of American Railroads, Transportation Technology Center, Pueblo, CO (October 1995).
3. Davis, J.C., Private Communication, Association of American Railroads, Transportation Test Center (2001).
4. Pellini, W.S., "Analysis of Tank Car Failures Related to Rail Burn Dents," Association of American Railroads, Report No. R-551, Chicago, Illinois (June 1983).
5. Beremin, F.M., "A Local Criterion for Cleavage Fracture of a Nuclear Pressure Vessel Steel," Metallurgical Transactions A, Vol. 14A, Nov. 1983, pp. 2277-2287.
6. Broberg, K.B., *Cracks and Fracture*, San Diego, Academic Press, pp. 596-603 (1999).
7. Whirley, R.G., and Engelman, B.E., "DYNA3D—A Nonlinear, Explicit, Three-Dimensional Finite Element Code for Solid and Structural Mechanics—User Manual," Report UCRL-MA-107254 Rev. 1, Lawrence Livermore National Laboratory (November 1993).
8. Owen, D.R.J., and Hinton, E., *Finite Elements in Plasticity: Theory and Practice,* Swansea, United Kingdom, Pineridge Press (1980).

9. Klopp, R.W., Kirkpatrick, S.W., and Shockey, D.A., "Damage Assessment of Tank Cars Involved In Accidents: Phase II–Modeling and Validation," Association of American Railroads, Transportation Technology Center, Pueblo, CO (October 1997).

10. Yagami, T., Hiyoshi, S., and Keio, J., "Effect of prestrain on fracture toughness and fatigue-crack growth of line pipe steels", *Journal of Pressure Vessel Technology*, Vol. 123, No. 3, pp. 355-361, (2001).

PVP-Vol. 458, Computer Technology and Applications

PVP2003-1921

SAFETY ANALYSIS AND RISK ASSESSMENT FOR PRESSURE SYSTEMS

Boris Blyukher
Health and Safety Department
Indiana State University
Terre Haute, Indiana 47809

ABSTRACT

There have been many instances where serious injuries and fatalities have resulted from over-pressurization, thermal stress, asphyxiation and other potential hazards associated with testing, handling and storage of compressed gases and pressure facilities at numerous production and research facilities. These hazards are major issues that should be addressed in system design and in materials selection appropriate for high pressure applications. Potential hazards may be mitigated through system analysis and design process which are the major factors in preventing thermal/pressure hazards caused by possible leaks and fragmentation, in the case of rupture.

This paper presents a conceptual model and framework for developing a safety analysis which will reduce potential hazards, accidents and legal liabilities. The proposed systematic approach allows to identify hazards provide timely documentation of potential hazards and risks associated with systems, facilities, and equipment. As a result of this hazard analysis process, provisions and actions for hazard prevention and control have been put in place, and all identifiable potential hazards can be reduced to a low risk level.

INTRODUCTION

The presence and use of pressure systems associated with storage, handling, and transfer of compressed gases is commonplace at most industries, R&D and test facilities. Number of recent accidents at pressure facilities (Blyukher, 2000) have served to focus public concern and legislative attention on the risks associated with the operations. While sharing the public concern about plant operational safety, plant management have the additional, financial incentive to implement an effective safety analysis system since the financial impacts of major incidents are large and appear to be growing.

Government and industry have responded to the need for improved system safety and hazard analysis with legislation and industry initiated guidelines (e.g., 29 CFR 1910.119; MIL-STD-882B, 1984, *System Safety Program Requirements*; DOE Order 5481.1B " Safety Analysis and Review System", U. S. Department of Energy, !984).

This paper demonstrates how the technology and methods used in the system safety and hazard analysis can be utilized and incorporated in Occupational Safety programs and in Emergency Preparedness plans for the projects and facilities containing pressure systems.

A proactive approach to safety should be reflected in hazard analysis. While safe operations are important, safe designs are a prerequisite to operational safety. System safety is an approach to accident prevention that involves the detection of deficiencies in system components that have an accident potential.

These methods are demonstrated in the example of safety analysis of pressure systems at Superconducting Super Collider Laboratory by developing Safety Analysis Report (SSC Laboratory, 1992) . Subsystem Safety Analysis shows how to analyze subsystem components interfaces and interrelationships to ensure that they meet specified safety criteria, identify failures and hazardous events that would create other hazards, and locate impacts on the safety of the total system. The individual components of the pressure subsystem and their hazard analysis are discussed in detail below. An application of risk assessment, control, and mitigation methods are also presented.

HAZARD ANALYSIS METHODOLOGY

The key element in system safety is hazard analysis that identifies, anticipates, and controls hazards. The hazard analysis extends over the life cycle of a system. Many kinds of controls extend from the hazard analysis. They may be engineering controls that modify a system to eliminate or reduce the hazards to acceptable levels. Controls include management policy and procedures, identification and implementation of training for system operators, maintainers and support staff, operating procedures, emergency response and other plans and application of many consensus standards and government standards and regulations for safety.

Hazard Analysis is developed using the techniques of teamwork. As safety analysis gains greater acceptance, cross-functional teams will increase in number within all organizations. A team system allows performance of the overall value-added pressure system to be analyzed by representatives from each functional area involved in the product or service. In addition, the voice of the external user has greater weight in the decision-making process. As member of these teams, the pressure safety specialist will contribute functional safety expertise to quality improvement efforts. Success in a team environment requires significant changes in traditional management roles, interpersonal skills, and compensation and reward systems. For example, a safety specialist may be part of a team comprised of design and manufacturing engineers, line managers, operators and marketing specialist. The team may be analyzing one pressure system--from design to final customer delivery and aftermarket servicing. Within this setting, the safety specialist can provide expertise in areas such as product liability, risk identification, process design, testing, and field service. This contribution will help identify areas caused by injury and safety management oversights that impact overall pressure system cost, quality, and safety.

In support of the safety analysis and review process, System Safety Engineering developed the methodology and approach for conducting detailed hazard analyses. It is recommended to develop the internal organizational procedure to augment and clarify MIL-STD-882B and to assist engineers in conducting hazard analyses and safety assessments of their system designs and operations.

The higher the risk category, the greater the importance of the input from the end user. This concept of delegating detailed hazard analysis tasks to the individuals closest to potential safety problems is the best method of identifying potential hazards and implementing hazard prevention, elimination, and mitigation measures.

HAZARD IDENTIFICATION

Hazard identification categorizes the hazards normally associated with industrial plants and large scientific laboratories: injury to personnel or damage to equipment along with the following grouping of potential hazards unique to pressure subsystems: pressure, noise, and environmental (oxygen deficiency hazard at confined spaces).

Pressure accidents stem from or is a result of failures of pressured equipment such as pressure vessels, piping, pressure relief valve, motor valves or associated control equipment, or an insulating vacuum jacket such as those encasing the magnet string test items. Such failures result from over-pressurization, thermal stress, and abuse. In some cases, combination of conditions and events can result in one pressure hazard leading to another.

Over-pressurization and thermal stress are major issues that are addressed in system design and on material selection that is appropriate for pressure applications. These two potential hazards are mitigated through design process which is the major factor in preventing thermal/pressure caused potential hazards with its possible fragmentation, in the case of a rupture. To a lesser extent, marginal pressure hazards will be present with the low pressure instrument air (Blyukher, 2000).

SAFETY ANALYSIS AND RISK ASSESSMENT ELEMENTS

The following elements should be considered at safety analysis stage: items pertaining to customer needs and satisfaction; safety and environmental compatibility; items pertaining to product specification and service requirements; specification of materials and components, including approved supplies; failure and fail –safe characteristics; failure modes and effects analysis; fault tree analysis; labels, warnings and user instructions; items pertaining to process specification and service requirements.

Safety aspect of product or service quality should be identified, with the aim of enhancing safety and minimizing liability. Steps should be taken to limit the risk of product liability and minimize the number of cases by: identifying element safety standards in order to enhance product or service specifications; conducting design evaluation test and prototype testing for safety and documenting test results; analyzing instructions and warnings to the user; maintenance manuals and labeling in order to remove misinterpretation; developing a means of traceability to facilitate products recall when features that compromise safety and discovered and allow investigation of faulty products / services.

Safety analysis is a comprehensive assessment and a documented process to:

- systematically identify systems and operational hazards;
- describe and analyze the adequacy of the measures taken to eliminate, control, or mitigate hazards, and evaluate potential accidents and their potential risks;

- confirm a facility planned and constructed to demonstrate that pressure systems can be installed, leak tested, and energized safely;
- reflect the potential hazards that have been identified, classified, analyzed, mitigated, and reclassified.

Preliminary Hazard List (PHL) (Blyukher, 1998)

PHL is prepared during an early concept definitions phase for the design of a system and is a list of possible hazards that may be inherent in the design of a system.

The PHL is prepared by informal conferencing and checklist review. The input required includes analytical trees, flow diagrams, sketches and drawings, and installation map, energy sources, external hazards, and lessons learned. The PHL report consists of a narrative, a list of hazards, and a risk assessment. This PHL is prepared by system safety working group (SSWG) consisting of representatives from the engineering, safety, and user communities and project management. This group also influences the level of the system safety effort by categorizing the proposed facility as low, medium, or high risk (see Risk Assessment section).

Preliminary Hazard Analysis (PHA)
(Blyukher, 1998)

PHA is generally performed by the designer during early design stage and includes hazard identification, an evaluation of hazards and identification of safety design criteria that will be used during design. Included in the analysis are hazardous components, safety related interface considerations among system elements, environmental constraints, operation, test, maintenance and emergency procedures, facilities, support equipment, training and safety-related equipment and safeguards.

Subsystem Hazard Analysis (SSHA)

As the system design progresses into the design of subsystems and components, the hazard analysis must move to greater detail. Hazards involving component failures, human error and relationships between components and equipment in each subsystem are identified and documented.

Failure Modes and Effect Criticality Analysis (FEMECA) (Blyukher, 1997)

The failure or malfunction of each pressure or cryogenic system component is identified, along with the mode of failure. The effects of the failure are traced through the system and the ultimate effect on task performance is evaluated. FEMECA starts with a listing comprised of all critical items identified as a result of performing PHA and SHA *(Critical Items List)*, and is an in-depth analysis of possible failures and their effects related to system functions (*functional FEMECA)* or system hardware and components (*hardware FEMECA)*. All hardware components are categorized on their *criticality* by the worst case potential. Direct effect of failure that results in a level of injury, damage , or loss of a magnitude that quick or total recovery would be possible although extremely difficult (e.g., personnel injuries, partial system loss, property or equipment damage). The pressure and temperature parameters for this categorization were established by designers in the system safety program plan as a policy-making documentation.

In assigning hardware criticality, the availability of redundancy modes of operation is considered. Assignment of functional criticality, however, assumes the loss of all redundant hardware elements.

Fault Tree Analysis (FTA)

Fault Tree is used as an analytical tree to determine fault and/or accident potential before one has occurred. FTA is a deductive system safety analysis method (top down) to evaluate fault or failure events and performed as an extension of the failure mode and effect analysis, that evaluates the overall effect of functional failures on other subsystems or the overall system itself.

Job Safety Analysis (JSA)

JSA is a generalized examination of the tasks associated with the performance of a given job and an evaluation of the hazards associated with those tasks and the controls used to prevent or reduce exposure to those hazards. A close relationship exists between standardization and job safety analysis (JSA). When coupled with a quality standard for a given task, JSA created a total job standard.

Occupational Health Hazard Assessment (OHHA)

A pressure subsystem may impose occupational health hazards to users, maintenance and other support people involved with operations. This assessment identifies those hazards and specifies what means will be used to control them.

RISK ASSESSMENT

Risk considerations must be given to risk related to the health and safety of users, dissatisfaction with goods and services, etc.

The safety analysis methods and techniques supporting the hazard analysis process are well recognized first steps used by high-tech industries, aerospace companies, defense contractors, and government agencies associated with large projects.

The in-depth system and subsystem hazard analyses supporting safety analysis are proven systematic approaches to identify hazards inherent in the system of hazards in its projected environment which will endanger the system. This approach includes risk assessments (Blyukher, 1997). To eliminate as many hazards as possible, and determine what corrective actions to take, all identified hazards were prioritized according to risk level criteria. Since not all pressure hazards are of equal magnitude or criticality to personnel safety. each identified hazard is assessed in terms of severity and probability. Hazard categorization involved the determination of the likelihood of the hazardous event actually occurring. For pressure equipment this likelihood can be estimated in numeric terms (quantitative method) These probabilities of failure (failure rates per hour and/or per demand) shall be determined by operating experience, otherwise data from similar systems elsewhere or other relevant values shall be used. Estimates of "spontaneous" equipment failure rates collected from reliability testing and assessments of historical data are presented in NASA and DOE sources (see also Blyukher, 2000). Use of this statistics is justified by noting that they are based on manufacture experience with specific components, and were gathered from a variety of sources including industrial plants and research pressure facilities, many of which were designed and built to the same codes.

Risk categorization process also involved qualitative analysis by utilizing non-numeric terms resulting in comparative hazard risk assessment. In accordance with risk assessment matrix (MIL-STD-882, 1984) all hazards were categorized by:
Frequency of Occurrence categories (A-frequent, B-probable, C-occasional, D-remote, and E-impossible), and *Hazard Severity* categories (1-catastrophic, 2-critical, 3-marginal, and 4-negligible).

An identified hazard assigned a hazard risk index of 1A, 1B, 1C, 2A, 2B, or 3A (suggested criteria are *unacceptable, high risk*) require immediate corrective action. A hazard risk index of 1D, 2C, 2D, 3B, or 3C (suggested criteria are *undesirable, medium risk*) would be tracked for possible corrective action.
A hazard risk index of 1E, 2E, 3D, 3B, 4A, 4B, 4C, 4D, or 4E (suggested criteria are *acceptable for review, low risk*) might have a lower priority for corrective action and may not warrant any tracking actions.
Low-risk facilities are those with low energy levels and those with the management has a considerable amount of trouble-free experience, such as basic administrative buildings and housing. The system safety effort for these facilities may consist primarily of the PHL, with no additional analysis required.
Medium- or high-risk facilities include those that have higher levels of energy associated with them and/or tend to be more unusual because of their function or design. The results of this PHA determine and drive the need for further hazard analysis in the form of SSHA, O&SHA, and/or OHHA.

For example, risk categories assigned to pressure equipment were also based on considerations of user involvement: *high risk* category is a result of heavy-, *medium-* of moderate, *low-* of low user involvement. Some results of risk categorization of main pressure facilities are presented below (SSC Laboratory, 1992):
High Risk- Refrigeration Plant, Main Cold Box, Saturated Vapor Nitrogen Generator, Piping System (helium, nitrogen), Compressor Skid, Cryogenic Valves (cooldown, warmup, relief), Cryostat Vacuum Vessel, Oxygen Monitoring and Audio Warning System;
Medium Risk- Liquid Nitrogen Dewar, Oil Processing System, Helium Vaporizer, Liquid Helium Storage Dewar, Helium High Pressure Piping, Pump & Compressor Noise, Compressed Air System (Storage Tank; Compressor Noise);
Low Risk- Main Cold Box Expansion Engines, Cold Compressor, Cooling Water System for Helium Compressor, Helium Storage Vessel.

Risk categorization process was implemented by using the supportive DOE regulations (e.g., DOE Order 5481.1B, 1984) as a guide.

HAZARD MITIGATION

1. *Design for minimum risk.*

- At design stage a formal, documented, comprehensive, systematic examination of a design to evaluate its requirements and capability to meet safety requirements shall be applied. Examination should also identify problems and propose solutions. Design capability of pressure equipment encompasses such things as fitness for purpose, feasibility, manufacture ability, measurability, performance, reliability, maintainability, safety and environmental aspects.
- Designing for minimum risk means that the first goal is to eliminate hazards.
- If the hazards cannot be eliminated, the associated risk should be reduced to acceptable levels through design selection.
- At design control stage suppliers of pressure equipment should establish procedures for evaluating safety performance and dependability incorporated in product design.

2. *Incorporate safety systems and safety devices.*
    - Safety devices are to be incorporated if identifies hazards cannot be eliminated or their associated risk adequately reduced through design selection.

- The risk is to be reduced through the use of fixed, automatic or other protective safety design features or devices.
- The design must include provisions for periodic functional checks of safety devices when applicable.

3. *Provide warning devices*
   - Warning devices are needed when neither design nor safety devices can effectively eliminate identified hazards or adequately reduce associated risk.
   - Devices are used to detect a hazardous condition and to produce an adequate warning signal to alert personnel to the hazard.
   - Warning signals and their application must minimize the probability of incorrect personnel reaction to the signals and must be standardized within like types of systems.

4. *Develop safety procedures*
   Provide safety training
   Provide safety equipment
   Develop administrative controls or restrictions.
   - Procedures and training are needed when the other three methods are impractical.
   - The standard requires that no warning, caution or other form of written advisory be used as the only risk reduction method for hazard severity categories 1or 2 (see the subsection entitled Risk Assessment) without a specific waiver.
   - Procedures may include the use of personal protective equipment. Precautionary notations shall be standardized and the managing activity must specify such standards.
   - The managing activity may specify that tasks and activities it judges critical require certification of personnel proficiency.

SUMMARY

1. The final result of the system safety analysis is a qualitative or quantitative value of the risk associated with the operation of each equipment item. All risk values are the result of hazard identification, system safety analysis, and measures of control and mitigation. The magnitude of the risk depends upon the damage criteria (equipment damage, personnel injury, or personnel fatality) chosen in the consequences analysis. Risk assessment provides a valuable screening tool for the assignment of priority in Occupational Safety program and Emergency Preparedness Plan.

2. By addressing the high risk items first, a more cost effective risk reduction program can be initiated and managed. This program could include reduction of failure probability by modification of process conditions or use of improved inspection methods/technologies or improved emergency procedures.

3. In conclusion, comprehensive system safety analysis is an effective and efficient tool for assessing the relative risks associated with various equipment items within a given facility and for assessing the comparative risk of similar facilities. It therefore provides a cost effective way of optimizing design, operation, test, inspection, and use procedures for the reduction of risk.

LIST OF REFERENCES

Blyukher, B., 2000, Probabilistic Risk Assessment of Environmental Hazards (with L. Ivanova, E. Malakhov, and V. Sitnikov), *Emerging Technologies: Risk Assessment, Computational Mechanics and Advanced Engineering Topics,* PVP-Vol.400, The American Society of Mechanical Engineers, United Engineering Center, New York, N.Y., pp. 69- 73.

Blyukher, B., 1998, Preliminary Hazard Analysis For Superconducting Test Facilities, *Risk Assessment Technologies, and Transportation, Storage, and Disposal of Radioactive Materials*, PVP-Vol.378, The American Society of Mechanical Engineers, United Engineering Center, New York, N.Y., pp. 21 - 27.

Blyukher, B., 1997, Determination of the Probability and Prediction of Accidents and Emergencies Consequences. Application of the Mathematical Theory of Catastrophes and Other Methods (with A.Tsykalo, V. Gogunsky, and V. Rodankov), *Emergency Situations and Civil Defense (Scientific and Informational Journal),* # 1, Odessa, Ukraine, pp. 79-80.

Code of Federal Regulations, 29 CFR Part 1910 (General Industry), U.S. Department of Labor.

DOE Order 6430.1A, 1989, *General Design Criteria*

DOE Order 5481.1B, 1984, *Safety Analysis and Review System,* U. S. Department of Energy (updated by Notice 1, 1987)

MIL-STD-882B, 1984, *System Safety Program Requirements*, U. S. Government Printing Office, Washington, DC.

SSC Laboratory, 1992, Accelerator Systems String Test. Safety Analysis Report, DMP-000001.

AUTHOR INDEX

PVP-Vol. 458
Computer Technology and Applications

Book Number: G01193